Encyclopedia of
Plant Physiology

New Series Volume 16 A

Editors

A. Pirson, Göttingen
M.H. Zimmermann, Harvard

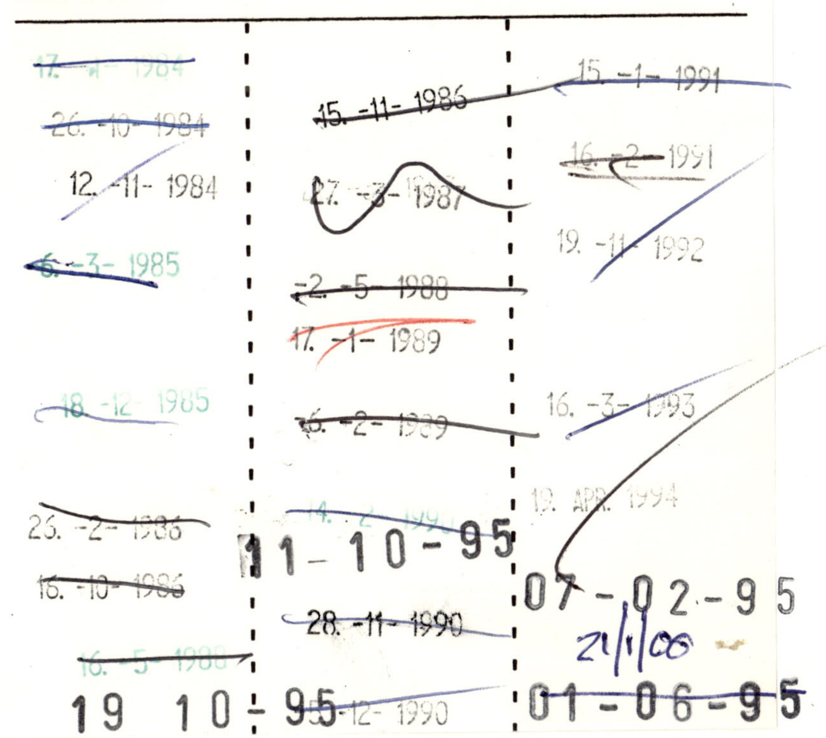

Photo-morphogenesis

Edited by

W. Shropshire, Jr. and H. Mohr

Contributors

K. Apel M. Black A.E. Canham J.A. De Greef M.J. Dring
H. Egnéus B. Frankland H. Frédéricq L. Fukshansky
M. Furuya V. Gaba A.W. Galston J. Gressel W. Haupt
S.B. Hendricks M.G. Holmes M. Jabben H. Kasemir
C.J. Lamb M.A. Lawton K. Lüning A.L. Mancinelli
H. Mohr D.C. Morgan L.H. Pratt P.H. Quail R.H. Racusen
W. Rau W. Rüdiger E. Schäfer H. Scheer J.A. Schiff
P. Schopfer S. D. Schwartzbach W. Shropshire, Jr.
H. Smith W.O. Smith R. Taylorson W.J. Van Der Woude
D. Vince-Prue H.I. Virgin E. Wellmann

With 173 Figures

Springer-Verlag
Berlin Heidelberg New York Tokyo 1983

W. Shropshire, Jr.
Smithsonian Institution
Radiation Biology Laboratory
12441 Parklawn Drive
Rockville, MD 20852/USA

H. Mohr
Biologisches Institut II der Universität
Lehrstuhl für Botanik
Schänzlestr. 1
D-7800 Freiburg/FRG

ISBN 3-540-12143-9 (in 2 Bänden) Springer-Verlag Berlin Heidelberg New York Tokyo
ISBN 0-387-12143-9 (in 2 Volumes) Springer-Verlag New York Heidelberg Berlin Tokyo

Library of Congress Cataloging in Publication Data. Main entry under title: Photomorphogenesis. (Encyclopedia of plant physiology; new ser., v. 16) Includes indexes. 1. Plants–Photomorphogenesis–Addresses, essays, lectures. I. Shropshire, Walter. II. Mohr, Hans, 1930. III. Apel, K. IV. Series. QK711.2.E5 vol. 16 581.1s [581.1′9153] 83-10615 [QK757] ISBN 0-387-12143-9 (U.S.).

Typesetting, printing and bookbinding: Universitätsdruckerei H. Stürtz AG, Würzburg.
2131/3130-543210

The Editors respectfully wish to dedicate this encyclopedia volume on Photomorphogenesis to STERLING B. HENDRICKS who died on January 4, 1981. His incisive and imaginative approach to research was a stimulus to all who had the pleasure of meeting or working with him. We are especially indebted to his critical and supportive friendship expressed over many years, beginning with the year 1956 as a supervisor at the Beltsville Plant Industry Station (H.M.) and as an examining Ph.D. committee member at the George Washington University in 1957 (W.S.).

We, in the field of photomorphogenesis, will miss his enthusiasm.

The Editors

Contents Part A

5 Models in Photomorphogenesis

L. FUKSHANSKY and E. SCHÄFER (With 5 Figures)

6 Phytochrome as a Molecule

W.O. SMITH (With 1 Figure)

9 Rapid Action of Phytochrome in Photomorphogenesis

P.H. QUAIL (With 9 Figures)

10 Photocontrol of Gene Expression

C.J. LAMB and M.A. LAWTON

11 Intracellular Photomorphogenesis

P. SCHOPFER and K. APEL (With 5 Figures)

12 Control of Plastid Development in Higher Plants

H.I. VIRGIN and H. EGNÉUS (With 5 Figures)

13 Control of Plastogenesis in Euglena

S.D. SCHWARTZBACH and J.A. SCHIFF

14 Pattern Specification and Realization in Photomorphogenesis

H. MOHR (With 10 Figures)

Contents Part B

List of Contributors Part A and B

K. Apel
Botanisches Institut
der Universität Kiel
Olshausenstraße 40–60
D-2300 Kiel/FRG

M. Black
Department of Biology
Queen Elizabeth College
(University of London)
Campden Hill Road
London W8 7AH/United Kingdom

A.E. Canham
University of Reading
Department of Agriculture and Horticulture
Earley Gate
Reading, Berks. RG6 2AT/
United Kingdom

J.A. De Greef
Department of Biology
University of Antwerpen
(UIA-RUCA)
Universiteitsplein 1
B-2610 Wilrijk/Belgium

M.J. Dring
Botany Department
Queen's University
Belfast BT7 1NN/United Kingdom

H. Egnéus
University of Göteborg
Botanical Institute
Department of Plant Physiology
Carl Skottsbergs Gata 22
S-413 19 Göteborg/Sweden

B. Frankland
School of Biological Sciences
Queen Mary College
(University of London)
Mile End Road
London EI 4NS/United Kingdom

H. Frédéricq
Laboratory of Plant Physiology
University of Gent
Ledeganckstraat, 35
B-9000 Gent/Belgium

L. Fukshansky
Institut für Biologie II/Botanik
Schänzlestraße 1
D-7800 Freiburg/FRG

M. Furuya
Department of Biology
Faculty of Science
University of Tokyo
Hongo, Tokyo, 113/
Japan

V. Gaba
Department of Biology
Queen Elizabeth College
(University of London)
Campden Hill Road
London W8 7AH/United Kingdom

A.W. Galston
Department of Biology
Yale University
P.O. Box 6666
New Haven, Connecticut 06511/
USA

J. Gressel
The Weizman Institute of Science
Department of Plant Genetics
Rehovot, 76100/Israel

W. Haupt
Institut für Botanik
und Pharmazeutische Biologie
der Universität Erlangen-Nürnberg
Schloßgarten 4
D-8520 Erlangen/FRG

S.B. Hendricks
(Deceased)

M.G. HOLMES
Smithsonian Institution
Radiation Biology Laboratory
12441 Parklawn Drive
Rockville, Maryland 20852/USA

M. JABBEN
Max-Planck-Institut
für Strahlenchemie
Stiftstraße 34–36
D-4330 Mühlheim/FRG

H. KASEMIR
Institut für Biologie II/Botanik
Schänzlestraße 1
D-7800 Freiburg/FRG

C.J. LAMB
The Salk Institute for Biological Studies
P.O. Box 85800
San Diego, California 92136/USA
and
Plant Biology Laboratory
10010 North Torrey Pines Road
La Jolla
San Diego, California 92138/USA

M.A. LAWTON
Department of Biology
Washington University
Campus Box 1137
St. Louis, Missouri 63130/USA

K. LÜNING
Biologische Anstalt Helgoland
Notkestraße 31
D-2000 Hamburg 52/FRG

A.L. MANCINELLI
Department of Biological Sciences
1108 Schermerhorn Hall
Columbia University
New York, NY 10027/USA

H. MOHR
Biologisches Institut II
der Universität
Lehrstuhl für Botanik
Schänzlestraße 1
D-7800 Freiburg/FRG

D.C. MORGAN
D.S.I.R.
Plant Physiology Division
Private Bag
Palmerston North/New Zealand

L.H. PRATT
Botany Department
University of Georgia
Athens, Georgia 30602/USA

P.H. QUAIL
Department of Botany
139 Birge Hall
University of Wisconsin-Madison
Madison, Wisconsin 53706/USA

R.H. RACUSEN
Department of Botany
University of Maryland
College Park, Maryland 20742/USA

W. RAU
Botanisches Institut der Universität
Menzinger Straße 67
D-8000 München 19/FRG

W. RÜDIGER
Botanisches Institut der Universität
Menzinger Straße 67
D-8000 München 19/FRG

E. SCHÄFER
Institut für Biologie II/Botanik
Schänzlestraße 1
D-7800 Freiburg/FRG

H. SCHEER
Botanisches Institut der Universität
Menzinger Straße 67
D-8000 München 19/FRG

J.A. SCHIFF
Brandeis University
Institute for Photobiology
of Cells and Organelles
South Street
Waltham, Massachusetts 02154/USA

P. SCHOPFER
Biologisches Institut II
der Universität
Lehrstuhl für Botanik
Schänzlestraße 1
D-7800 Freiburg/FRG

S.D. SCHWARTZBACH
Genetics, Cellular and Molecular
Biology Section
School of Life Sciences
348 Manter Hall
University of Nebraska-Lincoln
Lincoln, Nebraska 68588-0118/USA

W. SHROPSHIRE, JR.
Smithsonian Institution
Radiation Biology Laboratory
12441 Parklawn Drive
Rockville, Maryland 20852/USA

H. SMITH
Department of Botany
University of Leicester
University Road
Leicester LE1 7RH/
United Kingdom

W.O. SMITH
Smithsonian Institution
Radiation Biology Laboratory
12441 Parklawn Drive
Rockville, Maryland 20852/USA

R. TAYLORSON
U.S. Department of Agriculture
Bldg. 001 Rm. 40 BARC-West
Beltsville, Maryland 20705/USA

W.J. VANDERWOUDE
Light and Plant Growth Laboratory
Beltsville Agricultural
Research Center
Beltsville, Maryland 20705/USA

D. VINCE-PRUE
Glasshouse Crops Research
Institute
Worthing Road
Littlehampton
West Sussex, BN17 3PU/
United Kingdom

H.I. VIRGIN
University of Göteborg
Botanical Institute
Department of Plant Physiology
Carl Skottsbergs Gata 22
S-413 19 Göteborg/Sweden

E. WELLMANN
Biologisches Institut II
der Universität
Schänzlestraße 1
D-7800 Freiburg/FRG

1 Advice to the Reader

W. Shropshire, Jr. and H. Mohr

The following technical notes should be read before using this volume of the Encyclopedia.

In Appendix I at the end of the book is a brief list of abbreviations which have been used throughout the text without repeated explicit definition. Other abbreviations were introduced and defined at their first use in the different chapters.

In Appendix II are the most important units and symbols of the SI (Système Internationale d'Unités) as they apply to the present text. As far as possible SI units have been used in all chapters except where the author has explicitly and consistently used others for emphasis.

In Appendix III a description is given of the general characteristics of light fields employed in research on photomorphogenesis. The operational boundary conditions are listed and there is a brief discussion of recommended terminology.

A photoreceptor (or sensor pigment) is defined as a pigment which can transduce a light signal (see Fig. 6 in Chap. 3). Not every pigment is a photoreceptor. As an example, anthocyanin is a pigment, but not a photoreceptor (as far as is known at present).

A chapter on light and the physiological clock had to be deleted for technical reasons. Most of the pertinent questions related to clocks and photomorphogenesis are dealt with in the chapter on photomorphogenesis and flowering (Chap. 18).

Some overlap between the chapters was not only tolerated, but encouraged by the editors in an effort to make each individual chapter as autonomous as possible. Many readers will not study the book from cover to cover at one time, but rather prefer to consult particular sections. On the other hand, the sequence of the chapters is not arbitrary. To fully appreciate the arguments presented the present sequence is suggested.

When conflicting views and interpretations were presented by the authors, such as in Chapters 19 and 27, the decision was made to have both available to the reader rather than making an arbitrary editorial decision to exclude divergent opinions. Time and additional data will ultimately settle the issues. Even though the chapters have been edited, the responsibility for emphasis is the author's.

The editors wish to acknowledge the support and enthusiasm, even though sometimes left anonymous, received from many sources, especially the authors, in enabling the completion of this volume.

An attempt was made to exclude unpublished data. However, in a dynamically expanding field this was not always possible.

Similarly, the editors were very much aware that although the previous Encyclopedia series (1955) contained much data scattered in many volumes pertaining to photomorphogenesis, this is the first time a single volume with this emphasis has been compiled. Therefore, it was decided to include material aimed at giving a newcomer to the field a helpful perspective, for example, An Introduction to Photomorphogenesis for the General Reader (Chap. 3) and a chapter on Action Spectroscopy of Photoreversible Pigment Systems (Chap. 4). The historical completeness of each topic has been minimized and the reader is rather given key review references to pursue if desired. Thus, the aim has been to provide a stimulating entry into an exciting field of research with an emphasis upon current significant questions of interest, not only to the generalist, but to the specialist as well.

Finally, the editors decided to dedicate this volume on Photomorphogenesis to STERLING HENDRICKS and invited him to write a personal view of what he considered to be significant problems of the field. Unfortunately, he died shortly after the completion of the initial draft. An attempt was made by his co-author, WILLIAM VANDERWOUDE, to keep HENDRICKS' portions of the chapter as he wrote them, but the reader should keep in mind that HENDRICKS did not see the final version.

2 How Phytochrome Acts – Perspectives on the Continuing Quest

S.B. HENDRICKS and W.J. VAN DER WOUDE

1 Introduction

Knowledge about the influence of light on the form and function of plants developed over the last century. Phytochrome was recognized in 1949 as an essential absorber of light in these photomorphogenic processes. The recognition came entirely from logical deductions based on physiological responses of plants and their propagules. The discovery in 1952 of the photoreversibility of a potential response to light was the key factor leading to the present understanding of phytochrome action.

The photoreversibility was used in 1958 in the detection of phytochrome in extracts of plant tissue. It then served as an assay for isolation of the pigment which was realized in 1964. The isolation gave the necessary material for examination of molecular structure and function which is still in progress.

Phytochrome was the first protein recognized as having a regulatory function in plants other than by initial enzymatic action. Its immediate action thus is intracellular although by so acting it can influence release of material from cells to regulate intercellular action as expressed, for example, in flowering and material transport.

It was suggested in 1967 that phytochrome functions through action on membranes of cells and organelles (HENDRICKS and BORTHWICK 1967). This concept has withstood critical examination. The postulate here is that phytochrome action depends on chemiosmotic processes at the membrane (MITCHELL 1979). The concept still needs extensive examination.

2 Recognition of Photomorphogenesis

Prevention of etiolation, control of seed germination, and influence on flowering are three major aspects of the effect of light on plant growth and development, i.e., on photomorphogenesis. Each of these distinct actions has a specialized literature. The bibliography compiled by CORRELL et al. (1977) serves as a guide to the literature. Etiolation might well have been noted by prehistoric man in his grubbing for food. The differences between normal plants and ones grown in darkness were studied in detail by PRIESTLEY (1925) on *Vicia faba* seedlings. The differences are pronounced in plastid development, chlorophyll formation, stem lengthening, and leaf enlargement. The effectiveness of red light to prevent etiolation was noted by JACOBI (1914). TRUMPF (1924) found that exposure to light for times as short as 1 min profoundly affected the development of plants growing in darkness. Red light was used to suppress mesocotyl growth of *Avena* seedlings in studies on phototropism (NUERNBERGK 1927).

The recorded effects of light on seed germination preceded the finding on etiolation. Seeds of *Bullardia aquatica (Tillestrum aquaticum)* were observed by CASPARI (1860) to germinate better in light than darkness. The responsiveness of many kinds of seeds was studied by STEBLER (1881) and KINZEL (1908). Radiation in the 600 to 750 nm region was found most effective by FLINT and MCALISTER (1937) and RESÜHR (1939). Both increased and decreased germinations were noted. The literature is reviewed by TOOLE (1973).

Discovery of photoperiodic control of flowering by GARNER and ALLARD (1920) was the touchstone for studies on photomorphogenesis and biological rhythm. They observed that flowering of many plants responded to the relative length of day and night and accordingly had some responding light control. The literature has been reviewed by VINCE-PRUE (1975; see also Chap. 18, this Vol.).

3 Unity of Responses – Photoreversibility

Recognition of an underlying common cause for many of the responses to light came from studies of action spectra (PARKER et al. 1949, HENDRICKS and BORTHWICK 1954). Such studies were made in 1945 to 1948 on control of flowering of both long- and short-day plants by interruption of long nights with irradiation of their leaves for short periods (PARKER et al. 1946, BORTHWICK et al. 1948). The action spectra found with the two types of plants were closely similar in energy requirement and in the region, near 660 nm, of maximum effectiveness for their appropriate flowering responses. This also was the case for de-etiolation of *Pisum sativum* seedlings (PARKER et al. 1949), indicative of the same initial action of the light in control of flowering and etiolation.

In 1952 E.H. TOOLE suggested further measurements on germination control of lettuce seed by light which he earlier had prompted FLINT to undertake with MCALISTER. The action spectra in the 400 to 700 nm region were closely the same as those for other responses (BORTHWICK et al. 1954). A maximum for suppression of germination, moreover, was found at 730 nm. Photoreversibility of the potential germination response was recognized and was confirmed for short irradiances in sequence (BORTHWICK et al. 1952b). Reversibility of the potential responses with maxima for effectiveness at the same wavelengths, namely 730 and 660 nm, were also found for flowering (BORTHWICK et al. 1952a) and etiolation (DOWNS et al. 1957).

The ubiquity for the response to light was shown for germination of fern spores (BÜNNING and MOHR 1955) for which a response to light had been reported by HEALD (1898). HAUPT (1958) observed photoreversibility of chloroplast orientation in the alga *Mougeotia* which was known to respond to light (OLTMANNS 1922). The presence of a photoreversible pigment as a controlling agent for light action was thus indicated throughout plant phyla. It is thought to be ubiquitous for all chlorophyllous plants as well as their albino mutants.

Reversal of a potential response indicated control by the photoreversible change in form of an absorbing pigment with maxima of light absorption at the maxima of the action spectra in tissues of low chlorophyll content, thus:

$$P_{660} \underset{\text{far red}}{\overset{\text{red}}{\rightleftarrows}} P_{730} \quad \text{or} \quad P_r \underset{730\,\text{nm}}{\overset{660\,\text{nm}}{\rightleftarrows}} P_{fr}$$

The reversibility allowed measurement of quantum efficiency (ϕ) × molar extinction coefficents for the two final forms of the controlling pigment. This was equivalent to the method used by WARBURG and NEGELEIN (1929) in work on cytochrome oxidase. Absence of notable fluorescence at room temperature indicates that ϕ approaches 1.0. The molecular absorption coefficients were found to be the order of 5×10^4 cm^2 mol^{-1} at the maxima (HENDRICKS et al. 1956). The pigment was thus realized as being intensely colored. Failure to observe its blue color in physiologically responsive barley and maize albino plants indicated that the effective concentration must be very low. A temperature coefficient of 1.0 for several of the responses indicated that P alone is involved in the photoreactions.

P_{660} is thermodynamically stable. This conclusion followed from the germination responses of lettuce seed in the following sequence of treatments:

Imbibed in darkness (D) for 2 h at 20 °C
Saturating red radiation (R) 5 min
48 h at 35° (a temperature too high for germination)
48 h, D, 20° no germination, or alternatively
5 min R, 48 h, 20° full germination

These responses suggested that P_{fr} reverted to P_r within 48 h or less at 35 °C in the lettuce seed. The thermal reversion of P_{fr} to P_r led to the concept that timing in photoperiodism might be based on it as a sort of "hour-glass". This was too simplistic.

Several arguments indicated that P_{fr} is the active form of P. Thus a photon fluence of 1×10^{-8} mol m^{-2} near 660 nm sufficient to change $<10^{-3}$ P_r to P_{fr} caused a measurable enlargement of etiolated *Pisum sativum* leaves (PARKER et al. 1949). A hundredfold greater irradiance near 730 nm of tissue with predominant P_{fr} was ineffective.

The close similarity of the action spectra with the absorption spectrum of allophycocyanin indicated that the photochromic group of P is a fully conjugated phycobilin (PARKER et al. 1950). The similarities of the many features of the spectra indicate the P_{fr} is a protein as is allophycocyanin. The physiological responsiveness to very small amounts of P_{fr} indicated that the control acts through amplification.

4 Detection in Vitro and Isolation of Phytochrome

The knowledge about P (see BORTHWICK 1972 for another resume) was adequate in 1958 for formulation of an assay to guide isolation based on photoreversibility. This was accomplished by measuring differences in absorbancies in the 660 and 730 nm regions in a plaque of essentially achlorophyllous tissue with P in the P_r form relative to P in the P_{fr} form (BUTLER et al. 1959). A differential spectrometer, that came to be known as a ratio-spectrometer, was devised for the assay (BIRTH 1960). With the assay in hand, the concentration of P from *Avena sativa* seedlings to $>50\%$ purity was accomplished (SIEGELMAN and FIRER 1964). The properties of the separated P fully accorded with the physiological findings (BUTLER et al. 1965).

The P separated from the *Avena* seedlings had a molecular weight in the range of 60,000 to 120,000. Material separated from *Hordeum vulgare* dark grown seedlings gave molecular weights of 240,000 to 360,000 (CORRELL et al. 1968). Differences were shown to arise from proteolysis in extracts from *Avena.*

Phytochrome has also been isolated from the green algae *Mesotaenium* and the liverwort *Sphaerocarpus donnellii.* This verified the occurrence of P in phyla other than spermatophyta. P accordingly has an apparently primitive function in plants (TAYLOR and BONNER 1967). Pigments isolated from various phyla, however, are apt to differ somewhat in amino acid sequence while maintaining a close similarity in the region of the chromophore. This challenging possibility has not been examined.

5 Membrane Association of Phytochrome for Action

The first acts of light in photomorphogenesis are known. They are to change the ligand arrangement of the phycobilin chromophore of P and, in the 400 to 500 nm region, to excite an effective flavin. So attention is focused on the configuration of the P_{fr} chromophore and the immediate features of its action. The first of these involved molecular aspects of the subject. The second, the displays, is part of the physiology of the process. To know how P acts is to know the coupling between the initial molecular changes and the first steps of the displays. Evidence concerning membrane function in P action is discussed below. (See also a review by MARMÈ 1977).

5.1 Responses of Algae and Sporelings to Light

The control by phytochrome on the direction of outgrowth of fern and liverwort sporelings and the orientation of plastids in algae have been very informative about the action of P. Involvement of P in germination of fern spores (BÜNNING and MOHR 1955) led to observations on the direction of outgrowth of the protonema as influenced by polarized light (ETZOLD 1965). Irradiation with red light normal to the axis of the protonema leads to continued growth normal to the plane of polarization. ETZOLD suggested that the direction of maximum polarization of P_r is parallel to the cell surface while that of P_{fr} is normal to the surface. Outgrowth of germ tubes of the liverwort *Sphaerocarpus donellii* were similarly influenced by light (STEINER 1967).

Observations on the chloroplast orientation of the filamentous alga *Mougeotia* (HAUPT 1970, 1972) extended the findings from the fern and liverwort sporelings. It was known (OLTMANNS 1922) that the chloroplast was oriented normal to the light direction in dim light. HAUPT (1958) found from photoreversibility of the effect that the controlling action was due to P. By following changes in plastid orientation on irradiation with polarized microbeams at the cell surface HAUPT showed that part of the single plastid of a cell turns to the direction normal to the direction of maximum polarization of P_{fr}. The turning on irradiation with red light was initiated within minutes. The maximum position was reached in about 40 min. HAUPT suggested that the turning moment depends on the gradient of P_{fr} established by the radiation. The orientation results and the reversibility showed that both P_{fr} and P_r have fixed orientations with respect to the cell wall. The absorption vector of P_{fr} is normal to the cell wall while that of P_r is parallel.

5.2 The Structure of the Chromophore

The starting point on the structure of the P chromophore was the very close similarity between the action spectra dependent on P and the absorption spectrum of the biliprotein allophycocyanin (PARKER et al. 1950). FISHER, SEIDELL, and their students in the 1920 to 1940 period (See LEMBERG and LEGGE 1949 for the pertinent references) found the biliviolin to be an open chain tetrapyrole with full ligand conjugation (π bonding) through the four pyroles and the bridging methene groups. The structure approaches planarity.

The structure of the P chromophore has been examined by RÜDIGER and his associates by degradative methods (GROMBEIN et al. 1975). The photoisomerism involves ligand changes in ring A (Fig. 1). Binding of the chromophore to the apoprotein involves a thioether linkage through cysteine to the C_2 side chain of ring A. GROMBEIN et al. suggested the presence of a further ester linkage between the apoprotein and the propionic side chain of ring B. The change from P_r to P_{fr} was indicated as involving breakage of the (−C−S−) linkage and ligand rearrangement in ring A.

The structure of the biliviolin chromophore of phytochrome in the P_r form has also been examined in detail by LAGARIAS et al. (1980). The chromophore was cleaved from the apoprotein by pepsin and thermolysin which retain the polypeptide linkage through cysteine. The proposed structure does not have an ester linkage, through the propionic acid side chains, to the protein. This conclusion is based on the lack of ester hydrolysis by pepsin-thermolysin, which nevertheless gave a chromopeptide with linkage through cysteine for phytochrome P_r.

Breaking of the thioether linkage in the transformation of P_r to P_{fr} is unlikely since the chromophore would not be left with a covalent linkage to the protein in P_{fr}. LAGARIAS et al. suggest three possibilities for the structure of P_{fr}. These structures involve interaction of the phytochrome protein with the chromophore at the oxygen atom of ring A or ring D or the nitrogen atom of ring C. The appearance of 1.0 to 2.0 additional SH groups per P 120,000 molecular weight

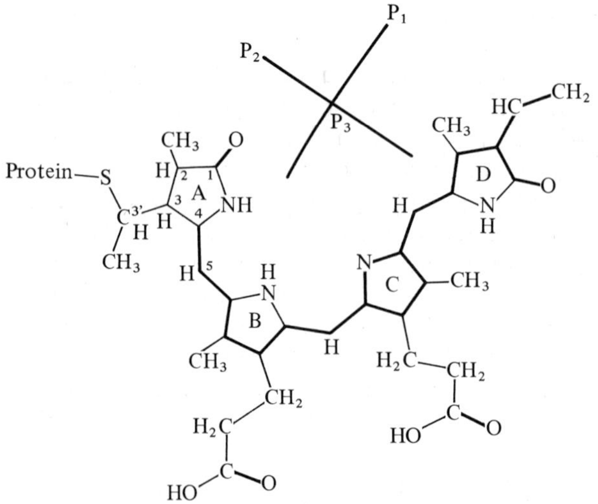

Fig. 1. The probable configuration of the phytochrome chromophore in the P_r form. P_1 and P_2 are the directions of maximum polarizations in the plane (approximate) of the chromophore. P_3, which is small compared to P_1 and P_2, is normal to the plane

is considered to arise by unmasking of existing groups upon transformation to P_{fr}. The detailed linkages in P_{fr}, however, must still be considered as unknown.

The high absorbancies of both forms of P at their absorption peaks are consistent with the conjugated π ligands of the chromophore and an approximately planar structure of the chromophore. SONG et al. (1979) calculated polarizabilities and positions of absorption maxima by the molecular orbital method for the possible biliviolin configurations. The observed and calculated absorbancies of P_r were in approximate agreement for the structure in Fig. 1. The directions of maximum polarizabilities are indicated.

5.3 Change of the Chromophore on Excitation

Transformation of P_r to P_{fr} shifts the absorption maximum from 660 to 730 nm corresponding to increased π conjugation or some type of charge displacement (see RÜDIGER, Chap. 7, this Vol.). The directions of calculated polarizabilities and the general configuration of the chromophore, particularly its approximate planarity, undergo little change. The protein moiety also apparently undergoes little conformational change. The activation of phytochrome resides in the chromophore.

In the polarotropic experiments with algae and ferns the shift of the absorption vector of P from parallel to the normal relationship to the plasma membrane upon transformation of P_r to P_{fr} suggests reorientation of the chromophore relative to the protein moiety. Details of such reorientation, its influence on P surface properties, and resultant changes in the membrane domain near P may be essential to understanding the action of P.

5.4 Membrane Charge and Transport

Further cogent observations for the present level of understanding of P_{fr} action were made by TANADA (1968). Segments of barley or of mung bean root tips (ca. 5 mm) suspended in an aqueous medium adhered to the bottom of the glass container when P was in the P_{fr} form. The segments were released upon transformation of P_{fr} to P_r. The response was repeatedly reversible in periods of <60 s for each transformation. A number of cofactors affected the display. The glass was exposed to phosphate or other multi-atom anions (see HUBBARD 1952 for ionic conditions at glass surfaces). The suspending solution contained Ca^{+2}, Mg^{+2}, and K^+ as 10^{-4} M salts, indoleacetic acid (10^{-10} M), ascorbic acid (10^{-6} M), and ATP (4×10^{-6} M). The rationale is that P_{fr} and some or all of these factors are involved in expression of a positive surface change in potential at the plasma membrane under control of P_{fr} associated with the membrane.

Attraction and repulsion of a cell of a teased mung bean root tip to a charged electrode on change in form of P demonstrated the positive charge on the cell surface (RACUSEN and ETHERTON 1975, see QUAIL Chap. 9, and RACUSEN and GALSTON Chap. 26, in this Vol. for treatments of the electric potential changes associated with P transformation). The potential across the plasma membrane of parenchyma cells of dark grown *Avena* coleoptiles depolarized by 5 to 10 mV in a few seconds upon change of P_r to P_{fr}. The change was reversible (RACUSEN 1976).

5.5 Turgor Change in Pulvini

The closure of legume leaves at dusk was commented on by Androsthenes on the march to India with Alexander's army in the 3rd century B.C. The influence of light and darkness on closure of *Mimosa pudica* pinnules was studied by BERT (1870). Detailed action spectra indicating phytochrome as the affecting pigment were made by FONDEVILLE et al. (1966). Involvement of phytochrome was also shown in pinnule closure of *Albizzia julibrissin* (HILLMAN and KOUKKARI 1967) and *Samanea saman* (SWEET and HILLMAN 1969). The essential findings of phytochrome action in the turgor control of legume-leaf movement were expressed by GALSTON and SATTER (1976) as follows:

"··· Because the electrical changes precede appreciable K^+ movement ··· anion or proton flux, possibly electrogenic, precedes and causes the K^+ flux that is important in leaflet movement".

The electrical changes are evident in seconds with *Samanea saman* pinnules. These findings are one of the indications for phytochrome action being a part of a protonmotive or cationmotive process.

5.6 Redox Potential and Cation Interplay

Phytochrome (P_{fr}) was shown to control expansion of bean-leaf disks cut from dark grown plants (LIVERMAN 1959). Reducing agents, including ascorbic acid, which was used by TANADA, were found to enhance expansion, and to act synergistically with P_{fr} (KLEIN and EDSALL 1966). Oxidizing agents such as 1–4 naphthaquinone and ferricyanide, repressed the expansion. The redox agents were of such a type as to be largely excluded from uptake by the leaf disks and thus were active across the plasma membrane. Their effects are consistent with proton or electron transfer across the membrane under P_{fr} influence.

The closing response of *Mimosa pudica* pinnules to touch requires the influx of Ca^{2+} (CAMPBELL and THOMSON 1977) as shown by high concentrations of EDTA preventing closure and by comparable results with lanthanum salts. A depolarization of the *Nitella* plasma membrane depended on the form of phytochrome and the ambient concentration of Ca^{2+} (WEISENSEEL and RUPPERT 1977). Irradiation of *Mougeotia* cells with red light for several seconds enhanced the Ca^{2+} uptake within 60 s as traced with $^{40}Ca^{2+}$ (DREYER and WEISENSEEL 1979). The enhanced rate was reversed by far-red radiation.

5.7 Membrane Fluidity

Germination at 40 °C of a lot of *Amaranthus retroflexus* seeds depended on the presence of P_{fr}. Cellular membrane fragments from the seeds show a transition in fluidity near 30 °C when tested with fluorescent probes (HENDRICKS and TAYLORSON 1979). The effectiveness of P_{fr} in causing germination varies with alternate temperature during periods <30 min. It is markedly enhanced if the temperature during this period is below 32 °C rather than above. The indication is that P_{fr} initial action is favored by the more organized membrane attained during the period at the lower temperature (HENDRICKS and TAYLORSON 1978).

A transition at 30 °C was observed for mitochondria isolated from hypocotyls of *Vicia faba* (MARX and BRINKMANN 1979). The transition is associated with energy conservation by the mitochondria as shown by its presence when NADH is the substrate under high phosphorylation (state 3) and absence when phosphorylation is low [state 4, absence of ADP + Pi or presence of uncouplers (phenyl hydrazones)].

6 Phytochrome and Cellular Organelles

Mitochondria isolated from etiolated pea epicotyls were observed by MANABE and FURUYA (1974) to contain significant amounts of phytochrome. The isolated mitochondria showed a reversible change in the reduction of NADP. Presence of $MgCl_2$ enhanced the rate of reduction by about twofold. The photoreversible effect and the reductive display were eliminated by swelling of the mitochondria in a hypotonic sucrose solution.

Phytochrome labeled with iodine bound in vitro to mitochondria isolated from etiolated *Avena sativa* seedlings (GEORGEVICH et al. 1977, CEDEL and ROUX

1980). The binding was enhanced by Mg^{2+}. P_{fr} probably binds to the outer mitochondrial-membrane surface. The activity of the antimycin A-insensitive NADH dehydrogenase is changed by the presence of P_{fr}. The enzyme is considered to be associated with the outer mitochondrial membrane.

Mitochondria isolated from etiolated *Avena* seedings varied by about two fold in (a) NADH-controlled oxygen consumption and (b) rates of ATP formation dependent on phytochrome change 5 to 10 min prior to isolation (HAMPP and WELLBURN 1979). Mitochondria from unirradiated tissue had no NADH-dependent ATP formation.

Etioplasts isolated from dark-grown plants contained photoreversible phytochrome (SMITH et al. 1978, HILTON and SMITH 1980). The isolated organelles increased two to three fold in gibberellin-type activity as measured by α-amylase action, following irradiation with red light. The enhanced potential activity was fully reversible.

The several findings demonstrate the functional association of P_{fr} with at least three types of membranes, namely mitochondria, etioplasts, and plasmalemma. The diversity of displays of P_{fr} action are probably connected in part with the effectiveness on various membranes (note PRATT 1979). Thus action at the plasmalemma is a part of cellular interplay and material transport. Action on the mitochondria is coupled with respiration and intermediary metabolism of a cell. The etioplast display is a control over the organization of an organelle, and with photosynthesis in the eventual chloroplast. The expectation, however, is that the associations of phytochrome with the several membranes are basically similar in affecting transfers across the membranes.

7 Pelletability of Phytochrome

Action of phytochrome through the control of membrane functions implies that P_{fr} might associate with a specific receptor on the membrane. An active search for biochemical evidence of such association has continued during the past decade. QUAIL et al. (1973) demonstrated that in vivo red irradiation of dark-grown corn coleoptiles enhanced the amount of particle-bound P in extracts obtained from such tissues in the presence Mg^{2+}. MARMÈ et al. (1973) showed that amounts of particle-bound P were increased by in vitro red irradiation of Mg^{2+}-containing extracts of dark-grown zucchini hypocotyl hooks. Both in vivo- and in vitro-induced pelletability phenomena have subsequently been actively investigated (for reviews see: MARMÈ 1977, PRATT 1978, PRATT 1979, QUAIL Chap. 9, this Vol.).

The significance of in vitro-induced pelletability was questioned by the demonstration that degraded 31S ribonucleoprotein material in the extracts bound P_{fr} (QUAIL and GRESSEL 1976). Definitive evidence that in vivo-induced pelletability, occurring with tissues and under conditions that minimize in vitro-induced pelletability, represents an in situ association of P with membranes is lacking. Support for this view, however, was provided by immunocytochemical studies of the light-induced, intracellular redistribution of P (PRATT et al. 1976, PRATT 1979). The speed of sequestering of P_{fr} at discrete, 1 μm regions within cells (completed in 1–2 min at 3 °C) was similar to that of in vivo-induced pelletability (40 s at 0.5 °C, 5–10 s at 25° C). Also, after a brief red–far-red irradiation sequence, the dispersion of phytochrome from the sequestered condition took place over the same time period (1–2 h at 24 °C) as in vivo loss of pelletability.

STONE and PRATT (1979) and JABBEN (1980) showed that although P may be destroyed in vivo in either the P_{fr} or the P_r forms, it must exist as P_{fr} for a short period of time for it to be subsequently destroyed as P_r. Although more detailed data are required, their findings indicated a half-time of a few minutes since FR given after 3 or 4 min of R inhibited destruction more than 50%. This period is greater than that required for in vivo induction of pelletability.

MACKENZIE et al. (1978) observed that the return of sequestered P to its diffuse distribution after a red–far-red irradiation occurred with kinetics parallel to those of in vivo destruction of P observed spectrophotometrically after a similar red–far-red irradiation. These observations suggest that the association of P_{fr} with membranes, indicated by sequestering and pelletability phenomena, may function in phytochrome destruction. They also suggest that a given phytochrome molecule has an increased probability of being conditioned for subsequent destruction as its photochemical lifetime as P_{fr} on the membrane exceeds 1 to 2 min. Section 9 discusses the possible significance of this in relation to "high irradiance responses".

The molecular nature of the binding of P to cellular membranes was examined by GRESSEL and QUAIL (1976). Phytochrome-containing membrane fractions from red-irradiated zucchini hypocotyl hooks were treated with phospholipase C. Although more than 80% of the choline was removed, no P was released. This finding suggested that P is not held by polar head groups of membrane phospholipids. Phytochrome was released by low concentrations of the detergent deoxycholate that did not cause substantial loss of integral membrane proteins or phospholipids. The findings indicated that P may associate with membrane proteins without forming nonpolar linkages. Studies by YAMAMOTO and FURUYA (1979) of the in vitro binding of P_{fr} to pea shoot membrane fractions treated with phospholipase C or trypsin suggested a similar conclusion.

Possible association of P with membrane steroids was examined in etiolated barley leaves by ROTH-BEJERANO and KENDRICK (1979). The pelletability of P in this tissue is not influenced by Mg^{2+} at 1 to 20 mM. Filipin, a polyene antibiotic which is known to combine with steroids, inhibited red light-induced pelletability of phytochrome in both in vivo and in vitro experiments. In vitro-induced pelletability was also inhibited by addition of cholesterol or stigmasterol to homogenates prior to irradiation. The combined addition of filipin and steroid at appropriate concentrations counteracted the inhibition of pelletability caused by either when added alone. The findings suggest that steroids as membrane components might also be involved in phytochrome binding.

Light-induced phytochrome pelletability has yet to be related to photomorphogenic responses. The pelletability might well represent a general rather than specific binding to membranes. That very low red fluences are effective in promoting some physiological responses (discussed below), is indicative of a specific binding.

8 Phytochrome Action at Very Low P_{fr} Levels

The use of a photomorphogenically active receptor for P_{fr} as a construction to examine mechanisms of physiological responses may focus the search for a receptor. The application of this approach to the difficulties of establishing quantitative relationships between the magnitudes of responses and photometrically estimated or spectrophotometrically measured levels of P_r and P_{fr} in tissues (HILLMAN 1967, 1972) is instructive. A response that exemplifies such difficulties is the R enhancement of phototropic sensitivity to blue light of dark-grown Zea coleptiles (BRIGGS and CHON 1966). Red light fluences sufficient to saturate the phototropic sensitivity were found to be two orders of magnitude lower than those required to induce measurable P transformation. Low FR fluences also increased phototropic sensitivity (CHON and BRIGGS 1966). Most significantly, promotion by R or low FR fluences could be inhibited by high FR fluences known to produce more P_{fr} than the promotive, low R fluences. The suggestion was advanced by HENDRICKS that photomorphogenically active receptors for P_{fr} exist and that their number in a cell is very low relative to the total number of phytochrome molecules. (See PRATT et al. 1976, VANDERHOEF et al. 1979). The P at these receptors remains bound even if photoconverted to P_r. Very low R or low FR fluences then produce P_{fr} molecules adequate to saturate the receptors. High FR fluences, near those adequate to establish photoequilibrium, are then required to photoconvert most receptor-bound P_{fr} to P_r and inhibit the response. The P_r held at the receptor prevents binding of FR-produced P_{fr}. Red fluences nearer the photoequilibrium level are then required to photoconvert receptor-bound P_r to P_{fr} and repromote the response. This hypothesis is consistent with the sensitivity changes of phototropic responses of *Zea* coleoptiles to brief R and FR irradiations.

RAVEN and SPRUIT (1973) proposed a similar hypothesis to explain the induction of rapid chlorophyll accumulation (under continuous white light) by preirradiations of dark-grown seedlings given many hours earlier. The fluence-response characteristics of this phenomena were determined by RAVEN and SHROPSHIRE (1975). The response was induced by very low R or low FR fluences and was not photoreversible by FR. A second red irradiation given several hours after the first required a 5,000-fold greater fluence for added stimulus and was fully inhibited by FR. To explain the absence of response photoreversibility by FR in the first irradiation, a major point of difference relative to the phototropic phenomena in *Zea,* RAVEN and SPRUIT postulated that the migration of P_{fr} and its binding to receptors might require several minutes or more. (Little P_r would then be formed by phototransformation at receptors and block subsequent binding of P_{fr}).

Very low P_{fr} levels are adequate to promote lettuce seed germination under some conditions. The fluence-response characteristics of normally non-light-requiring lettuce seeds (var. May Queen A and Noran) were examined by SMALL et al. (1979). A light requirement for germination was induced by either 48 h of continuous far-red irradiation or 48 to 72 h at 37 °C. Far red-dormant seeds required low R fluences. Thermodormant seeds were four orders of magnitude more sensitive to R. Their germination could also be promoted by FR fluences. Promotion by R or low FR fluences could be reversed by high FR fluences. This response pattern is similar to that of phototropic sensitization in *Zea* coleoptiles.

Low temperature treatment induces sensitivity to very low P_{fr} levels in Grand Rapids lettuce seeds (VAN DER WOUDE and TOOLE 1980). Seeds of the Grand Rapids variety were pretreated with brief far-red irradiation and a 24 h dark incubation to deplete initial levels of P_{fr} that would normally lead to high levels of dark germination. Several hours of low temperature incubation (termed prechilling) greatly enhanced the sensitivity of germination to subsequent brief far red irradiation compared to that of non-prechilled seeds. Enhancement of such sensitivity displayed temperature and time dependencies characteristic of membrane thermal adaptation. Prechilling did not alter subsequent rates of phytochrome action, but fluence-response characteristics were greatly changed. Seeds maximally sensitized by 24 h at 4 °C required R fluences that were four orders of magnitude less than those required by constant temperature controls for half maximal germination (VAN DER WOUDE 1982a, b). The R fluence-response characteristics of seeds prechilled for shorter periods were biphasic, having both a very low fluence component characteristic of maximally sensitized seeds and a low fluence component characteristic of unchilled seeds. Only the low fluence component of responses could be inhibited by far-red irradiation.

Evidence that phytochrome is a protein dimer (HUNT and PRATT 1980) suggests that the pigment may have the photointerconvertible forms $P_r{:}P_r$, $P_r{:}P_{fr}$ and $P_{fr}{:}P_{fr}$. An explanation of responses to very low "P_{fr}" levels based on the interaction of dichromophoric P with receptors, X, arose in studies of lettuce seed responses to prechilling (VAN DER WOUDE 1982a, b). Calculated fluence requirements for the formation of $P_r{:}P_{fr}$-X and $P_{fr}{:}P_{fr}$-X were very similar to those observed, respectively, for very low and low fluence responses. Comparisons of such fluence requirements suggested the cellular concentration of X to be about 10^{-3} that of total phytochrome dimer. In view of the apparent involvement, mentioned above, of membrane thermal adaptation in responses to prechilling, additional analyses suggested that the activity of $P_r{:}P_{fr}$-X is strongly influenced by membrane properties near X, and therefore by thermal or other factors that alter such properties.

In very low fluence responses, the probability may be low that a free $P_r{:}P_{fr}$ would be bound directly to a low number of receptors in time to produce the observed responses. This leads to the consideration that it is bound first to the membrane on which the receptor is located. Its probability of being bound to a receptor within a given time period would then be greatly increased, since movement would be limited by migration in the two-dimensional plane of the membrane surface. The phenomena of phytochrome pelletability and sequestering may in part reflect this possible membrane function.

9 High Irradiance Responses

Photomorphogenesis in plants often requires irradiation over long periods or frequent repetition for short times for significant display. (See FUKSHANSKY and SCHÄFER, Chap. 5, this Vol. for additional discussion of HIR responses). Three distinct types of photoreactions have been distinguished. They differ in the spectral regions involved. These regions and the probable effective pigments are:

1. 340 to 760 nm – phytochrome
2. 340 to 720 nm – chlorophyll or its degradation products
3. 340 to 500 nm – flavin.

Some essential features of the HIR with respect to phytochrome were shown by HARTMANN (1966) in studies on suppression of hypocotyl-lengthening of lettuce seedlings. Irradiation in the 600 to 800 nm region showed a maximum effectiveness near 716 nm. The degree of effectiveness depended on the energy. Many other photomorphogenic responses show the 716 nm peak of action and energy dependence (BORTHWICK et al. 1969, MOHR 1972). HARTMANN also found that an equal effectiveness could be attained by simultaneous irradiation with 658 and 768 nm bands both of which are ineffective when used alone. Both the single 716 nm band and the dual irradiance maintain about 3% P in the P_{fr} form. The conclusion from this work was that P_{fr} must be held at a relatively low level while repeatedly exciting both P_r and P_{fr}. The low level of P_{fr} was considered as reducing its destruction.

Action spectra for stem lengthening of green plants of *Beta vulgaris, Hyoscyamus niger* and *Spinacea oleracea* were obtained by interruption of long nights (SCHNEIDER et al. 1967, BORTHWICK et al. 1969). The plants were simultaneously irradiated in the 600 to 680 nm region adequately to maintain about 50% of total P as P_{fr}. The prominent action maxima nevertheless were at 716 nm. This was also the case for unfolding of *Mimosa pudica* leaflets (FONDEVILLE et al. 1967). These findings are not in apparent agreement with HARTMANN'S deductions about having a low P_{fr}/P value for the HIR. Possible explanations can be advanced: one is that total P_{fr}, rather than the ratio should be low. Such differences show the need for measurements of actual levels of P, P_r and P_{fr} in tissues being studied when the values are low or the plants are green. The possibility of measuring total P is now being attained with immunochemical methods (PRATT 1979).

HIR type actions with peaks for action in the 716 nm region also have peaks in the 340 to 500 region. The overall range of action has been reviewed by MANCINELLI and RABINO (1978). Examples of such actions are the suppression of stem lengthening for lettuce (MOHR and WEHRUM 1960, HARTMANN 1966) and anthocyanin synthesis in *Sinapis alba* (MOHR 1957) and turnip seedlings (SIEGELMAN and HENDRICKS 1957). The question as to whether phytochrome or a blue-light receptor is acting alone or together in various cases is open. The comparatively high responsiveness of the several HIR in the blue compared to the red, far-red regions is against phytochrome alone being effective.

In carotenoid synthesis in mycelia of *Neurospora crassa,* which do not contain phytochrome, MUNOZ and BUTLER (1975) found that radiation in the 400 to 500 nm region caused a reversible photoreduction of b-type cytochrome. The action spectrum for the reduction was similar to the absorbance of a flavin. The action spectrum in the 400 to 500 nm region for hypocotyl lengthening in lettuce seedlings corresponds to that of a flavin, although HARTMANN (1966) considered it to be due to P.

The action spectrum for anthocyanin synthesis in *Sorghum vulgare* cv. Wheatland seedlings (DOWNS and SIEGELMAN 1963) has a peak near 470 nm with less than 10% peak effectiveness over 530 nm. The action decreases to zero at 640 nm. While this is strictly a blue-light effect the extent to which it is displayed is, nevertheless, enhanced by conversion of P_r to P_{fr} at the termination of a light exposure. The interpretation is: The HIR involves the flavin-cytochrome system under modification by reversible phytochrome action at some stage in the overall process of forming anthocyanin. (See also SCHÄFER and HAUPT, Chap. 28, this Vol.).

Effectiveness of radiation in the 400 to 500 nm region on plastid orientation (HAUPT 1971) is informative about P_{fr} action. HAUPT and SCHÖNBOHM (1970) reviewed the pertinent findings over the 1938 to 1970 period. The organisms involved are the algae *Mougeotia, Mesotaenium,* and *Vaucheria sessilis,* and the club moss *Selaginella martensii.* The plastids can turn to positions space normal (+) or parallel (−) to the light at intensities low (L) or high (H).

Action maxima for the (H-) plastid movement of *Mougeotia* and *Mesotaenium* are near 480 and 450 nm respectively. Simultaneous irradiation of *Mougeotia* filaments with both blue (H) and red (L) light from different directions results in the (−) position with respect to red light which is photoreversible with far-red. The blue light is effective in changing the sign of the response irrespective of direction but does not control the orientation of the plastid except to the extent that it contributes to the $P_r \rightleftharpoons P_{fr}$ steady state.

Plastid orientation in *V. sessilis* is restricted to the 330 to 500 nm region and is not influenced by phytochrome. The action spectra are the same for (H) and (L) intensities, but the orientation responses are opposite. The plastid orientation in *V. sessilis* appears to arise from protoplasmic streaming that changes with light intensity in a manner that accords with the plastid orientation.

Anthocyanin synthesis in hypodermal cells of apple fruits is most effective in the 630 to 680 nm region with neglible action at >700 nm. 3-(3,4-dichlorophenyl) 1-,1-dimethylurea (DCMU) inhibits oxygen evolution similarly to the anthocyanin synthesis. DCMU is also an inhibitor of anthocyanin synthesis in strawberry leaf disks (CREASY 1968) and in *Spirodela polyrhiza* (MANCINELLI and RABINO 1978). The very low levels or absence of chlorophyll in dark-grown seedlings indicate that photosynthesis is not involved in their anthocyanin synthesis. Light-dependent phenylalanine ammonia lyase (PAL) activity in *Xanthium* leaf disks is also inhibited by DCMU (ZUCKER 1969). The activity of this enzyme in mustard seedling cotyledons has been shown to be a phytochrome-dependent HIR (MOHR 1972). PAL action is coupled to transcinnamic acid formation which is involved in the formation of the C9 moiety of anthocyanins.

High irradiance responses result from conditions of irradiance and spectral photon distribution that produce P_{fr} photochemical lifetimes less than 1 or 2 min (MANCINELLI and RABINO 1978). Consideration of the kinetics of P pelletability, sequestering, redistribution and destruction, discussed in Section 7, suggests that short P_{fr} lifetimes maximize the quantity of P, both P_r and P_{fr}, on membranes (VAN DER WOUDE 1982a). The resultant alteration of membrane properties and functions may, at least in part, underlie high irradiance responses.

10 How Phytochrome Acts

JOHN MUIR wrote in *My First Summer in the Sierras:* "When we try to pick out anything by itself, we find it hitched to everything in the universe." JOHN MUIR's comment is true for phytochrome. P_{fr} associated with a membrane would be expected to interact with other factors affecting the membrane. These factors include change in temperature, presence of hormones, inorganic ions, and compounds such as anesthetics which influence membrane fluidity. Eventually transport and metabolism become involved as a result of membrane functioning.

The manifold displays have meant complexity of action to some observers, but simplicity to others. Ubiquitous reversal of potential responses point to a unique initial action which is clearly identified as functional association of P_{fr} with membranes. Possibilities include the plasma, mitochondria, and plastid membranes as previously discussed. Others are unexamined rather than excluded. The manner in which these three membranes act is at the core of MITCHELL's (1979) chemiosmotic theory. Involvement of temperature change, phytochrome, and other membrane-modifying substances in movement of protons, i.e., proticity, is expected even though the details of the involvement are only beginning to be known.

The H_2, O_2 fuel cell (HABER and FLEISCHMANN 1906) illustrates concepts of electricity and proticity (MITCHELL 1979) associated with the reaction $H_2 + \frac{1}{2}O_2 \rightarrow H_2O + \text{energy}$. In an equivalent way the reduction of oxygen $2\,NADH + 2\,H^+ + O_2 \rightarrow 2\,NAD + 2\,H_2O + \text{energy}$ in the several steps of the respiratory chain generates flow of charge (LUNDEGARDH 1945). Much remains unknown about how protons or other cations cross the membranes. This unknown part, possibly involving carriers and aqueous channels, is where phytochrome is likely to influence the process by membrane modification.

The charge appearing on the root tip surface with transformation of P_r to P_{fr} (TANADA 1968, RACUSEN and ETHERTON 1975) in the chemiosmotic sense is an expected result of proticity. The positive charge appearing on the outside of the membrane is in the Donnan equilibrium space of the plasma membrane outer surface which includes the amphoteric phosphatidyl choline and serine groups. The Ca^{2+}, Mg^{2+}, and H^+ ions are preferentially held at this surface and thus allow the surface to be positive and be attacted to the negative fixed charge space at a glass surface. The experiment of WEISENSEEL and RUPPERT (1977) showing depolarization of *Nitella* plasma membrane by P_{fr} in the presence of Ca^{2+} is an expression of the same generation of proticity. This is also seemingly the case for Ca^{2+} movement into *Mougeotia* cells under control of P_{fr} by producing H^+ counter transport to Ca^{2+} (DREYER and WEISENSEEL 1979). The influx of Ca^{2+} in the response to touch of *Mimosa pudica* (CAMPBELL and THOMSON 1977) is a further expression as also could be the expression of legume leaf movement under P_{fr} control as previously noted by GALSTON and SATTER (1976). The detailed coupling-energy source (ATP) remains unknown in these several cases.

The P_{fr} control is expressed by association with the inner surface of the plasma membrane through the mediation of the P_{fr} chromophore as discussed for plastid movement in *Mougeotia*.

Mitochondria, etioplasts, and plastids have the common feature of inner and outer membranes. Phytochrome associated with the outer surface of organelles may be lost during isolation and the possible retention of P_{fr} at inner membranes and its action in isolated systems deserves consideration. Herein might lie the significance of the necessity of irradiation prior to isolation of the mitochondria noted by HAMPP and WELLBURN (1979). The effects observed by them on rates of NADH-dependent O_2-consumption and ATP formation under some degree of control of P_{fr} would be responsive to the functioning of the cristae as a chemiosmotic system. Reduction of NADP as noted by MANABE and FURUYA (1974) would appear to be remote from the immediate action at the cristae, where the system is oxidative for NADH. A similar situation exists for the phytochrome interplay with NAD kinase (YAMAMOTA and TEZUKA 1972).

Phytochrome as P_{fr} exerts a control both over plastid development in vivo (BRADBEER and MONTES 1976) and plastid action in vitro (SMITH et al. 1978, DOWNS and SIEGELMAN 1963). (See also LAMB and LAWTON Chap. 10, and SCHÖPFER and APEL Chap. 11, this Vol.). Both of these responses may involve P_{fr} inside the organelle where action could be influenced by proticity at inner membranes. In a third type of response, i.e., P_{fr} influence on anthocyanin synthesis associated with photosynthesis, P_{fr} may exert its effect on the outer membrane of the plastid where protons are appearing or an oxidation reduction-loop could be functioning.

References

Bert P (1870) Recherches sur les mouvements de la sensitive (*Mimosa pudica* L.). Mem Soc Sci Phys Nat Bordeaux 8:1–58

Birth GS (1960) Agricultural applications of the dual-monochrometer spectrometer. Agric Eng 41:432–435

Borthwick HA (1972) History of phytochrome. In: Mitrakos K, Shropshire W Jr (eds) Phytochrome. Academic Press, London New York, pp 3–23

Borthwick HA (1972) The biological significance of phytochrome. In: Mitrakos K, Shropshire W (eds) Phytochrome. Academic Press, London New York, pp 27–44

Borthwick HA, Hendricks SB, Parker MW (1948) Action spectrum for photoperiodic control of floral initiation of a long-day plant, Wintex barley (*Hordeum vulgare*). Bot Gaz 110:103–118

Borthwick HA, Hendricks SB, Parker MW (1952a) The reaction controlling floral initiation. Proc Natl Acad sci USA 38:929–934

Borthwick HA, Hendricks SB, Parker MW, Toole EH, Toole VK (1952b) A reversible photoreaction controlling seed germination. Proc Natl Acad Sci USA 38:662–666

Borthwick HA, Hendricks SB, Toole EH, Toole VK (1954) Action of light on lettuce seed germination. Bot Gaz 115:205–225

Borthwick HA, Hendricks SB, Schneider MJ, Taylorson RB, Toole VK (1969) The high energy light action controlling plant responses and development. Proc Natl Acad Sci USA 64:479–486

Bradbeer JW, Montes G (1976) The photocontrol of chloroplast development – ultrastructural aspects and photosynthetic activity. In: Smith H (ed) Light and plant development. Butterworth, London, pp 213–227

Briggs WR, Chon HP (1966) The physiological versus the spectrophotometric status of phytochrome in corn coleoptiles. Plant Physiol 41:1159–1166

Bünning E, Mohr H (1955) Das Aktionsspektrum von Lichteinfluß auf die Keimung von Farnsporen. Naturwissenschaften 42:212

Butler WL, Norris KH, Siegelman HW, Hendricks SB (1959) Detection, assay, and preliminary purification of the pigment controlling photoresponsive development of plants. Proc Natl Acad Sci USA 45:1703–1708

Butler WL, Hendricks SB, Siegelman HW (1965) Purification and properties of phytochrome. In: Goodwin TW (ed) Biochemistry of plant pigments. Academic Press, London New York, pp 197–210

Campbell NA, Thomson WW (1977) Effects of lanthanum and ethylenediamine tetraacetate on leaf movement of *Mimosa*. Plant Physiol 60:635–639

Caspari R (1860) *Bullardia aquatica*. Schriften Kon Physikal-Oekon Ges Königsberg 1:66–91

Cedel TE, Roux SJ (1980) Further characterization of the in vitro binding of phytochrome to a membrane fraction enriched for mitochondria. Plant Physiol 66:696–703

Chon HP, Briggs WR (1966) Effect of red light on the phototropic sensitivity of corn coleoptiles. Plant Physiol 41:1715–1724

Correll DL, Steers E, Towe KM, Shropshire W Jr (1968) Phytochrome in etiolated annual rye. IV. Physical and chemical characterization of phytochrome. Biochim Biophys Acta 168:46–57

Correll DL, Edwards JL, Shropshire W Jr (1977) Phytochrome. Smithson Inst Press, Washington DC

Creasy L (1968) The significance of carbohydrate metabolism in flavonoid synthesis in strawberry leaf disks. Phytochemistry 7:1743–1749

Downs RJ, Siegelman HW (1963) Photocontrol of anthocyanin synthesis in milo seedlings. Plant Physiol 38:25–30

Downs RJ, Hendricks SB, Borthwick HA (1957) Photoreversible control of elongation of Pinto beans and other plants under normal conditions of growth. Bot Gaz 118:199–208

Dreyer EM, Weisenseel MH (1979) Phytochrome-mediated uptake of calcium in *Mougeotia* cells. Planta 146:31–39

Etzold H (1965) Der Polarotropismus und Phototropismus der Chloronemen von *Dryopteris filix-mas*. Planta 64:254–280

Flint LH, McAlister ED (1937) Wave lengths in the visible spectrum inhibiting the germination of light-sensitive lettuce seeds. Smithson Misc Collect 96:1–8

Fondeville JC, Borthwick HA, Hendricks SB (1966) Leaflet movement of *Mimosa pudica* L. I. Identification of phytochrome action. Planta 69:357–364

Fondeville JC, Schneider MJ, Borthwick HA, Hendricks SB (1967) Photocontrol in *Mimosa pudica* L. leaf movement. Planta 75:228–238

Galston AW, Satter RL (1976) Light, clocks and ion flux: An analysis of leaf movement. In: Smith H (ed) Light and plant development. Butterworth, London, pp 159–184

Garner WW, Allard HA (1920) Effect of the relative length of day and night and other factors of the environment on growth and reproduction in plants. J Agric Res 18:553–606

Georgevich G, Cedel TE, Roux SJ (1977) Use of ^{125}I-labeled phytochrome to quantitate phytochrome binding to membranes of *Avena sativa*. Proc Natl Acad Sci USA 74:4439–4443
Gressel J, Quail PH (1976) Particle-bound phytochrome: differential pigment release by surfactants, ribonuclease and phospholipase C. Plant Cell Physiol 17:925–940
Grombein S, Rüdiger W, Zimmerman H (1975) The structure of the phytochrome chromophore in both photoreversible forms. Hoppe-Seyler's Z Physiol Chem 356:1709–1714
Haber F, Fleischmann F (1906) The oxyhydrogen cell. I. Z Anorg Chem 51:245–288
Hampp R, Wellburn AR (1979) Control of mitochondrial activity by phytochrome during greening. Planta 147:229–235
Hartmann KM (1966) A general hypothesis to interpret "high energy phenomena" of photomorphogenesis on the basis of phytochrome. Photochem Photobiol 5:349–366
Haupt W (1958) Hellrot-dunkelrot-Antagonismus bei der Auslösung der Chloroplastenbewegung. Naturwissenschaften 45:273–274
Haupt W (1970) Über den Dichroismus von Phytochrome 660 und Phytochrome 730 bei *Mougeotia*. Z Pflanzenphysiol 62:287–298
Haupt W (1971) Schwachlichtbewegung des *Mougeotia*-Chloroplasten im Blaulicht. Z Pflanzenphysiol 65:248–265
Haupt W (1972) Localization of phytochrome within the cell. In: Mitrakos K, Shropshire W Jr (eds) Phytochrome. Academic Press, London New York, pp 553–569
Haupt W, Schönbohm E (1970) Light-oriented plastid movements. In: Halldal P (ed) Photobiology of microorganisms. Wiley-Interscience, New York, pp 282–307
Heald F De F (1898) Conditions for the germination of the spores of bryophytes and pteridophytes. Bot Gaz 26:25–45
Hendricks SB, Borthwick HA (1954) Photoresponsive growth. In: Rudnik D (ed) Aspects of synthesis and order in growth. Princeton Univ Press, Princeton, pp 149–169
Hendricks SB, Borthwick HA (1967) The function of phytochrome in regulation of plant growth. Proc Natl Acad Sci USA 58:2125–2130
Hendricks SB, Taylorson RB (1978) Dependence of phytochrome action in seeds on membrane organization. Plant Physiol 61:17–18
Hendricks SB, Taylorson RB (1979) Dependence of thermal responses of seeds on membrane transitions. Proc Natl Acad Sci USA 76:778–781
Hendricks SB, Borthwick HA, Downs RJ (1956) Pigment conversion in the formative responses of plants to radiation. Proc Natl Acad Sci USA 42:19–26
Hillman WS (1967) The physiology of phytochrome. Annu Rev Plant Physiol 18:301–324
Hillman WS (1972) On the physiological significance of in vivo phytochrome assay. In: Mitrakos K, Shropshire W Jr (eds) Phytochrome. Academic Press, London New York, pp 573–584
Hillman WS, Koukkari WL (1967) Phytochrome effects on the nyctinastic movements of *Albizzia julibrissin* and some other legumes. Plant Physiol 42:1413–1418
Hilton JR, Smith H (1980) The presence of phytochrome in purified barley etioplasts and its in vitro regulation of biologically active gibberellin levels in etioplasts. Planta 148:312–318
Hubbard DH (1952) Heterogeneous equilibria at the glass electrode-solution interface. J Res Nat Bureau Standards 48:428–437
Hunt RE, Pratt LH (1980) Partial characterization of undegraded oat phytochrome. Biochemistry 19:390–394
Jabben M (1980) The phytochrome system in light-grown *Zea mays* L. Planta 149:91–96
Jacobi H (1914) Wachstumsreaktionen von Keimlingen, hervorgerufen durch monochromatisches Licht. 1. Rot. Sitzungsber Akad Wiss Wien Math Naturwiss 123:617–631
Kinzel W (1908) Die Wirkung des Lichtes auf die Keimung. Ber Dtsch Bot Ges 26:105–115, 631–645, 654–665
Klein RM, Edsall PC (1966) Substitution of redox potential for radiation in phytochrome mediated photomorphogenesis. Plant Physiol 41:949–952
Lagarias JC, Glazer AN, Rapoport H (1980) Chromopeptides from C-phycocyanin, struc-

ture and linkage of a phycocyano-bilin bound to the β subunit. J Am Chem Soc 101:5030–5037

Lemberg R, Legge JW (1949) Hematin compounds and bile pigments. Wiley-Interscience, New York

Liverman JL (1959) Control of leaf growth by interaction of chemicals and light. In: Withrow RB (ed) Photoperiodism and related phenomena in plants and animals. AAAS Washington, pp 161–180

Lundegardh H (1945) Absorption, transport, and exudation of inorganic ions by roots. Arkiv Bot 32A, 12:1–58

Mackenzie JM Jr, Briggs WR, Pratt LH (1978) Phytochrome photoreversibility: empirical test of the hypothesis that it varies as a consequence of pigment compartmentation. Planta 141:129–134

Manabe K, Furuya M (1974) Phytochrome dependent reduction of nicotinamide nucleotides in the mitochondrial fraction isolated from etiolated pea epicotyls. Plant Physiol 53:343–347

Mancinelli AL, Rabino I (1978) The "high irradiance responses" of plant photomorphogenesis. Bot Rev 44:129–180

Marmè D (1977) Phytochrome: membranes as possible sites of primary action. Annu Rev Plant Physiol 28:173–198

Marmè D, Boisard J, Briggs WR (1973) Binding properties in vitro of phytochrome to a membrane fraction. Proc Natl Acad Sci USA 70:3861–3865

Marx R, Brinkmann K (1979) Effect of temperature on the pathway of NADH-oxidation in broad-bean mitochondria. Planta 144:359–365

Mitchell P (1979) Keilin's respiratory chain concept and its chemiosmotic consequences. (Nobel Lecture 1978) Science 206:1148–1159

Mohr H (1957) Der Einfluß monochromatischer Strahlung auf das Längenwachstum des Hypocotyls und auf die Anthocyanbildung bei Keimlingen von *Sinapis alba* L. Planta 49:389–405

Mohr H (1972) Lectures on photomorphogenesis. Springer, Berlin Heidelberg New York

Mohr H, Wehrung M (1960) Die Steuerung des Hypokotylwachstums bei den Keimlingen von *Lactuca sativa* L. durch sichtbare Strahlung. Planta 55:438–450

Munoz V, Butler WL (1975) Photoreceptor pigment for blue light in *Neurospora crassa*. Plant Physiol 55:421–426

Nuernbergk E (1927) Untersuchungen über die Lichtverteilung in *Avena*-Koleoptilen und anderen phototropisch reizbaren Pflanzenorganen bei einseitiger Beleuchtung. Bot Abhand 12:5–162

Oltmanns F (1922) Morphologie und Biologie der Algen. Fischer, Jena

Parker MW, Hendrick SB, Borthwick HA, Scully NJ (1946) Action spectrum for the photoperiodic control of floral initiation of short-day plants. Bot Gaz 108:1–26

Parker MW, Hendricks SB, Borthwick HA, Went FW (1949) Spectral sensitivity for leaf and stem growth of etiolated pea seedlings and their similarity to action spectra for photoperiodism. Am J Bot 36:194–204

Parker MW, Hendricks SB, Borthwick HA (1950) Action spectrum for the photoperiodic control of floral initiation of the long-day plant *Hyoscyamus niger*. Bot Gaz 111:242–252

Pratt LH (1978) Molecular properties of phytochrome. Photochem Photobiol 27:81–105

Pratt LH (1979) Phytochrome: function and properties. Photochem Photobiol Rev 4:59–124

Pratt LH, Coleman RA, Mackenzie JM Jr (1976) Immunological visualization of phytochrome. In: Smith H (ed) Light and plant development. Butterworth, London, pp 75–94

Priestley JH (1925) Light and growth. I. The effect of bright light exposure on etiolated plants. II. On the anatomy of etiolated plants. New Phytol 24:271–283, 25:145–170

Quail PH, Gressel J (1976) Particle-bound phytochrome: interaction of the pigment with ribonucleoprotein material from *Cucurbita pepo* L. In: Smith H (ed) Light and plant development. Butterworth, London, pp 111–128

Quail PH, Marmè D, Schäfer E (1973) Particle-bound phytochrome from maize and pumpkin. Nat New Biol 245:189–191

Racusen RH (1976) Phytochrome control of electrical potentials and intercellular coupling in oat-coleoptile tissue. Planta 132:25–29

Racusen RH, Etherton B (1975) Role of membrane-bound fixed charges in phytochrome mediated mung bean tip adherence phenomena. Plant Physiol 55:491–495

Raven CW, Shropshire W Jr (1975) Photoregulation of logarithmic fluence-response curves for phytochrome control of chlorophyll formation in *Pisum sativum* L. Photochem Photobiol 21:423–429

Raven CE, Spruit CJP (1973) Induction of rapid chlorophyll accumulation in dark-grown seedlings. III. Transport model for phytochrome action. Acta Bot Neerl 22:135–143

Resühr B (1939) Beiträge zur Lichtkeimung von *Amaranthus caudatus* L. und *Phacelia tanacetifolia* Benth. Planta 30:471–506

Roth-Bejerano N, Kendrick RE (1979) The effects of filipin and steroids on phytochrome pelletability. Plant Physiol 63:503–506

Schneider MJ, Borthwick HA, Hendricks SB (1967) Effects of radiation on flowering of *Hyoscyamus niger*. Am. J Bot 54:1241–1249

Siegelman HW, Firer EM (1964) Purification of phytochrome from oat seedlings. Biochemistry 3:418–423

Siegelman HW, Hendricks SB (1957) Photocontrol of anthocyanin formation in turnip and red cabbage seedlings. Plant Physiol 32:393–398

Small JGC, Spruit CJP, Blaauw-Jansen G, Blaauw OH (1979) Action spectra for light-induced germination in dormant lettuce seeds. I. Red region. Planta 144:125–131

Smith H, Evans A, Hilton JR (1978) An in vitro association of soluble phytochrome with a partially purified organelle fraction from barley leaves. Planta 141:71–76

Song PS, Chae Q, Gardner JD (1979) Spectroscopic properties and chromophore conformation of the photomorphogenic receptor phytochrome. Biochim Biophys Acta 576:479–495

Stebler FG (1881) Über die Einwirkung des Lichtes auf die Keimung. Bot Centrl 2:157–158

Steiner AM (1967) Phytochrome action elicited by short wave length irradiation in polarotropism of germlings of a fern and a liverwort. Action Spectra. Proc Eur Ann Symp Plant Photomorphogenesis, Hvar, pp 113–116

Stone HJ, Pratt LH (1979) Characterization of the destruction of phytochrome in the red-absorbing form. Plant Physiol 63:680–682

Sweet HC, Hillman WS (1969) Phytochrome control of nyctinasty in *Samanea* as modified by oxygen, submergence, and chemicals. Physiol Plant 22:776–786

Tanada T (1968) Substances essential for a red, far-red light reversal attachment of mung bean root tips to glass. Plant Physiol 43:2070–2071

Taylor AO, Bonner BA (1967) Isolation of phytochrome from the alga *Mesotaenium* and the liverwort *Sphaerocarpus*. Plant Physiol 42:762–766

Toole VK (1973) Effects of light, temperature, and their interactions on the germination of seed. Seed Sci Tech 1:339–396

Trumpf C (1924) Über den Einfluß intermittierender Belichtung auf die Etiolation der Pflanzen. Bot Arch 5:381–410

Vanderhoef LN, Quail PH, Briggs WR (1979) Red light-inhibited mesocotyl elongation in maize seedlings II. Kinetic and spectral studies. Plant Physiol 63:1062–1067

Van Der Woude WJ (1982a) Mechanisms of photothermal interactions in phytochrome control of seed germination. In: Meudt W (ed) Strategies of plant reproduction. Beltsville Symp Agric Res 6 (In press)

Van Der Woude WJ (1982b) A dichromophoric model for the action of phytochrome: Evidence from photothermal interactions in lettuce seed germination. Proc Natl Acad Sci USA (In press)

Van Der Woude WJ, Toole VK (1980) Studies of the mechanism of enhancement of phytochrome-dependent lettuce seed germination by prechilling. Plant Physiol 66:220–224

Vince-Prue D (1975) Photoperiodism in Plants. McGraw-Hill, New York
Warburg O, Negelein E (1929) Über das Absorptionsspektrum des Atmungsferments. Biochem Z 214:64–100
Weisenseel MH, Ruppert HK (1977) Phytochrome and calcium ions are involved in light-induced depolarization in *Nitella*. Planta 137:225–229
Yamamoto KT, Furuya M (1979) Effects of enzymatically digested microsome fractions on red-light-enhanced pelletability of pea phytochrome in vitro in the presence of calcium ion. Plant Cell Physiol 20:1591–1601
Yamamoto Y, Tezuka T (1972) Regulation of NAD kinase by phytochrome and control of metabolism by variation of NADP level. In: Mitrakos K, Shropshire W Jr (ed) Phytochrome. Academic Press, London New York, pp 408–429
Zucker M (1969) Induction of phenylalanine ammonia-lyase in *Xanthium* leaf disks. Photosynthetic requirement and effect of daylength. Plant Physiol 44:912–922

3 An Introduction to Photomorphogenesis for the General Reader

H. MOHR and W. SHROPSHIRE JR.

1 Aim and Scope of this Volume

In photosynthetic green plants light is the decisive environmental factor. The terrestrial green plant is organized almost ideally in a way so as to maximize absorption and processing of photosynthetic light quanta. The *genetic* adaptation to the factor, light, has taken place in the course of the genetic evolution (phylogeny) of terrestrial plants. However, light also affects the individual development (ontogeny) *specifically* insofar as the genes which control development of plants can only express themselves fully in the presence of light. Thus, the development of plants ("photomorphogenesis") is characterized by the obligatory interaction between genes and environment, specifically the light environment. Our present knowledge about the mechanisms of light absorption, transduction of the light signal, and final expression of the photoresponses, i.e., the biophysical, molecular and physiological events during photomorphogenesis, are described in this present volume.

2 Photomorphogenesis in Seedlings and Sprouts

Photomorphogenesis is essential to all terrestrial plant life. Normal development in higher plants *is* photomorphogenesis. This is most conspicuously exhibited by seedlings of spermatophytes or sporelings of ferns, although the photomorphogenic light effect is detected at all major stages of plant development, e.g., seed germination, later stages of vegetative growth, transition to reproductive development and senescence.

For a specific example, let us consider mustard seedlings (*Sinapis alba*), which illustrate the basic phenomena (Fig. 1). All three seedlings have essentially the same genes, the same chronological age, and all three were grown on the same medium. The conspicuous differences in appearance (morphogenesis) are thus due to light. Since the development of the seedling under white light (which allows formation of chlorophyll and photosynthesis) does not significantly differ from the development of the seedling under continuous far-red light (which allows only traces of chlorophyll to be formed and does not support photosynthesis), the effect of light on morphogenesis is not a consequence of photosynthesis.

Photomorphogenesis is not restricted to seedlings, but occurs in structures of vegetative reproduction as well (Fig. 2). The two potato plants (sprouts)

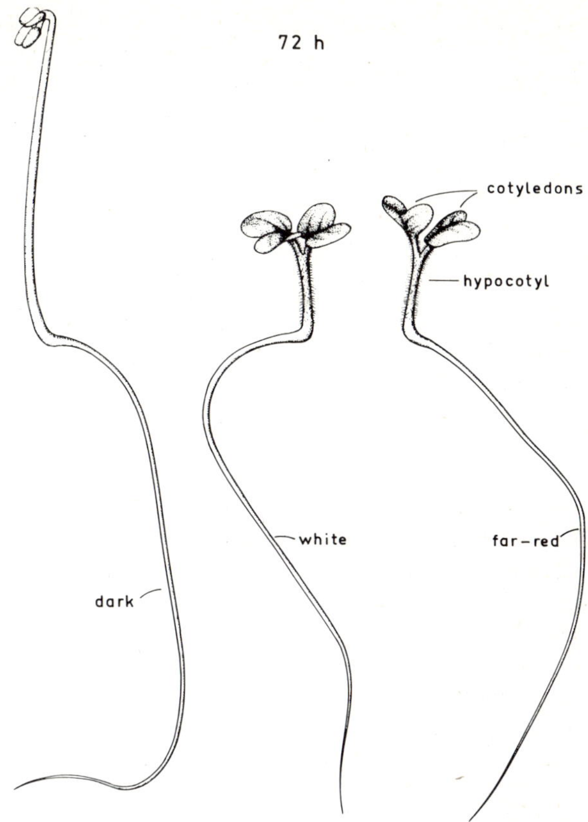

Fig. 1. Seedlings of white mustard (*Sinapis alba*) 72 h after sowing at 25 °C. The seedlings were either kept in the dark, in white light (3800 lx) or in far-red light (3.5 Wm^{-2})

are genetically identical since they originate from tubers produced by the same mother plant. The traditional description of the phenomenon is that sprouts or seedlings will "etiolate" in darkness, whereas in the light normal development will take place. The biological or ecological significance of etiolation is obvious. As long as a plant has to survive in darkness it uses the limited supply of storage material predominantly for axis growth. In this way the probability is highest that the tip of the plant (including the rudimentary leaves) will reach the light before the storage material is exhausted.

A slightly different description of the phenomena is the following: the seedling or the sprout of a terrestrial plant is genetically endowed with the ability to follow two different strategies of development, depending on the ambient light conditions. These are "photomorphogenesis" and "skotomorphogenesis" (from the Greek skotos = dark). Photomorphogenesis is the strategy of development if and as long as light is available, while skotomorphogenesis (etiolation) is the developmental strategy of choice as long as light is not yet, or no longer, available. During skotomorphogenesis the reserve material of a seed or a tuber is predominantly invested in rapid axis growth (in an effort to reach the light

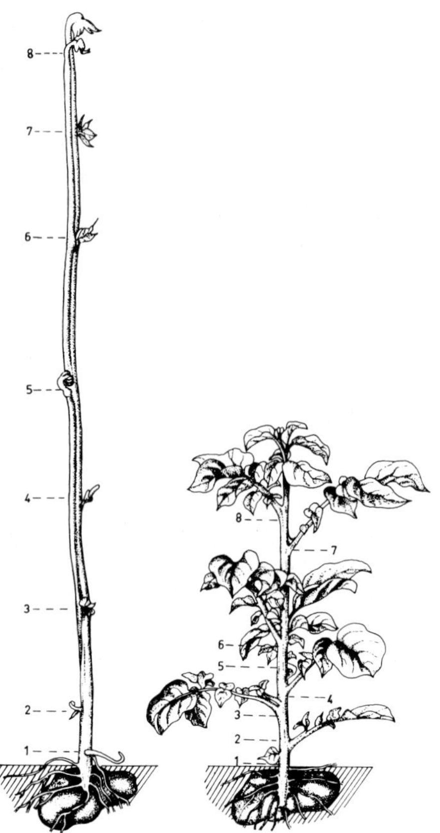

Fig. 2. Genetically identical potato plants (*Solanum tuberosum*) grown in the dark (*left*) or in natural daylight (*right*). (After PFEFFER 1904). *Numbers* indicate the position of the corresponding leaves along the main axis

as rapidly as possible!), while in the case of photomorphogenesis, the seedling or sprout invests the storage material predominantly in the construction of the photosynthetic apparatus, including expansion of leaves, differentiation of conducting tissues in the axis and formation of chloroplasts in the mesophyll and (to a limited extent) in the cortex of the stem.

The transition from skotomorphogenesis to photomorphogenesis has been investigated intensively since the early days of quantitative plant physiology (see PFEFFER 1904, KLEBS 1913). Many plant physiologists have been intrigued by the phenomenon that a single, well-defined environmental factor, light, can cause a plant to switch over from one developmental strategy to another, i.e., from one pattern of gene expression to another. Modern biophysical and molecular methods have greatly fostered the causal analysis of the transition from skoto- to photomorphogenesis. Moreover, it has been recognized recently that the transition from skoto- to photomorphogenesis can be considered a particularly useful model process in the present efforts to understand in *molecular* terms the control of gene expression during development in multicellular eukaryotic organisms.

A caveat should be kept in mind, however. Even though the difference between skoto- and photomorphogenesis is striking, it is clear that light does not carry any information with regard to the specificity of morphogenesis. Light can only be considered as an "elective factor" with regard to that genetic information which is present in the particular organism. The developmental *potential* of a plant is, of course, determined by its genes, not by light. This is dramatically illustrated by the phyllotactic pattern of the potato plants in Fig. 2. It is obvious that *specification* of the phyllotactic pattern is independent of light even though pattern *realization* (see Chap. 14, this Vol.), i.e., actual leaf development, takes place only in the light.

3 Photomorphogenesis in Sporelings of Ferns

The sporelings (=young gametophytes) of the common male fern (*Dryopteris filix-mas*) may be considered as being representative for the sporelings of many leptosporangiate ferns. The gonospores (the spores resulting from meiosis) of *Dryopteris* germinate only in the light. This germination is a typical phytochrome response (see Sect. 4.1) brought about by red light only. The sporelings, however, can develop "normally" from the very transient filamentous one-dimensional

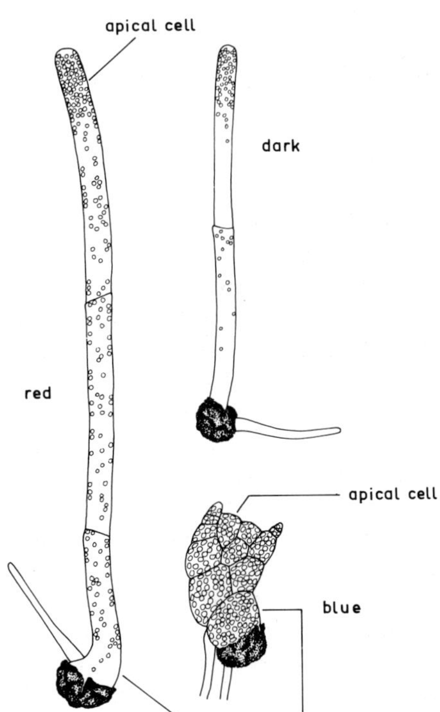

Fig. 3. Typical sporelings of *Dryopteris filix-mas* after 6 days culture on inorganic nutrient solution. The sporelings were either grown in darkness or in continuous red or blue light of equal photon fluence rate (approximately 1 Wm^{-2} in blue light). The blue-light-grown and the red-light-grown sporelings have about the same dry mass. (After MOHR and OHLENROTH 1962)

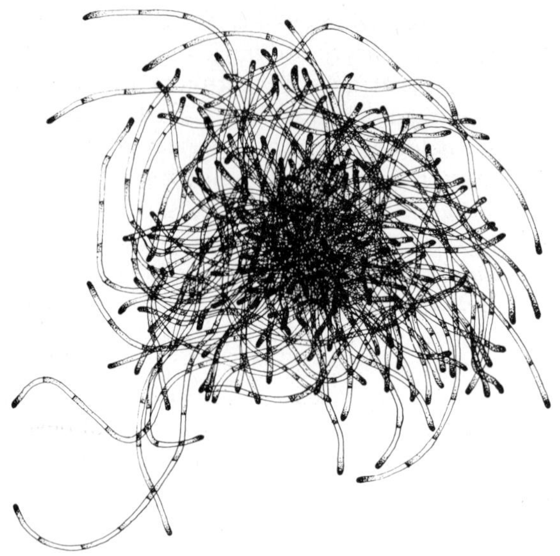

Fig. 4. A manifold-branched fern protonema (*Dryopteris filix-mas*) grown on inorganic liquid medium in red light about $2^1/_2$ months after spore germination. The protonema originated from a single spore. (Courtesy H. MAY)

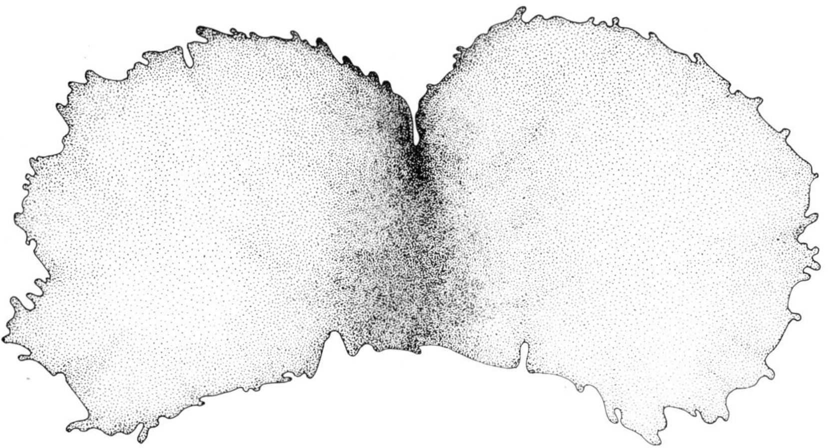

Fig. 5. A typical prothallus of the fern *Dryopteris filix-mas* grown on inorganic liquid medium in blue light about 2 months after spore germination. (Courtesy H. MAY)

stage, called the protonema, to the two- or three-dimensional stage, called the prothallus, only if they receive enough short-wavelength visible light below 500 nm (Fig. 3). Under long-wavelength visible light, e.g., red or far-red light of any fluence rate tested, these sporelings will continue to grow for many days as cellular filaments which are similar to the filaments of the dark controls (Fig. 3). The characteristic differences in morphogenesis will remain if the culture of the sporelings is continued over long periods (Figs. 4, 5).

It is apparent in the present case that only short-wavelength visible light ($\lambda < 500$ nm) can cause the sporelings to switch over from skotomorphogenesis (filamentous growth) to photomorphogenesis which we consider the normal development of a fern gametophyte (heart-shaped prothallus). While in the case of a seedling (Fig. 1), long-wavelength light is capable of mediating photomorphogenesis, the sporelings respond only to blue (or UV) light with photomorphogenesis. Obviously at least two different photoreceptors are involved in terrestrial plants in the switch from skoto- to photomorphogenesis.

In conclusion, during development of the fern gametophyte we observe *alternative* strategies of growth depending on the light conditions (filamentous vs. heart-shaped growth). Remember that a fern plant will also exhibit *different sequential* strategies of development on the basis of the same genome, referred to as alternation of generations. While the gametophyte grows as a thallus, the sporophyte develops as a cormus, in accordance with the "Bauplan" of higher plants. Thus, the ferns are well suited to illustrate the point that plants are genetically endowed with the potential to follow different strategies of development, depending on the (light) environment and on the nature of the germ cell (in the present case a gonospore or a zygote).

4 Photoreceptors in Photomorphogenesis

Research in photomorphogenesis aims at understanding the signal-response chains in photomorphogenesis in a preferentially biophysical and molecular terminology. Consider the plant (e.g., a seedling) as a black box (Fig. 6). The input is light which can be applied and measured relatively easily and accurately. The output (terminal responses) can be measured as final photoresponses, such as light-induced growth of the leaves, light-induced inhibition of axis etiolation, light-induced formation of plastids, or light-induced synthesis of pigments (e.g.,

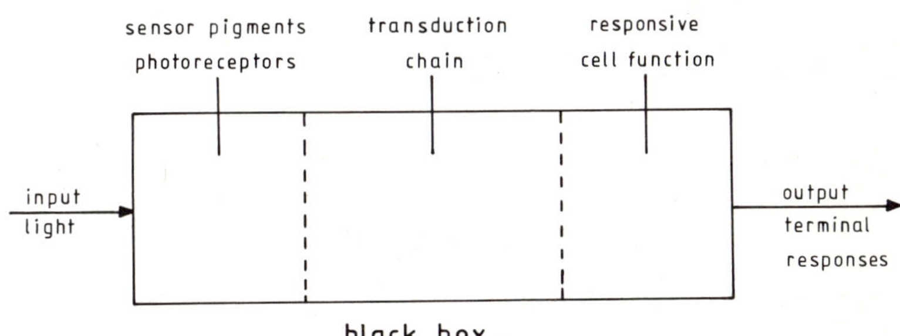

Fig. 6. Scheme to illustrate the experimental approach to photomorphogenesis. As an example, consider light-mediated accumulation of anthocyanin in the vacuoles of the upper epidermal cells of mustard seedling cotyledons (see Fig. 1). Accumulation of anthocyanin (a protective pigment without significant turnover during the experimental period) is considered a terminal response. The immediate questions are: what kind of photoreceptor pigment absorbs the effective light, how is this light signal transduced to that cell function which produces anthocyanin after the arrival of the signal, and how does the responsive cell function bring about synthesis and accumulation of anthocyanin?

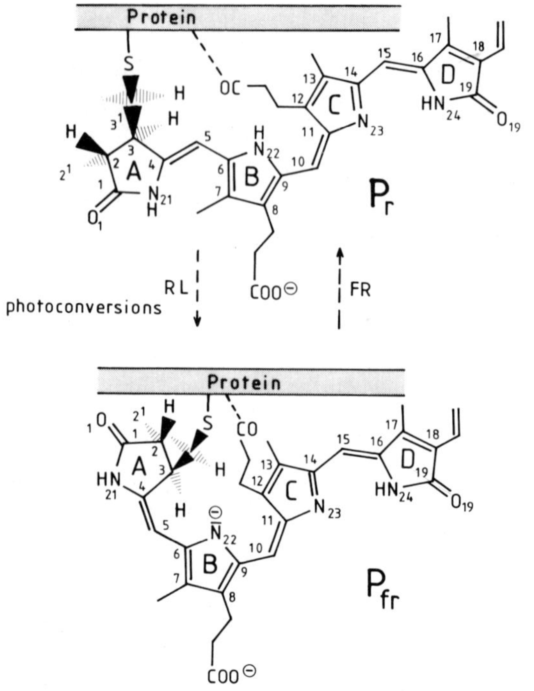

Fig. 7. Suggested structures of the P_r and P_{fr} chromophores and their binding to the apoprotein. *RL* red light; *FR* far-red light. (W. RÜDIGER personal communication; for details see Chap. 7, this Vol.)

anthocyanins, chlorophylls or carotenoids). Biochemical responses are preferred by modern researchers because of their apparent simplicity. The scheme (Fig. 6) does not account for the basic question of how the *integration* of the large number of photoresponses can be understood. This integration of the photoresponses of the different cells, tissues, and organs is a characteristic and essential feature of photomorphogenesis. Since photomorphogenesis is a harmonic process, the integration of the different photoresponses must be achieved precisely in space and time (see Chap. 14, this Vol.). For the present purpose development is assumed to be primarily the consequence of an orderly sequence of changes in the enzyme complement of an organism. Therefore, an examination of those responses is preferred in which changes in enzyme levels have a well-defined causal role in well-defined developmental steps. Returning to the photoreceptor side (Fig. 6) the primary question becomes what kind of photoreceptor pigment(s) absorbs the effective light?

4.1 Phytochrome

At wavelengths longer than 500 nm the light effective in photomorphogenesis is absorbed by the sensor pigment phytochrome. Phytochrome is a bluish chromoprotein with a molecular weight of 124,000 and an open chain tetrapyrrole as a chromophore. The phytochrome system consists of two interconvertible forms (Fig. 7), P_r with an absorption maximum in the red spectral range at

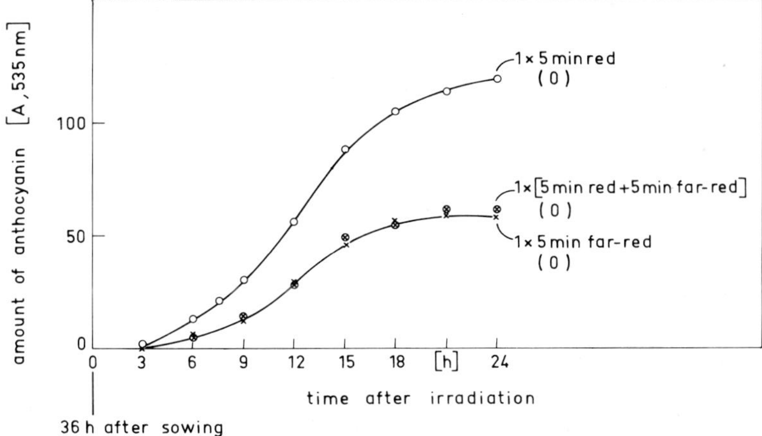

Fig. 8. Time course of anthocyanin accumulation in a mustard (*Sinapis alba*) seedling following brief light treatments with red and/or far-red light (5 min each). Light pulses were given at 36 h after sowing (time zero). (After LANGE et al. 1971)

660 nm and P_{fr} with an absorption maximum in the far-red spectral range at 730 nm. In a dark-grown seedling only P_r, the physiologically inactive form, is present. The physiologically active P_{fr} originates under the influence of light. If a dark-grown plant is irradiated with red light, a large part of its P_r is transformed into P_{fr}. This is a complex reaction. However, the only photochemical event is the excitation of P_r which leads to an intermediate. This intermediate gives rise in the dark to others of short life-times and eventually to P_{fr}.

In turn, P_{fr} can be excited by far-red light and subsequently relaxes via a different set of short-lived intermediates to P_r.

The absorption spectra of P_r and P_{fr} overlap throughout the visible range. This is the reason why, under conditions of saturating irradiations, photoequilibria are established. The P_{fr}/P_{tot} ratio in photoequilibrium (abbreviated by the symbol φ) is a function of the wavelength of light (φ_λ). As an example, if the photoequilibrium is established with a saturating red light (660 nm) treatment, φ_r is of the order of 0.8, whereas the establishment of the photoequilibrium with far-red (718 nm) light leads to a φ_{fr} of the order of 0.025. In red and far-red light the photoequilibrium of phytochrome can readily be established since the extinction coefficients of both forms of phytochrome and the relative quantum efficiencies of both photochemical transformations are high. A minute of irradiation with an energy fluence rate about 10 Wm^{-2} suffices to establish the photoequilibrium. In brief, only light pulses are required to establish photoequilibria of phytochrome in the red and far-red spectral range.

Therefore, a simple operational criterion for the involvement of phytochrome as the responsible photoreceptor is defined as follows: if a photoresponse can be induced by a red light pulse, and if this inductive effect of the red light pulse can be fully reversed by immediately following with a saturating far-red light pulse, then the photoreceptor involved is phytochrome. "Full reversibility" means that the extent of the response following the irradiation sequence, red pulse followed by a far-red pulse, must be identical with the extent of the response following a far-red pulse alone.

This operational criterion has been verified in many instances (Fig. 8, as an example).

If the operational criterion in the above sense cannot be met, phytochrome is not excluded necessarily as the photoreceptor. Failure to obtain reversibility can be due to the fact, for example, that the signal transduction process is fast (of the order of seconds, see Chap. 9, this Vol.) or that a plant requires a pretreatment with blue/UV light before phytochrome can become effective (see Chap. 28, this Vol.).

Phytochrome operates in green algae, mosses, pteridophytes and spermatophytes. There is no unequivocal experimental evidence that phytochrome is involved in light-mediated responses in other classes of the plant kingdom. In particular, it is apparently lacking in fungi (see Chap. 23, this Vol.).

In the course of evolution of green plants the significance of phytochrome has increased steadily. In the most advanced taxa (as far as evolutionary progress is concerned) phytochrome plays the decisive role in mediating photomorphogenesis while in the cryptogams its role is of minor importance compared to an evolutionarily ancient photoreceptor which absorbs only blue/UV light (see SENGER 1980).

4.2 Cryptochrome

As a case study, consider light-mediated carotenoid synthesis in the mycelium of the fungus *Fusarium aquaeductuum*.

This photoresponse can be considered the prototype of a fungal photoresponse, including morphogenic photoresponses. The action spectrum, more precisely, the effectiveness spectrum (Fig. 9), shows that carotenoid synthesis is caused only by light less than 500 nm, i.e., by blue/UV light. The action spectrum has a maximum in the near UV and three maxima (or shoulders) between 400 and 500 nm. Action spectra very similar to the one in Fig. 9 were determined for a multiplicity of photoresponses in fungi (see Chap. 23, this Vol), in algae (see Chap. 21, this Vol.), mosses, liverworts, pteridophytes and spermatophytes. As an example, the action spectrum for photomorphogenesis in fern gametophytes (see Fig. 3) shows the same characteristics. In higher plants the action spectrum of phototropism has the same structure (Fig. 10). Clearly, a second photomorphogenic photoreceptor pigment besides phytochrome is operating throughout the plant kingdom. This photoreceptor is particularly important in the life of lower plants (cryptogams), whereas phytochrome seems to dominate in spermatophytes (phanerogams). In the course of evolution the relative significance of the blue/UV light-absorbing photoreceptor has decreased whereas the significance of phytochrome has correspondingly increased. However, the blue/UV light photoreceptor is still present probably in *all* higher plants even though in some evolutionarily progressive taxa (e.g., Brassicaceae, to which *Sinapis alba* belongs) the operation of a blue/UV light photoreceptor cannot readily be detected in photomorphogenesis. On the other hand, with regard to the

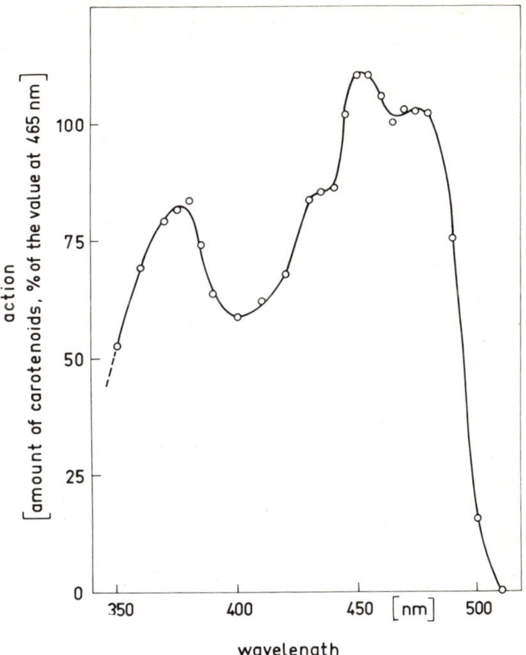

Fig. 9. "Action spectrum" of light-induced carotenoid synthesis in the mycelium of the imperfect fungus *Fusarium aquaeductuum*. The amounts of carotenoids which can be induced by 4.2×10^{-3} mol photon m^{-2} is given as a function of wavelength (effectiveness spectrum). (After RAU 1967)

phototropic response, *all* higher plants depend on the blue/UV light photoreceptor (see Fig. 10).

Following a suggestion by J. Gressel (SENGER 1980), the ubiquitous blue/UV light photoreceptor is now called "cryptochrome" to acknowledge its particular role in the lower plants (cryptogams).

The operational criterion for the involvement of cryptochrome in eliciting a photoresponse is an action spectrum with a peak near 370 nm and three peaks or shoulders in the blue spectral range between 400 and 500 nm. In contrast to phytochrome the inductive effect of a blue/UV light treatment operating through cryptochrome is *not* reversed by a subsequent treatment with light of longer wavelengths.

Regarding the molecular nature of cryptochrome only two candidates for the cryptochrome chromophore have been discussed seriously, namely flavins and carotenoids (see SENGER 1980). While the precise molecular nature of the cryptochrome chromophore is not yet known, there can be hardly any doubt that it is some kind of a flavin (Fig. 11). The best-known flavins are FMN and FAD (see Chap. 7, this Vol.). Both have broad absorption maxima around 370 nm and in the blue spectral range (see Fig. 10). Even the fine structure of the action spectrum in the blue range can be accounted for by flavins in a particular environment (see Chap. 7, this Vol.).

Carotenoids as candidates for the cryptochrome chromophore were convincingly eliminated by two different approaches. First, "carotenoid-less" mutants of *Phycomyces, Neurospora* and *Euglena* display normal light sensitivity and yet, in the case of *Phycomyces,* may have less than 0.004% of the wild-type

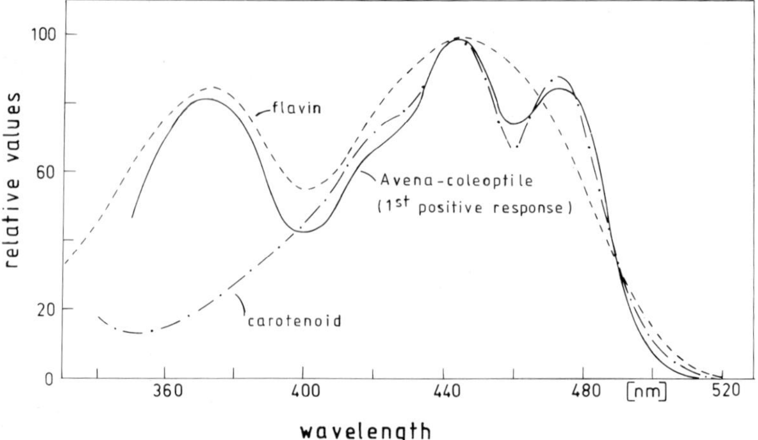

Fig. 10. A comparison of the action spectrum for phototropism (*Avena* coleoptile; first positive curvature) with the absorption spectra of carotenoids and flavins. Only a comparison of the general characteristics is intended. (After Fig. 9 in SHROPSHIRE 1958)

Fig. 11. A suggestion for the structure of the chromophore of cryptochrome. The chromophore is believed to be non-covalently bound to an apoprotein. The flavin as present in cryptochrome is thought to be similar to riboflavin (6,7-dimethyl-9-ribityl-isoalloxazin) rather than FMN or FAD. However, the nature of the side chain on N-9 and the functional side groups on C-6 and C-7 remain undetermined at present (M. DELBRÜCK 1976 personal communication)

carotenoid content (see VIERSTRA and POFF 1981, for recent review of relevant literature).

In higher plants where carotenoid-free mutants are not readily available, the herbicide Norflurazon has been used to inhibit carotenoid synthesis (BARTELS and MCCULLOUGH 1972). This herbicide interferes with the desaturation of cis-phytoene, thus blocking the accumulation of colored carotenoids. Norflurazon does not seem to affect general metabolism or other biosynthetic pathways (including synthesis of flavins) significantly. For example, corn seedlings retain normal light sensitivity for the photomorphogenic responses mediated through phytochrome (JABBEN and DEITZER 1979). Using these facts, virtually carotenoid-free corn (*Zea mays*) seedlings were found to retain their ability to respond phototropically (VIERSTRA and POFF 1981). The reduction of their sensitivity toward blue light can be accounted for by the role carotenoids play as screening pigments in coleoptile phototropism. In another approach using Norflurazon-treated tomato (*Lycopersicon esculentum*) seedlings with no detectable carotenoid content, it is now found that the action of blue/UV light on photomorphogenesis in general, and on anthocyanin synthesis in particular, was not impaired (DRUMM-HERREL and MOHR 1982). Since the response is linearly dependent on the blue/UV light fluence rate, it has been concluded that at least "bulk" carotenoids (i.e., detectable carotenoids) are not involved in blue/UV light-dependent photomorphogenesis of the tomato seedling. Moreover, the presently available evidence strongly suggests that the idea that a carotenoid is the chromophore of cryptochrome be abandoned. The thrust of current research in several laboratories is in favor of a flavin (or flavins) as chromophore of cryptochrome.

5 Photomodulations

While photomorphogenic responses such as growth of a leaf, synthesis of anthocyanin, development of plastids or formation of flowers are irreversible and obviously related to differential gene expression, a second category of light effects – photomodulation – is readily reversible and apparently does not involve gene expression. As an example, while light-dependent leaf growth is a typical photomorphogenic response, light-dependent leaf movements may be considered as typical photomodulations. At present, photomodulations are usually attributed to light effects on membranes (see Chap. 26, this Vol.). Some categories of photomodulations, namely light-controlled movements of parts of plants including phototropism have been covered in a previous volume of this Encyclopedia (Vol. 7). Photomodulations must be considered as an essential feature of the behavioral strategies of plants since at least the photoautotrophic plant depends throughout its life span on an *optimum* and *rapid* adaptation to the *actual* light conditions at the site where it grows (see Chap. 19, this Vol.).

A convenient case study is "chloroplast migration in cells". Chloroplasts commonly undergo two types of movements in response to light (Fig. 12). The first response is called the weak light response (or low intensity movement). It leads to a diastrophic orientation of the chloroplasts and thus allows the chloroplast to absorb maximum light. The second response is called strong light response (or high intensity movement). It protects the chloroplasts from damage if the level of incident light becomes too high. This movement leads to a parastrophic orientation of the plastids.

The usual statement "the plastids migrate in the cell" is not correct insofar as the plastids are passively moved by forces in the cytoplasm. Moreover, the light absorption which acts as a signal for chloroplast migration takes place in the cytoplasm rather than in the plastid itself. Most light-induced chloroplast migration responses are mediated by blue/UV light, i.e., through cryptochrome. The action spectrum for the induction of the weak light response in "leaf" cells of the moss *Funaria hygrometrica* (Zurzycki 1967) is a typical cryptochrome spectrum (see Fig. 9).

In two related green algae, *Mougeotia* (Haupt 1959) and *Mesotaenium* (Haupt and Thiele 1961) the weak light response is mediated by red light acting through phytochrome. Filaments of *Mougeotia* (Fig. 13) are made up of long cylindrical cells each of which contains a single, large, flat plate-like chloroplast. The chloroplast is able to turn inside the cell. The light-induced movement is a response to light absorbed in the cytoplasm. The weak light

Fig. 12. Diastrophic (*left*) and parastrophic (*right*) orientation of the chloroplasts in the "leaf" cells of a moss (*Funaria hygrometrica*). The cells were irradiated either with weak (*left*) or with strong (*right*) white light. (After MOHR 1972)

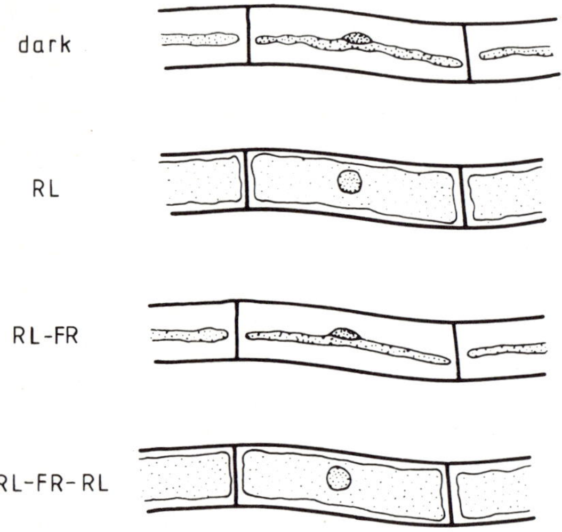

Fig. 13. An experiment which demonstrates the involvement of phytochrome in the weak light response (profile position → face position) of the *Mougeotia* chloroplast. *Dark*, starting position; *RL, RL-FR, RL-FR-RL*, orientation of the chloroplast about 30 min after brief light pulses (1 min each) with red or far-red light. (After HAUPT 1970)

response is mediated by phytochrome since the operational criteria for the involvement of this photoreceptor are clearly fulfilled (Fig. 13).

P_{fr}-mediated responses, such as formation of anthocyanin which can be traced back to phytochrome-induced gene transcription and phytochrome-mediated responses such as the movement of the *Mougeotia* chloroplast, obviously do not share the same signal-response chain (see Fig. 6). However, in spite of some effort, it is not clear yet whether the *initial* action of P_{fr} is always the same. Different views about this problem are discussed at the end of Chap. 9, this Vol.

6 Biochemical Model System of Photomorphogenesis

In recent years several laboratories (Wellmann, Hahlbrock; see Chap. 29, this Vol.) have succeeded in using plant cell suspension cultures obtained from parsley (*Petroselinum hortense*) to elucidate the molecular mechanism of light-induced synthesis of flavonoid compounds. These cultures offer the advantage that they can be handled like yeast or microbial suspension cultures and that light- and water-soluble compounds can readily be applied to the cells. The use of cell suspension cultures has considerably increased our knowledge about light-induced transcription of the mRNA of key enzymes involved in biogenesis of flavonoid compounds. At present, there is no doubt that light-induced synthesis of flavonoids can be understood in terms of light-mediated gene expression (see Chap. 10, this Vol). A "black box", however, still exists between the formation of P_{fr} and the onset of differential gene transcription (see Chap. 14, this Vol.).

Of course, biochemical model systems of the photomorphogenic process will not provide us with a *complete* answer to the key question which the phenomena of photomorphogenesis pose. What is the relationship between changing enzyme levels and the appearance of specific growth and form? However, these studies of biochemical model systems will increase our understanding of the formal rules and molecular mechanisms underlying developmental processes in general. Thus, a solution of the problem is approached which is prevalent throughout this volume; i.e., what is the relationship between control of protein synthesis and development of structural specificity – form and pattern – in space and time.

References

Bartels PG, McCullough C (1972) A new inhibitor of carotenoid synthesis in higher plants: 4-chloro-5-(dimethylamino)-2-α,α,α-(trifluoro-*m*-tolyl)-3(2H)-pyridazinone (Sandoz 6706). Biochem Biophys Res Commun 48:16–22

Drumm-Herrel H, Mohr H (1982) Effect of blue/UV light on anthocyanin synthesis in tomato seedlings in the absence of bulk carotenoids. Photochem Photobiol 35:233–236

Haupt W (1959) Die Chloroplastendrehung bei *Mougeotia* I. Über den quantitativen und qualitativen Lichtbedarf der Schwachlichtbewegung. Planta 53:484–501

Haupt W (1970) Localization of phytochrome in the cell. Physiol Veg 8:551–563

Haupt W, Thiele R (1961) Chloroplastenbewegung bei *Mesotaenium*. Planta 56:388–401

Jabben M, Deitzer G (1979) Effects of the herbicide SAN 9789 on photomorphogenic responses. Plant Physiol 63:481–485

Klebs G (1913) Über das Verhältnis der Außenwelt zur Entwicklung der Pflanzen. In: Sitzungsber Heidelberger Akad Wiss Math Naturwiss Klasse, Abt B, Jg 1913, 5. Abhandl

Lange H, Shropshire W Jr, Mohr H (1971) An analysis of phytochrome-mediated anthocyanin synthesis. Plant Physiol 47:649–655

Mohr H (1972) Lectures on photomorphogenesis. Springer, Berlin Heidelberg New York

Mohr H, Ohlenroth K (1962) Photosynthese und Photomorphogenese bei Farnvorkeimen von *Dryopteris filix-mas*. Planta 57:656–664

Pfeffer W (1904) Pflanzenphysiologie. Engelmann, Leipzig

Rau W (1967) Untersuchungen über die lichtabhängige Carotenoidsynthese. I. Das Wirkungsspektrum von *Fusarium aquaeductuum*. Planta 72:14–28

Senger H (ed) (1980) The blue light syndrome. Springer, Berlin Heidelberg New York

Shropshire W Jr (1958) A detailed action spectrum of the first positive phototropic tip-curvature of *Avena*. PhD Thesis, George Washington Univ, Washington DC

Vierstra RD, Poff KL (1981) Role of carotenoids in the phototropic response of corn seedlings. Plant Physiol 68:798–801

Zurzycki J (1967) Properties and localization of the photoreceptor active in displacements of chloroplasts in *Funaria hygrometrica*. I. Action spectrum. Acta Soc. Bot Pol 31:489–538

4 Action Spectroscopy of Photoreversible Pigment Systems

E. SCHÄFER, L. FUKSHANSKY, and W. SHROPSHIRE, JR.

1 Introduction

Action spectroscopy is a non-destructive method for analyzing the properties of a functional pigment. It uses the natural agent light, to which plants have adjusted in the course of evolution. Another advantage of the method is the absence of any indeterminate delay in action due to transportation of the acting factor to the place of action.

Since ENGELMANN'S first brilliant studies (1882), a general method of action spectroscopy has been developed by WARBURG (1949) and described in a series of publications (LOOFBOUROW 1948, SETLOW and POLLARD 1964, DUYSENS 1970, SHROPSHIRE 1972). The purpose of classical action spectroscopy is to identify the functional pigment (photoreceptor) of a response by comparing the absorption spectrum of any would-be photoreceptor to the action spectrum of the response. The action spectrum is defined as a plot of relative responsivity of a biological system (with respect to the response) to light of different wavelengths. As we will see below, such a problem is relevant when one studies either a single pigment or a mixture of pigments working additively and having constant concentrations. In photomorphogenesis such a situation is very probable when we deal with cryptochrome.

In the case of phytochrome [and any photochromic (photoreversible) system in general], the problem of identifying a photoreceptor cannot be solved by means of action spectroscopy. On the other hand, it becomes irrelevant since the photoreversibility itself is an excellent tool to identify the functional pigment. This does not mean, however, that the experimental approach, based on the investigation of action spectra, will not be useful. The same properties of a photoreversible photoreceptor, which make the basic problem of classical action spectroscopy irrelevant, create new problems to be solved by this approach. These problems are mainly connected with the dynamics of the processes in the photoreceptor and the dynamics of signal transition from photoreceptor to the transduction chain. Their solution requires not only the measurements of action spectra but also the use of mathematical models, as has been pointed out by HARTMANN and COHNEN UNSER (1972) and HARTMANN (1977), who suggested a separate name for this field – analytical action spectroscopy.

In this chapter we will first present classical action spectroscopy, discuss limits of its applicability and then analytical action spectroscopy, as applied to the phytochrome system, will be considered. First, we discuss some general problems which are based on the principle of equivalent light action and are not bound to any particular model. Second, we discuss problems which require particular models for their solution but these are treated only briefly in this chapter and as one could expect, are closely related to Chapter 5, this Volume.

Complex optical properties of plant tissue impose additional obstacles in solving problems of action spectroscopy. Two main causes of optical artifacts are fluence rate gradients within a tissue and distortion of in vivo absorption (and difference) spectra as compared to in vitro spectra (cf. FUKSHANSKY 1981). To provide a clear presentation of principal problems involved in action spectroscopy, we ignore optical artifacts initially and deal with them separately in the last part of the chapter.

2 Classical Action Spectroscopy

2.1 The Grotthus-Draper Law and the Rate of the Primary Reaction

Action spectroscopy is based on Grotthus-Draper's law (the first law of photochemistry) which states that only absorbed light can cause photochemical reactions and, therefore, photobiological responses. An absorption process does not necessarily lead to a molecular transformation of interest with respect to a certain photobiological response. Hence, under monochromatic radiation with a photon fluence rate N_λ the number of molecules transformed per second (i.e. the rate of a definite phototransformation) is:

$$v = P N_\lambda \varepsilon_\lambda \phi_\lambda, \tag{1}$$

where P is the concentration of the pigment in the ground state, ε_λ is the molar absorption cross-section of pigment, ϕ_λ is the quantum yield of transformation, i.e., the probability that the absorption of a photon leads to a phototransformation. The product

$$\varepsilon_\lambda \phi_\lambda = \sigma_\lambda \tag{1a}$$

is called cross-section of the phototransformation. Considering Eq. (1) from the point of view of the mass action law one can see that the rate constant of the phototransformation, k, which has a dimension $[k] = s^{-1}$ can be expressed as:

$$k = N_\lambda \varepsilon_\lambda \phi_\lambda. \tag{2}$$

The dimension of the right side of (2) is also s^{-1} since:

$$[N_\lambda] = mol_{(photons)}\, m^{-2} s^{-1}; \quad [\varepsilon_\lambda] = \left[\frac{mol_{(pigm)}}{m_3}\right]^{-1} m^{-1};$$

$$[\phi_\lambda] = \frac{mol_{(pigm)}}{mol_{(photons)}}.$$

Correlation (2) can be easily extended to the case of polychromatic irradiation, where we do not deal with discrete numbers N_λ, ε_λ, ϕ_λ but with continuous functions $\varepsilon(\lambda)$, $\phi(\lambda)$ and density probability of the photon fluence rate spectrum $n(\lambda)$, which, when integrated over a relevant spectral interval, produces a total polychromatic photon fluence rate:

$$\int_{\lambda_1}^{\lambda_2} n(\lambda)\,d\lambda = N.$$

The analogy of Eq. (2) for polychromatic irradiation is:

$$k = \int_{\lambda_1}^{\lambda_2} n(\lambda)\,\varepsilon(\lambda)\,\phi(\lambda)\,d\lambda \tag{2a}$$

with the following dimensional correlations:

$$[\varepsilon(\lambda)] = [\varepsilon_\lambda]; \quad [\phi(\lambda)] = [\phi_\lambda]; \quad [n(\lambda)] = [N_\lambda]\,m^{-1}.$$

The concept of the quantum yield and the idea of the primary reaction as presented above need additional discussion. The absorption of a photon leads, in the case of visible light, to a transition of the pigment molecule from the ground state to some excited state (Fig. 1). This non-stable configuration is a junction from which different directions of excitation energy can start with different probabilities: return to the ground state either through fluorescence or by means of non-radiative transfer (internal conversion), transition to a metastable state (intersystem conversion) followed either by phosphorescence or internal conversion to the ground state. Besides this, in each excited or metastable state the molecule can, in principle, take part in an interaction.

One of the possible interactions, which leads to the response of interest, we call the *primary reaction*. For example, in photosynthesis the excited molecule of chlorophyll a (in photosystem II) reduces the intermediate acceptor of electrons and after this is reduced itself by a donor. One could consider as a primary reaction either the reduction of the acceptor or the whole result of this interaction, i.e., the transition of an electron from the donor to the acceptor. In other cases, for example, in phytochrome- and rhodopsin-induced processes, the primary reaction is the transformation of a pigment molecule.

Thus in all cases the ideal primary reaction is the first molecular change promoted by the absorbed light quantum. The quantum yield is defined as a probability that this change occurs. Even in an ideal case, when the primary reaction immediately follows the molecular excitation, the quantum yield can be λ-dependent. As shown in Fig. 2, if the primary reaction starts from the second excited state, only the photons from the "short wavelength" part of the absorption spectrum will contribute to the action spectrum, because only these photons have enough energy to populate the second excited state. This leads to λ-dependence of the quantum yield. Another cause of the λ-dependence of the quantum yield has been illustrated by SETLOW and POLLARD (1964,

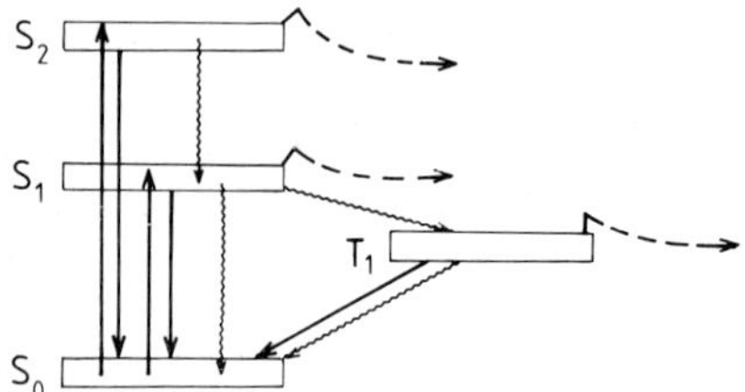

Fig. 1. Schematic presentation of the changes in the state of a molecule after photon absorption. S and T symbolize different (electronic) energy levels. S ground state; S_1 and S_2 first and second excited states, respectively; T_1 metastable state. *Arrows of type* →: S_0 S_1, S_0 S_2 show transfer from the ground state to corresponding excited states after a photon absorption; S_1 S_0, S_2 S_0 show transfer back to the ground state via fluorescence; T_1 S_0 shows transfer to the ground state via phosphorescence. *Arrows of type* ⇝: S_2 S_1, S_1 S_0 show non-radiative transfer to lower energy levels via internal conversion; S_1 T_1, T_1 S_0 show non-radiative transfer to lower energy levels via intersystem conversion. *Arrows of type* ---→ show possible ways of coupling of an excited molecule to external reactions. Since S_2 is a very short-living state, S_1 a relatively short-living state and T_1 a rather long-living state, the largest probability of coupling is in T_1, while coupling in S_2 is very unlikely. Finite thickness of (electronic) energy levels indicates that they are superimposed by different levels of vibrational energy

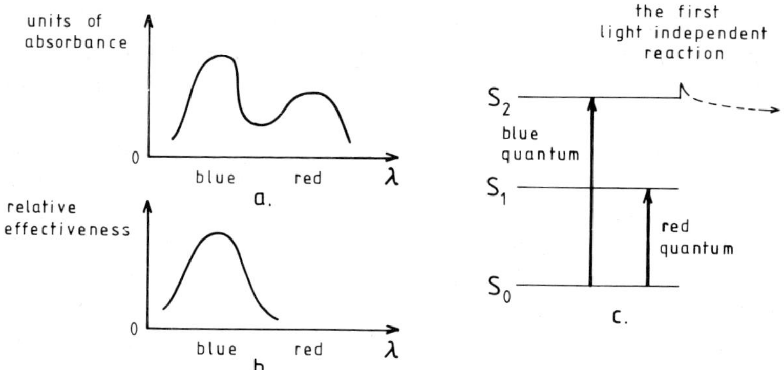

Fig. 2 a–c. Scheme which visualizes the appearance of a fluence rate dependent quantum yield. **a** absorption spectrum **b** action spectrum **c** the first light-independent reaction occurs from the excited state S_2 which can be populated only by "blue" photons

pp. 283–285) using the inactivation of dry trypsin by ultraviolet light. The action spectrum of this effect is not parallel to the absorption spectrum of trypsin, indicating that the quantum yield depends on λ. The reason is that the absorption spectra of different amino acids constituting the protein molecule do not coincide, and, in addition, that different amino acids contribute to a different degree to the effect. The action spectrum can be presented as the sum of a highly efficient component (with cystine as a photoreceptor) and a low efficiency component (with aromatic amino acids as a photoreceptor). In fact, one has here two additively acting photoreceptors.

In these examples the value of the quantum yield and its dependence on λ can be affected by the choice of the primary reaction. Moreover, this choice can be very difficult because the first steps in the light-induced chain of reactions cannot always be resolved well. It is clear, however, that the primary reaction has to be chosen as close as possible to the absorption process for several reasons. Firstly, the quantum yield of an event which is separated from the absorption process by many biochemical steps will be very sensitive to changes in temperature, ph, etc. Secondly, these intermediate steps can provide amplification of the initial action promoted by the absorption of a photon; this deprives the quantum yield of its probabilistic interpretation, which is important as a characteristic of the efficiency of a photoreceptor. Thirdly, a primary reaction placed close to the absorption process, as we will see below, is very useful in comparing the action of light of different wavelengths, even in the case when conditions for the application of classical action spectroscopy are violated.

2.2 The Principle of Equivalent Light Action and the Basic Equation of Classical Action Spectroscopy

Let us consider two imaginary experiments having the same type of light-induced response. If we have defined the primary reaction and there are no other light-dependent steps except those which bring about this primary reaction, the following statement is valid. The action of light on the measured response will be equivalent in any two experiments when and only, when the rates of the primary reactions are equal during the experiments. This simple statement we will call the principle of equivalent light action. It is the foundation for classical and, as we shall, see, analytical action spectroscopy. The significance of this principle is based on the fact that it does not contain any record of the light quality. This enables one to compare the effectiveness of photons of different wavelengths in bringing about a certain level of response.

Of course, the magnitude of response depends on many other external and internal parameters. To elucidate the influence of light one should keep all these parameters equal in the experiments to be compared (but not necessarily constant). The measured response R is a function of the internal parameters of the system and external characteristics of the light action and the procedure of measurements:

$$R = R[t_0, \sigma_\lambda \cdot N_\lambda(t), \lambda, t_f, P(t), t^0, r_1, \ldots, r_i, c_1, \ldots, c_j, t_m]. \tag{3}$$

Here: t_0 is the time of the beginning of irradiation; $N_\lambda(t)$ contains the whole time schedule of the irradiation; $P(t)$ is the current pigment concentration (it is not necessarily constant and also P can be a vector if more than one pigment is involved); r_i, c_j are characteristics of reactions in photoreceptor and transduction chain, respectively; t_f is the time of the end of the irradiation; t_m is the time of the measurement of the response. Expression (3) describes a general case. Not all of the presented parameters in (3) are necessary for each experiment. For example, if the response is measured simultaneously with the irradiation (as the rate of O_2 production in photosynthesis) parameter t_m is superfluous.

It must be emphasized that in general parameters of internal processes, r_i, c_j, may depend on N_λ and on the current level of the response.

According to the principle of equivalent light action any two experiments with equal rates of primary reactions will show equal responses:

$$v_a(t) = v_b(t) \rightarrow R_a = R_b, \tag{4}$$

providing the initial states of the system and all the parameters from (3) are equal. The only prerequisite of this is that there is no direct action of light except through the primary reaction considered. If, in addition, the concentration of the pigment in the ground state is constant, [P(t) = const], then v(t) = const. and we can substitute the rates in (4) by the rate constants k defined in (2):

$$k_a = k_b \rightarrow R_a = R_b. \tag{4a}$$

Let us introduce one more restriction for the experiments to be compared. *Responses are only considered in which the light action is rate limiting,* i.e., R(v) is always a monotonously growing function. This enables expression (4a) to be reversed:

$$R_a = R_b \rightarrow k_a = k_b. \tag{5}$$

Using Eqs (2) and (1a) we can rewrite Eq. (5):

$$R_a = R_b \rightarrow N_{\lambda a}\, \sigma_{\lambda a} = N_{\lambda b}\, \sigma_{\lambda b} \tag{5a}$$

and

$$R_a = R_b \rightarrow \frac{\sigma_{\lambda a}}{\sigma_{\lambda b}} = \frac{1}{N_{\lambda a}} : \frac{1}{N_{\lambda b}}. \tag{6}$$

This is the basic equation of classical action spectroscopy which states: if in two comparable experiments monochromatic light of different wavelengths brings about an equal level of response, then the effective cross-sections of the photoreceptor (with respect to primary reaction) are related as reciprocals of photon fluence rates. The conditions of validity of this rule are mentioned above and later we will consider what happens if any of these conditions are violated.

The practical use of Eq. (6) is shown in Fig. 3a. Curves $R_a(N_\lambda)$, $R_b(N_\lambda)$, called fluence rate-response curves, contain results of many experiments with the same time schedule of irradiation and measurements but different fluence rates. Each tested wavelength has its fluence rate-response curve. If many wavelengths have been tested, one of them may be chosen to be a reference wavelength. The normalized plot of reciprocals of fluence rates bringing about a fixed level of response is called the action spectrum. According to Eq. (6), the action spectrum also presents the relative spectrum of cross-sections of

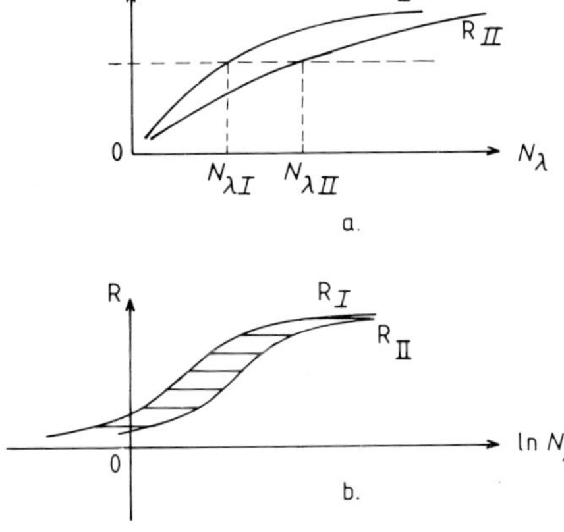

Fig. 3 a, b. On the derivation of the basic equation of classical action spectroscopy. **a** Photon fluence rate-response curves R_I and R_{II} have been obtained in two experiments, *I* and *II*, with light irradiation using wavelengths λI and λII, respectively; a certain level of response has been reached at photon fluence rates $N_{\lambda I}$ and $N_{\lambda II}$, respectively. **b** The same curves as in **a** have been plotted against ln N_λ instead of N_λ. Curves R_I and R_{II} are parallel (i.e., they can be made to coincide by a shift along the axis ln N_λ) under the same conditions which facilitate the derivation of the basic equation of classical action spectroscopy

the photoreceptor (with respect to the chosen primary reaction). If the quantum yield of the primary reaction is known (or known to be λ-independent) then, according to Eq. (1 a), the action spectrum also provides the relative absorption spectrum of the photoreceptor. On the other hand, if an independent estimation of cross-sections is possible, the action spectrum provides the quantum yield of the primary reaction.

2.3 The Parallelism of Fluence Rate-Response Curves

If the fluence rate-response curves are not plotted against N_λ but against ln N_λ such a graph possesses one useful property. Curves for different λ must be parallel, i.e., any two curves can be made to coincide by shifting them along the axis, ln N_λ. This can be seen from the following consideration. Since ln X is a monotonous function of X:

$$N_{\lambda a}\sigma_{\lambda a} = N_{\lambda b}\sigma_{\lambda b} \leftrightarrow \ln(N_{\lambda a}\sigma_{\lambda a}) = \ln(N_{\lambda b}\sigma_{\lambda b}).$$

Together with Eq. (5a) this gives:

$$R_a = R_b \rightarrow \ln(N_{\lambda a}\sigma_{\lambda a}) = \ln(N_{\lambda b}\sigma_{\lambda b})$$

and further

$$R_a = R_b \rightarrow \ln N_{\lambda a} - \ln N_{\lambda b} = \ln\frac{\sigma_{\lambda b}}{\sigma_{\lambda a}}. \tag{7}$$

Expression (7) says that if we have two curves "response vs. ln N_λ" then the value ln N_λ from one curve, which brings about *any* fixed level of response will be shifted against a corresponding value from another curve in the interval

$\left|\ln \frac{\sigma_{\lambda b}}{\sigma_{\lambda a}}\right|$, which is independent of fluence rate. In Fig. 3b two fluence rate-response curves from Fig. 3a are shown in this logarithmic presentation. The shift $\left|\ln \frac{\sigma_{\lambda b}}{\sigma_{\lambda a}}\right|$ depends on λ and contains all the information which is available from the action spectrum. It should be noted that the curves "response vs. N_λ" contain the same information which exists in curves "response vs. $\ln N_\lambda$". The parallelism of curves from Fig. 4b has a corresponding (but not so obvious) feature in curves from Fig. 4a: the fluence rates which bring about equal responses in different curves are proportional to each other with the same proportionality factor along the whole curve (cf. p. 11 in DUYSENS 1970).

Since the derivation of Eq. (7) did not require any additional assumptions and limitations as compared with Eq. (6), the parallelism of curves "response vs. $\ln N_\lambda$" is an indication that all conditions for the application of the basic equation of classical action spectroscopy are fulfilled. On the contrary, the non-parallelism means that some of these conditions have not been fulfilled.

2.4 The Bunsen-Roscoe Law of Reciprocity

In general the magnitude of a response depends on N_λ and on the duration of irradiation, Δt, in a very complicated way. Omitting those parameters which are equal in all compared experiments we can rewrite Eq. (3):

$$R = R(\sigma_\lambda \cdot N_\lambda, \Delta t, \dots). \tag{8a}$$

Often, however, the magnitude of a response may be dependent only on the product $N_\lambda \cdot \Delta t$, i.e., the magnitudes of a response will be equal for all pairs N_λ and Δt satisfying:

$$N_\lambda \cdot \Delta t = \text{const.} \tag{8b}$$

This simplifies the correlation (8a):

$$R = R(N_\lambda \cdot \sigma_\lambda \cdot \Delta t, \dots). \tag{8c}$$

This property is called reciprocity and a correlation between two experiments

$$N_{\lambda a}\, \Delta t_a = N_{\lambda b}\, \Delta t_b \rightarrow R_a = R_b \tag{8d}$$

is called the Bunsen-Roscoe law of reciprocity. Reciprocity enables one to compare experiments with different durations of irradiation. Instead of fluence rate-response curves one can apply in this area (which has to be previously checked) fluence-response curves.

The validity of the Bunsen-Roscoe law is, of course, not mandatory for the application of Eqs. (6) and (7), i.e., classical action spectroscopy may be feasible outside the area of reciprocity.

3 Limitation of Classical Action Spectroscopy

1. The Photoreceptor Contains More Than One Pigment and These Pigments Act Additively, i.e., They Contribute to a Unique Primary Reaction. Analogously to Eq. (1) we can write for two pigments

$$v = P_1\, N_\lambda \sigma_{1\lambda} + P_2\, N_\lambda\, \sigma_{2\lambda} \quad (P_1 = \text{const};\ P_2 = \text{const}), \tag{9a}$$

where indexes 1, 2 mean first and second pigment, respectively. Introducing the total concentration of functional pigments $P = P_1 + P_2$ and the fractional concentrations of single pigments $\frac{P_1}{P} = \gamma$, $\frac{P_2}{P} = 1 - \gamma$ we obtain from Eq. (9a):

$$v = P \cdot N_\lambda \cdot \sigma_{\text{ef}\,\lambda} \quad (P = \text{const, ef} = \text{effective}), \tag{9b}$$

which provides the analogy of Eq. (2):

$$k = N_\lambda\, \sigma_{\text{ef}\,\lambda} \tag{9c}$$

with

$$\sigma_{\text{ef}\,\lambda} = \gamma \cdot \sigma_{1\lambda} + (1 - \gamma)\sigma_{2\lambda}.$$

From this point considerations completely analogous to those in Section 2.2 lead us, however, to the conclusion that the basic problem of classical action spectroscopy can only be solved with respect to an "effective pigment" composed of one and/or more than one real pigment with unknown fractional concentrations and partial cross-sections of the primary reaction. The Bunsen-Roscoe law remains, of course, valid since the properties of the photoreceptor and transduction chain do not change with time. We cannot conclude from fluence rate-response curves whether we are dealing with a single pigment or with an "effective pigment".

2. Adaptation (Sensitization) of the Transduction Chain. In this case some of the characteristics c_j from Eq. (3) may change with time and become functions of N_λ and/or of R(t) (through a feedback reaction). However, since only one primary reaction exists, the influence of N_λ on c_j will occur as an influence of v as described by Eq. (1). Furthermore, since the concentration of the pigment is constant, the influence of N on c_j will occur as an influence of k as described by Eq. (2). This means that the response will be a function of k only (the existence of feedback reactions does not affect this statement). Since we work only within the area of monotony of the function $R(N_\lambda \cdot \sigma_\lambda, \ldots)$ all considerations in Section 2.2 which lead to Eq. (6) are valid and classical action spectroscopy is applicable.

On the contrary, the Bunsen-Roscoe law (8d) has been violated. The violation of Eq. (8d) occurs in both cases: when the c_j are affected by feedback reactions from the response and when they are functions of N_λ. This can be shown unequivocally (FUKSHANSKY unpublished).

3. Variable Concentration of the Pigment in the Ground State. We will distinguish between two cases which lead to different consequences.

3a. The concentration of the pigment either changes independently of light action or it depends on light action with the same quantum yield as the primary reaction. In this case some of the characteristics r_i from Eq. (3) will change with time and may become functions of N_λ. A discussion completely analogous to that in the case of adaptation (sensitization) of the transduction chain leads to the following statements. Classical action spectroscopy remains applicable but the Bunsen-Roscoe law is violated.

3b. The concentration of the pigment changes due to light action but with a quantum yield which is not equal to the quantum yield of the primary reaction leading to the response. This is the same as if we have two different primary reactions and Eq. (3) becomes:

$$R = R[N_\lambda \cdot \sigma_\lambda, \dots P(N_\lambda \cdot \sigma'_\lambda, t), \dots)],$$

where σ'_λ is the cross-section of the changes in pigment concentration and $\sigma'_\lambda \neq \sigma_\lambda$. One can easily see that under these conditions the monotonous dependence of R on N_λ *does not mean* that R depends monotonously on $k = N_\lambda \sigma_\lambda$. Therefore, we *cannot* make the initial logical step

$$R_a = R_b \rightarrow k_a = k_b$$

in the derivation of Eq. (6). Thus, classical action spectroscopy is not applicable (and, of course, the Bunsen-Roscoe law is not valid).

4. Two (or More) Pigments Are Involved in Bringing the Response About and Their Contributions Are Non-Additive, i.e., the Contribution of One Pigment Can Be Affected by Another Pigment in an Arbitrary Way. The interaction between different pigments can take place either on the level of the photoreceptor or on the level of the transduction chain (and, also on both levels simultaneously). In the first case the situation discussed under number 3b occurs. In the second case some of the characteristics c_j become functions of N_λ, however, with a separate primary reaction(s) having spectral sensitivity which differs from that of primary reaction(s) leading to the response. In fact, one cannot always say which primary reaction(s) is (are) leading to the response and which affect(s) the transduction chain. In all these cases the reasoning identical to that in number 3b leads to the conclusion: neither is the Bunsen-Roscoe law valid, nor is classical action spectroscopy applicable.

For completeness it should be mentioned that classical action spectroscopy is still applicable if condition 1 and 2, or 2 and 3a occur simultaneously but *not* for 1 and 3a.

From the measured fluence rate-response curves and action spectra one can conclude whether the Bunsen-Roscoe law is valid and whether the fluence rate-response curves are parallel in logarithmic presentation, i.e., whether Eq. (6) is applicable (WITHROW et al. 1957, SCHÄFER et al. 1982). Which effect is responsible for the violation of the Bunsen-Roscoe law and non-applicability of Eq. (6) should be found out independently if possible.

4 Analytical Action Spectroscopy of a Single Photoreversible Pigment System

4.1 The Problem

As we have seen, the identification of a photoreceptor on the basis of an action spectrum is, in many cases, impossible in principle. The properties of photoreceptors which forbid the classical application of action spectra are mainly consequences of two facts: (1) the parameters of a photoreceptor cease to be constant; (2) more than one primary reaction exists.

On the other hand, some new problems, irrelevant for objects of classical action spectroscopy, are concerned with these complex photoreceptors: for example, the elucidation of the structure of a complex photoreceptor, of dynamics of the concentration of pigment(s), and dynamics of signal transduction and of the interplay between different sets of photochemical and non-photochemical reactions. Well-known examples of complex photoreceptors are: the photoreceptors of photosynthesis in higher plants containing two active photosystems and different types of antenna pigments, photoreversible pigments such as phytochrome and rhodopsin.

The intricate behaviour of complex photoreceptors alone provides some additional tools for solving the new problems. Among these tools are changes in action spectra in specially constructed experiments, i.e., comparison of action spectra under inductive conditions (WITHROW et al. 1957) with those under prolonged irradiation (BEGGS et al. 1980, HOLMES and SCHÄFER 1981) or comparison of action spectra at low and high fluence rates (HOLMES et al. 1982, SCHÄFER et al. 1982). The construction and interpretation of such experiments is often connected with the analysis of mathematical models (HARTMANN 1966, GAMMERMANN and FUKSHANSKY 1974, SCHÄFER 1975). Frequently a photoreceptor, which is initially studied by means of classical action spectroscopy, reveals more and more its complex structure and becomes an object of the new approach. This was the case in the studies of photosynthesis, while another, simpler example, is provided by the inactivation of trypsin as described by SETLOW and POLLARD (1964) and discussed in Section 2.1.

It is highly improbable that a general theory of analytical action spectroscopy can be developed which embraces all kinds of complex photoreceptors, for example, photoreversible pigments, sets of pigments acting successively and in parallel, etc. Even for a given type of photoreceptor this seems unlikely because of the great diversity of associated non-photochemical reactions providing strongly different descriptions of dynamics (mathematical models) of photoreceptors.

In the framework of photomorphogenesis, we are concerned below with photoreversible pigment systems only. Some general properties of photoreversible systems can be outlined which are independent of the kind of associated non-photochemical reactions. These properties are discussed in Section 4.2. A detailed study of particular systems requires accounting for specific reactions associated with a photoreceptor, i.e., is bounded by certain mathematical models. Such a treatment of the phytochrome system, is intimately related to Chapter 5, this Volume.

4.2 Elements of General Analytical Action Spectroscopy of Photoreversible Systems

4.2.1 Extension of the Principle of Equivalent Light Action to Photoreversible System

A single photoreversible pigment system must have two primary reactions leading to the formation of two different forms of the pigment. Such a system is presented in Fig. 4 where two forms of the pigment are designated P_r, P_{fr} as in the phytochrome system which does not diminish the generality of the conclusions. Primary reactions with rates v_1, v_2 and rate constants k_1, k_2 lead to the formation of P_{fr}, P_r, respectively, from the ground states P_r, P_{fr}, respectively. For the following discussion it is irrelevant what kind and number of non-photochemical reactions (shown by dashed lines in Fig. 4) are associated with the pigment system; important is that only two photochemical reactions take place. Equation (3) can be written as

$$R = R[\ \ldots, \sigma_{r\lambda} \cdot N_\lambda, \sigma_{fr\lambda} \cdot N_\lambda, P_r(t), P_{fr}(t), \ldots\ ]$$

which reflects our assumption that the primary reactions of pigment phototransformations are also primary reactions of the response.

Applying the principle of equivalent light action to this system we require the assumption that both v_1, v_2 should be equal in the two experiments (a and b) which are being compared. This provides equal responses if other characteristics of the compared experiments are equal:

$$v_{1a} = v_{1b},\ v_{2a} = v_{2b} \rightarrow R_a = R_b \tag{10a}$$

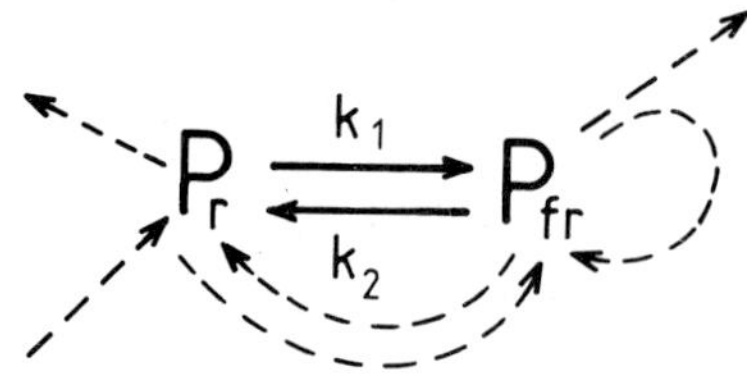

Fig. 4. A photoreversible system which possesses two photochemical reactions having rate constants k_1 and k_2 and a number of non-photochemical reactions (see Chap. 5, this Vol) indicated by dashed lines (two forms of the photoreversible pigment are symbolized by P_r, P_{fr} as in the phytochrome system)

Since these two photochemical reactions are primary reactions for both response and pigment phototransformations, the dynamics $P_r(t)$, $P_{fr}(t)$ are completely determined by the pair $v_1(t)$, $v_2(t)$. Therefore, we can rewrite Eq. (10a) using the rate constants

$$k_{1,2} \quad \left(k_1 = \frac{v_1(t)}{P_r(t)},\ k_2 = \frac{v_2(t)}{P_{fr}(t)}\right)$$

instead of rates $v_{1,2}$; the only precondition for this transition is that initial concentrations $P_r(o)$, $P_{fr}(o)$ should be equal in the compared experiments

$$k_{1a} = k_{1b}, k_{2a} = k_{2b} \rightarrow R_a = R_b \tag{10b}$$

A schematic plot of response as a function of light action is now not two-dimensional (as the corresponding plot for a one-pigment photoreceptor as shown in Fig. 3a) but three-dimensional as shown in Fig. 5a. This illustrates once more that we cannot now reverse the logical correlation (10b) as we did with the correlation (4) which gave us (5) and furthermore the basic equation (6) of classical action spectroscopy.

Before discussing results of the application of the principle of equivalent light action to systems with two primary reactions we introduce instead of k_1, k_2 another pair of parameters – φ, ϑ, which also completely describe the light action and is in the following one-to-one correlation to k_1, k_2:

$$\varphi = \frac{k_1}{k_1 + k_2};\ \vartheta = k_1 + k_2; \tag{11a}$$

$$k_1 = \varphi\vartheta;\ k_2 = (1 - \varphi)\vartheta. \tag{11b}$$

Parameters φ, ϑ are very convenient for a description of a photoreversible system. If we consider phototransformations taken alone [see model (M1) in Chap. 5, this Vol.) φ is equal to the fractional photostationary P_{fr} concentration and ϑ is equal to the rate constant with which the system approaches its photostationary state. In an actual situation, when a set of non-photochemical reactions takes place in the photoreceptors, parameters *φ and ϑ lose their meaning as characteristics of pigment concentration and absolute rate of phototransformation* (Symbol "photostationary concentration" *has no meaning* for such a complex system). But φ and ϑ preserve their important meaning as *characteristics of the irradiation with respect to the photoreceptor* (cf. GAMMERMAN and FUKSHANSKY 1971, SCHÄFER 1981). Under monochromatic irradiation φ is independent of the photon fluence rate and presents the quality of the light as a contributory factor for P_{fr} formation. On the contrary, ϑ is proportional to the fluence rate and is symmetrical regarding phytochrome forms; it presents the power of light in promoting phytochrome phototransformations. In the new coordinates φ, ϑ, the three-dimensional surface from Fig. 5a takes another form (Fig. 5b); between these two surfaces a one-to-one correlation exists because transition from k_1, k_2 to φ, ϑ causes only a permutation of different points on the surface of 5a without changing their heights.

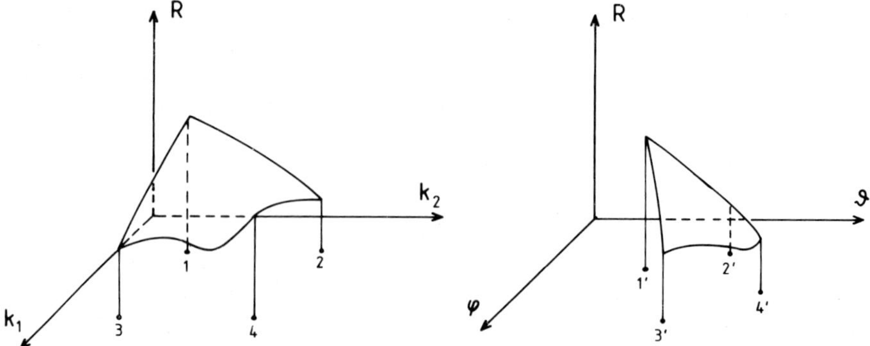

Fig. 5 a, b. A three-dimensional plot of a response as a function of light action which is characterized by two parameters. The time schedule of irradiation is fixed. **a** The light action is represented by rate constants of primary reactions k_1, k_2. **b** The light action is represented by characteristics φ, ϑ. The function $R(k_1, k_2)$ is converted into $R(\varphi, \vartheta)$ by the non-linear one-to-one coordinate transformation $k_1, k_2 - \varphi, \vartheta$ (for example, points 1, 2, 3, 4 are transferred to 1′, 2′, 3′, 4′ respectively)

It should be remembered that each of the surfaces like those presented in Fig. 5 corresponds to a certain time schedule of irradiation and can undergo drastic changes if this time schedule is changed.

Using parameters φ, ϑ, we will discuss some practical consequences of the principle of equivalent light action which are applicable to any response and any duration of irradiation provided that in the compared experiments only φ and ϑ have been varied.

4.2.2 The Plot "Response vs. φ with ϑ=const"

Let us consider two responses, R_a and R_b, obtained in experiments a and b, where irradiation of two different wavelengths, λ_a and λ_b have been used. If the light applied in experiments I and II has the same characteristics φ, ϑ

$$\varphi_a = \varphi_b, \ \vartheta_a = \vartheta_b$$

then according to the principle of equivalent light action

$$R_a = R_b.$$

If the response was measured in many comparable experiments using different λ one can plot these response levels against φ. If one additional condition is fulfilled, namely, all these experiment have the same value ϑ, then all points for the same φ must coincide on this graph *irrespective of* λ.

Such a coincidence is the consequence of the principle of equivalent light action and of our assumption that *two known primary reactions underlie the response*. Therefore, if some points say from the blue spectral area, deviate from the curve drawn through the points provided by light from other spectral areas, one should conclude that an additional pigment(s) absorbing in the blue area takes part in bringing the response about (see Chap. 28, this Vol., Fig. 4a, b; SCHÄFER 1981, SCHÄFER et al. 1981b). The same reasoning is, of course, relevant for polychromatic irradiation. For example, daylight ($\varphi = 0{,}61$) can be substituted by monochromatic irradiation with either $\lambda = 687$ nm or $\lambda = 525$ nm, shade light ($\varphi = 0.27$) by irradiation with $\lambda = 701$ nm or $\lambda = 390$ nm or $\lambda = 450$ nm (HOLMES et al. 1982, SCHÄFER 1981).

One practical difficulty in applying this plot "response vs. φ with $\vartheta = \text{const}$" is concerned with the estimation of the value of ϑ at the place of light action. Possible correction for different optical artifacts will be discussed below. Another way to estimate ϑ is to measure the rate constant of the phototransformation, $k_1 + k_2 = \vartheta$, directly. This is possible for phytochrome in etiolated and Norfluorazon-treated (see Chap. 8, this Vol. and JABBEN et al. 1982) plants.

4.2.3 Limitation for Application of the Plot "R vs. φ with $\vartheta = \text{const}$" and the Theory of Dichromatic Irradiation

If one applies only monochromatic irradiation the application of plot "R vs. φ with $\vartheta = \text{const}$" underlies the following severe restriction. Unlike the case of one primary reaction, where the rate constants in two experiments could have been made equal ($k_a = k_b$) by varying N_λ, in a photoreversible system we have to adjust two rate constants ($k_{1a} = k_{1b}$, $k_{2a} = k_{2b}$) or, which is the same, two parameters ($\varphi_a = \varphi_b$, $\vartheta_a = \vartheta_b$) by varying the single parameter N_λ. One should expect that this is not always possible because of the limitations imposed by the known spectra $\sigma_{r\lambda}$, $\sigma_{fr\lambda}$. Indeed, from the left side of Eq. (10b) and the definition (2) we obtain the following necessary condition for equivalent light action:

$$k_{1a} = N_{\lambda a}\,\sigma_{r\lambda a} = k_{1b} = N_{\lambda b}\,\sigma_{r\lambda b}$$

$$k_{2a} = N_{\lambda a}\,\sigma_{fr\lambda a} = k_{2b} = N_{\lambda b}\,\sigma_{fr\lambda b}$$

and further:

$$\frac{\sigma_{r\lambda a}}{\sigma_{fr\lambda a}} = \frac{\sigma_{r\lambda b}}{\sigma_{fr\lambda b}}. \tag{12}$$

Expression (12) means that we can achieve equivalent light action ($\varphi_a = \varphi_b$, $\vartheta_a = \vartheta_b$) only in those pairs of experiments with monochromatic irradiation in which the applied wavelengths λ_a, λ_b provide the same ratio σ_r/σ_{fr}.

To avoid this limitation one should be able to vary more than one characteristic of irradiation in experiments to be compared. Such a possibility occurs when dichromatic (or polychromatic) irradiation instead of monochromatic irradiation is applied. In this case the fluence rate $N_{\lambda 1 + \lambda 2}$ is composed of two monochromatic components $N_{\lambda 1}$ and $N_{\lambda 2}$, each of which can be varied. On the basis of *known* $\sigma_{r\lambda}$, $\sigma_{fr\lambda}$ a mixture of fluence rates $N_{\lambda 1}$ and $N_{\lambda 2}$ can be calculated to satisfy any desired effective values φ_{ef}, ϑ_{ef}. The only remaining

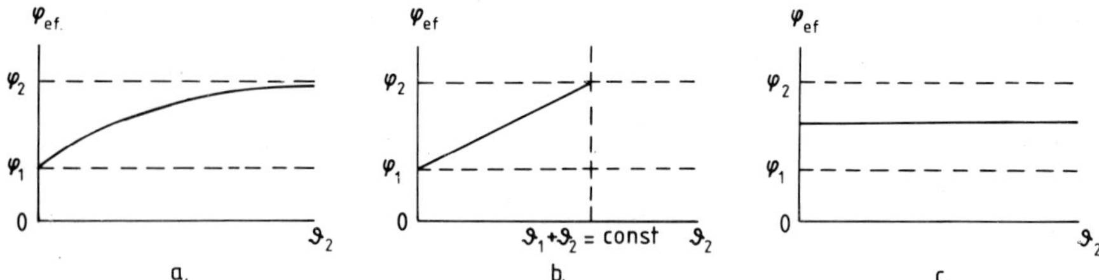

Fig. 6 a–c. Three types of dichromatic experiments. Pairs of characteristic φ_{ef} and ϑ_{ef}; φ_1 and ϑ_1; φ_2 and ϑ_2 describe the dichromatic photon flux, the partial monochromatic photon flux at λ 1, the partial monochromatic photon flux at λ 2, respectively. The characteristic φ_{ef} is shown as a function of varied ϑ_2 for three types of dichromatic experiments. **a** $\vartheta_1=\text{const}$, ϑ_2 is varied; **b** both ϑ_1 and ϑ_2 are varied under condition $\vartheta_1+\vartheta_2=\text{const}$; **c** both ϑ_1 and ϑ_2 are varied under condition ${}^{\vartheta_1}/_{\vartheta_2}=\text{const}$

limitation is that φ_{ef} varies *between* values φ_1 and φ_2 (see Fig. 6). The application of dichromatic photon fluxes has been introduced in HARTMANN'S pioneer work (1966).

The theory of dichromatic irradiation appears very simple in terms of parameters φ, ϑ (FUKSHANSKY unpublished). In a dichromatic experiment using simultaneously two photon fluence rates $N_{\lambda 1}$, $N_{\lambda 2}$ with parameters φ_1, ϑ_1 and φ_2, ϑ_2 respectively, one obtains

$$k_{1\,ef}=k_{11}+k_{12},\; k_{2\,ef}=k_{21}+k_{22}$$

and further, substituting the last equations in (11):

$$\vartheta_{ef}=\vartheta_1+\vartheta_2;$$

$$\varphi_{ef}=\frac{\varphi_1\,\vartheta_1+\varphi_2\,\vartheta_2}{\vartheta_1+\vartheta_2}. \tag{13}$$

Expression (13) correlates the characteristics of the dichromatic flux with those of monochromatic fluxes. Now, choosing λ_1 and λ_2 one establishes the parameters φ_1, φ_2 and with the changing of $N_{\lambda 1}$ and $N_{\lambda 2}$ produces desired values of ϑ_{ef}, φ_{ef}, including those which are forbidden by Eq. (12). There are three different ways of performing a dichromatic experiment.

1. $\vartheta_1=\text{const}$; ϑ_2 is varied; In this case ϑ_{ef} changes with the same increments as ϑ_2. The φ_{ef} is changed as shown in Fig. 6a, being a monotonous but not linear function of ϑ_2 (in Fig. 6 we have taken $\varphi_1<\varphi_2$ for definitiveness which does not affect generality). A dichromatic fluence rate-response curve obtained in this way is a projection of the three-dimensional function R from Fig. 5b on a rather complicated surface, which is perpendicular to the plane $\varphi 0 \vartheta$. For very large values of $\vartheta_2 (\vartheta_2 \to \infty)$ this surface approaches a plane $\varphi=\text{const}=\varphi_2$ asympthotically, which is parallel to the axis ϑ. This is exactly the type of experiments applied by HARTMANN (1966).

2. Both ϑ_1 and ϑ_2 are varied but in such a way that $\vartheta_1+\vartheta_2=\text{const}$. Here φ_{ef} is a linear function of ϑ_2 (and, of course, of ϑ_1) as shown in Fig. 6b. A

dichromatic fluence rate response curve obtained in this way is a projection of the three-dimensional function R from Fig. 5b on a plane which is perpendicular to the plane $\varphi 0 \vartheta$ and parallel to the axis φ. This more cumbersome procedure has some advantages as compared with the first type of dichromatic experiment. For example, it enables one to compare responses which are functions of light quality (i.e., φ_{ef}) only, whereas the "cycling" of the photoreversible pigment (i.e., φ_{ef}) remains constant.

3. All three magnitudes ϑ_1, ϑ_2, ϑ_{ef} are varied under the condition

$$\vartheta_1/\vartheta_2 = \text{const.}$$

providing $\varphi = \text{const.}$ (Fig. 6c). This is an exact imitation of a sequence of monochromatic irradiations with the same λ and different values of N_λ. A dichromatic fluence-rate response curve obtained in this way is a projection of the three-dimensional function R from Fig. 5b on a plane which is perpendicular to the plane $\varphi 0 \vartheta$ and parallel to the axis ϑ.

This type of dichromatic irradiation is a cumbersome imitation of a monochromatic fluence rate-response curve. The advantage of this procedure is that it is then possible to simulate a monochromatic irradiation at a wavelength where strong screening has to be expected by simultaneous irradiation with two wavelengths having no, or at least different, static screening (see SCHÄFER and FUKSHANSKY 1983, SCHÄFER et al. 1982). Furthermore, the effect of dynamic light-induced screening (JOHNSON 1980) can also be analyzed by using this type of dichromatic irradiation. For example, the possible role of protochlorophyll-chlorophyll transformation can be checked by comparing monochromatic irradiation at λ_1 (690 nm) with dichromatic irradiation at 650+750 nm with $\varphi_{ef} = \varphi_{690}$. Changes in the fluence rate response curves are probably due to influences of chlorophyll formation.

4.2.4 Additional Remarks Concerning General Analytical Action Spectroscopy

1. Both of the procedures described above – "plot R vs. φ with $\vartheta = \text{const.}$" and dichromatic experiments of all three types – lead to the following conclusions: If discrepancies with theoretical predictions have been found, then additional photoreceptor(s) must be involved. If no discrepancies have been found, then a process induced by the photoreversible pigment is the limiting step in the mechanism of response and perhaps no other pigments are involved.

This analysis is based on known spectra $\sigma_{r\lambda}$, $\sigma_{fr\lambda}$. Any attempt *to determine* $\sigma_{r\lambda}$, $\sigma_{fr\lambda}$ from the action spectra without additional information cannot be successful because the number of unknown variables will always exceed the number of equations.

2. The question whether there is more than one primary reaction can be raised in general, i.e., outside the scope of an established photoreversible photoreceptor. The non-parallelism of the fluence rate-response curves discussed in Section 2.3 is a useful indication of a complex photoreceptor, however, the whole problem is more complicated.

There are two different types of photobiological responses with respect to this problem. Type one responses do not require the existence of fluence rate gradients or other spatial features of photon fluxes. They are functions of the number of absorbed photons and their distribution in time (for example, photosynthesis, inhibition of elongation, anthocyanin production). Type two responses are functions of spatial properties of photon fluxes like fluence rate gradients or state of polarisation (for example, phototropism or polarotropism).

Since spatial properties of the photon flux are wavelength-dependent, the action of the photoreceptor of type two response can be modified in a way so that it will imitate the presence of an additional photoreceptor, for example, fluence rate-response curves may become non-parallel and, it is possible that they may even have different saturating levels. To visualize such an effect one can imagine that a response depends on both a local fluence rate and the size of a light spot produced by a lens-effect which is different for λ_a and λ_b; the number of light-sensitive areas in action may then depend on the whole process of light propagation. This can cause different (for different λ) rates of the increase of response with an increasing fluence rate. Similar conclusions can be drawn with the help of a λ-dependent fluence rate gradient and a gradient-dependent response.

The parallelism of fluence rate-response curves of type 1 responses will not be affected by spatial rearrangements of the photon fluxes. As we shall see below (Sect. 5.3) only fluence rate gradients which vary with time and are themselves fluence-rate-dependent (non-stationary screening) will affect the parallelism of these curves.

3. One more example of a problem concerned with an indication of a complex photoreceptor is given by the extrema of dichromatic fluence rate-response curves (HARTMANN 1966). One can ask whether the existence of an extremum is a necessary and/or sufficient condition for two primary reactions or even for a definite type of a photoreceptor with two primary reactions, for example, a photoreversible pigment? Since one can easily imagine a fluence rate-response curve of a simple photoreceptor with one primary reaction which has an extremum, we conclude that an extremum by itself is not a sufficient condition for two primary reactions.

From the discussion in Section 4.2.3 we know that by changing $N_{\lambda 2}$ in a dichromatic experiment we take our system through different values φ, ϑ and can encounter a local extremum of the function R (φ, ϑ); this is, however, not the property of the photoreceptor alone but of the whole set of processes bringing the response about.

In fact, the extremum only accompanies the following property (of a dichromatic fluence rate-response curve) which seems to be bound to a complex photoreceptor: if neither the monochromatic fluence rate $N_{\lambda 1}$, nor the monochromatic fluence rate $N_{\lambda 2}$ but only the dichromatic fluence rate $N_{\lambda 1+\lambda 2}$ provides a significant response this means that more than one primary reaction exists. This property causes an extremum of a dichromatic curve independently since this curve starts and ends in a monochromatic condition, where the response is nullified. Whether an extremum should always exist when two primary reactions exist (i.e., is a necessary condition) is unknown. This general problem, as some others in this area, needs investigation.

What can be said with respect to this discussion in the particular case of a phytochrome system? For phytochrome controlled responses up to now *no extrema* for $R_{\varphi=\text{const}}$ (ϑ) have been observed either under continuous irradiation or in pulse experiments. For $R_{\vartheta=\text{const}}$ (φ) extrema have been observed (HARTMANN 1977, BEGGS et al. 1980, SCHÄFER 1981, HOLMES and SCHÄFER 1981) under continuous irradiation dark-grown seedlings. For light-grown seedlings these extrema seem to disappear (BEGGS et al. 1980). In models by GAMMERMAN and FUKSHANSKY (1971, 1974) and SCHÄFER (1975, 1976) it has been shown that the phytochrome dynamics alone *could* provide, under prolonged irradiation, extrema which, when the time of irradiation is varied, move along the spectrum in the same way as the extrema of action spectra under HIR conditions do. These models, which lie within the framework of the reduced problem of phytochrome action (see Chap. 5, this Vol.), say nothing about the possibility that the extrema of action spectra depend on the properties of the transduction chain (this can be studied only using the dynamics of the signal transition from the photoreceptor within the framework of the full problem of photoreceptor action). Thus, we can say that extrema in plots $R_{\vartheta=\text{const}}$ (φ) and in dichromatic fluence rate-response curves reflect properties of phytochrome dynamics and perhaps of the mechanism, which transduces these dynamics into response. Of course, these extrema have nothing to do with the fact itself that the dichromatic technique is applied.

4.3 Model-Bounded Analytical Action Spectroscopy of Phytochrome-Induced Responses

The photoreversible pigment phytochrome (see Fig. 7) is shown to be the photoreceptor of many photomorphogenic responses. This has been done without applying action spectra. The reversibility of the response (cf. MOHR 1972, SMITH 1975) is a much more convenient tool to do this in the case of pulse experiments (induction conditions). In the case of prolonged irradiation (HIR conditions) the involvement of phytochrome was shown in the dichromatic experiments by HARTMANN (1966) which are based on the principle of equivalent light action as discussed in Sections 4.2.3 and 4.2.4. At the same time the involvement of a blue light photoreceptor in some photomorphogenical responses was shown by means of "plot R vs. φ with ϑ=const." (Figs. 4a, b in Chap. 28, this Vol.). Thus, in various responses phytochrome is either the single photoreceptor or one among others, and the problem of involvement of a photoreceptor was solved either using photoreversibility of the response or methods of general analytical action spectroscopy.

However, as it has been noted in Section 4.1, other problems specific for complex photoreceptors appear here. The connection between the characteristics of irradiation and those of the signal which arise on the input of the transduction chain is much more complicated than in the case of a single pigment. This

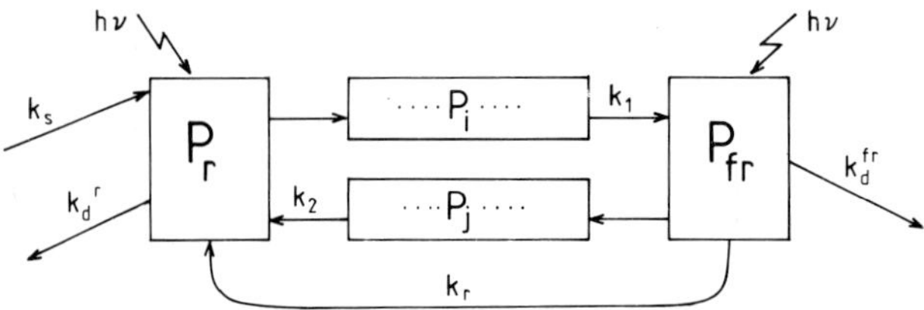

Fig. 7. The network of reactions in the photoreversible pigment system phytochrome, according to current knowledge. P_r, P_{fr} are forms of phytochrome having different absorption spectra (shown in Fig. 1 Chap. 6, this Vol.). The transition of phytochrome from the stable form, P_r, to a rather non-stable form, P_{fr}, by a red light pulse causes photomorphogenic responses which can be prevented if the red pulse is followed by an almost infrared pulse ($\simeq$760 nm) converting P_{fr} to P_r. These pulse experiments (induction conditions) show action spectra corresponding to the effective cross-sections of phototransformations $P_r \rightarrow P_{fr}$, $P_{fr} \rightarrow P_r$, respectively. Action spectra of photomorphogenic responses under prolonged irradiation show no resemblance either with cross-sections of photoequilibrium in the phytochrome system (when a rather high fluence rate is applied such spectra are called-HIR-action spectra reflecting the accepted abbreviation of "high irradiance conditions"). The scheme is based on the data obtained mainly by the spectrophotometrical assay (see Chap. 6, this Vol.) and reflects interconversions between *total amounts* of phytochrome having the chromophore in a definite form. It is very likely that the pools, P_r, P_{fr}, are subdivided into subpools (see Chap. 6 for detailed consideration). Some of the reactions shown are not found in all objects

connection is influenced by the total *dynamics of photochemical and non-photochemical reactions within the photoreceptor*.

The total result of the whole network of these reactions is to some degree measurable: one can measure spectrophotometrically (see Chap. 5, this Vol.) interconvertions between overall pools of phytochrome forms P_r and P_{fr} which have the chromophore in different states. These measurements are, however, feasible only in objects which either are etiolated or in which the chlorophyll is bleached after a special pretreatment (see Chap. 25, this Vol.).

The second measurable dynamic is that of the signal transition from the photoreceptor to the transduction chain. This dynamic can be measured as the rate with which the reversibility of a response (induced by a red pulse) disappears in the darkness which follows (cf. FUKSHANSKY and MOHR 1980). Obviously, the rate of a signal transition may depend on some characteristics of the photoreceptor (for example, P_{fr} for $t = t_e$, real time) and on some characteristics of the transduction chain (for example, the current number at time t of specific receptor sites available for P_{fr}). Direct measurement of the loss of reversibility is possible only for induction conditions but not for HIR. Using the techniques of intermittent irradiation developed by Mancinelli and coworkers (cf. MANCINELLI and RABINO 1978 and Chap. 24, this Vol.), one can substitute the action of prolonged continuous irradiation with a series of pulses (SCHÄFER et al. 1981 b, HEIM and SCHÄFER 1982) and then estimate the dynamics of loss of reversibility of contributions to the response which have been made by certain pulses. It must be stressed that the dynamics of signal transduction under intermittent irradiation is not necessarily equal to that under continuous irradiation even when the levels of response are equal (HEIM and SCHÄFER 1982).

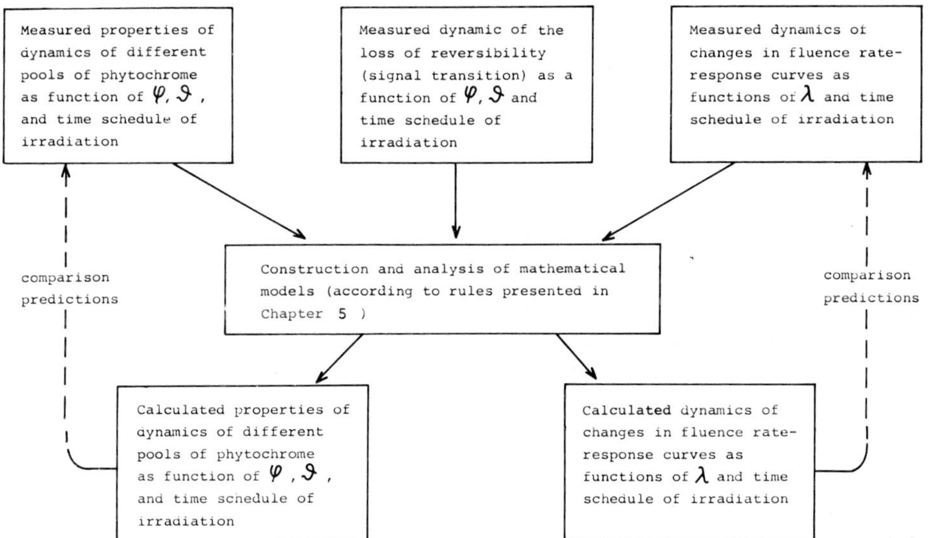

Fig. 8. A scheme illustrating the procedure of model-bounded action spectroscopy

The level of response and the dynamics of signal transduction are both functions of the dynamics in the photoreceptor which may be different in to experiments with the same total amount of photons applied but different time schedules of the irradiation.

The third measured dynamic is that of action spectra which depends on the time schedule of irradiation and on different light pretreatments (MANCINELLI and WALSH 1979).

Using these three sources of experimental data on can apply *the method of mathematical modeling* to study the following main problems of a complex photoreceptor:

1. What do the dynamics (and, therefore, the structure) of the complex photoreceptor look like, and how do they depend on the characteristics of irradiation?
2. How does phytochrome act on the transduction chain?
3. Which characteristics of the phytochrome dynamics correlate to the response and what are the correlations?

The construction and analysis of mathematical models must be performed according to the rules presented in Chapter 5, this Volume where the expected results and the limitations of the method are shown. In fact, this is a cyclic procedure in which one selects assumptions about the mechanism of photoreception which have consequences consistent with experimental results (see Fig. 8).

The current stage in the modelling of the phytochrome system is presented in detail in Chapter 5, this Volume. In the course of modeling one obtains, however, not only a concrete analysis of sets of assumptions but also some general principles applicable to the studied system, irrespective of the particular model used.

5 Optical Artifacts

5.1 The Problem

The core of action spectroscopy is the comparison of absorption and action spectra (and also the comparison of in vivo and in vitro absorption spectra). All these spectra depend not only on spectral properties of the functional pigment but also on the whole process of light propagation within the sample. Two kinds of optical artifacts disturb the procedure of action spectroscopy.

1. Gradients of the photon fluence rate occur within the sample. Therefore, the real photon fluxes which hit the functional pigment may have drastic discrepancies with the incident photon fluxes (especially if the pigment itself has a non-homogeneous spatial distribution). A crucial consequence is that the discrepancies are wavelength-dependent, which distorts the action spectra.
2. Absorption spectra measured in vivo differ from corresponding spectra measured in vitro.

Both fluence rate gradients and distortion of the absorption spectra occur on a biological tissue due to self- and mutual screening of different pigments, multiple scattering, multiple mirror reflection on the boundaries of areas having different refractive indexes, sieve-effect and screen fluorescence (see FUKSHANSKY 1981). Therefore, corrections for optical artifacts means describing the influence of these effects on the process of light propagation.

We consider the optical artifacts and their correction in Sections 5.3 and 5.4 after a general discussion of the influence of fluence rate gradients on the fluence rate-response curves.

5.2 The Influence of Fluence Rate Gradients on Fluence Rate-Response Curves

A non-even spatial distribution of the photon fluence rate within the tissue, whatever its cause may be, provides that the real "acting" monochromatic fluence rate $N_{\lambda\xi}$ at a certain depth ξ differs from the incident fluence rate $N_{\lambda 0}$ by a correction factor $f(\lambda,\xi)$ which depends on both λ and ξ:

$$N_{\lambda\xi} = f(\lambda, \xi) N_{\lambda 0}. \tag{14}$$

The factor f causes *a parallel shift* of the logarithmic fluence rate-response curve as compared to the curve which would have been produced by the incident photon fluence rate $N_{\lambda 0}$. This can be easily shown following the same reasoning which led to Eq. (7) in Section 2.3. The difference of values $\ln N_{\lambda\xi}$ at λa and λb which bring about the same value of response will be:

$$\ln N_{\lambda a\xi} - \ln N_{\lambda b\xi} = \ln \frac{\sigma_{\lambda b}}{\sigma_{\lambda a}} + \ln \frac{f(\lambda b, \xi)}{f(\lambda a, \xi)} = \ln N_{\lambda ao} - \ln N_{\lambda bo} + \ln \frac{f(\lambda b, \xi)}{f(\lambda a, \xi)}, \tag{15}$$

i.e., the additional shift of the fluence rate-response curve b against curve a is $\ln[f(\lambda b, \xi)/f(\lambda a, \xi)]$.

It can be shown (FUKSHANSKY unpublished) that after a transition from the local consideration at depth ξ to the overall response by integration over the entire thickness of the object the expression (15) holds. It holds also when the pigment is non-evenly distributed in depth and when the concentration of the pigment changes either independently of irradiation or in a light-induced process with the same primary reaction as that of the response. The only difference between these more general situations and that considered above is that the rate constant k will be substituted by some effective constant, which is a function of both fluence rate gradients and the spatial distribution of the pigment.

The reason why the parallelism of fluence rate-response curves is not affected by the fluence rate gradients (14) is that the response in the entire object appears as a result of the total action of all the photons absorbed (the overall rate of the primary reaction appears as an integral of the local rates over all the depths). In other words, we have a type one response as discussed in Section 4.2.4. In the case of a type two response not the integral but some parameters of spatial distribution of fluence rates should be significant for the response (for example, the difference between $N_{\lambda\xi}$ at different dephts ξ). Therefore, the fluence rate-response curves at different λ presented as functions of the logarithm of the *total number of photons absorbed per second may not be parallel.*

The parallelism of type 1 responses will be lost only if the screening itself is a function of the fluence rate (as, for example, in the course of greening). The photon fluence rate at the depth ξ will be in this case

$$N_{\lambda\xi} = f[\lambda, \xi, N_{\lambda 0}] \cdot N_{\lambda 0}$$

which, unlike Eq. (14) does not allow one to derive an expression similar to Eq. (15).

The discussion in this section has only been concerned with the parallelism of fluence rate-response curves. Even if the parallelism holds the information about the functional pigment will be lost through the shifts of fluence rate-response curves due to gradients (14) if these gradients are unknown. The estimation of fluence rate gradients in a tissue which (as well as correcting the absorption spectra) is the problem of heterogeneous optics will be briefly discussed in the following section.

5.3 Fluence Rate Gradients in a Tissue

The fluence rate gradients occur, obviously, even in a homogeneously absorbing sample where the light propagation can be described by Beer's law:

$$T = e^{-\alpha\rho L}, \qquad A = -\ln T = \alpha\rho L, \tag{16}$$

where A is absorbance, α is molar absorbance, P is concentration of the pigment, L is thickness of the object, $T = N_{\lambda L}/N_{\lambda 0}$ is the transmittance, $N_{\lambda 0}$ is the incident fluence rate, $N_{\lambda\xi}$ is the fluence rate at the depth ξ.

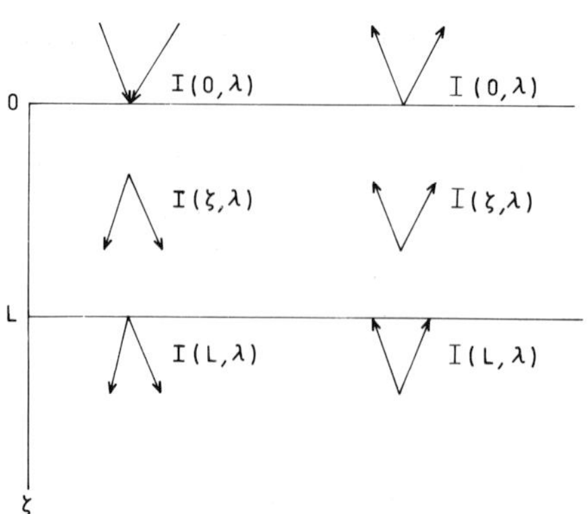

Fig. 9. Graphical presentation of the idealization underlying the K-M theory. ξ spatial coordinate; L thickness of the object; λ wavelength; I and J photon fluxes propagating in opposite directions

Expression (16) accounts for absorption only. Biological objects are usually highly scattering. Scattering affects the light propagation in two ways.

Firstly, parts of the transmitted and remitted photon fluxes may *not reach the measuring device*. This leads to an overestimation of absorbance. This "instrumental" error can be almost completely corrected by applying ULBRICHT's sphere or partly corrected by using opal glasses. Corresponding methods are considered in KORTÜM's excellent monograph (1969) and we will not discuss them here.

Secondly, scattering causes an increase in the average pathlength of photons which also leads to an increase of the measured absorbance. The increase of the pathlength can be up to factor 20 (BUTLER 1962) and cannot be directly measured or calculated *in an intact object*.

To account for both absorption and scattering in a turbid media *phenomenological theories* of light propagation must be applied (see FUKSHANSKY 1981). The simplest and most commonly used is the two-flows Kubelka-Munk (K-M) theory which is, in fact, a two-parametrical generalization of the one-parametrical Beer's law (see FUKSHANSKY 1981).

In the idealized scheme underlying the K-M theory (Fig. 9) the overall photon flux N_λ within the object is thought to consist of two diffuse fluxes I and J — propagating in opposite directions (in Fig. 11 I is directed from above as incident flux I_0, J is directed from below, the object is presented as an infinite strip with parallel sides). Photons may be absorbed with the unknown propability k (phenomenological absorption coefficient) which is the same for both fluxes. Scattering manifests itself in the transfer of photons from I to J and vice versa, which occurs with the unknown probability s (phenomenological coefficient of scattering) and is the same for both fluxes. This scheme leads to a system of two differential equations, which, when combined with any two of four values of I, J measured at the boundaries of the object (boundary conditions), provides solutions for fluence rates $I(\xi)$, $J(\xi)$ as functions of the depth ξ. The remaining pair of boundary values of I, J, if substituted in these solutions, gives the unknown phenomenological optical parameters k and s.

The important difference between scattering and non-scattering samples reflected in the difference between the K-M theory and Beer's law is concerned with the additivity of absorbance as a function of thickness. In a non-scattering sample with thickness $L = L_1 + L_2$ the absorbance of the whole sample is a sum of absorbances of its parts (having thickness L_1 and L_2) taken separately. Correspondingly, the fluence rate at the output of part one (having thickness L_1) will be equal to the fluence rate within the intact sample at depth L_1. This is not true for scattering samples because here the fluence rate at a certain depth is established as a result of the contributions from all the layers constituting the sample, including those which are cut off when the transmittance of a part of the sample is measured.

The conditions of applicability of the K-M theory are discussed in detail by HARTMAN and COHNEN UNSER (1972), FUKSHANSKY (1981) and FUKSHANSKY and KAZARINOVA (1980). It is essential that a diffused light is applied, that the object is a plane layer much larger than the measuring beam's cross-section, or that it is placed in a cuvette with non-selective reflecting walls. The reflection at the boundaries of the object is not accounted for in the standard K-M theory. This can be done by combining it with a (modified) classical procedure originated by STOKES (1862) which derives the overall remission and transmission of a pile of planar plates from remissions and transmissions of individual plates. We do not consider this combined theory here which is presented briefly in FUKSHANSKY (1981). Using developing cotyledons of *Cucurbita pepo,* detailed measurements and calculations showed that transmission and remission and therefore the fluence rate gradient depend strongly on the age of the seedling and light pretreatments (SEYFRIED, FUKSHANSKY and SCHÄFER unpublished). If the object is not macrohomogenous, for example, if it consists of layers with different absorption and scattering, the application of the K-M theory yields only "effective" values of k and s and does not provide a reliable estimation for the fluence gradients. There are no methods to determine fluence rate gradients in a layered intact plant tissue (cf. FUKSHANSKY 1981, FUKSHANSKY and KAZARINOVA 1980).

In summary, accounting for scattering results in completely different fluence rate gradients as compared with those obtained by means of Beer's law and this discrepancy increases when the reflection at the boundaries is accounted for. These more precise calculations reveal, in particular, that plants' organs have developed properties of light traps to utilize light more efficiently (SEYFRIED and FUKSHANSKY unpublished).

Recently the K-M theory had been extended to account also for screen fluorescence (FUKSHANSKY and KAZARINOVA 1980). The application of this rather complicated theory to phytochrome screened by chlorophyll in the special case of daylight irradiation and the 0.018 quantum yield of chlorophyll fluorescence led to the following conclusions: The absolute contribution of Chl fluorescence is quite enough to cause a loss of detection of the phytochrome-specific changes in absorbance measured at 660 nm and 725 nm. It has been supposed that the measurements in the neighborhood of 645 nm could prove more sensitive by slightly green objects. The influence of the Chl fluorescence on the phytochrome phototransformations can be ignored. These and some other statements are valid only for the light source and the value of the quantum yield used.

The theory allows one, however, to test what will happen under any other conditions, for example, for a higher value of fluorescence quantum yield and under red and far-red irradiation typical for photomorphogenic experiments.

Another problem to be tested by the extended theory is the estimation of the influence of protochlorophyll fluorescence on the phytochrome specific signal in etiolated plant tissues (cf. PRATT 1978, FUKSHANSKY and KAZARINOVA 1980).

5.4 Distortion of Absorption (Difference) Spectra

Absorption spectra can be distorted in vivo (as compared with in vitro spectra) due to the spatial inhomogeneity of the pigment, which implies that different parts of the photon flux intersect different amounts of the pigment. This phenomenon is called sieveeffect or absorption statistics.

Let us consider the overall photon flux as consisting of such small fractions that, if we further subdivide these fractions into subfractions, each subfraction from the same fraction intersects the same amount of pigment. This means that for each fraction the object is homogeneous (however, with different effective amounts of pigment for different fractions) and Beer's law can be applied[1]. This reasoning leads to the following law of light absorption (FUKSHANSKY unpublished)

$$T = \int_0^\infty f(L, x)\, e^{-\alpha x}\, dx, \qquad A = -\ln T, \tag{17}$$

where L is the thickness of the sample, x is the amount of pigment intersected by a small (differential) fraction of the photon flux (x is a random variable), f(L, x) is the density probability of the distribution of the differential fractions of the photon flux with respect to the amount of pigment intersected, A, T and α are as in Eq. (16). Expression (17) is a generalization of Beer's law (16), and converts into Eq. (16) when all the fractions intersect the same amount of pigment.

Obviously, expression (17) can provide a different value of T for fixed α and L depending on the spatial distribution of the pigment [i.e., on f(L, x)]. This is why the universal characterization of a pigment must be related to a fixed spatial pattern and, naturally, a homogeneous pattern is generally chosen. However, in a tissue the heterogeneous spatial distribution of a pigment occurs because the pigment is part of the biological structure. Therefore, one is interested in absorption spectra which correspond to the situation described in Eq. (17). To correct absorption spectra for changes caused by spatial reorganization of the pigment the sieve-effect factor β is applied, which is defined as

[1] This is not quite correct. The scattering within the sample causes a distribution of different parts of the photon flux with respect to pathlength which can vary between infinitely small (for photons remitted by the infinitely thin layer below the irradiated surface) and infinitely large values. The reasoning must be applied not to the over all photon flux but to such a part of the flux and then the second step should be made for transfer to the overall flow

the relative change of absorbance caused by transition from homogeneous to inhomogeneous spatial pattern:

$$\beta = \frac{A_0 - A}{A_0}. \tag{18}$$

Here A_0 is from Eq. (16), A is from Eq. (17). Analogously, the sieve-effect factor between two different heterogeneous patterns is defined:

$$\beta_{ij} = \frac{A_j - A_i}{A_j}. \tag{18a}$$

Here A_j, A_i are from Eq. (17) with functions f_j, f_i ($f_j \neq f_i$), respectively. Of course, the total amount of pigment in the compared objects must be equal. One should remember that β, β_{ij} are not constants but functions which also depend on α (and, therefore, on λ).

The properties of β, e.g., correlations between β and α, dependence on parameters of pigments spatial distribution and on sample thickness are studied using the sieve-effect theory. This theory consists of two parts (FUKSHANSKY unpublished):

1. general theory, i.e., statements which can be made in the absence of any information about f(L, x) and are valid for any type of spatial pattern;

2. statistical theory, i.e., more detailed statements which can be made when the type of f(L, x) [and, perhaps, even parameters of f(L, x)] is known on the basis of a probabilistic reasoning. Examples of statements of general theory: (1) Any heterogeneity can only diminish the absorbance of a homogeneous object (FUKSHANSKY 1978). (2) The sieve-effect is independent of the object thickness L (FUKSHANSKY unpublished).

The simplest objects of the statistical theory are homogeneous suspensions and powders. Investigators were confronted with such objects in photobiology (for example, the flattening of absorption spectra of suspensions (DUYSENS 1956) or optical changes caused by chloroplasts movement (ZURZYCKI 1961) and also in the spectroscopy of films, poweders and protective coatings (DUYCKAERTS 1959, GLEDHILL and JULIAN 1963, FELDER 1964).

Theories for suspensions and powders [the most advanced have been developed by DUYSENS (1956) and FELDER (1964)] have much in common and are based on the discontinuous model of the distribution of particles proposed in the fundamental paper by DUYSENS (1956): the idealized suspension is described as a number of regularly oriented cubes randomly placed in a spatial lattice having cubical compartments of the same size.

A more realistic (one postulate less) semi-continuous model (FUKSHANSKY 1978) provides some improvements in calculating the sieve-effect in suspensions and powders and, importantly, allows one to extend the treatment beyond these simple objects.

Spatial distribution of pigments in tissues is usually more complicated than those in suspensions due to a layered structure containing different cell types in different layers, the absence of pigment in some cells, vacuolarization and sequestering. The theory for such objects (FUKSHANSKY unpublished) deals predominantly with hierarchical models: absorbing particles are gathered into specific regions – clusters – which are randomly distributed over the object, while particles are randomly distributed over a cluster.

It has been shown (FUKSHANSKY unpublished) that scattering does not affect the contribution of the sieve-effect to the distortion of absorption spectra.

The current results of the theory of the sieve-effect can be described as follows. For any known stochastic spatial pattern of the pigment an exact absorption law can be derived. Any two spatial patterns can be compared with respect to the sieve-effect. If the absorption spectrum has been measured for any known spatial pattern, one can predict the absorption spectrum of any other spatial pattern, provided the difference in light scattering between these two spatial patterns is known. One can give quantitative estimates of the changes in absorbance caused by such alterations in the spatial structure of the object such as a subdivision of particles, dilution of their content, splitting or sticking of the clusters, squeezing of the particles or clusters, reorientation of non-symmetrical particles and/or clusters, etc.

In order to apply the hierarchical theory of the sieve-effect to phytochrome studies the theory has been extended to account for a photoreversible system and difference spectra (FUKSHANSKY, MOHR and SCHÄFER, unpublished). The parameters of the spatial distribution of phytochrome used in calculations have been taken from immunochemical analysis made by PRATT et al. (1976) and MACKENZIE et al. (1974).

The sieve-effect caused by phytochrome sequestering appears small (2–4%), however, the influence of changes in vacuole size is rather high. This effect, if neglected, may lead to a misinterpretation of the changes in measured difference of absorption as a consequence of the changes in the total amount of phytochrome. This danger is restricted to absolute spectra. Distortions of the relative difference spectra due to the sieve-effect have been found negligible. All these calculations can only be made for etiolated objects.

A theory of the sieve-effect in two-component systems (i.e., when, for example, both screening and sensory pigments have heterogeneous and different spatial patterns) does not exist. Nor does a unified theory exist for both the sieve-effect and scattering.

The sieve-effect of the screen (i.e., chlorophyll) in a green leaf can provide different conditions of illumination for phytochrome at different points at the same depth as has been acknowledged by HANKE et al. (1969). The authors attempted an approximate empirical estimation of these fluctuations. A statistical calculation of such an effect has not been feasible up till now since it requires the combined consideration of the sieve-effect and scattering.

In summary: Useful analytical action spectroscopy can be carried out only on the basis of a reliable model of phytochrome dynamics. The data must be corrected for "optical artifacts" in order to analyze the dynamics of signal transduction and in order to analyze fully the dynamics of the phytochrome system. From these data the mechanism of phytochrome action (Fig. 8) may be solved.

References

Beggs CJ, Holmes MG, Jabben M, Schäfer E (1980) Action spectra for the inhibition of hypocotyl growth by continuous irradiation in light and dark-grown *Sinapis alba* L. seedlings. Plant Physiol 66:615–618

Butler WR (1962) Absorption of light by turbid materials. J Opt Soc Am 52:292–299

Duyckaerts G (1959) The infrared analysis of solid substances. (A review.) Analyst 84:201–214

Duysens LNM (1956) The flattening of the absorption spectrum of suspensions as compared to that of solutions. Biochim Biophys Acta 19:1–12

Duysens LNM (1970) Photobiological principles and methods. In: Haldall P (ed) Photobiology of microorganisms. Wiley-Interscience, London

Engelmann TW (1882) Über Sauerstoffausscheidung von Pflanzenzellen im Mikrospektrum. Bot Z 40:419–426

Felder B (1964) Über die Teilchengrößenabhängigkeit der Lichtabsorption in heterogenen Systemen. I. Theoretische Betrachtungen. Helv Chim Acta 47:488–497

Fukshansky L (1978) On the theory of light absorption in non-homogeneous objects: the sieve-effect in one-component suspensions. J Math Biol 6:177–196

Fukshansky L (1981) Optical properties of plants. In: Smith H (ed) Plants and the daylight spectrum. Academic Press, London New York, pp 21–40

Fukshansky L, Kazarinova N (1980) Extension of the Kubelka-Munk theory of light propagation in intensely scattering materials to fluorescent media. J Opt Soc Am 70:1101–1111

Fukshansky L, Mohr H (1980) Boundary conditions for mathematical models in photomorphogenesis. In: De Greef J (ed) Photoreceptors and plant development. Antwerpen Univ Press, Antwerpen, pp 135–144

Gammermann A, Fukshansky L (1971) Theory and calculation of the transformation dynamics of phytochrome in a green leaf. Physiol Rastenii 18:661–667

Gammermann A, Fukshansky L (1974) A mathematical model of phytochrome — the receptor of photomorphogenic processes in plants. Ontogenez 5:108–114

Gledhill RJ, Julian DB (1963) Light absorption in heterogeneous systems with application to photographic dye images. J Opt Soc Am 53:239–246

Hanke J, Hartmann KM, Mohr H (1969) Die Wirkung von „Störlicht" auf die Blütenbildung von *Sinapis alba* L. Planta 86:235–249

Hartmann KM (1966) A general hypothesis to interpret the high energy phenomena of photomorphogenesis on the basis of phytochrome. Photochem Photobiol 5:349–366

Hartmann KM (1977) Aktionsspektroskopie. In: Hoppe W, Lohmann W, Markl H, Ziegler H (eds) Biophysik. Springer, Berlin Heidelberg New York, pp 197–222

Hartmann KM, Cohnen Unser I (1972) Analytical action spectroscopy with living systems: photochemical aspects and attenuance. Ber Dtsch Bot Ges 85:481–555

Heim B, Schäfer E (1982) Light-controlled inhibition of hypocotyl growth of *Sinapis alba* L. seedlings. Fluence rate dependence of hourly light pulses and continuous irradiation. Planta 154:150–155

Holmes MG, Schäfer E (1981) Action spectra for changes in the "high irradiance reaction" in hypocotyls of *Sinapis alba* L. Planta 153:267–272

Holmes MG, Beggs CJ, Jabben M, Schäfer E (1982) Hypocotyl growth in *Sinapis alba* L.: the role of light quality and light quantity. Plant Cell Environ 5:45–51

Jabben M, Beggs CJ, Schäfer E (1982) Dependence of P_{fr}/P_{tot} ratios on light quality and light quantity. Photochem. Photobiol 35:709–712

Johnson CB (1980) The effect of red light in the high irradiance reaction of phytochrome: evidence for an interaction between P_{fr} and a phytochrome cycling driven process. Plant Cell Environ 3:45–51

Kortüm G (1969) Reflectance spectroscopy: principles, methods, applications. Springer, Berlin Heidelberg New York

Loofbaourow JR (1948) Effects of ultraviolet radiation on cells. Growth Symp 12:75–143

Mackenzie Jr JM, Coleman RA, Briggs WR, Pratt LH (1974) Reversible redistribution of phytochrome within the cell upon conversion to its physiologically active form. Proc Natl Acad Sci USA 72:799–803

Mancinelli AL, Rabino I (1978) The "high irradiance responses" of plant photomorphogenesis. Bot Rev 44:129–150

Mancinelli AL, Walsh L (1979) Photocontrol of anthocyanin synthesis in young seedlings. Plant Physiol 55:251–257

Mohr H (1972) Lectures on photomorphogenesis. Springer, Berlin Heidelberg New York

Pratt LH (1978) Molecular properties of phytochrome (Review article) Photochem Photobiol 27:81–105

Pratt LH, Colman RA, Mackenzie Jr JM (1976) Immunological visualisation of phytochrome. In: Smith H (ed) Light and plant development. Butterworth, London, pp 75–94

Schäfer E (1975) A new approach to explain the "high irradiance responses" of photomorphogenesis on the basis of phytochrome. J Math Biol 2:41–56

Schäfer E (1976) The "high irradiance reaction". In: Smith H (ed) Light and plant development. Butterworth, London, pp 45–59

Schäfer E (1981) Phytochrome and daylight. In: Smith H (ed) Plants and the daylight spectrum. Academic Press, London New York, pp 461–480

Schäfer E, Fukshansky L (1983) Action spectroscopy. In: Smith H (ed) Methods in photomorphogenesis. (in press)

Schäfer E, Beggs CJ, Fukshansky L, Holmes MG, Jabben M (1981a) A comparative study of the responsivity of *Sinapis alba* L. seedlings to pulsed and continuous irradiation. Planta 153:258–261

Schäfer E, Beggs CJ, Fukshansky L, Holmes MG, Jabben M (1981b) A method to check the involvement of additional photoreceptors to phytochrome in photomorphogenesis. Eur Symp Light Mediated Plant Dev 9. 14 Bischofsmais, FRG

Schäfer E, Lassig T-U, Schopfer P (1982) Phytochrome controlled extension growth of *Avena sativa* L. seedlings: Fluence rate relation ship and action spectra of mesocotyl and coleoptile responses. Planta 154:231–240

Setlow RB, Pollard EC (1964) Molecular biophysics. Addison-Wesley, Reading

Shropshire W Jr (1972) Action spectroscopy. In: Mitrakos K, Shropshire W Jr (eds) Phytochrome. Academic Press, London New York, pp 161–181

Smith H (1975) Phytochrome and photomorphogenesis. McGraw-Hill, London New York

Stokes GG (1860/1862) On the intensity of the light reflected from or transmitted through a pile of plates. Proc R Soc Lond 11:545–557

Warburg O (1949) Heavy metal prostetic groups and enzyme action. Oxford Univ Press, Oxford

Withrow RB, Klein WH, Elstad V (1957) Action spectra of photomorphogenic induction and its photoinactivation. Plant Physiol 32:453–462

Zurzycki J (1961) The influence of chloroplast displacements on the optical properties of leaves. Acta Soc Bot Pol 30:505–527

5 Models in Photomorphogenesis

L. FUKSHANSKY and E. SCHÄFER

1 General Uses and Limitations

Before discussing models in photomorphogenesis it is necessary to give a brief outline of the general uses and limitations of mathematical models in investigating the mechanisms of biological phenomena (see also FUKSHANSKY and MOHR 1980).

The use of mathematical models is an attempt to apply quantitative descriptions of phenomena to situations which lack precise information. This method can become very successful within a combined study of a complex phenomenon, however, only if applied correctly.

The basic properties of mathematical models are presented in Fig. 1. First of all, the analysis of a mathematical model must give consequences of underlying statements, which are not obvious, i.e., cannot be obtained on the basis of the same statements by means of qualitative reasoning only.

In order to build a model one must have a set of firmly established facts (set F in Fig. 1). This set has to be divided into two subsets (F_1 and F_2). Subset F_1, with the possible addition of supplementary statements, underlies the model; subset F_2 is to be compared with results of its analysis [i.e., with non-obvious (N.C.) and obvious (O.C.) consequences]. In photomorphogenesis F_1 usually contains data of spectrophotometrical in vivo and in vitro measurements, while F_2 contains results concerning physiological responses (such subdivision is, however, not obligatory). The statements to be added to subset F_1 can be conditionally divided into two classes: plausible assumptions (P.A.) and arbitrary postulates (A.P.) (the first being statements about properties which are likely, but not firmly established, the second group statements which cannot be supported by experimental findings to any degree).

The third property can be formulated half-quantitatively:

$$|\text{P.A.}| + |\text{A.P.}| \ll |\text{N.C.}| + |\text{O.C.}| \,,$$

i.e., the number of statements underlying the model must be considerably less than that of statements derived from the model. This requirement is obvious in an extreme case: nobody is permitted to make an assumption for each "explained fact". Here occurs, in a natural way, the concept of a minimal model, i.e., the model containing a minimal number of P.A. and A.P.

Results of model analysis can contain not only statements about facts from F_2, but also other statements which cannot be confronted with known facts. These statements about supposed new properties, or predictions (P) are most

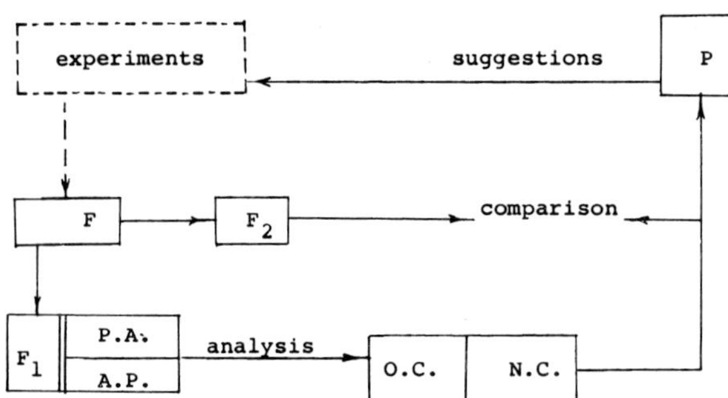

I. There should be N.C. of P.A. and A.P.
II. The set F should be divided into subsets F_1 and F_2 having no common parts.
III. $|\text{P.A.}| + |\text{A.P.}| \ll |\text{N.C.}| + |\text{O.C.}|$
IV. Predictions are desirable
V. Only relative comparison of models is possible.

Fig. 1. Diagram illustrating the application of a mathematical model and its basic features. *F* is the set of firmly established experimental facts which is to be divided into subsets F_1 and F_2 in an arbitrary way; *P.A.* and *A.P.* are plausible assumptions and arbitrary postulates, respectively; *O.C.* and *N.C.* are obvious and non-obvious consequences of the underlying model statements; *P* are predictions

valuable. Even the design of special experiments based on the model analysis must be regarded in itself as an achievement.

The above-mentioned properties of mathematical models enable the relative comparison of different models. Among the models with an equal number of P.A. and A.P. it is natural to consider as more useful that one which permits more N.C. to be drawn. Among the models with equal numbers of drawn conclusions doubtless the one which has at least one underlying assumption less will be much more interesting.

The common properties of models presented in Fig. 1 show rules which must always be fulfilled. Point I describes the peculiarity of the method, points I–III give conditions of its applicability, points IV–V show criteria of its utility. In addition, two problems specific for the use of models should be mentioned. Firstly, one must remember that within the limits of accuracy when presenting experimental data, a good fit of a given set of points can be achieved by different mathematical expressions. An excellent illustration of this statement was given by W. FELLER (1940) in the example of growth curves. Secondly, one must remember that two phenomena described by the same mathematical expressions are not necessarily related. Therefore, a good fit of experimental points by a theoretical curve does not prove that the scheme underlying the model represents the real mechanism.

After this not very optimistic statement one could ask: what do we still achieve with modeling?

When no discrepancies arise between results of model analysis and elements of F_2 or between predictions and results of supplementary experiments, we know that statements added to F_1 do not contradict the possible mechanism of the phenomenon. Otherwise, we come to know that these statements are incompatible and we repeat the whole procedure after changing the statements. After each round of modeling we select and accumulate compatible statements and get some ideas for new experiments. A useful model can disappear at once after finishing the analysis. On the contrary, a useless model can remain for a long time unchanged with statements which cannot be disproved or strengthened.

2 Models for Cryptochrome-Controlled Processes

The main differences between mathematical modeling in phytochrome and cryptochrome research are based on the fact that up to now cryptochrome cannot be measured directly. Even the nature of the molecule is still under debate (DEFABO 1980, SHROPSHIRE 1980). Therefore, modeling in cryptochrome research is restricted to cybernetic models, i.e., the "black box" consideration. This type of model has been developed for the light-growth response of *Phycomyces* (DELBRÜCK and REICHARDT 1956). Based on white noise analysis LIPSON (1975) has developed a complex cybernetic scheme which describes the dynamics of the light-growth response. Unfortunately, the molecular basis of the elements and reactions in this scheme is poor.

Up till now, mathematical modeling, in the sense as described by FUKSHANSKY and MOHR (1980), has appeared impossible for cryptochrome research. Therefore, we shall confine our discussion to phytochrome.

3 Models for Phytochrome-Controlled Processes

Here the quantitative description of the phytochrome pigment system is considered within the scope of general rules and aims as presented in Section 1. We shall proceed from the simplest schemes to more sophisticated models, dealing with increasing amounts of experimental data, assumptions and predictions. We shall not attempt to elucidate particular models. Our purpose is to describe trends and discuss general principles which become clear through model analysis. In addition, we shall try to indicate where typical mistakes can occur and outline future prospects.

3.1 Description of Phytochrome Phototransformations

According to procedures developed in BUTLER et al. (1959) the dynamics of interconvertions between two forms of phytochrome—P_r and P_{fr}—can be studied experimentally, following light induced absorbance changes (see Chap. 8, this Vol. for detailed description). When the fluence rate of the actinic light is not too low the phototransformations are much faster than some dark reactions which also take place in the phytochrome system (see Chap. 8, this Vol.). Hence, for exposures that are not too long the influence of dark reactions on the P_r and P_{fr} concentrations will be negligible.

If the above-mentioned conditions are fulfilled, the qualitative scheme for the model of phototransformations appears as

$$P_r \underset{k_2}{\overset{k_1}{\rightleftarrows}} P_{fr}, \tag{M1}$$

whereas the set of experimental facts (F_1) is presented in the following statement:

Both photoreactions are of first order with respect to their substrates.

Translation of this statement into a formal description gives the following linear system of differential equations:

$$\begin{aligned} \dot{P}_r &= k_2 P_{fr} - k_1 P_r \\ \dot{P}_{fr} &= k_1 P_r - k_2 P_{fr} \end{aligned} \tag{1}$$

with arbitrary initial conditions

$$\left.\begin{aligned} P_r(t=0) &= P_{r0} \\ P_{fr}(t=0) &= P_{fr0} \end{aligned}\right\} \tag{1a}$$

which in the case of a dark-grown object will be

$$\begin{aligned} P_r(t=0) &= P_{tot}(t=0) \\ P_{fr}(t=0) &= 0, \end{aligned} \tag{1b}$$

where $P_{tot}(t) \equiv P_{fr}(t) + P_r(t)$ is the total concentration of P_r and P_{fr}.

Equations (1) are not independent because we are considering a *closed* system, with a conservation law:

$$P_r(t) + P_{fr}(t) \equiv P_{tot}(t) = \text{const.} \equiv P_{tot}. \tag{2}$$

To find the complete dynamics of phototransformations one needs either of the two Eqs. (1) and the condition (2).

For example:

$$\dot{P}_{fr} = k_1 (P_{tot} - P_{fr}) - k_2 P_{fr}, \tag{3}$$

which together with the second initial condition from (1a) has the solution:

$$\left.\begin{aligned} P_r(t) &= (P_{r0} = P_{r\infty})\, e^{-(k_1+k_2)t} + P_{r\infty} \\ P_{fr}(t) &= (P_{fr0} - P_{fr\infty})\, e^{-(k_1+k_2)t} + P_{fr\infty} \end{aligned}\right\}, \tag{4}$$

where

$$\left.\begin{aligned} P_{fr\infty} &\mathrel{\widehat{=}} P_{fr}(t)_{t\to\infty} = \frac{k_1}{k_1+k_2} P_{tot} \\ P_{r\infty} &\mathrel{\widehat{=}} P_r(t)_{t\to\infty} = \frac{k_2}{k_1+k_2} P_{tot} \end{aligned}\right\}, \tag{4a}$$

or, introducing parameters φ, ϑ [cf. Eq. (11) in Chap. 4]

$$\vartheta \hat{=} k_1 + k_2, \qquad \varphi \hat{=} \frac{k_1}{k_1 + k_2} = \frac{P_{fr\infty}}{P_{tot}}, \tag{5}$$

we can write (4), (4a) as

$$\left.\begin{aligned} P_r(t) &= (P_{r0} - P_{r\infty})\,e^{-\vartheta t} + P_{r\infty} \\ P_{fr}(t) &= (P_{fr0} - P_{fr\infty})\,e^{-\vartheta t} + P_{fr\infty} \end{aligned}\right\} \tag{6}$$

$$\left.\begin{aligned} P_{fr\infty} &= \varphi\, P_{tot} \\ P_{r\infty} &= (1-\varphi)\, P_{tot} \end{aligned}\right\}. \tag{6a}$$

The first conclusion of the model (M1) is that a plot

$$\ln \frac{P_{fr}(t) - P_{fr\infty}}{P_{fr0} - P_{fr\infty}} = -\vartheta t$$

should be a straight line.

Other conclusions could be drawn from the correlations between k_1, k_2 and spectral properties of light and pigment [see Eqs. (2), (2a) in Chap. 4]. Under monochromatic irradiation the rate constant of the phototransformation is proportional to N_λ and $\sigma_{r\lambda} + \sigma_{fr\lambda}$. The photostationary concentrations of both P_r and P_{fr} are independent of N_λ. For the polychromatic irradiation a more restricted statement is valid: if the photon fluence rate is changed without alterations in the spectral composition the rate constant of the phototransformation will be changed proportionally, but the photostationary concentrations of P_r and P_{fr} remain unchanged.

These conclusions are in agreement with experimental data obtained under conditions where (M 1) is applicable (BUTLER et al. 1964, PRATT and BRIGGS 1966, SCHMIDT et al. 1973).

The applicability of (M 1) is very limited because all the dark reactions of phytochrome are neglected as well as the formation of the intermediates of phototransformations (see Chap. 8, this Vol.). When using short light flashes or very bright light the kinetics of the intermediates and even two-photon absorption should be accounted for. It should be mentioned that in (M 1), as well as in all the following models, the quantity N_λ is related to the place of action of light. The transition from the measured incident photon fluence rate $N_{o\lambda}$ to N_λ is discussed in Chapters 4 and 8, this Volume.

We would like to stress that even such a simple model facilitates the introduction of a general principle concerned with photochromic systems. This is the principle of equivalent light action in a photochromic pigment together with its two consequences—the plot "response vs. φ with $\vartheta = \text{const}$" and the theory of dichromatic fluence-response curves as presented in Chapter 4.

3.2 The Basic Model of Phytochrome Dynamics

A short time after the in vivo spectroscopy of phytochrome became feasible several dark reactions in the phytochrome system have been elucidated (cf.

FRANKLAND 1972). The extension of (M 1) to the following model scheme

$$\xrightarrow{{}^0k_s} P_r \underset{k_2}{\overset{k_1}{\rightleftarrows}} P_{fr} \xrightarrow{{}^1k_d} \qquad (P_{fr} \xrightarrow{{}^1k_r} P_r) \tag{M2}$$

has been based on the firmly established zero-order P_r accumulation (SCHÄFER et al. 1972), first-order P_{fr} destruction (KENDRICK and FRANKLAND 1968, MARMÉ et al. 1971) and first-order dark reversion (MARMÉ et al. 1971).

The mathematical description

$$\left.\begin{aligned} \dot{P}_r &= -k_1\, P_r + (k_2 + {}^1k_r) P_{fr} + {}^0k_s \\ \dot{P}_{fr} &= k_1\, P_r - (k_2 + {}^1k_r + {}^1k_d) P_{fr} \end{aligned}\right\}, \tag{7}$$

corresponding to (M 2) has a solution (SCHÄFER and MOHR 1974) consisting of two exponents for both P_r and P_{fr}:

$$\left.\begin{aligned} P_r(t) &= c_1\, e^{\lambda_1 t} + c_2\, e^{\lambda_2 t} + (\vartheta(1-\varphi) + {}^1k_r + {}^1k_d) \frac{{}^0k_s}{\vartheta\, \varphi\, {}^1k_d} \\ P_{fr}(t) &= \frac{\vartheta\, \varphi + \lambda_1}{\vartheta\,(1-\varphi) + {}^1k_r}\, c_1 \cdot e^{\lambda_1 t} + \frac{\vartheta\, \varphi + \lambda_2}{\vartheta\,(1-\varphi) + {}^1k_r}\, c_2\, e^{\lambda_2 t} + \frac{{}^0k_s}{{}^1k_d} \end{aligned}\right\} \tag{8}$$

with eigenvalues of the system (7):

$$\lambda_{1,2} = -\frac{\vartheta + {}^1k_r + {}^1k_d}{2} \pm \sqrt{\left(\frac{\vartheta + {}^1k_r + {}^1k_d}{2}\right)^2 - \varphi\, \vartheta\, {}^1k_d} \tag{8a}$$

and c_1, c_2-parameters determined by initial conditions.

Before discussing consequences of (M 2) we shall make one simple but useful remark. According to Eq. (8), the dynamics of a substance acting within a system of first-order reactions can be presented as a sum of exponents, whereas the number of exponents is *equal to the number of independent participants* (pools of substances) in the whole system, but *not to the number of reactions* in which the considered substance takes part. This almost trivial remark is still relevant since the following *wrong* reasoning sometimes appears. The measured dynamics of a substance consists of two parts—"fast" and "slow" exponents—therefore, the substance should take part in two reactions and the corresponding rate constants could be determined from slopes of the logarithmic plot. In the model (M 2) both P_r and P_{fr} participate in four reactions but the corresponding dynamics contain two exponents.

The analysis of (M 2) led to the following conclusions.

1. Under prolonged irradiation the phytochrome system approaches the following steady state values.

$$P_{fr\infty} = \frac{{}^0k_s}{{}^1k_d}; \qquad P_{r\infty} = \frac{{}^0k_s}{{}^1k_d}\left(\frac{1-\varphi}{\varphi} + \frac{{}^1k_r + {}^1k_d}{\varphi\, \vartheta}\right);$$

$$P_{tot\infty} = \frac{{}^0k_s}{{}^1k_d}\left(\frac{1}{\varphi} + \frac{{}^1k_r + {}^1k_d}{\varphi\, \vartheta}\right). \tag{8b}$$

One can see that $P_{fr\infty}$, unlike $P_{r\infty}$ and $P_{tot\,\infty}$, is fluence-rate and wavelength-independent.

2. In the area of very large N_λ ($N_\lambda \rightarrow \infty$) the total amount of phytochrome approaches asymptotic value $\frac{{}^0k_s}{{}^1k_d}\frac{1}{\varphi}$, whereas the rate of approaching this value is inversely proportional to the photon fluence rate:

$$\frac{P_{tot\infty}(N_\lambda) - P_{tot\infty}(N_\lambda \rightarrow \infty)}{P_{tot\infty}(N_\lambda \rightarrow \infty)} = \frac{{}^1k_r + {}^1k_d}{\vartheta} = \frac{1}{N_\lambda}\frac{{}^1k_r + {}^1k_d}{\sigma_r + \sigma_{fr}} \tag{9}$$

This prediction has been verified experimentally (SCHÄFER and MOHR 1974).

All restrictions for applicability of (M 2) concerned with photochemical intermediates and optical artifacts remain the same as for (M 1).

3.2.1 Two General Principles Which Can Be Eluciated Within the Framework of the Basic Model

1. Completely New Phenomena Can Be Brought About Solely by Interplay Between Light and Dark Reactions. In the (M 1) model the rate of approaching steady state depends on N_λ, but the quantities $P_{r\,\infty}$, $P_{fr\,\infty}$ are fluence rate independent. On the contrary, in (M 2) the complete dynamics $P_{fr}(t)$, $P_r(t)$ and the quantities $P_{r\,\infty}$, $P_{tot\,\infty}$ are fluence rate-dependent. When comparing (M 1) and (M 2) one can see that the phytochrome system under a saturating light pulse is described equally in both models. Indeed, during the pulse the slow dark reactions can be neglected and (M 2) is reduced to (M 1), whereas during the subsequent darkness the exponential decrease (destruction + dark reversion) of P_{fr}

$$P_{fr}(t) = P_{fr\,0}\, e^{-({}^1k_r + {}^1k_d)t}$$

and corresponding increase (dark reversion) of P_r have been brought about by dark reactions only. Thus, we conclude: under prolonged irradiation the fluence-rate dependence of the dynamics and steady-state levels of phytochrome forms can be caused by an interplay between light and dark reactions.

2. The Quasistationary Approximation. As we have seen above, the light pulse separates experimentally the light and dark reactions leading to a simpler description of phytochrome dynamics. Attempts have been made to transfer this simplified description to continuous irradiation, referring to the fact that normally light reactions are much faster than dark reactions (HARTMANN 1966, KENDRICK, FRANKLAND 1968, SCHÄFER et al. 1971, FRANKLAND 1972, HARTMANN and UNSER 1972, SCHÄFER et al. 1975, 1976, HARTMANN 1977). This is a situation where a reasonable approximation, known as a quasistationary description, can facilitate the analysis of the model. The quasistationary approximation can be very useful. However, if applied outside the restricted area of its validity it can lead to misinterpretations of the data (cf. FUKSHANSKY and

MOHR 1980). This can be shown in the example of the description of P_{tot}-destruction under continuous irradiation within the framework of (M 2). The dynamics of P_{tot}, on the basis of Eq. (7) are:

$$P_{tot}(t)=\frac{P_{tot}(0)\,{}^1k_d\,\varphi\,\vartheta+{}^ok_s\,\lambda_2}{\lambda_1(\lambda_2-\lambda_1)}\,e^{\lambda_1 t}-\frac{P_{tot}(0)\,{}^1k_d\,\varphi\,\vartheta+{}^ok_s\,\lambda_1}{\lambda_2(\lambda_2-\lambda_1)}\,e^{\lambda_2 t}+P_{tot\infty}, \tag{10}$$

where the eigenvalues λ_1, λ_2 are from (8a), the steady-state concentration $P_{tot\,\infty}$ is from (8b), $P_{tot}(0)$ is the initial concentration at the beginning of the exposure $t=0$ and φ, ϑ are parameters of light action correlated with rate constants of photoconversion by (5). The solution (10) consists of two exponents and is *fluence-rate-dependent*. If the fluence rate is high enough to establish the unequality, $k_1, k_2 \gg {}^1k_d, {}^1k_r, {}^ok_s$, the photochemical part of (M 2) can be considered as a fast subsystem in a quasistationary state. This means that within each point of a certain time interval (between some $t_1>0$ and $t_2>t_1$) this subsystem, being removed from its steady state, returns to the steady state so fast that the contributions of slow reactions to the concentration of any substance during this transient process are negligible. This assumption separates *artificially* light and dark reactions in the phytochrome system: the light reactions taken alone establish correlation between P_r and P_{fr} for each value $P_{tot}(t)$ on the background of slow changes of $P_{tot}(t)$ due to dark reactions taken alone. Mathematically this means that within the time interval (t_1, t_2) the complete description (7) can be approximated by

$$P_{tot}={}^ok_s-{}^1k_d\,P_{fr}(t),$$

where $P_{fr}(t)$ is found as the stationary solution of (1). i.e., $P_{fr}(t)=\varphi\,P_{tot}(t)$, which leads to

$$\dot{P}_{tot}={}^ok_s-{}^1k_d\,\varphi\,P_{tot}(t)$$

with solution

$$P_{tot}(t)=\left[P_{tot}(0)-\frac{{}^ok_s}{{}^1k_d\,\varphi}\right]e^{-{}^1k_d\varphi t}+\frac{{}^ok_s}{{}^1k_d\,\varphi} \tag{11}$$

or

$$\ln\frac{P_{tot}(t)-P_{tot}(\infty)}{P_{tot}(0)-P_{tot}(\infty)}=-{}^1k_d\,\varphi\,t,$$

where

$$P_{tot}(\infty)=\frac{{}^ok_s}{{}^1k_d\,\varphi}. \tag{11a}$$

Solution (11), which shows some formal resemblance to the pure dark destruction of P_{fr}

$$P_{fr}(t) = P_{fr\,0}\, e^{-{}^{1}k_{d}t},$$

is the quasistationary approximation of (10) and *has a restricted area of application*. Two following conditions determine the boundaries of this area: (1) If N_λ or $\sigma_{r\lambda}$, $\sigma_{fr\lambda}$ are insufficiently large, the rates of photoconversion become comparable to the rates of P_{fr} destruction and P_{fr} dark reversion and Eq. (11) cannot be applied. (2) If the quantity $P_{tot}(t)$ is very low the rates of photoconversion become comparable to P_r resynthesis and Eq. (11) cannot be applied. Outside this area of applicability discrepancies occur between *fluence-rate-dependent* experimental dynamics $P_{tot}(t)$ and steady-state value $P_{tot}(\infty)$ on the one hand and *fluence-rate-independent* theoretical expressions for these magnitudes Eq. (11), (11a) on the other hand. If somebody mistakingly holds Eqs. (11), (11a) for a correct reflection of (M 2) he would require a change in the model in order to account for fluence-rate dependence of $P_{tot}(t)$ and $P_{tot}(\infty)$. This, however, may be wrong because the fluence rate dependence of $P_{tot}(t)$ and $P_{tot}(\infty)$ is incorporated into the correct reflection of the model [Eqs. (10) and (8b)] and the discrepancy mentioned occurs only on account of using simplified descriptions (11), (11a) outside the range of their validity.

Considerations similar to those presented in this section exist in different areas and have different modifications. For example, the classical Michaelis-Menten theory of enzyme kinetics is based on the quasistationary approximation which has similarities but also differences to the approximation used above.

3.3 The Modified Basic Model of Phytochrome Dynamics

3.3.1 Difficulties of the Basic Model as Concerned With HIR

It is obvious that photomorphogenic responses under induction conditions are neither controlled by a decrease in P_r nor by the photoconversion reaction itself. They are controlled in some way by the formation of P_{fr}. This can be demonstrated firstly by the fact that for induction fascinatingly small amounts of $P_r \rightarrow P_{fr}$ conversions are necessary (BRIGGS and CHON 1966, DRUMM and MOHR 1974) whereas for the reversion of an inductive light pulse higher photon fluence rates of about three orders of magnitude are necessary (HENDRICKS 1960, GORTON and BRIGGS 1980). Secondly, the inductive red light pulse can be reverted by a subsequent far-red light pulse. Full reversibility can be obtained over a period as short as some seconds (JABBEN and MOHR 1975) and as long as several hours (BORTHWICK et al. 1959, HOPKINS and HILLMAN 1966). Thirdly, the mesocotyl growth of *Avena* after light flashes is only a function of the amount of P_{fr} and not a function of the way in which this P_{fr} has been established (flash duration, color, and number of flashes have been varied; KRAML and SCHÄFER unpublished).

On the other hand, under induction conditions it is, in principle, impossible to distinguish which characteristic(s) of the P_{fr} dynamics is (are) significant for the signal transition (see Fig. 2): photostationary concentration $P_{fr}(t^*)$, the total

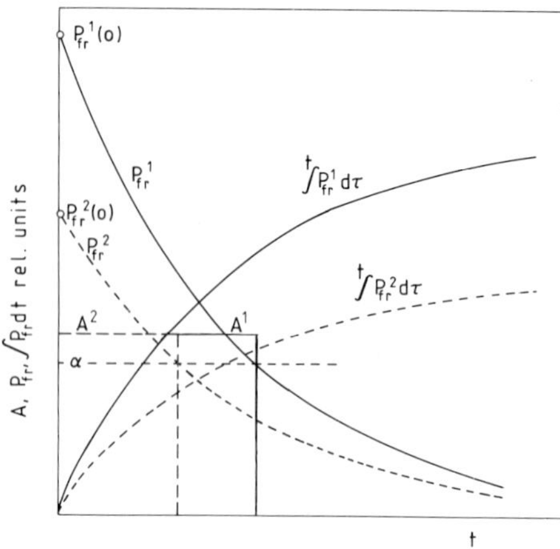

Fig. 2. Dynamics of various functions of P_{fr} in the dark after two light pulses establishing P_{fr}^1 (0) and P_{fr}^2 (0), respectively. The chosen functions of P_{fr} are:

1. $P_{fr}(t)$
2. $\int_0^t P_{fr}(t)\,dt$
3. A for $P_{fr} > \alpha$; 0 for $P_{fr} \leqq \alpha$

The functions for light pulse *1* are shown by the *full lines*, for the light pulse *2* by the *dashed lines*

amount of P_{fr} in darkness $\int_{t^*}^{t} P_{fr}(t)\,dt$, time interval with P_{fr} level above some fixed value or some other.

In each pair of induction experiments any such characteristic will behave as the photostationary concentration: if in the first experiment the photostationary concentration is larger than in the second, then any characteristic from the first experiment exceeds the corresponding characteristic of the second experiment. Therefore, in induction experiments we are not confronted with the full problem of the mechanism of photoreception in photomorphogenesis (see FUKSHANSKY and MOHR 1980) which consists of two parts:

1. Elucidation of significant features of dynamics in the photoreceptor.
2. Establishing correlations between processes in photoreceptor and response.

Here we are only confronted with a so-called "reduced problem": establishing a resemblance between N_λ- and λ-dependence of phytochrome dynamics and those of the response.

Only under prolonged irradiation do the characteristics of P_{fr} dynamics diverge. For example, the total amount of P_{fr}, which is present during a certain time interval, or the length of a time interval, when the P_{fr} concentration exceeds a fixed level, can appear *larger* in that experiment which has a *lower photostationary* concentration of P_{fr}. This divergence of characteristics of P_{fr} dynamics has a direct connection to the problem of HIR. After it became clear that phytochrome is involved in HIR the following problem occurs. Can the P_{fr} dynamic have a characteristic which, with respect to spectral sensitivity and fluence rate dependence, behaves like physiological responses? To wit, does any characteristic of P_{fr} dynamic have a larger magnitude under red light and is it fluence-rate-independent in an induction experiment and, at the same time, does it have a larger magnitude under far-red light and is it fluence-rate-dependent in an HIR experiment? To find such a characteristic is equivalent to proposing a unified description of phytochrome action in photomorphogenesis. Of course,

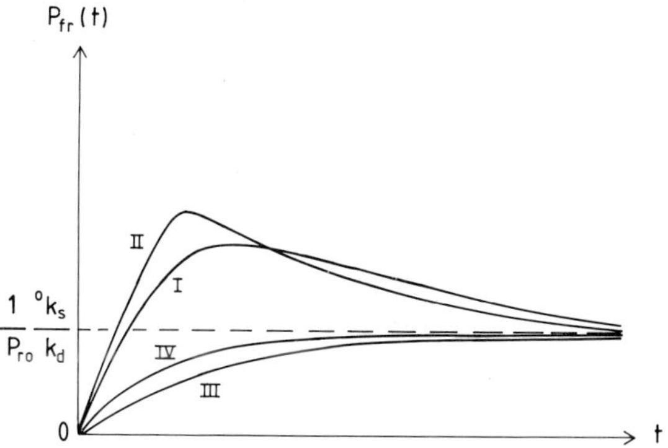

Fig. 3. Necessary condition for intersection of curves presenting P_{fr} dynamics $[P_{fr}(t)]$ within the basic model. Curves which lie below the straight line indicating the stationary level P_{fr} will not intersect (curves *III* and *IV*)

it may be that such a universal characteristic does not exist at all. However, even in this case, to show that current knowledge about the reactions in the photoreceptor exclude such a universal characteristic would be desirable.

What is the necessary condition for a unified description of phytochrome action within the framework of a model? Some $P_{fr}(t)$ curves, which express P_{fr} dynamics for different pairs φ, ϑ, *must* intersect as curves I and II shown in Fig. 3 do. Will the curves $P_{fr}(t)$, calculated from (M2), intersect? Yes, but only those which have parameters satisfying the following condition (we omit the cumbersome proof of this statement):

$$P_{r0}\,\varphi\,\vartheta+\lambda_2\frac{{}^{o}k_s}{{}^{1}k_d}>0, \tag{12}$$

where $P_{r0}=P_{tot}(0)$ is the initial concentration of phytochrome. For relatively large fluence rates $k_1+k_2 \gg k_d$, k_r one has $\lambda_2 \approx -\vartheta$. Hence, the condition (12) means that the crossing of curves appears only under irradiation which provides photoequilibrium φP_{r0} exceeding the stationary value $\frac{{}^{o}k_s}{{}^{1}k_d}$ as shown in Fig. 3. In other words, in a spectral interval, where the experiment shows a strong HIR the basic model forbids crossing!

Another problem of the basic model is the fluence rate dependence of HIR. According to Eq. (8) the stationary concentration, $P_{fr\,\infty}$, is in (M 2) wavelength and fluence-rate-independent. Under prolonged irradiation with a rather high fluence rate the current concentration, $P_{fr}(t)$, approaches the level $P_{fr\,\infty}$ in a way presented by the "slower" exponent from Eq. (8) (cf. quasistationary approximation) and the kinetics of this approach should be almost fluence rate independent because of the continuity of the dependence of $P_{fr}(t)$ on the parameter ϑ.

3.3.2 Construction and Consequences of the Modified Basic Model

The modified basic model was analyzed (GAMMERMAN and FUKSHANSKY 1971, NEUSYPINA et al. 1972, GAMMERMAN and FUKSHANSKY 1974) even before the

(M 2) model, and illustrates that the same mathematical description can reflect different qualitative ideas (c.f. Sect. 1 and also FUKSHANSKY and MOHR 1980).

CLARKSON and HILLMAN (1967) observed an apparent synthesis of phytochrome in *Pisum* which seemed to be a function of P_r current concentration. They suggested that the rate of P_r synthesis is a function of the amount of destroyed P_{fr} or of the difference between the initial and current value of P_r concentration. Using this suggestion GAMMERMAN and FUKSHANSKY (1971, 1974) analyzed the following scheme:

$$\xrightarrow{\dot{P}_r=\alpha(P_{r0}-P_r)} P_r \underset{k_2}{\overset{k_1}{\rightleftarrows}} P_{fr} \xrightarrow{{}^1k_d} \qquad P_{fr} \xrightarrow{{}^1k_r} P_r \tag{M*}$$

where α is the rate constant of the *first-order* P_r resynthesis and the rate of resynthesis is proportional to the decrease of the current value, $-P_{r0}-P_r$. The use of the additional parameter, P_{r0}, makes no difference between this and any other model because this parameter is identical to the initial concentration $P_{r0}=P_{tot}(0)$ before the first irradiation. One can simply put $P_{r0}=1$ which only affects those units in which all concentrations are measured. The scheme (M*) implies equations:

$$\left.\begin{aligned} \dot{P}_r &= \alpha(P_{r0}-P_r)-k_1\,P_r+(k_2+{}^1k_r)P_{fr} \\ \dot{P}_{fr} &= k_1\,P_r-(k_2+{}^1k_d+{}^1k_r)\,P_{fr} \end{aligned}\right\} \tag{*}$$

or

$$\left.\begin{aligned} \dot{P}_r &= {}^0k_s-(k_1+{}^1k_{dr})P_r+(k_2+{}^1k_r)P_{fr} \\ \dot{P}_{fr} &= k_1\,P_r-(k_2+{}^1k_d+{}^1k_r)P_{fr}) \end{aligned}\right\} \tag{13}$$

with

$${}^0k_s \mathrel{\hat{=}} \alpha P_{r0};\ {}^1k_{dr} \mathrel{\hat{=}} \alpha.$$

On the other hand, the system (13) can be interpreted as a description of the scheme

$$\xrightarrow{{}^0k_s} P_r \underset{k_2}{\overset{k_1}{\rightleftarrows}} P_{fr} \xrightarrow{{}^1k_d} \qquad P_r \xrightarrow{{}^1k_{dr}} \qquad P_{fr} \xrightarrow{{}^1k_r} P_r \tag{M 3}$$

which contains the basic model (M 2) and additionally one slow first-order reaction of P_r destruction. We shall refer all results of the analysis of (13) to (M 3) because this is the scheme verified later. Using the technique of density labeling, QUAIL et al. (1973b) were able to show that P_r has been synthesized in vivo in *Cucurbita pepo* cotyledons and that the measured P_r value is a steady state between P_r synthesis and P_r destruction (turnover). The rate of P_r destruction was found to be at least one order of magnitude smaller than that of

P_{fr} destruction in the same system. It should be noted at this point that the slow P_r turnover, as found in *Cucurbita* and later in *Amaranthus* (QUAIL et al. 1973b, SCHÄFER 1978, HEIM et al. 1981), should not be confused with the P_{fr}-induced P_r destruction found in *Avena, Zea, Cucurbita, Brassica* and *Amaranthus* (CHORNEY and GORDON 1966, DOOSKIN and MANCINELLI 1968, STONE and PRATT 1979, JABBEN 1980, ZIPFEL and SCHÄFER unpublished). This latter reaction has not been considered in (M3) and will be discussed later (see Sect. 4.1).

The general solution of (13) with arbitrary initial conditions $P_{fr}(0)$, $P_r(0)$ is:

$$\left.\begin{aligned} P_{fr}(t) &= P_{fr\infty} + G_1 c^{\lambda_1 t} + G_2 e^{\lambda_2 t} \\ P_r(t) &= P_{r\infty} + \frac{\lambda_1 + {}^1k_r + {}^1k_d + (1-\varphi)\vartheta}{\varphi\vartheta} G_1 e^{\lambda_1 t} \\ &\quad + \frac{\lambda_2 + {}^1k_r + {}^1k_d + (1-\varphi)\vartheta}{\varphi\vartheta} G_2 e^{\lambda_2 t} \end{aligned}\right\} \tag{14}$$

where λ_1, λ_2 are eigenvalues of (13)

$$\lambda_{1,2} = -\tfrac{1}{2}(\vartheta + {}^1k_{dr} + {}^1k_r + {}^1k_d) \pm \tfrac{1}{2}\sqrt{(\vartheta + {}^1k_{dr} + {}^1k_r + {}^1k_d)^2 - 4\lambda_1\lambda_2}$$
$$\lambda_1 \cdot \lambda_2 = {}^1k_r\,\varphi\,\vartheta + {}^1k_{dr}[(1-\varphi)\vartheta + {}^1k_r + {}^1k_d];$$

the function G_1 is defined as

$$G_1 = \frac{{}^0k_s\,\varphi\,\vartheta}{\lambda_1} + \varphi\,\vartheta\,P_r(0) - [\lambda_2 + {}^1k_r + {}^1k_d + (1-\varphi)\vartheta]\,P_{fr}(0),$$

G_2 is obtained from G_1 after substitution λ_2 for λ_1 and vice versa; $P_{fr\infty}$, $P_{r\infty}$ are steady-state concentrations under prolonged irradiation

$$\left.\begin{aligned} P_{r\infty} &= \frac{{}^0k_s\left(1-\varphi+\dfrac{{}^1k_r+{}^1k_d}{\vartheta}\right)}{\varphi\left({}^1k_r-{}^1k_{dr}\right)+{}^1k_{dr}+\dfrac{1k_r+{}^1k_d}{\vartheta}} \\ P_{fr\infty} &= \frac{{}^0k_s\,\varphi}{\varphi({}^1k_r-{}^1k_{dr})+{}^1k_{dr}+\dfrac{{}^1k_r+{}^1k_d}{\vartheta}} \end{aligned}\right\} \tag{14a}$$

Under continuous irradiation of a dark-grown object with conventional light sources, the concentration of P_{fr} increases rapidly approximately to the photostationary level, $\varphi P_{tot}(0)$, and then decreases slowly to the level $P_{fr\infty}$ which depends on φ and ϑ and lies in the range of 3%–6% of $P_{tot}(0)$. The concentration of P_r monotonously approaches $P_{r\infty}$, which fluctuates between 1% and 35% of $P_{tot}(0)$ and is strongly dependent on φ and ϑ. *The remarkable non-obvious consequence of the slow P_r destruction, supplemented in the basic model, is the*

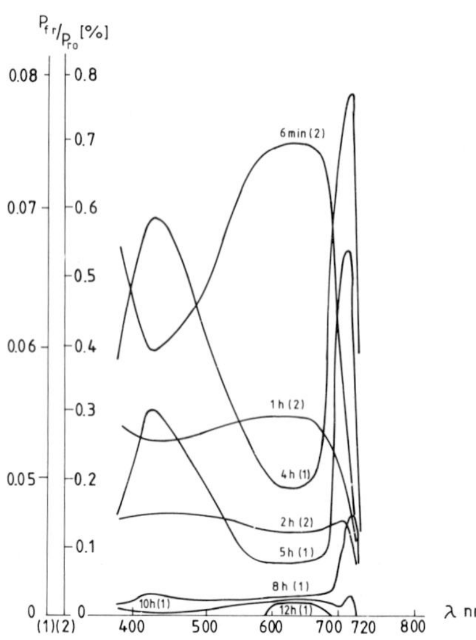

Fig. 4. The relative amount of P_{fr} $\left(\frac{P_{fr}}{P_{r0}}\%\right)$ calculated within the modified basic model as a function of λ for different durations of continuous irradiation. The index *1* or *2* in brackets indicates which of two ordinate scales has to be used for a certain curve

wavelength and fluence rate dependence of P_{fr} dynamics which become more pronounced with the increase of ${}^{1}k_{dr}$. Besides this, the crossing of curves $P_{fr}(t)$ discussed in Section 3.31 becomes possible for wavelengths with a strongly expressed HIR. The relative P_{fr} concentrations at different time points calculated from Eq.(14) as functions of the wavelength (Fig. 4) show some resemblance to action spectra for corresponding times of irradiation. The primary new result of (M 3) is the exposed possibility of a unified description of inductive and HIR responses in the framework of P_{fr} dynamics.

Of course, one should also check whether (M 3) possesses the experimentally confirmed properties of simpler models. It can be shown that the inverse proportionality between N_λ and $P_{tot\,\infty}$, as shown by (9), holds in (M 3) too.

3.3.3 The Principle of Phytochrome Savings

The red → far-red shift of "action spectra" with an increasing time or irradiation and the fluence-rate dependence of a response, as predicted on the basis of mathematical analysis, could be understood qualitatively on the basis of a general phenomenon which we shall call the principle of phytochrome savings. This is also a result of the interplay between light and dark reactions. Assuming that a higher P_{fr} concentration maintained in the system during a certain time interval brings about a higher response, one can see that under continuous irradiation a contradiction must be solved. With increasing P_{fr} production (higher φ) the P_{fr} destruction increases which implies a decreased value of P_{tot} and, therefore, of P_{fr}. On the other hand, decreasing P_{fr} production (lower φ) prevents some part of phytochrome from contributing to the photoreceptor action. This typical compromise situation implies an optimal value of φ (with respect to response) which, obviously, may be different for different times of irradiation. Increasing N_λ for fixed φ will increase the response.

The principle of saving has been suggested several times to explain HIR. With this principle in mind the protonation of P_{fr} (HENDRICKS et al. 1959) and a limited amount of a receptor for P_{fr} (HARTMANN 1966) have been suggested. In contrast to these suggestions the saving in the modified basic model is a consequence of interplay between light and dark reactions but not a new specific assumption (cf. Sect. 1).

3.4 The Cyclic Models of Phytochrome Dynamics

3.4.1 Analysis of New Spectrophotometric Data

Since the development of the modified basic model our knowledge of the phytochrome system has been increased by the use of new techniques—immunocytology, pelletability—and by refined kinetic analysis of the interplay of the dark reactions and of dark and light reactions. Since first observing the interplay of P_{fr} destruction and $P_{fr} \rightarrow P_r$ dark reversion some inconsistancies have become apparent. The reappearance of P_r (measured as $P_{tot} - P_{fr}$) seemed to be, on the one hand, rapid, but on the other hand, had stopped when P_{fr} was still spectrophotometrically measurable (cf. FRANKLAND 1972). The problem was solved by assuming two phytochrome fractions. Anaylzing the temperature dependence of dark reactions SCHÄFER and SCHMIDT (1974) demonstrated that there is an interplay between destruction and dark reversion. Based on M2 and M3 one has for the dark kinetics after a light pulse:

$$\begin{aligned} \dot{P}_{fr} &= -({}^1k_r + {}^1k_d)P_{fr} \\ P_{fr} &= P_{fr}(0)\, e^{-({}^1k_r + {}^1k_d)t} \end{aligned} \tag{15}$$

and, if the time interval is short enough that P_r synthesis can be neglected,

$$\begin{aligned} \dot{P}_r &= {}^1k_d\, P_{fr} \\ P_r(t) - P_r(\infty) &= [P_r(0) - P_r(\infty)]\, e^{-({}^1k_r + {}^1k_d)t}, \end{aligned} \tag{16}$$

whereby

$$P_r(\infty) - P_r(0) = \frac{{}^1k_r}{{}^1k_r + {}^1k_d}\, P_{fr}(0).$$

In words: the rate constant of the approach of the final level of P_r is ${}^1k_r + {}^1k_d$, the same as that of the disappearance of P_{fr}(!!). The amount of reverted P_{fr} to P_r is proportional to $P_{fr}(0)$ and to ${}^1k_r/({}^1k_r + {}^1k_d)$. This prediction was *always* in contradiction to the fact that the rate of P_r reaccumulation $[P_r(\infty) - P_r(t)]$ is much faster than the rate of P_{fr} disappearance. To overcome this problem SCHÄFER and SCHMIDT (1974) predicted, on the basis of temperature jump experiments, a further dark reaction intervening between P_{fr} formation and P_{fr} destruction.

The analysis of the intracellular localization of phytochrome by immunochemical techniques introduced by PRATT and co-workers indicated that other dark reactions of phytochrome existed which were not included in the modified basic model. After a red light treatment, a concentration of the stain was observed, indicating a sequestering of phytochrome in its P_{fr} form in special areas of the cell (COLEMAN and PRATT 1974). The sequestering was only found if the plants had been irradiated with red light. This effect was only reversible if red was followed immediately by far-red light (MACKENZIE et al. 1975). A sequence of 5 min red + 5 min far-red light leads to sequestered P_r which slowly—within 2 h—reverts to a homogenous distribution (MACKENZIE et al. 1975).

This reaction cycle—visualized by sequestered and homogenous states of P_r and P_{fr}—is dynamically very similar to the reaction cycle of light-induced phytochrome pelletability (QUAIL et al. 1973a, QUAIL and SCHÄFER 1974). Although this field of phytochrome has been frequently debated since the pioneering work of QUAIL et al. (1973a) we want to point out some facts which appear to be relevant for mathematical modeling in phytochrome research. Irrespective of whether the measuring device of the Mg^{2+}-dependent pelletability is an in vitro artifact the data clearly show: (1) There is a fast dark reaction starting from P_{fr} (QUAIL and SCHÄFER 1974, LEHMANN and SCHÄFER 1975, 1978, PRATT and MARMÉ 1976, QUAIL 1975) which leads to a changed system. (2) After this reaction a reversion of P_{fr} to P_r does not seem to change the state of the system (QUAIL et al. 1973a). (3) This changed system reverts to the dark state if the phytochrome is in the P_r form (QUAIL et al. 1973a). The half-life times of these two reactions at 25 °C are: dark to light state (P in P_{fr} form) about 2–5 s (LEHMANN and SCHÄFER 1978, QUAIL 1978), and light to dark state (P in P_r form) about 20–50 min (QUAIL et al. 1973a, BOISARD et al. 1974, PRATT and MARMÉ 1976).

On the other hand, there still seems to be some discussion whether the transition from a dark to light state of the system is a one- or two-step process (QUAIL and SCHÄFER 1974, LEHMANN and SCHÄFER 1978, QUAIL 1978).

3.4.2 Construction and Analysis of a Cyclic Model

Based on the discussions at the end of the previous section a simple cyclic model can be developed which accounts for most of the phenomena of P dynamics known up to now (SCHÄFER 1975, 1976).

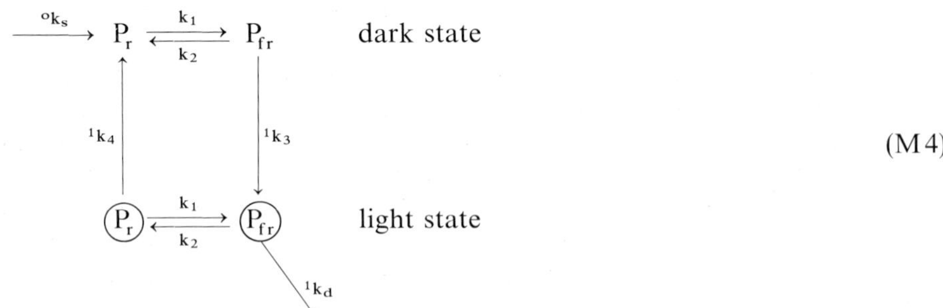

(M4)

The new rate constants, ${}^1k_3$, ${}^1k_4$, can either be obtained on the basis of pelletability experiments and/or by analyzing phytochrome sequestering (LEHMANN and SCHÄFER 1978, QUAIL 1978, MACKENZIE et al. 1975). In 1975 this model was analyzed with respect to the wavelength and photon fluence-rate dependence of the various pools in steady state. The rate constants of photoconversion were assumed to be the same in the dark and light state (QUAIL 1975).

A mathematical description of M 4 with four pools and six different rate constants is a set of four coupled linear differential equations:

$$\begin{aligned}
\dot{P}_r &= -k_1 P_r + k_2 P_{fr} + {}^1k_4 \textcircled{P_r} + {}^0k_s \\
\dot{P}_{fr} &= k_1 P_r - (k_2 + {}^1k_3) P_{fr} \\
\dot{\textcircled{P_r}} &= \quad = (k_1 + {}^1k_4) \textcircled{P_r} + k_2 \textcircled{P_{fr}} \\
\dot{\textcircled{P_{fr}}} &= {}^1k_3 P_{fr} + k_1 \textcircled{P_r} - (k_2 + {}^1k_d) \textcircled{P_{fr}}
\end{aligned} \tag{17}$$

Under steady-state conditions this set of differential equations will be reduced to a set of four linear equations with four variables which can easily be solved.

The primary predictions of this analysis are the following:

1. $\textcircled{P_{fr}} = {}^0k_s/{}^1k_d$ a photon fluence rate and wavelength independent value [see (M 2)]

2. $\dfrac{P^{ss}_{tot}(N_\lambda) - P^{ss}_{tot}(N_\lambda \to \infty)}{P^{ss}_{tot}(N_\lambda \to \infty)} \sim \dfrac{1}{N_\lambda}$ [see (9)]

3. The pool P_{fr}—and therefore the flux through the cycle, $k_3 \cdot [P_{fr}]$,—is photon fluence-rate-dependent, having its maximum in the far-red spectral range. Both pool size and photon fluence-rate dependence are very small in the red area of the spectrum. This model has some properties similar to the modified basic model, (M 3); for example, the saving of P, as a system characteristic based on the interplay between light and dark reactions. In contrast to the other models, one must assume in M 4 two different sites of action for induction and HIR. Since under short irradiations the P_{fr} concentration and the flux through the cycle is maximal under red light one has to assume that $\textcircled{P_{fr}}$ is the effect site for induction responses. This is primarily based on the argument that P_{fr} is only a transient pool and, being a site of action, this pool implies the loss of reversibility would be faster than experimentally verified (cf. SMITH 1975).

A computer analysis of Eq. (17) has shown that the time course of different pools of phytochrome depends, in a very complex way, on the photon-fluence rate and wavelength (GRUBER and SCHÄFER unpublished). After an 18 h irradiation period, (the time chosen by HARTMANN 1966) the P_{fr} pool and the flux through the system show all the characteristics of a candidate for an HIR controlling element. At medium and high photon fluence rates the maximal amount of P_{fr} is achieved by far-red light, while at very low photon fluence rates, by red light (Fig. 5).

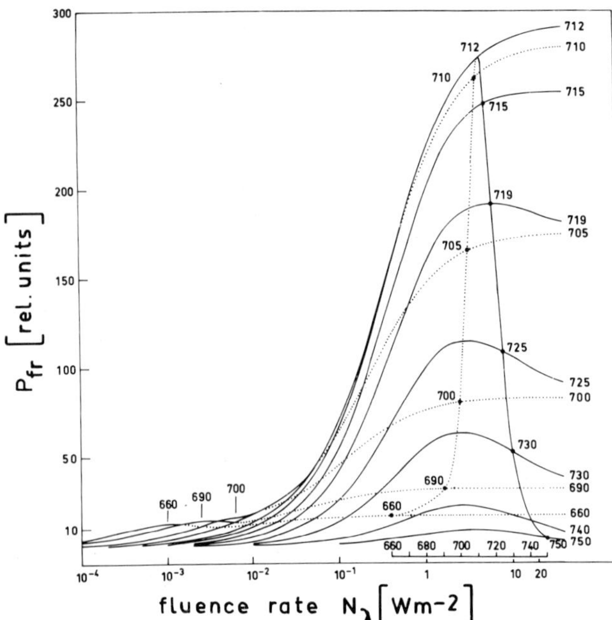

Fig. 5. Semilogarithmic plot of the pool size of P_{fr} (or the cycling through the system) after an 18-h irradiation period vs. fluence rate

4 General Principles and Future Aims in Model-Related Phytochrome Research

Some general principles learned from modeling have already been discussed, such as equivalent light action (Chap. 4, this Vol.; Sect. 3.1), quasi-stationary description (Sect. 3.2.1), phytochrome savings (Sect. 3.3.3), and divergence of characteristics of phytochrome dynamics (Sect. 3.3.1). Due to the divergence under HIR conditions we are confronted with the full problem of the photoreceptor action.

In photomorphogenesis one deals, generally speaking, with an analysis of the input–output relationship, whereby light acts on the input and the output is the measurable response. Various research groups are looking for very fast responses in order to come as close as possible to the input photoreceptor. Other groups are investigating the photoreceptor dynamics and their interaction with processes taking place in different cell organelles. The purpose of mathematical modeling in photomorphogenesis is to establish a model of the photoreceptor dynamic *and* a model for the "translation" of one or more characteristics of this dynamic to the response dynamic. The two-step problem, as formulated in this way, is called "the full problem" of photoreceptor action (see FUKSHANSKY and MOHR 1980 and also Sect. 3.3.1) as distinct from the "reduced problem", where the purpose is to describe photoreceptor dynamics and discuss whether these dynamics have features resembling measured responses. The reduced problem does not require an explicit description of the response as a function of phytochrome dynamics.

In the case of induction conditions the full problem seems to be beyond the means of modeling since all the characteristics of phytochrome dynamics coincide (see Sect. 3.3.1). The divergence of the characteristics under prolonged irradiation allows one to rule out some ways of action (and support others) on the basis of a non-suitable dependence of characteristics on N_λ, λ, duration of irradiation, etc. We should not forget, however, that any such reasoning is confined to a definite model scheme, i.e., we follow two independent arbitrary steps, which cannot be tested directly. We postulate the model scheme and assume which characteristic(s) is (are) significant. For example, HARTMANN (1966), in one of the earlier models containing the quasi-stationary description of (M2) as a submodel, realized that he also had to postulate a significant characteristic, (the current concentration of the product of a reaction driven by P_{fr}). Furthermore, he also realized that, in order to describe the response he had to postulate the correlation (proportionality) between this characteristic and the response.

In (M3) the question has been raised whether a unified characteristic exists which possesses an appropriate dependence on λ and N_λ for inductive and HIR conditions simultaneously. It has been found that such a characteristic may exist and can be an arbitrary function of $P_{fr}(t)$ [or the $P_{fr}(t)$ itself]. In (M4) the same question has been raised, however, with many new spectrophotometrically obtained data concerned with phytochrome dynamics. The new property of this model—two pools of both P_r and P_{fr}—is a consequence of spectrophotometrical data and their analysis and not an arbitrary assumption in order to get a more adaptable model with two possible sites of action. Neither (M3) nor (M4) contain an explicit description of the response, i.e., they are not concerned with the full problem.

The urge to apply $P_{fr}(t)$ as a significant characteristic is understandable, because of the great resemblance between the action spectra of the inductive response and those of the photoconversion, $P_r \rightarrow P_{fr}$, as well as between the action spectra of the reversion of the inductive response and those of the photoconversion, $P_{fr} \rightarrow P_r$. The complexity of the HIR action spectra stimulated other proposals. The suggestion that some photochemically excited state of P_{fr} can act as a specific HIR effector has been contradicted by the fundamental experimental results of Mancinelli and coworkers (cf. MANCINELLI and RABINO 1978), which demonstrated the equivalence of continuous and intermittent irradiation in a wide range of conditions. A suggestion that the process of photoconversion itself (i.e., cycling of phytochrome molecules) may initiate a photomorphogenetic response was made by SMITH (1970). This course of action was visualized in a scheme where phytochrome molecules act as carriers for an important metabolite. This helpful working hypothesis remains on the qualitative level.

Recently JOHNSON and TASKER (1979, see also JOHNSON 1980) attempted a quantitative analysis concerned with the site of action. They proposed that both P_{fr} and cycling are directly involved in the photoreceptor action and developed a quantitative model which should provide testable consequences of this assumption. According to this model an explicit description of the response was written as a function of two effectors – phytochrome concentration and rate of cycling – in two variants:

either (additive interaction of effectors)

$$R = \int_0^T [\bar{k}_1 P_{fr}(t) - \bar{k}_2 C_\lambda P_{tot}(t)]\, dt \tag{18}$$

or (multiplicative interaction of effectors)

$$R = \int_0^T \bar{k}\, P_{fr}(t)\, P_{tot}(t)\, C_\lambda\, dt, \tag{18a}$$

where

$$P_{tot}(t)\, C_\lambda \equiv \bar{k}\, P_{tot}(t)\, [(1-\varphi)\,\sigma_r + \varphi\,\sigma_{fr}] \tag{19}$$

is called "the rate of cycling".

Unfortunately, the whole discussion about the involvement of cycling within the framework of the Johnson-Tasker model was based on a misunderstanding. In fact, there are two different and contradictory models – a mathematical model (18), (19) examined in a computer experiment and a verbal model used in discussions.

1. The model (18), (19) does not contain any description of cycling but a construction (19) which is difficult to interpret. The real rate of cycling is

$$[P_{fr}(t)\sigma_{fr} + P_r(t)\sigma_r] \cdot N_\lambda. \tag{20}$$

The model (18), (19) has N_λ-dependent P_{fr} dynamics (for reason, see below) and shows in a computer experiment a N_λ-dependent response R with λ-dependence resembling the HIR action spectra.

2. The verbal model is presumably based on the interaction between P_{fr} and some product of cycling. The result of the computer experiment is supposed to be a property of the verbal model. The physiological data are discussed in a simple way: the dependence of a response on N_λ indicates a significant action of cycling while, on the contrary, the N_λ-independence indicates the pure action of P_{fr}. Such a substitution contradicts the scheme of phytochrome reactions used by the authors (for reason, see below).

The verbal model cannot claim support from the computer experiment until both models coincide. But can these models be made to coincide? Inserting in Eqs. (18) or (18a) the right cycling (20) instead of (19) results in contradictions between the model and simple pulse experiments and, in addition, causes drastic changes in the results of computer simulation of the HIR experiments. On the other hand, adjustment of the verbal model to the mathematical model will imply a new (and very difficult) edition of the type of photoreceptor action under consideration.

We owe the readers an explanation of two statements made above in points (1) and (2). Both these statements have the same explanation. The core of Johnson and Tasker's model (mathematical and verbal, as well) is the underlying scheme of reactions in the phytochrome system which, after translation into differential equations, gives the dynamics $P_{fr}(t)$, $P_r(t)$, $P_{tot}(t)$. The scheme used by Johnson and Tasker is nothing more than (M3) [compare (M3) to the scheme on p. 261 in JOHNSON and TASKER 1979]. Hence, though the authors never present their equations for phytochrome dynamics, this can only be Eq. (13), with solutions (14). This means that in Johnson and Tasker's model the phytochrome concentration is N_λ-dependent (!) due to savings and the P_{fr} dynamics show some resemblance to the HIR action spectra as has been discussed in Section 3.3.2. The multiplication of the exponential solution with $P_{tot}(t) \cdot C_\lambda$ and the following integration should be expected to cause only minor changes in the basic properties of the model's output, and thus provide an approximative numerical simulation of the HIR action spectra.

The direct action of cycling has been neither ruled out nor supported and awaits analysis.

None of the transduction models has been analyzed in detail up to now.

4.1 The Dynamics of Loss of Reversibility as a Tool in Approaching the Full Problem of Photoreceptor Action

Our present understanding of induction responses is based on the photochromic nature of phytochrome. Photoreversibility has also been used in studying HIR (HARTMANN 1966). According to action spectroscopy of a photochromic pigment (see Chap. 4, this Vol.) two parameters – φ and ϑ – completely determine the response initiated by phytochrome. This implies the role of dichromatic experiments.

An even more important test to understand the dynamics in the photoreceptor is the analysis of the kinetics of the loss of reversibility (see FUKSHANSKY and MOHR 1980). One can ask how long after the end of a light pulse P_{fr} can remain in the system before reversion to obtain 100% or 50% reversibility. The rate of the loss of reversibility is nothing more than the rate of the signal transition from the photoreceptor. The full problem of photoreceptor action becomes well formulated if it contains the requirement that the model should satisfy two sets of conditions simultaneously – one on the input of the photoreceptor (phytochrome dynamics) and another on the output of the photoreceptor (dynamics of signal transition). The benefits of this approach become obvious if the response itself lies within the photoreceptor and is an event in phytochrome dynamics. Two examples of such benefits are given below.

For phytochrome pelletability a half-life time of 40 s at 0 °C (PRATT and MARMÉ 1976) and 2–5 s at 25 °C (LEHMANN and SCHÄFER 1978, QUAIL 1978) was measured. The loss of reversibility [red pulse + Δt dark + far-red pulse] yields the same value, indicating that no further step (besides that leading with a half-life time of 2–5 s to pelletable phytochrome) needs to be included between P_{fr} formation and the P_{fr} light state (M4).

In contrast to this the analysis of P_r destruction in *Amaranthus* yields the following (ZIPFEL and SCHÄFER in SCHÄFER 1981): The P_r destruction is a slow process having a half-life similar to the P_{fr} destruction. Because P_r destruction requires a previous formation of P_{fr} it was possible to test how long P_{fr} has to exist to induce P_r destruction. The measurement of the loss of reversibility showed a half-life for this reaction almost two orders of magnitude smaller than the P_r destruction. Based on these and some other experiments we have to propose a modified cyclic model:

$$
\begin{array}{ccccc}
\xrightarrow{{}^0k_s} & P_r & \underset{k_2}{\overset{k_1}{\rightleftarrows}} & P_{fr} & \\
& \uparrow {}^1k_4 & & \downarrow {}^1k_3 & \\
& \textcircled{P_r} & \underset{k_2}{\overset{k_1}{\rightleftarrows}} & \textcircled{P_{fr}} & \\
\swarrow {}^1k_d^r & & & & \searrow {}^1k_d^{fr}
\end{array}
\qquad {}^1k_d^r \approx {}^1k_d^{fr} \qquad \text{(M5)}
$$

It is worth mentioning that with this modification one should expect the N_λ dependence of both P_{fr} and $\textcircled{P_{fr}}$. Though the (M5) has not been investigated yet this conclusion follows from noting that (M5) is roughly approached by (M3) as the dashed line shows.

4.2 Substitution of the HIR by Light Pulses

Another helpful tool for the further understanding of HIR is the use of intermittent light. This method has been used by MANCINELLI and RABINO (1975) and was the basis of their model of phytochrome dynamics. This is also a qualitative model and unfortunately the number of variables taken was larger than the number of equations necessary to describe the scheme. Nevertheless, the use of light-dark cycles will give us a tool to measure a further parameter of phytochrome dynamic by varying the dark interval between the light pulses (MANCINELLI and RABINO 1975). Furthermore, this method also seems to allow one to measure the dynamic of coupling between the phytochrome system and the transduction chain under HIR conditions. Continuous light can be substituted by light pulses if the dark interval is short enough and it seems to be possible to measure the loss of reversibility for these light pulses. Such types of experiment may give us a direct answer whether induction and HIR have the same or different coupling to the transduction chain for a given photomorphogenic response.

5 General Problems in Further Research

Irrespective of our progress in the understanding of phytochrome dynamics and especially of HIR, one should not forget that several general problems exist and only little effort has been made to overcome them.

5.1 The Role of Phytochrome Intermediates

It is well known that several intermediates exist in the $P_r \rightarrow P_{fr}$ and $P_{fr} \rightarrow P_r$ pathway. KENDRICK and SPRUIT (1973a, b) established a useful reaction scheme for photoconversion in both directions and it was shown that these intermediates can accumulate under continuous white light (KENDRICK and SPRUIT 1972) and in dehydrated tissues (KENDRICK 1974). That intermediates play an important role in seed germination and under bright white light and for short light pulses simply by saving phytochrome from destruction is obvious. Whether intermediates also play a direct role in the pathway of signal transduction is very doubtful because of the measurements of dynamics of reversibility. If the intermediates are photoconvertible as suggested (KENDRICK and SPRUIT 1973a, b) their contribution can be analyzed by comparing the plotts' responses vs. φ for various levels of ϑ.

5.2 Bulk and Active, Old and New Phytochrome

The problem of both bulk and active phytochrome, as well as that of old and new P_{fr}, has been discussed previously (HILLMAN 1967, HILLMAN 1972, SMITH 1975). Both concepts raise fundamental questions for mathematical modeling in photomorphogenesis.

Both concepts, generally speaking, are based on the problem that some physiological effects, especially the so-called phytochrome paradoxa, cannot obviously be understood on the basis of direct phytochrome measurements or the "known" phytochrome dynamic. The best-known paradoxa are the *Zea* paradox and the *Pisum* paradox. To explain these effects either a small number of phytochrome receptors with high specificity for P_{fr} and/or two populations of phytochrome with different photoconversion rates and/or different destruction rates have been suggested. Because the *Zea* paradox occurs at immeasurably low P_{fr} levels, mathematical modeling seems to be almost meaningless. All predictions about the active fraction of phytochrome have to be based on the assumption that the dynamic of this immeasurable pool is the same as or similar to that of the measurable "bulk" phytochrome.

In the case of the *Pisum* paradox a larger amount of phytochrome seems to be involved, making phytochrome measurements possible. Recently it has been observed that the P_{fr} destruction in several dicotyledonous tissues cannot be described by a simple first-order kinetic (HEIM et al. 1981, BROCKMANN and SCHÄFER 1982). It was shown that destruction kinetics an *Amaranthus* and *Sinapis* can be described by assuming a large fraction (97–99%) having a fast destruction and a small fraction (1–3%) having a very slow destruction (BROCKMANN and SCHÄFER 1982). Data indicate that the pool size of the small fraction remains almost constant during development irrespective of the pool size of total phytochrome. It may be possible that these two pools show a different physiological responsiveness. This would imply that future mathematical modeling has to start from *two different* phytochrome pools with *different* kinetic parameters and probably a *different* responsiveness.

5.3 Sensitization and Adaptation

It should be mentioned that the transduction chain is also probably a dynamic system. It is, therefore, not surprising that a plant which has been exposed to light before will respond differently to another light stimulus than a dark-grown seedling. Evidence has been recently accumulated for sensitization of induction responses by previous inductive pulses (TANADA 1972, RAVEN and SHROPSHIRE 1975, STEINITZ et al. 1976, MOHR et al. 1979). Recently the sensitization of induction responses by HIR has been reported (BEGGS et al. 1981). Although adaptation has mainly been discussed in blue-light-controlled responses, this type of feedback may also occur in phytochrome controlled responses. In all cases of adaptation and sensitization it should be checked whether this is due to changes in the transduction chain and/or the phytochrome system. Both models (M4) and (M5) predict that phytochrome distribution will be different in dark-grown and light-grown or in preirradiated seedlings. If the concept of two phytochrome populations is used, the effects of preirradiation on the phytochrome system may be even more complex.

In conclusion, it is obvious that our knowledge of the phytochrome system has increased greatly and thus makes the use of mathematical modeling more *necessary* and *successful*. New ways to improve our knowledge of phytochrome dynamics and coupling to the transduction chain are available. A more comprehensive use of mathematical models will help us to test our conclusions and to devise experiments with predictable results for future research.

References

Beggs CJ, Geile W, Holmes MG, Jabben M, Jose AM, Schäfer E (1981) High irradiance response promotion of a subsequent light induction response in *Sinapis alba* L. Planta 151:135–140

Boisard J, Marmé D, Briggs WR (1974) In vivo properties of membrane-bound phytochrome. Plant Physiol 54:272–276

Borthwick HA, Hendricks SB, Toole EH, Toole VK (1959) Action of light on lettuce-seed germination. Bot Gaz 115:205–225

Briggs WR, Chon HP (1966) The physiological versus the spectrophotometric status of phytochrome in corn coleoptiles. Plant Physiol 41:1159–1166

Brockmann J, Schäfer E (1982) Analysis of P_{fr} destruction in *Amaranthus caudatus* L. Evidence for two pools of phytochrome. Photochem Photobiol 35:555–558

Butler WL, Norris KH, Siegelman HW, Hendricks SB (1959) Detection, assay, and preliminary purification of the pigment controlling photoresponsive development of plants. Proc Natl Acad Sci USA 45:1703–1708

Butler WL, Hendricks SB, Siegelman HW (1964) Action spectra of phytochrome in vitro. Photochem Photobiol 3:521–528

Chorney W, Gordon SA (1966) Action spectrum and characteristics of the light activated disappearance of phytochrome in oat seedlings. Plant Physiol 41:891–896

Clarkson DT, Hillman WS (1967) Apparent phytochrome synthesis in *Pisum* tissue. Nature 213:468–470

Coleman RA, Pratt LH (1974) Subcellular localization of the red-absorbing form of phytochrome by immunocytochemistry. Planta 121:119–131

De Fabo E (1980) On the nature of the blue light receptor: Still an open question. In: Senger H (ed) The blue light syndrome. Springer, Berlin Heidelberg New York, pp 187–197

Delbrück M, Reichardt W (1956) System analysis for the light growth reactions of Phycomyces. In: Rudnick H (ed) Cellular Mechanisms in Differentiation and Growth. Princeton Univ Press, Princeton, pp 3–44

Dooskin RH, Mancinelli AL (1968) Phytochrome decay and coleoptile elongation in *Avena* following various light treatments. Bull Torrey Bot Club 95:474–487

Drumm H, Mohr H (1974) The dose-response curve in phytochrome-mediated anthocyanin synthesis in the mustard seedlings. Photochem Photobiol 20:151–157

Feller W (1940) On the logistic law of growth and its empirical verifications in biology. Acta Biotheor 5:51–65

Frankland E (1972) Biosynthesis and dark transformations of phytochrome. In: Mitrakos K, Shropshire W, Jr (eds) Phytochrome. Academic Press, London New York, pp 195–225

Fukshansky L, Mohr H (1980) Boundary conditions for mathematical models in photomorphogenesis. In: De Greef J (ed) Photoreceptors and Plant Development. Antwerpen Univ Press, Antwerpen, pp 135–144

Gammerman AY, Fukshansky L (1971) Theory and calculation of dynamics of phytochrome transformations in the green leaf. Fiziol Rast 18:661–667 (See Consultants Bur Eng Transl Plant Physiol (1972) 557–562)

Gammerman AY, Fukshansky L (1974) A mathematical model of phytochrome – the receptor of photomorphogenetic processes in plants. Ontogenez 5:122–129

Gorton H, Briggs WR (1980) Phytochrome responses to end-of-day irradiations in light-grown corn grown in the presence and absence of Norflurazon. Plant Physiol 66:1024–1026

Hartmann KM (1966) A general hypothesis to interpret 'high energy phenomena' of photomorphogenesis on the basis of phytochrome. Photochem Photobiol 5:349–366

Hartmann KM (1977) Aktionsspektrometrie. In: Hoppe W, Lohmann W, Markl H, Ziegler H (eds) Biophysik: Ein Lehrbuch. Springer, Berlin Heidelberg New York, pp 197–222

Hartmann KM, Unser IC (1972) Analytical action spectroscopy with living systems: photochemical aspects and attenuance. Ber Dtsch Bot Ges 85:481–551

Heim B, Jabben M, Schäfer E (1981) Phytochrome destruction in dark- and light-grown *Amaranthus caudatus* seedlings. Photochem Photobiol 34:89–93

Hendricks SB (1960) Rates of change of phytochrome as an essential factor determining photoperiodism in plants. Cold Spring Harbor Symp Quant Biol 25:245–258

Hendricks SB, Toole EH, Toole VK, Borthwick HA (1959) Photocontrol of plant development by the simultaneous excitations of two interconvertible pigments. III. Control of seed germination and axis elongation. Bot Gaz 121:1–8

Hillman WS (1967) The physiology of phytochrome. Annu Rev Plant Physiol 18:301–324

Hillman WS (1972) On the physiological significance of in vivo phytochrome assays. In: Mitrakos K, Shropshire W, Jr (eds) Phytochrome. Academic Press, London New York, pp 573–584

Hopkins WG, Hillman WS (1966) Relationship between phytochrome state and photosensitive growth of *Avena* coleoptile segments. Plant Physiol 41:593–598

Jabben M (1980) The phytochrome system in light-grown *Zea mays* L. Planta 149:91–96

Jabben M, Mohr H (1975) Stimulation of the shibata shift by phytochrome in the cotyledons of the mustard seedling *Sinapis alba* L. Photochem Photobiol 22:55–58

Johnson CB (1980) The effect of red light on the high irradiance reaction of phytochrome. Plant Cell Environ 3:45–51

Johnson CB, Tasker R (1979) A scheme to account quantitatively for the action of phytochrome in etiolated and light-grown plants. Plant Cell Environ 2:259–265

Kendrick RE (1974) Phytochrome intermediates in freeze-dried tissue. Nature 250:159–161

Kendrick RE, Frankland B (1968) Kinetics of phytochrome decay in *Amaranthus* seedlings. Planta 82:317–320

Kendrick RE, Spruit CJP (1972) Light maintains high levels of phytochrome intermediates. Nat New Biol 237:281–282
Kendrick RE, Spruit CJP (1973a) Phytochrome intermediates in vivo. I. Effects of temperature, light intensity, wavelength and oxygen on intermediate accumulation. Photochem Photobiol 18:139–144
Kendrick RE, Spruit CJP (1973b) Phytochrome intermediates in vivo. III. Kinetic analysis of intermediate reactions at low temperature. Photochem Photobiol 18:153–159
Lehmann U, Schäfer E (1975) Kinetic analysis of phytochrome pelletability. In: Smith H (ed) Light and Plant Development. Butterworth, London, pp 92A
Lehmann U, Schäfer E (1978) Kinetics of phytochrome pelletability. Photochem Photobiol 27:767–773
Lipson E (1975) White noise analysis of *Phycomyces* light-growth response system. I. Normal intensity range. Biophys J 15:989–1011
Mackenzie JM Jr, Coleman RA, Briggs WR, Pratt LH (1975) Reversible redistribution of phytochrome within the cell upon conversion to its physiological active form. Proc Natl Acad Sci USA 72:799–803
Manchinelli AL, Rabino I (1975) Photocontrol of anthocyanin synthesis. IV. Dose dependence and reciprocity relationships. Plant Physiol 56:351–355
Manchinelli AL, Rabino I (1978) The "high irradiance responses" of photomorphogenesis. Bot Rev 44:129–180
Marmé D, Marchal B, Schäfer E (1971) A detailed analysis of phytochrome decay and dark reversion in mustard cotyledons. Planta 100:331–336
Mohr H, Drumm H, Schmidt R, Steinitz B (1979) The effect of light pretreatments on phytochrome-mediated induction of anthocyanin and of phenylalanine-ammonialyase. Planta 146:369–376
Neusypina TA, Pumpjanskaja SL, Fukshansky L (1972) A mathematical model of plant photoperiodism. Probl Cybern 25:28–57
Pratt LH, Briggs WR (1966) Photochemical and non photochemical reactions of phytochrome in vivo. Plant Physiol 41:467–474
Pratt LH, Marmé D (1976) Red-light-enhanced phytochrome pelletability. Re-examination and further characterization. Plant Physiol 58:682–692
Quail PH (1975) Particle-bound phytochrome: spectral properties of bound and unbound fractions. Planta 118:345–355
Quail PH (1978) Irradiation-enhanced phytochrome pelletability. Plant Physiol 62:773–778
Quail PH, Schäfer E (1974) Particle-bound phytochrome: A function of light dose and steady-state level of the far-red absorbing form. J Membr Biol 15:393–404
Quail PH, Marmé D, Schäfer E (1973a) Particle-bound phytochrome from maize and pumpkin. Nat New Biol 245:189–191
Quail PH, Schäfer E, Marmé D (1973b) Turnover of phytochrome in pumpkin cotyledons. Plant Physiol 52:128–131
Raven CW, Shropshire W Jr (1975) Photoregulation of logarithmic fluence response curves for phytochrome control of chlorophyll formation in *Pisum sativum* L. Photochem Photobiol 21:423–429
Schäfer E (1975) A new approach to explain the "high irradiance responses" of photomorphogenesis on the basis of phytochrome. J Math Biol 2:41–56
Schäfer E (1976) The "high irradiance reaction". In: Schmith H (ed) Light and Plant Development. Butterworth, London, pp 45–59
Schäfer E (1978) Variation in the rates of synthesis and degradation of phytochrome in cotyledons of *Cucurbita pepo* L. during seedlings development. Photochem Photobiol 27:775–780
Schäfer E (1981) Phytochrome and daylight. In: Smith H (ed) Plants and the Daylight Spectrum. Academic Press, London New York, pp 461–480
Schäfer E, Mohr H (1974) Irradiance dependency of the phytochrome system in cotyledons of mustard (*Sinapis alba* L.). J Math Biol 1:9–15
Schäfer E, Schmidt W (1974) Temperature dependence of phytochrome dark reactions. Planta 116:257–266

Schäfer E, Marchal B, Marmé D (1971) On the phytochrome phototransformation kinetics in mustard seedlings. Planta 101:265–276

Schäfer E, Marchal B, Marmé D (1972) In vivo measurements of the phytochrome photostationary state in far-red light. Photochem Photobiol 15:457–464

Schäfer E, Lassig TU, Schopfer P (1975) Photocontrol of phytochrome destruction in grass seedlings. The influence of wavelength and irradiance. Photochem Photobiol 22:193–202

Schäfer E, Lassig TU, Schopfer P (1976) Photocontrol of phytochrome destruction and binding in dicotyledonous vs. monocotyledonous seedlings. The influence of wavelength and irradiance. Photochem Photobiol 24:567–572

Schmidt W, Marmé D, Quail P, Schäfer E (1973) Phytochrome: first-order phototransformation kinetics in vivo. Planta 111:329–336

Shropshire W Jr (1980) Carotinoids as primary photoreceptors in blue-light responses. In: Senger H (ed) The blue light syndrome. Springer, Berlin Heidelberg New York, pp 172–186

Smith H (1970) Phytochrome and photomorphogenesis in plants. Nature 227:665–668

Smith H (1975) Phytochrome and photomorphogenesis. McGraw-Hill, London

Steinitz B, Drumm H, Mohr H (1976) The appearance of competence for phytochrome-mediated anthocyanin synthesis in the cotyledons of *Sinapis alba* L. Planta 130:23–31

Stone HJ, Pratt LH (1979) Characterisation of the destruction of phytochrome in the red absorbing form. Plant Physiol 63:680–682

Tanada T (1972) Phytochrome control of another phytochrome-mediated process. Plant Physiol 49:560–562

6 Phytochrome as a Molecule

W.O. Smith

1 Introduction

Action spectra studies of photomorphogenic responses in plants imply that several photoreceptor molecules probably exist (see Chap. 2, this Vol.). A necessary part of gaining an understanding of the mechanisms of these light-mediated responses in plants is the identification and characterization of the photoreceptor molecules. To date, the only photomorphogenically active molecule to be unequivocally identified and isolated from plants is phytochrome. The detection and isolation of phytochrome (Butler et al. 1959) was made possible by its unique photoreversible absorbance changes in the red and far-red regions of the spectrum which matched action spectra for photoreversible physiological responses in plants (Borthwick et al. 1952). These reversible spectral changes in phytochrome provided the basis for a photometric assay (Chap. 8, this Vol.) and made possible the subsequent purification and characterization of phytochrome.

The purpose of this chapter is to summarize the progress that has been made in purifying and characterizing phytochrome. The ultimate goal of these studies, as yet unattained, is to gain an understanding of the biochemical action of phytochrome.

2 Purification of Phytochrome

2.1 Sources

Phytochrome appears to be present in all higher plants and at least in some mosses, liverworts and algal genera. However, relatively few plants have proved to be convenient sources for large-scale purification of this protein. It has been known for some time that dark-grown seedlings of cereals and legumes contain high quantities of phytochrome. These plants have been convenient sources also because of the ready availability of seeds, their ability to produce large amounts of tissue in the absence of light and the lack of chlorophyll in dark-grown seedlings.

2.2 Extraction Conditions

All phytochrome purifications have been performed on protein which is readily soluble in neutral to slightly alkaline buffer. Initial extraction from fresh

(CORRELL et al. 1968a, RICE et al. 1973), frozen (PRATT 1982a) or lyophilized tissue (SMITH and DANIELS 1981) have all been successful. Methods of preference for tissue breakage have usually involved grinding with mortar and pestle (CORRELL et al. 1968a) or mechanical blenders (PRATT 1982a). The former method probably produces less surface denaturation of protein but is laborious and time-consuming. Lyophilization allows considerable flexibility in the handling of the initial extraction of phytochrome. Once the tissue is dry the phytochrome is quite stable, even at room temperature in the light. The dry material can be pulverized to a fine powder and stored until needed. The actual extraction of phytochrome from the powder is accomplished by simply stirring the powder into a suitable buffer. This considerably shortens the duration of the crude homogenate stage of the purification. Furthermore, the dry powders can be pretreated in a variety of ways to remove unwanted substances from the tissue before bringing phytochrome into solution. For example, excess chlorophyll or other lipids can be greatly reduced by extraction of the powders with anhydrous acetone, hexane, ether, or petroleum ether (TAYLOR and BONNER 1967, SMITH and DANIELS unpublished results). As long as the organic material is completely removed prior to addition of aqueous buffer there is no apparent denaturation of phytochrome. Caution should be exercised, however, as some polar solvents such as methanol and dimethylsufoxide have been found to cause spectral denaturation of the phytochrome (SMITH and DANIELS unpublished results). One further advantage of lyophilized powders is a high recovery of phytochrome with a minimum of agitation of the initial homogenate.

Phytochrome, being a protein, is susceptible to the same difficulties which are encountered in many protein preparations from higher plants. It is well documented that phytochrome is susceptible to proteases occurring in crude extracts. Phenolic compounds are also known to be deleterious to proteins. Other reactive compounds are always present in crude extracts depending upon the plant species. A variety of conditions have been utilized to minimize the effect of these factors. The P_r form of phytochrome is generally thought to be more stable than the P_{fr} form. For this reason essentially all purifications are carried out in darkness or 'safe-light' conditions. All chemical interactions are reduced by performing procedures at low temperature, generally at 2–4 °C. The usual method of minimizing proteolytic activity is to work as fast as possible and to plan the purification so that the most effective and least time-consuming steps are performed first. Proteolytic inhibitors such as phenylmethylsulfonyl fluoride, benzamidine and ω-amino caproic acid have been used (CORDONNIER and PRATT 1982) in extractions of phytochrome. The usefulness of those inhibitors should probably be further explored in future purification attempts. Reducing agents have been utilized extensively, although the exact nature of their action is not clear. Probably their most important function is to counteract oxidizing conditions which occur due to aeration of the protein solutions. Additionally, they might prevent oxidation of polyphenols to forms which covalently bind to proteins. The most commonly used reducing agents in crude extracts have been mercaptoethanol (PRATT 1982a) and bisulfite (ROUX et al. 1975). Dithiothreitol has been used at late stages of purification when smaller volumes are involved. Other additives such as EDTA, glycerol or polyvinylpyrollidone have been used in initial extracts.

2.3 Precipitants

Among the fastest and generally most useful purification techniques are those involving the addition of precipitation agents and the centrifugation of the mixture to separate insoluble materials.

Calcium Chloride, when added to crude extracts removes pectic substances by precipitation. In some procedures this has been done to prevent interference with subsequent brushite chromatography (BRIGGS et al. 1972).

Streptomycin Sulfate has been utilized (ROUX et al. 1975) to precipitate nucleic acids and protein from partially purified phytochrome.

Polyethylenimine has been reported to precipitate nucleic acids and considerable protein from crude phytochrome extracts (BOLTON and QUAIL (1981 b) without affecting the phytochrome. Perhaps of more importance, it was also reported to drastically reduce the chlorophyll in a crude extract from green tissue to a level comparable to extracts from etiolated tissue.

Ammonium Sulfate has been used extensively to purify phytochrome in crude extracts as well as to concentrate the protein at various stages of the purification. There have been reports of instability of phytochrome, especially P_{fr}, in ammonium sulfate solutions; however, these difficulties were probably of a technical nature rather than due to an inherent instability of phytochrome. It has been observed that the apparent photoreversibility of phytochrome can sometimes be repressed on resolubilization of ammonium sulfate pellets (BRIGGS et al. 1968). This is a reversible repression of phytochrome absorbance in the 730 nm range due to the presence of residual ammonium sulfate. On desalting, the spectral properties usually return to normal (SMITH unpublished results).

Polyethylenglycols have been used to purify a number of proteins by precipitation. They have not been used extensively with phytochrome, but a recent report indicates that these polymers can be useful for phytochrome purification (LITTS 1980). One advantage of this nonionic polymer was pointed out – that redissolved phytochrome could be applied directly to ion exchange columns without an intervening desalting step being necessary.

2.4 Adsorption Chromatography

Brushite ($CaHPO_4 \cdot 2H_2O$) was utilized in the first phytochrome purification (SIEGELMAN and FIRER 1964) and has been an important component of most procedures since. It is generally used early in the purification and usually involves binding of phytochrome from a low molarity phosphate buffer followed by batch elution of phytochrome by a stepwise increase in phosphate concentration. This is usually a very efficient procedure and commonly gives tenfold purification of the phytochrome (Tables 1 and 2). The power of this technique can readily be seen by the use of a second brushite column in the procedure in which a further twofold purification can be achieved (Table 1). The major disadvantage of this material probably arises from the fact that brushite is usually prepared in the laboratory instead of purchased. Seemingly slight variations in preparation of the material can result in alteration of the binding and elution behavior of the column. Inconsistencies in particle size can result in uneven flow of the columns. Also, overloading the column with protein and other constituents of crude extracts can cause localized shrinkage and channeling of the column. The user is well advised to read carefully the published methods and precautions concerning preparation of brushite (BRIGGS et al. 1972) and exercise careful control of all aspects of preparation such as temperature of solutions, rate of brushite precipitation and extent of incubation in buffer before use.

Hydroxyapatite is a form of calcium phosphate prepared by treatment of brushite with strong base. Its mechanism of binding and purification of phytochrome is apparently quite similar to brushite (ROUX et al. 1975) but, it is not as well suited for preliminary purifications. It generally has a higher binding capacity for proteins than brushite, but a much reduced flow rate.

2.5 Ion Exchange Chromatography

Cation Exchange media were used in earlier purification procedures (HOPKINS and BUTLER 1970) in which a 60,000 molecular weight phytochrome species was obtained. It has since been found that this species was a proteolytic fragment of phytochrome (GARDNER et al. 1971) and procedures which are designed to minimize or avoid proteolysis produce a species of phytochrome which is a dimer of 120,000 molecular weight polypeptides (PRATT 1982b). This species, which is presumed to be native, does not bind to carboxymethyl cellulose at pH values greater than 6.5. In buffers of lower pH the aggregate precipitates from solution. This limits the usefulness of cation exchangers in designing purification of phytochrome, but does not rule them out, as they might prove valuable in "pass through" purification steps in which unwanted contaminants might bind to the column.

Anion Exchange media in the form of the DEAE-moiety immobilized on either cellulose, agarose or dextran polymers has been used successfully in many phytochrome purifications, although considerable losses of phytochrome have occurred in some cases (BRIGGS et al. 1972). DEAE cellulose has also been used as a "flow through" purification for phytochrome extracted from chlorophyll containing material (TAYLOR and BONNER 1967). When crude extract was passed through a column equilibrated in 0.15 M potassium phosphate buffer, the phytochrome was not bound but a large quantity of chlorophyll remained on the column.

2.6 Gel Filtration Chromatography

Gel filtration is the procedure of choice for desalting phytochrome during purification procedures. Media such as Sephadex G-50 allow very rapid desalting as compared to dialysis. This is very important for most phytochrome preparations in order to minimize exposure to contaminating proteases.

Gel filtration is also useful at later stages of purification as a means of separating phytochrome from other proteins. Undegraded phytochrome normally exists in solution as a 240,000 molecular weight dimer. However, it migrates through gel filtration media as if it were a larger protein of about 400,000 molecular weight (PRATT 1978). This phenomenon, which will be discussed further in Section 3, is apparently due to the fact that phytochrome is not a spherical protein. This makes it necessary to select a medium of large pore size in order for phytochrome to elute in the fractionation range of the column. Media used have included Sephadex G-200 (RICE et al. 1973), Biogel P-300 (PRATT 1982a) and Biogel A-1.5 m (SMITH and CORRELL 1975). Phytochrome elutes from the latter at a volume far enough removed from the void volume to allow separation from proteins larger than phytochrome, as well as those smaller, This is not the case with Sephadex G-200.

2.7 Ultracentrifugation

Preparative ultracentrifugation of phytochrome in sucrose density gradients has been used as a final purification step (SMITH and CORRELL 1975). Because of technical limitations and lengthy procedures, its use has been limited. However, this technique utilized in tandem with gel filtration takes advantage of the unique shape of phytochrome to provide an excellent purification. A protein such as phytochrome which is non-spherical will migrate on gel filtration columns as a larger protein, but sediment in the ultracentrifuge as a smaller protein. Utilizing both procedures should theoretically separate phytochrome from all other proteins in the solution except those that have both the same size and shape. The use of this combination is limited to phytochrome preparations that are stable and protease free as they are quite lengthy. The gel filtration column usually requires a few hours, while ultracentrifugation requires 24 to 30 h to complete.

2.8 Electrophoretic Procedures

Preparative electrophoresis methods have been utilized in attempts to purify native phytochrome (MUMFORD and JENNER 1966) as well as denatured subunits from SDS-containing systems (STOKER et al. 1978). There has been only one report of the use of preparative isoelectric focusing (BALANGÉ and ROLLIN 1973). This technique has not been further exploited, but potentially it could be very useful.

2.9 Affinity Chromatography

Affinity chromatography utilizes specific attraction of a protein for a biological ligand such as a substrate, cofactor, effector, or antibody. This procedure provides in principle a one-step purification, but in practice very few proteins can be purified from plant extracts in one step. The presence of phenolic compounds, proteases etc. leads to fouling of columns, non-specific interference with binding to the column and denaturation of the protein during the sometimes lengthy column runs. These problems can be reduced in the case of phytochrome by a preliminary partial purification through a brushite column.

An affinity column for phytochrome has been developed utilizing agarose immobilized anti-phytochrome immunoglobulins (HUNT and PRATT 1979). This procedure has the advantage of being highly specific for phytochrome and yielding homogeneous phytochrome in a matter of only a few hours. Phytochrome is selectively adsorbed from a brushite eluate by agarose-immobilized antiphytochrome immunoglobulins and the contaminating proteins washed out with 1 M NaCl. Phytochrome is recovered by elution with 3 M $MgCl_2$. The elution step is the source of one of the drawbacks of this method. Release of phytochrome requires strenuous eluting conditions so that recovery of spectrally undenatured protein is difficult. The most successful elution procedure, indicated above, yields only about 15% of the phytochrome applied. However, the remaining phytochrome can be recovered by a 1 M formic acid wash. This material is denatured, but the polypeptide chain is still intact and suitable for chemical characterization.

Another affinity procedure has been developed based on the binding of the dye, Cibacron blue 3 GA, to phytochrome (SMITH 1981). This dye and its dextran conjugate, blue dextran, have a high affinity for specific binding domains on many proteins. These dyes appear to be acting as analogs of the biological ligands of the proteins and by immobilizing them on a solid support such as agarose they have been utilized as affinity media for the purification of many proteins (STELLWAGEN 1977). These affinity media are somewhat different from that discussed above in that they are not specific for a single protein, but for a class of proteins that have a binding domain compatible with the dye. Additionally, some proteins bind in a non-specific manner through hydrophobic or ionic interactions with the dye. The selectivity of these media is greatly enhanced, however, by varying the method of elution of the column. Proteins can usually be eluted by increasing the ionic strength or addition of non-polar substances such as ethylene glycol. Proteins that are bound specifically through binding domains for cofactors, substrates, effectors, or other natural ligands can usually be eluted by washing the column with a solution of the natural ligand.

No natural ligands are known for phytochrome. Of the cofactors known to be analogs of the dye, FMN and FAD were found to elute phytochrome from blue agarose columns with a high degree of purification. The structural implications of this phenomenon will be discussed in Section 3.7. The utilization of this phenomenon in purification of phytochrome from dark-grown rye seedlings (SMITH and DANIELS 1981) has resulted in a procedure which yields homogenous phytochrome with a high yield (25%–30% of the phytochrome in the crude extract). Just as in the case of the immunoaffinity procedure, phytochrome is partially purified before application to the affinity column. For this procedure, the preliminary purification is probably more important than in the case of the immunoaffinity column as a large number of proteins might be expected to bind to the blue agarose. However, rye phytochrome is bound so tightly by the blue agarose that it is not eluted by a 0.5 M KCl wash. Most other proteins that bind to the agarose, whether specifically or not, are washed out by this level of salt. It might be anticipated also that the use of this procedure in purifying phytochrome from plant species distant from rye would require adjustment of buffer conditions such as pH and ionic strength to obtain comparable purification. The conditions reported for rye (0.1 M potassium phosphate, pH 7.8) might not be optimal for a phytochrome species which has a different isoelectric point or aggregation state and the possibility remains that these properties might vary considerably throughout the plant kingdom. Among the advantages of this purification procedure are the ability to readily scale up the purification and the general availability of the affinity matrix. One report utilizing an adaptation of this method suggests that lumichrome, a common contaminant of FMN preparations, will bind tightly to phytochrome (SONG et al. 1981). This possibility can be prevented by using purified FMN in the procedure.

2.10 Summary of Purification Procedures

Tables 1, 2 and 3 illistrate in a stepwise fashion three procedures for purification of phytochrome from dark-grown seedlings. All three take advantage of a unique property of phytochrome to insure complete separation from other protein in the extraction. The first procedure (SMITH and CORRELL 1975, SMITH unpublished results) utilizes classical techniques, with the final two sizing steps taking advantage of the nonspherical shape of phytochrome to insure purity.

Table 1. Purification of phytochrome from 100 g lyophilized powder of dark-grown rye by conventional techniques[a]

	Total phytochrome mg	Specific activity A_{667}/A_{280}	Yield %
Buffer extraction	28.9	0.004[b]	100
$(NH_4)_2SO_4$ precipitation			
Sephadex G-50	23.3	0.018[b]	81
Brushite	18.8	0.108[b]	65
DEAE Agarose	14.6	0.207	51
Brushite	6.4	0.350	22
Biogel A 1.5 m	4.3	0.671	15
Ultracentrifugation	2.3	0.833	8

[a] Adapted from SMITH, WO (1975), SMITH and CORRELL (1975) and DANIELS and SMITH (unpublished results)
[b] Values of A_{280} estimated from protein concentration

Table 2. Purification of phytochrome from 100 g lyophilized powder of dark-grown rye on blue agarose[a]

	Total phytochrome mg	Specific activity A_{667}/A_{280}	Yield %
Buffer extraction	34.7	0.004[b]	100
$(NH_4)_2SO_4$ precipitation	28.9	0.012[b]	83
Sephadex G-50	28.1	0.014[b]	81
Brushite	20.8	0.121	60
Blue agarose	13.3	0.93[b]	38
Biogel A 1.5 M	8.7	0.96	25

[a] From SMITH and DANIELS (1981)
[b] Values of A_{280} estimated from protein concentration

Table 3. Purification of phytochrome from 1 kg of dark-grown oats by immunoaffinity methods[a]

	Total phytochrome mg	Specific activity A_{667}/A_{280}	Yield %
Buffer extraction	28.9[b]	–	100
Brushite	–	–	–
$(NH_4)_2SO_4$ precipitation	–	–	–
Sephadex G-25	18.8	0.044	65[b]
Immunoaffinity column	2.7 (1.8)[c]	0.83	9.1 (6.2)[c]

[a] Adapted from HUNT and PRATT (1979) and PRATT (1982a)
[b] These represent estimates based on the assumption that the first three steps result in yields comparable to these obtained with rye in Tables 1 and 2
[c] The values represent denatured phytochrome that can be recovered from in immunoaffinity column by washing with 1 M formic acid

The second is the immunoaffinity procedure of HUNT and PRATT (1979) utilizing the selective binding of antiphytochrome immunoglobulin. The third procedure utilizes the blue agarose affinity column capitalizing on the selective binding and subsequent elution of phytochrome by FMN (SMITH and DANIELS 1981). All three of these procedures have been reported to yield phytochrome preparations that are homogeneous as judged by SDS gel electrophoresis with apparent particle mass of about 120,000 molecular weight.

3 Properties of Purified Phytochrome

3.1 Background

Characterization of phytochrome has been underway since its first detection in crude extracts (BUTLER et al. 1959), but progress in these investigations has been hindered by the presence of proteolytic enzymes in extracts and by difficulties encountered in obtaining homogeneous phytochrome. There are several excellent reviews of these studies (see BRIGGS and RICE 1972, PRATT 1978, PRATT 1979, RÜDIGER 1980 and references therein) and the reader should refer to them for complete details. The properties discussed in this chapter will deal mostly with highly purified phytochrome preparations which are believed to be undegraded by proteolysis except where noted.

The first significant finding that led to the realization that proteolysis of phytochrome was occurring during purification resulted from the selection of rye seedlings to purify phytochrome (CORRELL et al. 1968a). At that time the preferred plant source for obtaining phytochrome was dark-grown oat seedlings and the phytochrome species obtained consisted of a single 60,000 molecular weight peptide. CORRELL et al., screened a variety of plant tissues for in vivo content of phytochrome, as well as yield and stability after an initial extraction and ammonium sulfate precipitation. Rye seedlings seemed superior to other readily available plant materials and was used for all subsequent work. As it turned out they found a much larger phytochrome molecule in their preparations (CORRELL et al. 1968b). It was later discovered by BRIGGS and coworkers (GARDNER et al. 1971, BRIGGS and RICE 1972) that this large species ($2 \times 120{,}000$ m.w.) was apparently the native form of phytochrome with the 60,000 molecular weight species, being a relatively stable core of phytochrome remaining after partial proteolysis during extraction. The rye extracts seemed to have lower proteolytic activity than oat extracts, making possible the discovery of the larger form. With proper precautions taken to limit proteolysis it has been possible to obtain the same large form of phytochrome from oats, pea, zucchini, and lettuce (PRATT 1982b).

3.2 Chemical Composition

The amino acid composition of phytochrome has been reported by several laboratories. There is considerable variation in the results of these analyses. Much of it is probably due to a combination of factors including low purity and partial proteolysis of the phytochrome. Table 4 contains analyses for preparations of phytochrome from three plant species, as well as a comparison of large and small phytochrome from one species. An estimation of the polarity of these proteins according to the method of CAPALDI and VANDERKOOI (1972) is accomplished by summing the mole percentages of polar amino acids (the first seven in Table 4). Large phytochrome from zucchini, rye, and oats had values of 47%, 42% and 46%, respectively. These values fall within the range that has been observed for most soluble proteins. Small oat phytochrome had a value of 0.3% lower than large phytochrome. Of the 240 Asp and Glu residues listed for large oat phytochrome, 114 have been estimated to be in the carboxylate form with the remainder presumably being amides.

Oat phytochrome has been reported to contain 6 to 8 disufides along with 10 to 14 free sulfhydryl groups (HUNT and PRATT 1980). This is in contrast to another report of 1 disulfide along with 11 free sulfhydryl groups (ROUX

Table 4. Amino acid analyses of phytochrome from three plant species

Amino acid	Zucchini[a]	Rye[b]	Oat[c]	Oat[d]
Lys	64	55	64	33
His	26	26	34	17
Arg	50	47	51	30
Asp	104	97	118	59
Thr	56	43	38	24
Ser	82	71	73	40
Glu	126	120	122	54
Pro	58	83	45	36
Gly	80	73	72	37
Ala	77	103	93	50
Cys	18	25	27	11
Val	74	84	79	36
Met	29	30	26	4
Ile	59	51	51	25
Leu	110	105	119	54
Tyr	25	22	23	18
Phe	42	41	45	25
Trp	–	–	8	7
Total residues	1080	1076	1085	560

[a] From CORDONNIER and PRATT (1980) for 120,000 m.w. peptide
[b] From RICE and BRIGGS (1973a) for 120,000 m.w. peptide
[c] From HUNT and PRATT (1980) for 120,000 m.w. peptide
[d] From MUMFORD and JENNER (1966) for 60,000 m.w. peptide

et al. 1982). These reports are interesting in light of the general trend which occurs in proteins in that they are either sulfhydryl-containing proteins or disulfide-containing proteins, but usually not mixed (CECIL 1963).

Oat phytochrome has been reported to contain carbohydrate residues (ROUX et al. 1975), but this work could not be confirmed by other workers (PRATT 1982b).

One analysis of phosphate content for oat phytochrome has been reported with the indication of one mole of phosphate being present for each mole of phytochrome monomer (HUNT and PRATT 1980). The significance of phosphorylation of phytochrome is unknown at this time.

The only other non-amino acid constituent that is known to be a part of phytochrome is an open chain tetrapyrrole chromophore. The exact number of these chromophores is not known, but most probably there is only one per 120,000 molecular weight monomer. There are two lines of evidence to support this view. One comes from a comparison of the extinction of the phytochrome chromophore to that of other biliproteins of similar structure (GARDNER and BRIGGS 1974). The other comes from proteolysis experiments in which chromopeptides were separated and analyzed (LAGARIAS and RAPAPORT 1980). In that case all chromopeptides seem to be a part of the same unique amino acid sequence indicating that most likely one chromophore exists per peptide. The chromophore structure and properties are covered in Chapter 7, this Volume.

3.3 Primary Structure

Knowledge of the amino acid sequence of phytochrome is limited to about 1% of the length of the protein. FRY and MUMFORD (1971) reported a partial characterization of an undecapeptide of oat phytochrome with a tetrapyrrole chromophore attached. LAGARIAS and RAPAPORT (1980) confirmed and completed the analysis of the same region. The structure is as follows:

```
Leu—Arg—Ala—Pro—His—Ser—Cys—His—Leu—Gln—Tyr
                         |
                         S
                         |
                    Chromophore
```

Although only one covalent linkage between the chromophore and protein has been confirmed, the possibility exists that a second covalent linkage might exist (LAGARIAS and RAPAPORT 1980, Chap. 7, this Vol.). HUNT and PRATT (1980) attempted to sequence the amino terminal of immunoaffinity purified oat phytochrome by manual Edman degradation and dansylation with the following results:

NH_2—(Lys; Ala; 60%, 40%)—(Ala, Leu; 42%, 58%)—
(Leu, Val; 52%, 48%).

If it is assumed that a terminal Lys is missing from one of the phytochrome polypeptides then the sequence would be:

NH_2—Lys—Ala—Leu—Val
NH_2—Ala—Leu—Val

The only other report of amino terminal analysis was that of RICE and BRIGGS (1973a) in which they found using the Sanger DNP method two major DNP residues from rye phytochrome, Glu and Asp along with trace amounts of Gly and Ser. Clearly, the primary structure of the amino terminal portion of phytochrome is only tentatively identified and the meaning of this heterogeneity remains unknown.

3.4 Secondary and Tertiary Structure

Based on circular dichroism TOBIN and BRIGGS (1973) estimated large rye phytochrome to contain 20% α-helix, 30% β-structure and 50% random coil. They reported an α-helix content for small rye phytochrome of about 10%–13%, but did not extend the analysis to β-structure and random coil. HUNT and PRATT (1980) have analyzed CD spectra of large oat phytochrome and estimated 35% α-helix, 23% β-structure and 42% random coil.

As indicated in Section 3.2, there appears to be one or more disulfide bridges in the phytochrome molecule. These are probably intrachain links (HUNT and PRATT 1980) and therefore contribute to the stability of the secondary and tertiary structure of the protein.

3.5 Quaternary Structure

The particle mass of purified phytochrome has been found to be in the range of 240,000 molecular weight existing in solution as a dimer of identical or almost identical polypeptides of about 120,000 (BRIGGS and RICE 1972, PRATT 1982b). There have been reports of the possibility of larger aggregates (PRATT 1973, GROMBEIN and RÜDIGER 1976) and dissociation to monomer in solution (SMITH and CORRELL 1975), but the predominant species at pH values in the range of 7.0 to 8.0 is a dimer which is held together by noncovalent interactions (HUNT and PRATT 1980).

3.6 Three-Dimensional Structure

In solution phytochrome does not exhibit properties of an ideal spherical protein (BRIGGS and RICE 1972). On gel filtration of the dimer it appears to have a particle mass much greater than molecular weight 240,000 – as much as 400,000. This is apparently due to phytochrome being non-spherical. Gel filtration is a better indication of molecular size than molecular mass and this technique has been used to estimate a Stokes radius for the phytochrome dimer. Under conditions which should minimize aggregation of phytochrome, a value of 6.5 nm was estimated for rye phytochrome (SMITH, WO 1975) and 7.0 nm for oat phytochrome (LITTS 1980). Assuming a Stokes radius of 6.5 nm correlates to a particle mass of 240,000 molecular weight, a frictional ratio (f/fo) can be estimated for phytochrome to be 1.58 (SMITH, WO 1975). This is the ratio of the radius of phytochrome as measured by gel filtration to that of a theoretical non-hydrated spherical protein. The value for phytochrome is considerably greater than that of a typical globular protein, but less than that of a typical fibrous protein. The interpretation of this apparent non-globular property of phytochrome awaits further study.

Attempts have been made to directly observe the three-dimensional structure of phytochrome by electron microscopy. CORRELL et al. (1968b) first published negative contrast images of rye phytochrome preparations containing tetrameric structures that they interpreted as aggregates of 42,000 molecular weight subunits. Subsequent to this work McKenzie and Briggs investigated rye phytochrome preparations that were demonstrated to be dimers of a 120,000 molecular weight subunit. They also obtained negative contrast images of tetrameric structures (report in SMITH, H 1975). These images have been interpreted to represent dumbbell-shaped dimers (SMITH, H 1975, RÜDIGER 1980). SMITH and CORRELL (1975) reexamined rye phytochrome and found that the tetrameric structures in their phytochrome preparations were not even phytochrome, but a contaminant which remained in preparations of greater than 90% purity. No interpretable images could be obtained once this contaminant was removed from the phytochrome. Although the double-dumbbell interpretation of electron micrographs appears to be premature, there is indirect evidence that the phytochrome monomer might consist of two domains. This arises from limited proteolysis studies which indicate that the 120,000 molecular weight monomer is readily cleaved to about half that size (see reviews by BRIGGS and RICE 1972 and PRATT 1982b). One interpretation of these results is that two domains of the protein are joined by a segment that is readily cleaved by proteases.

3.7 Properties of the Functional Chromoprotein

3.7.1 Phytochrome as Photoreceptor Molecule

3.7.1.1 Light and Dark Transformations

The absorption spectra of phytochrome as found in etiolated seedlings (P_r) and after saturating irradiation with red light (P_{fr}) are shown in Fig. 1. Because of overlapping absorption of P_r and P_{fr} below 720 nm, it is impossible to photochemically convert all of P_r to P_{fr} so the recording obtained after saturating red light actually represents the absorbance of a mixture of 16% P_r and 84% P_{fr} (YAMAMOTO and SMITH 1981c). This photoequilibrium state will be further discussed below.

The extinction coefficient of P_r from rye has been reported to be 7×10^4 l mol^{-1} cm^{-1} in the red region based on a colorimetric assay for protein (TOBIN and BRIGGS 1973). ROUX et al. (1982) have estimated the extinction coefficient for P_r from oats to be 10.2×10^4 l mol^{-1} cm^{-1} using amino acid analysis to estimate protein concentration. The spectral characteristics of phytochrome are dependent on the interaction of the open chain tetrapyrrole chromophore with the protein. This is demonstrated by the fact that denaturation of the protein under conditions which should not chemically alter the chromophore lead to a loss of photoreversibility, a large bathochromic shift in the visible absorbance peak and a reduction in extinction coefficient (Chap. 7, this Vol.). The photoconversions of P_r and P_{fr} are first order with respect to phytochrome (BUTLER 1961, GARDNER and BRIGGS 1974). A number of transient spectral intermediates have been detected using techniques of flash photolysis (LINSCHITZ et al. 1966, LINSCHITZ and KASCHE 1967, PRATT and BUTLER 1970, BRASLAVSKY et al. 1980, SHIMAZAKI et al. 1981), stabilization of intermediates by low temperature (CROSS et al. 1968, PRATT and BUTLER 1968) or dehydration (KENDRICK 1974, KENDRICK and SPRUIT 1977), and spectral analysis during continuous actinic irradiations (BRIGGS and FORK 1969a, b). There are points of disagreement among all of these studies which remain to be resolved. Among the remaining questions are the actual number of intermediates, whether some occur sequentially or in parallel, what molecular changes are accompanying these spectral changes and possibly of greatest importance, do any of the inter-

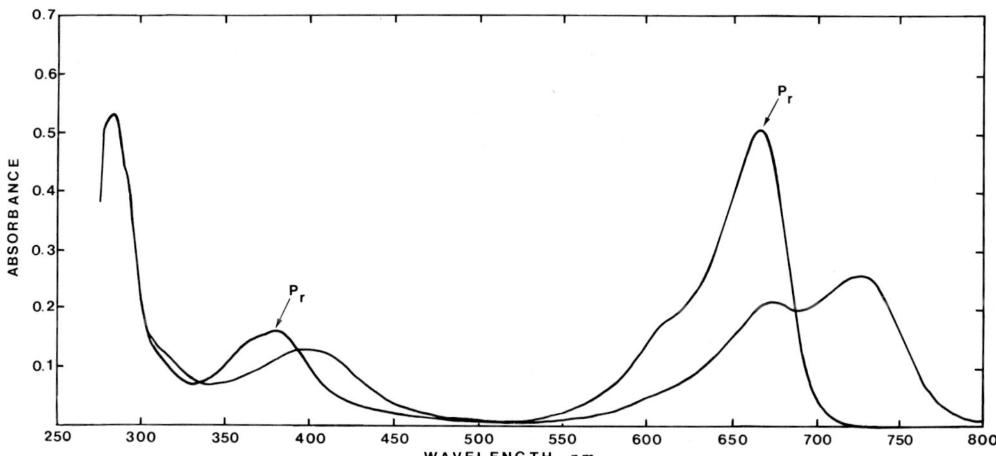

Fig. 1. Absorbance spectra of purified rye phytochrome after saturating far-red light (P_r) or red light. (SMITH and DANIELS 1981). Spectra were recorded at 7 C in 0.1 M potassium phosphate, pH 7.8 in Aminco DW-2a spectrophotometer using 1-cm-light path quartz cuvette. The absorption maximum for P_r was at 667 nm and for P_{fr} at 730 nm. The A_{667}/A_{280} was 0.96

mediates have biological activity? This last question is prompted by the observation that continuous irradiation of phytochrome leads to the steady-state accumulation of significant levels of spectral intermediates (BRIGGS and FORK 1969a, b, KENDRICK and SPRUIT 1977). The general picture which has emerged is that light activation of P_r is followed by dark relaxation via a number of intermediates to P_{fr}. P_{fr} in turn can be light activated and subsequently relaxes via a different set of intermediates to P_r.

In addition to the phototransformations there is a dark reversion of P_{fr} to P_r (see PRATT 1978, PIKE and BRIGGS 1972 for discussions). This reaction is readily detected in vivo in some dicots, but not in monocots but on extraction and purification, phytochrome, regardless of its source, will undergo dark reversion in solution (FRANKLAND 1972). The time course of this process is quite long in comparison to the dark reactions involved in the phototransformation pathways, with a half-life of minutes to hours depending on the conditions of the study (see for example YAMAMOTO and SMITH 1981b, c). Dark reversion has been found to follow complex kinetics during the early part of the dark period following photoconversion to P_{fr} (CORRELL et al. 1968c, PIKE and BRIGGS 1972), followed by first-order decay until reversion is complete. The rate of reversion as well as the relative contribution of the early non-first-order kinetics have been reported to be affected by a number of factors including reducing agents (PIKE and BRIGGS 1972), metal ions (PRATT and CUNDIFF 1975, NEGBI et al. 1975), relative content of ions that salt-in or salt-out proteins (SMITH 1978) and ionic strength (YAMAMOTO and SMITH 1981b). These observations cannot be reconciled with one another because the inherent reversion rates of the different phytochrome preparations varied considerably. The complexity of dark reversion is not understood, but one factor is certainly that a photosensitive component(s) other than P_r and P_{fr} occurs in many phytochrome preparations on irradiation (PRATT 1975, YAMAMOTO and SMITH 1981b). This component(s) reverts rapidly to P_r in the dark. Whether this component(s) is an altered form of P_{fr} or an intermediate is unclear at the present time, nor can its occurrence be readily correlated to any molecular changes in the protein of phytochrome. It does appear that a fast reversion rate can many times be correlated with the presence of this component. Although dark reversion could potentially play an important role in photomorphogenesis our present knowledge of its significance both in vitro and in vivo is somewhat confused. In fact, its physiological significance has been questioned in at least one study (OELZE-KAROW et al. 1976).

3.7.1.2 The Photoequilibrium State

The relative ratio of P_r and P_{fr} under a given set of conditions will be determined by the quality of the impinging light, the extinction coefficients of P_r and P_{fr} at those wavelengths of light and the quantum yields for phototransformation of P_r and P_{fr}. BUTLER et al. (1964a) first devised a method for determining the photoequilibrium state of a P_r/P_{fr} mixture under a given monochromatic light regime. This was done with 60,000 molecular weight oat phytochrome and it was determined that the highest ratio of P_{fr} to P_r could be obtained by irradiation with monochromatic light in the region of 660 nm giving a photoequilibrium of 81% P_{fr}. PRATT extended these studies to 120,000 molecular weight (presumably undegraded) phytochrome from oats, rye and pea tissue (PRATT and CUNDIFF 1975, PRATT 1975). The proteolytic chromopeptide which BUTLER et al. (1964a) analyzed showed a high absorbance in the far-red region at the red-light-induced photoequilibrium state, and the change in absorbance in the red region was accompanied by an almost equivalent change in the far-red region. The P_{fr} form would slowly revert to P_r in darkness – again with a $\Delta A_{fr}/\Delta A_r$ of about unity. Similar spectral properties have been reported in large phytochrome from rye (RICE and BRIGGS 1973a). In the large phytochrome preparations that PRATT analyzed, however, there was reduced absorbance in the far-red region at the red-light-induced photoequilibrium, $\Delta A_{fr}/\Delta A_r$ was much less than unity and there was initially a faster dark reversion of P_{fr} in which the absorbance decrease in the far-red region was not accompanied by an equivalent increase in the red region. PRATT (1975) proposed an explanation of these spectral properties based on the existence of an altered P_{fr} in these preparations which exhibited a reduced far-red extinction and rapid dark reversion. It was further found that EDTA or mercaptoethanol would inhibit these anomolous spectral properties (PRATT and CUNDIFF 1975). As the method of BUTLER et al. (1964a) is only applicable to systems containing two components; i.e., P_r and P_{fr}, the red-light-induced photoequilibrium was determined under these conditions and found to be 75% P_{fr}. YAMAMOTO and SMITH (1981b) have also studied these anomolous spectral properties in undegraded pea phytochrome. They found the presence of a photosensitive component(s) other than P_r and P_{fr} during phototransformations as well as during the first few minutes in the dark after phototransformations. This pea phytochrome was very sensitive to the conditions of the solvent and at high ionic strength became a two-component system at the red-light-induced photoequilibrium, but the other component(s) was still evident during phototransformations. Pea phytochrome under these conditions was found to contain 80% P_{fr} at the red-light-induced photoequilibrium (YAMAMOTO and SMITH 1981b). They also analyzed undegraded rye phytochrome which had no anomolous spectral features and found it to contain 84% P_{fr} at the red-light-induced photoequilibrium (YAMAMOTO and SMITH 1981c). The molecular nature of this anomolous behavior remains unknown, but it does appear to be due to some alteration of the protein which occurs on extraction and purification. This is indicated by the fact that the phenomenon has been found to be present at times, but not at others by researchers working with one plant material (see example in SMITH 1978 vs. YAMAMOTO and SMITH 1981b, c). If one examines published absorption spectra for undegraded phytochrome (for example, RICE and BRIGGS 1973a, PRATT 1978, SMITH and DANIELS 1981, YAMAMOTO and SMITH 1981b, c), one can see that the ratio of absorbance in the far-red region to that in the red region varies considerably at the red-light-induced photoequilibrium. If this is interpreted as due to varying degrees of presence of these anomolous spectral properties, one might also wonder what other properties of this protein are altered.

3.7.1.3 Dynamics of Phytochrome in the Plant Cell

The chief purpose for studying the light and dark transformations of phytochrome is to gain an understanding of how these processes occur in vivo and how they are related to the mechanism of action of phytochrome. There are two other processes which contribute to the dynamic state of phytochrome in the cell, biosynthesis and destruction. The quantitative aspects of these processes are an active area of research and are reviewed elsewhere (Chap. 5, this Vol.). Only a brief description of these processes will be given here.

The phytochrome protein is synthesized in growing seedlings de novo as P_r (QUAIL et al. 1973a, b). There is essentially nothing known of the biosynthetic origin of the linear tetrapyrrole chromophore. BOLTON and QUAIL (1981a, 1982) have reported the cell-free synthesis of the phytochrome polypeptide in both wheat germ and rabbit reticulocyte lysate systems that were primed with poly(A) RNA from oats. The translation product had the same apparent molecular mass on SDS gels as native phytochrome extracted from oats. The fact that the mRNA contained poly(A) sequences and was translated in two eukaryotic cell-free protein synthesis systems and that there was no apparent intracellular processing of the translation product were interpreted as indicative of phytochrome being a nuclear encoded soluble protein.

Although there is a turnover of P_r in dark-grown seedlings (QUAIL et al. 1973b) relatively large amounts of the protein accumulate in meristematic tissues – as much as 0.5% of the extractable protein in etiolated oat and rye seedlings. Exposure of dark-grown seedlings to light, i.e., conversion of P_r to P_{fr} initiates a destruction process (FRANKLAND 1972) which is apparently a proteolytic phenomenon (PRATT et al. 1974). This destruction process appears to be an important part of the mechanism by which the level of the active form of phytochrome is controlled in the cell. It has been demonstrated, however, that at least some of the P_r in the cell is also destroyed by this process (STONE and PRATT 1979). Under continuous light the level of phytochrome eventually reaches a photo-steady state which represents only a small fraction of that occurring in dark-grown material (FRANKLAND 1972, Chap. 27, this Vol.). It is believed that this level is supported either by a biosynthetic rate which matches the destruction rate (CLARKSON and HILLMAN 1968), or a population of phytochrome which is resistant to destruction such as reported in cauliflower florets (BUTLER et al. 1963) and amaranthus seedlings (HEIM et al. 1981).

3.7.2 Phytochrome as a Biologically Active Protein

The general properties of the photochrome molecule as described in earlier sections offer no clues as to the biochemical action of phytochrome. If it is accepted that P_{fr} is the biologically active form of phytochrome, then it would seem reasonable to expect phototransformation of P_r to P_{fr} to produce a change in the protein that is directly related to its activity. There have been a number of comparisons of P_r and P_{fr}, several of which were performed on preparations that had undergone proteolytic degradation. These earlier studies are still useful to some extent in that they were studying a large fragment of the protein containing the photochemically active portion of the molecule. The findings with preparations of these type were that P_r and P_{fr} did not differ on gel filtration, sedimentation in sucrose gradients, electrophoresis or brushite chromatography (BRIGGS et al. 1968). P_{fr} was more readily spectrally denatured by urea, p-mercuribenzoate, proteases (BUTLER et al. 1964b), ammonium sulfate (BRIGGS et al. 1968) and metal ions (LISANSKY and GALSTON 1974), and P_r was more reactive to the lysine reagents, glutaraldehyde and trinitrobenzene sulfonic acid than P_{fr} (ROUX 1972). Ultraviolet difference spectra, CD spectra and immunological comparison were interpreted to indicate a small difference between P_r and P_{fr} (HOPKINS and BUTLER 1970). Differential sedimentation velocity studies indicated P_r to have a slightly greater sedimentation coefficient than P_{fr} (HOPKINS 1971). Another CD study found to difference in P_r and P_{fr} that could be attributed to protein changes (ANDERSON et al. 1970). Calculations of entropies of activation of phototransformation intermediates also gave no indication of major protein conformational changes (PRATT and BUTLER 1970).

A number of investigations of purified and undegraded phytochrome have indicated no differences in the P_r and P_{fr} forms of the protein. These include CD spectra (TOBIN and BRIGGS 1973, HUNT and PRATT 1981), protein fluorescence (TOBIN and BRIGGS 1973), immunochemical reactivity (CUNDIFF and PRATT 1975, RICE and BRIGGS 1973b), and isoelectric focusing (HUNT and PRATT 1981). Small differences were noted in UV difference spectra that could be attributed to changes in the protein (TOBIN and BRIGGS 1973). As in degraded phytochrome, spectral denaturation by metal ions was more rapid in P_{fr} (PRATT and CUNDIFF 1975). P_{fr} was also found to be more susceptible to permanganate oxidation than was P_r (HAHN et al. 1980).

It was reported that the hydrophobicity of phytochrome increased on conversion of P_r to P_{fr} as determined by partitioning in an aqueous two-phase system (TOKUTOMI et al. 1981). Salting out of P_{fr} was reported to occur at a lower concentration of ammonium sulfate than P_r and irradiation of phytochrome in solutions containing divalent cations led to precipitation of phytochrome as well as binding to microsomal fractions (YAMAMOTO et al. 1980). Another study reported that P_{fr} had a higher affinity than P_r for DEAE agarose and certain alkyl and amino alkylagaroses (YAMAMOTO and SMITH 1981a). A higher affinity of P_{fr} than P_r for liposomes was also observed (KIM and SONG 1981) and the hydrophobic fluorescence probe, ANS (HAHN and SONG 1981).

As described in Section 2, phytochrome binds to the dye Cibacron blue 3 GA (SMITH 1981, SMITH and DANIELS 1981) with P_{fr} exhibiting a greater affinity for the dye than P_r. It was shown that in the case of the dextran conjugate of the dye the P_r form had essentially no affinity for the dye while the P_{fr} form would bind (SMITH 1981). As this dye has been found to act as an analog of a number of biological ligands (STELLWAGEN 1977) attempts were made to determine if any of the reported analogs of the dye would competitively elute phytochrome from agarose beads that contained the covalently linked blue dye. FMN was found to be very effective in eluting phytochrome from the beads. It was suggested that this might indicate a flavin binding site on phytochrome (SMITH 1981), but subsequent studies of the eluted phytochrome indicated that no flavin remained bound to the protein (DANIELS and SMITH unpublished results). Although others have reported interactions of flavins and phytochrome in solution (SARKAR and SONG 1982), a direct demonstration of complex formation has not been given.

A greater reactivity of P_{fr} for the sulfhydryl reagent, N-ethyl maleimide was found (GARDNER et al. 1974), as well as dithionitrobenzoic acid (HUNT and PRATT 1981). In the latter case, it was concluded that P_{fr} had one more exposed cysteine than P_r. The same conclusion was reached for histidine as judged by reactivity of P_r and P_{fr} for diethyl pyrocarbonate. P_r and P_{fr} had identical reactivity toward the carboxyl-modifying reagent, 3-[(dimethylamino)-propyl]carbodiimide and the tyrosine-modifying reagent, tetranitromethane (HUNT and PRATT 1981).

All of these findings which have been innumerated above are not amenable to far-reaching conclusions, but do suggest certain facts. They can be interpreted as indicating that photoconversion of phytochrome to its active form results in a localized change on the protein surface that makes a binding domain more available to the exterior of the protein. This domain has hydrophobic properties and contains a cysteine and histidine residue.

SONG et al. (1979) have proposed a working model for light activation of oat phytochrome which states that the chromophore of P_r is located on the surface of the protein and on conversion to P_{fr} becomes reoriented away from the surface, thereby exposing a hydrophobic area of the protein. This is then the active site for binding of phytochrome to some receptor in the plant cell. Their spectroscopic data (SONG et al. 1979) and in vitro comparisons of P_r and P_{fr} (HAHN et al. 1980, KIM and SONG 1981) are consistent with this model, but do not prove it. Part of their argument for the location of the hydrophobic site immediately adjacent to the chromophore is that binding of substances such as ANS (HAHN et al. 1980) interfere with the photoconversions between P_r and P_{fr}. The findings of others are not in direct agreement with this (PRATT 1982b). For example, chemical modification of one histidine and one cysteine, each made more accessible to the surface by conversion of P_r to P_{fr}, did not alter subsequent phototransformations (HUNT and PRATT 1981). TOKUTOMI et al. (1981) found P_{fr} of pea phytochrome to be more hydrophobic than P_r in undegraded protein while no difference was found between P_r and P_{fr} in the 60,000 molecular weight proteolytically derived chromopeptide. The spectral properties of the chromopeptide are apparently unchanged from native phytochrome so presumably the area in the immediate vicinity of the chromophore is undisturbed by proteolysis and according to Song's model would also exhibit differences in hydrophobicity. Similar comparisons were also made in the binding of P_{fr} to microsomal fractions from pea tissue (YAMAMOTO et al. 1980) in that the P_{fr} form of the 60,000 molecular weight chromopeptide would not bind while the undegraded form would, indicating that the binding site on phytochrome was destroyed by proteolysis. One possibility that cannot be ruled out at this time is that more than one binding domain is altered on conversion of P_r to P_{fr} but our understanding of these phenomena is still too preliminary to know and, therefore, it is still too early to know their biological significance.

4 Conclusions

At the time of publication of this volume it is approaching a quarter of a century since the first isolation of phytochrome. Even though significant progress has been made in our knowledge of the phytochrome molecule during that time, the data base from which we work today is still very limited. The inability to purify the protein in an undenatured state has been the chief obstacle in these endeavors. The recent advances in purification of phytochrome which have been reviewed in this chapter are resulting in a greater availability of purified protein now and investigations of the molecular properties of phytochrome are significantly increasing in numbers.

Even now we are being presented with the probability that many of the studies presented in this chapter on putatively undegraded phytochrome were actually using a preparation which consisted of a mixture of proteolytically altered phytochrome peptides. Many phytochrome preparations which are free of contamination by other proteins still exhibit some heterogeneity relative to isoelectric point, amino-terminus and electrophoretic migration (see PRATT 1982b for discussion). VIERSTRA and QUAIL (personal communication) have recently obtained evidence that this heterogeneity is due to limited proteolysis during extraction. This could certainly affect many of the observations reported in this chapter, but the extent of this remains to be seen.

Our knowledge of phytochrome has been further limited in that studies have concentrated on preparations from two or three species of plants grown under the very unnatural condition of complete darkness. Possibly future studies of phytochrome from other more diverse plant species and from plant material grown under natural conditions will reveal information which has been unavailable to us under our past restrictions.

References

Anderson GR, Jenner EL, Mumford FE (1970) Optical rotatory dispersion and circular dichroism spectra of phytochrome. Biochim Biophys Acta 221:69–73

Balangé AP, Rollin P (1973) Purification of photoreversible phytochrome from *Avena* seedlings by isoelectric focusing. Plant Sci Lett 1:59–64

Bolton GW, Quail PH (1981a) Cell-free synthesis of *Avena* phytochrome. Plant Physiol 67 (suppl):130

Bolton GW, Quail PH (1981b) A method for preparing green plant tissue extracts for spectrophotometric measurement of phytochrome. Plant Physiol 67 (suppl):587

Bolton GW, Quail PH (1982) Cell-free synthesis of phytochrome apoprotein. Planta 155:212–217

Borthwick HA, Hendricks SB, Parker MW, Toole EH, Toole VK (1952) A reversible photoreaction controlling seed germination. Proc Natl Acad Sci USA 38:662–666

Braslavsky SE, Matthews JI, Herbert HJ, DeKok J, Spruit CJP, Schaffner K (1980) Characterization of a microsecond intermediate in the laser flash photolysis of small phytochrome from oat. Photochem Photobiol 31:417–420

Briggs WR, Fork DC (1969a) Long-lived intermediates in phytochrome transformation I: in vitro studies. Plant Physiol 44:1081–1088

Briggs WR, Fork DC (1969b) Long-lived intermediates in phytochrome transformation II: in vitro and in vivo studies. Plant Physiol 44:1089–1094

Briggs WR, Rice HV (1972) Phytochrome: chemical and physical properties and mechanism of action. Annu Rev Plant Physiol 23:293–334

Briggs WR, Zollinger WD, Platz BB (1968) Some properties of phytochrome isolated from dark-grown oat seedlings (*Avena sativa* L.). Plant Physiol 43:1239–1243

Briggs WR, Gardner G, Hopkins DW (1972) Some technical problems in the purification of phytochrome. In: Mitrakos K, Shropshire W (eds) Phytochrome. Academic Press, London New York, pp 145–158

Butler WL (1961) Some photochemical properties of phytochrome. In: Christensen B, Buchmann B (eds) Progress in photobiology. Elsevier, Amsterdam, pp 569–571

Butler WL, Norris KH, Siegelman HW, Hendricks SB (1959) Detection, assay, and preliminary purification of the pigment controlling photoresponsive development of plants. Proc Natl Acad Sci USA 45:1703–1708

Butler WL, Lane HC, Siegelman HW (1963) Nonphotochemical transformation of phytochrome in vivo. Plant Physiol 38:514–519

Butler WL, Hendricks SB, Siegelman HW (1964a) Action spectra of phytochrome in vitro. Photochem Photobiol 3:521–528

Butler WL, Siegelman HW, Miller CO (1964b) Denaturation of phytochrome. Biochemistry 3:851–857

Capaldi RA, Vanderkooi G (1972) The low polarity of many membrane proteins. Proc Natl Acad Sci USA 69:930–932

Cecil R (1963) Role of sulfur in proteins. In: Neurath H (ed) The proteins, 2nd edn, Vol 1. Academic Press, New York, pp 379–476

Clarkson DT, Hillman WS (1968) Stable concentrations of phytochrome in *Pisum* under continuous illumination with red light. Plant Physiol 43:88–92

Cordonnier M-M, Pratt LH (1980) Preparation, characterization, and utilization of antiserum against zucchini phytochrome. In: DeGreef J (ed) Photoreceptors and plant development, Antwerpen Univ Press, Antwerpen, pp 69–78

Cordonnier M-M, Pratt LH (1982) Immunopurification and initial characterization of dicotyledonous phytochrome. Plant Physiol 69:360–365

Correll DL, Edwards JL, Klein WH, Shropshire W (1968a) Phytochrome in etiolated annual rye. III. Isolation of photoreversible phytochrome. Biochim Biophys Acta 168:36–45

Correll DL, Steers E, Towe KM, Shropshire W (1968b) Phytochrome in etiolated annual rye. IV. Physical and chemical characterization of phytochrome. Biochim Biophys Acta 168:46–57

Correll DL, Edwards JL, Shropshire W (1968c) Multiple chromophore species in phytochrome. Photochem Photobiol 8:465–475

Cross DR, Linschitz H, Kasche V, Tenenbaum J (1968) Low-temperature studies on phytochrome: light and dark reactions in the red to far-red transformation and new intermediate forms of phytochrome. Proc Natl Acad Sci USA 61:1095–1101

Cundiff SC, Pratt LH (1975) Phytochrome characterization by rabbit antiserum against high molecular weight phytochrome. Plant Physiol 55:207–211

Frankland B (1972) Biosynthesis and dark transformations of phytochrome. In: Mitrakos K, Shropshire W (eds) Phytochrome. Academic Press, London New York, pp 195–225

Fry KT, Mumford FE (1971) Isolation and partial characterization of a chromophore-peptide fragment from pepsin digests of phytochrome. Biochem Biophys Res Commun 45:1466–1473

Gardner G, Briggs WR (1974) Some properties of phototransformation of rye phytochrome in vitro. Photochem Photobiol 19:367–377

Gardner G, Pike CS, Rice HV, Briggs WR (1971) "Disaggregation" of phytochrome in vitro – a consequence of proteolysis. Plant Physiol 48:686–693

Gardner G, Thompson WF, Briggs WR (1974) Differential reactivity of the red- and far-red-absorbing forms of phytochrome to [^{14}C]N-ethyl maleimide. Planta 117:367–372

Grombein S, Rüdiger W (1976) On the molecular weight of phytochrome: a new high molecular phytochrome species in oat seedlings. Hoppe-Seyler's Z Physiol Chem 357:1015–1018

Hahn T-R, Song P-S (1981) Hydrophobic properties of phytochrome as probed by 8-anilinonaphthalene-1-sulfonate fluorescence. Biochemistry 20:2602–2609

Hahn T-R, Kang S-S, Song P-S (1980) Difference in the degree of exposure of chromophores in the P_r and P_{fr} forms of phytochrome. Biochem Biophys Res Commun 97:1317–1323

Heim B, Jabben M, Schäfer E (1981) Phytochrome destruction in dark- and light-grown *Amaranthus caudatus* seedlings. Photochem Photobiol 34:89–93

Hopkins DW (1971) Protein conformational changes of phytochrome. PhD dissertation, Univ California, San Diego

Hopkins DW, Butler WL (1970) Immunochemical and spectroscopic evidence for protein conformational changes in phytochrome transformations. Plant Physiol 45:567–570

Hunt RE, Pratt LH (1979) Phytochrome immunoaffinity purification. Plant Physiol 64:332–336

Hunt RE, Pratt LH (1980) Partial characterization of undegraded oat phytochrome. Biochemistry 19:390–394

Hunt RE, Pratt LH (1981) Physicochemical differences between the red- and the far-red-absorbing forms of phytochrome. Biochemistry 20:941–945

Kendrick RE (1974) Phytochrome intermediates in freeze-dried tissue. Nature 250:159–161

Kendrick RE, Spruit CJP (1977) Phototransformations of phytochrome. Photochem Photobiol 26:201–214

Kim I-S, Song P-S (1981) Binding of phytochrome to liposomes and protoplasts. Biochemistry 20:5482–5489

Lagarias JC, Rapaport H (1980) Chromopeptides from phytochrome. The structure and linkage of the P_r form of the phytochrome chromophore. J Am Chem Soc 102:4821–4828

Linschitz H, Kasche V (1967) Kinetics of phytochrome conversion: multiple pathways in the P_r to P_{fr} reaction, as studied by double-flash technique. Proc Natl Acad Sci USA 58:1059–1064

Linschitz H, Kasche V, Butler WL, Siegelman HW (1966) The kinetics of phytochrome conversion. J Biol Chem 241:3395–3403

Lisansky SG, Galston AW (1974) Phytochrome stability in vitro. I. Effect of metal ions. Plant Physiol 53:352–359

Litts JC (1980) Phytochrome: A purification and partial characterization. PhD dissertation, Univ Minnesota

Mumford FE, Jenner EL (1966) Purification and characterization of phytochrome from oat seedlings. Biochemistry 5:3657–3662

Negbi M, Hopkins DW, Briggs WR (1975) Acceleration of dark reversion of phytochrome in vitro by calcium and magnesium. Plant Physiol 56:157–159

Oelze-Karow H, Schäfer E, Mohr H (1976) On the physiological significance of dark reversion of phytochrome in the mustard seedling. Photochem Photobiol 23:55–59

Pike CS, Briggs WR (1972) The dark reactions of rye phytochrome in vivo and in vitro. Plant Physiol 49:514–520

Pratt LH (1973) Comparative immunochemistry of phytochrome. Plant Physiol 51:203–209

Pratt LH (1975) Kinetic analysis of a very rapidly reverting population of high-molecular-weight phytochrome. Photochem Photobiol 21:99–103

Pratt LH (1978) Molecular properties of phytochrome. Photochem Photobiol 27:81–105

Pratt LH (1979) Phytochrome: functions and properties. Photochem Photobiol Rev 4:59–124

Pratt LH (1982a) Phytochrome purification. In: Smith H (ed) Techniques in photomorphogenesis. Academic Press, London New York (in press)

Pratt LH (1982b) Phytochrome: the protein moity. Annu Rev Plant Physiol 33:557–582

Pratt LH, Butler WL (1968) Stabilization of phytochrome intermediates by low temperature. Photochem Photobiol 8:477–485

Pratt LH, Butler WL (1970) The temperature dependence of phytochrome transformations. Photochem Photobiol 11:361–369

Pratt LH, Cundiff SC (1975) Spectral characterization of high-molecular-weight phytochrome. Photochem Photobiol 21:91–97

Pratt LH, Kidd GH, Coleman RA (1974) An immunochemical characterization of the phytochrome destruction reaction. Biochim Biophys Acta 365:93–107

Quail PH, Schäfer E, Marmé D (1973a) De novo synthesis of phytochrome in pumpkin hooks. Plant Physiol 52:124–127

Quail PH, Schäfer E, Marmé D (1973b) Turnover of phytochrome in pumpkin cotyledons. Plant Physiol 52:128–131

Rice HV, Briggs WR (1973a) Partial characterization of oat and rye phytochrome. Plant Physiol 51:927–938

Rice HV, Briggs WR (1973b) Immunochemistry of phytochrome. Plant Physiol 51:939–945

Rice HV, Briggs WR, Jackson-White CJ (1973) Purification of oat and rye phytochrome. Plant Physiol 51:917–926

Roux SJ (1972) Chemical evidence for conformational differences between the red- and far-red-absorbing forms of oat phytochrome. Biochemistry 11:1930–1936

Roux SJ, Linsansky SG, Stoker BM (1975) Purification and partial carbohydrate analysis of phytochrome from *Avena sativa*. Physiol Plant 35:85–90

Roux SJ, McEntire K, Brown WE (1982) Determination of extinction coefficients of oat phytochrome by quantitative amino acid analyses. Photochem Photobiol 35:537–543

Rüdiger W (1980) Phytochrome, a light receptor of plant photomorphogenesis. Struct Bonding 40:101–140

Sarkar HK, Song P-S (1982) Blue light induced phototransformation of phytochrome in the presence of flavin. Photochem Photobiol 35:243–246

Shimazaki Y, Inoue Y, Yamamoto KT, Furuya M (1981) Phototransformation of the red-light-absorbing form of undegraded pea phytochrome by laser flash excitation. Plant Cell Physiol 21:1619–1625

Siegelman HW, Firer EM (1964) Purification of phytochrome from oat seedlings. Biochemistry 3:418–423

Smith H (1975) Phytochrome and photomorphogenesis. McGraw-Hill, London

Smith WO (1975) Purification and physicochemical studies of phytochrome. PhD Dissertation, Univ Kentucky

Smith WO (1978) The effects of neutral salts on the dark reversion of phytochrome. Plant Physiol 61 (Suppl):61

Smith WO (1981) Probing the molecular structure of phytochrome with immobilized Cibacron blue 3GA and blue dextran. Proc Natl Acad Sci USA 78:2977–2980

Smith WO, Correll DL (1975) Phytochrome: a re-examination of the quaternary structure. Plant Physiol 56:340–343

Smith WO, Daniels SM (1981) Purification of phytochrome by affinity chromatography on agarose-immobilized Cibacron blue 3GA. Plant Physiol 68:443–446

Song P-S, Chae Q, Gardner JD (1979) Spectroscopic properties and chromophore conformations of the photomorphogenic receptor: phytochrome. Biochim Biophys Acta 576:479–495

Song P-S, Kim I-S, Hahn T-R (1981) Purification of phytochrome by affi-gel blue chromatography; an effect of lumichrome on purified phytochrome. Anal Biochem 117:32–39

Stellwagen E (1977) Use of blue dextran as a probe for the nicotinamide adenine dinucleotide domain in proteins. Acc Chem Res 10:92–98

Stoker BM, Roux SJ, Brown WE (1978) Evidence for symmetry in the phytochrome subunit. Nature 271:180–182

Stone HJ, Pratt LH (1979) Characterization of the destruction of phytochrome in the red-absorbing form. Plant Physiol 63:680–682

Taylor AO, Bonner BA (1967) Isolation of phytochrome from the alga *Mesotaenium* and liverwort *Sphaerocarpus*. Plant Physiol 42:762–766

Tobin EM, Briggs WR (1973) Studies on the protein conformation of phytochrome. Photochem Photobiol 18:487–495

Tokutomi S, Yamamoto KT, Furuya M (1981) Photoreversible changes in hydrophobicity of undegraded pea phytochrome determined by partition in an aqueous two-phase system. FEBS Lett 134:159–162

Yamamoto KT, Smith WO (1981a) Alkyl and ω-amino alkyl agaroses as probes of light-induced changes in phytochrome from pea seedlings (*Pisum sativum* cv. Alaska). Biochim Biophys Acta 668:27–34

Yamamoto KT, Smith WO (1981b) Effect of neutral salts on spectral characteristics of undegraded phytochrome partially purified from etiolated pea shoots. Plant Cell Physiol 22:1149–1158

Yamamoto KT, Smith WO (1981c) A re-evaluation of the mole fraction of P_{fr} at the red-light-induced photostationary state of undegraded rye phytochrome. Plant Cell Physiol 22:1159–1164

Yamamoto KT, Smith WO, Furuya M (1980) Photoreversible Ca^{2+}-dependent aggregation of purified phytochrome from etiolated pea and rye seedlings. Photochem Photobiol 32:233–239

7 Chromophores in Photomorphogenesis

W. RÜDIGER and H. SCHEER

1 Introduction

Chromophores in photomorphogenesis are those parts of the photoreceptor molecules which absorb the light responsible for the physiological response. Absorption spectra of the chromophores should therefore principally correspond to the action spectra of photomorphoses. However, the absorption of isolated chromophores can strongly deviate from physiological action spectra due to several reasons (e.g., perturbation by the environment, dichroitic effects of ordered structures, shading by bulk pigments). Therefore, we restrict our discussion here to those chromophores on which at least some complementary information is available.

The chromophore of phytochrome has previously been treated in several books and reviews (MITRAKOS and SHROPSHIRE 1972, SMITH 1975, BRIGGS and RICE 1972, SMITH and KENDRICK 1976, KENDRICK and SPRUIT 1977, PRATT 1978, RÜDIGER 1980). A comprehensive bibliography on the literature prior to 1975 is available (CORRELL et al. 1977). Phycochrome and adaptochrome chromophores have been discussed by BOGORAD (1975) and BJÖRN and BJÖRN (1980). For a recent survey on cryptochrome (the blue light receptor) the reader is referred to the book edited by SENGER (1980).

2 Phytochrome Chromophores

2.1 P_r Structure

Because of spectral similarity of P_r and PC[1], a bile pigment structure was suggested for the phytochrome chromophore at an early stage of phytochrome research (PARKER et al. 1950). Subsequently, the biliproteins PC, APC and PE, and their chromophores phycocyanobilin and phycoerythrobilin which are readily available, have been used extensively as model compounds for phytochrome and its chromophores.

2.1.1 Degradation Studies

Chromic acid degradation of bile pigments and biliproteins under carefully controlled conditions leads to well-defined oxidation products, namely maleimids

[1] Abbreviations: PC = phycocyanin, PE = phycoerythrin, APC = allophycocyanin

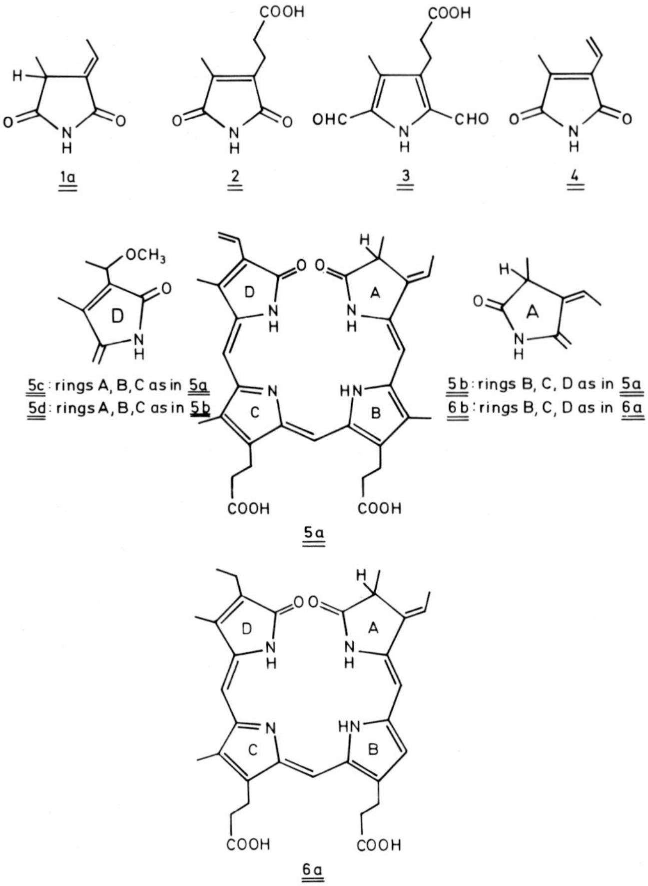

5c: rings A, B, C as in 5a
5d: rings A, B, C as in 5b
5b: rings B, C, D as in 5a
6b: rings B, C, D as in 6a

Fig. 1. Structure of phytochromobilin, related tetrapyrroles and degradation products thereof

and succinimides with typical substitution patterns. These products can be identified by thin layer chromatography and specific staining (RÜDIGER 1969, 1970). Porphyrins and chlorophylls yield the same or similar oxidation products, but bile pigments can be distinguished from these tetrapyrrols by oxidation at pH 0–1. Under these conditions, only bile pigments (and biliproteins) are degraded but no other tetrapyrrols.

Investigation of phytochrome with this method proceeded in several steps. With the first (denatured) sample, the bile pigment nature of the P_r chromophore was unequivocally confirmed (RÜDIGER and CORRELL 1969). Furthermore, the true degradation products from pyrrole rings B and C [(2) and (3); see Fig. 1] were obtained, whereas other products probably derived from rings A and D later turned out to be artifactual. The true degradation product from ring D (4) was only obtained 3 years later (RÜDIGER 1972). The key product from ring A (1a) was only obtained by modified degradation procedure (chromic acid – ammonia degradation, KLEIN et al. 1977, KLEIN and RÜDIGER 1978) which also cleaved the covalent linkage between ring A and the protein (see Sect. 2.1.5). In summary, the hypothetical structure 5a for free phytochromobilin was derived from these studies. Additional evidence for the protein binding was also derived from these studies (see Sect. 2.1.5). It should be kept in mind that degradation studies only allow the deduction of chromophore side chains. Structure (5a) differs from that of phycocyanobilin (6a) only by a formal exchange of an ethyl group for the vinyl group at ring D. The side chains of (5a) are identical with those of phycoerythrobilin, but the conjugated system is interrupted between rings C and D in the latter whereas the conjugation comprises all four rings in (5a) according to spectral studies.

2.1.2 Spectral Studies

Electronic spectra of free bile pigments consist of one broad band in the visible and possibly a second band in the near UV range. Mainly the visible band has been used extensively for classification and characterization of bile pigments (RÜDIGER 1971). Not only the position of this band, but also the shift induced

Table 1. Visible absorption maxima (nm) of some bile pigments and biliprotein chromophores related to phytochrome P_r

	Cation	Base	Zinc complex	References
Biliverdin (19a)[a]	700	653	715	RÜDIGER et al. (1968)
Mesobiliverdin[a]	685	630–655	688	KÖST et al. (1975)
Octaethylbiliverdin (26)[a]	693	657	691	SCHEER (1976)
Phytochromobilin (5a)[a]	690	–	–	SIEGELMAN et al. (1966)
	708	610	–	WELLER and GOSSAUER (1980)
Phycocyanobilin (6a)	687	603	628, 673	KÖST et al. (1975)
Phytochrome P_r (15)[b]	675–689	620–625	650 (590)	GROMBEIN et al. (1975)
Phycocyanin PC (28)[b]	665–670	610 (590)	640 (590)	GROMBEIN et al. (1975)
A-dihydrobiliverdin (20)[a]	665	594	638	SCHEER (1976)
A-dihydrobiliverdin (20)[c]	–	617+566	–	
Phycocyanobilin (6a)[c]	–	641+587	–	RÜDIGER et al. (1980)
Methanoladduct (5c)[c]	–	636+582	–	
Phytochromobilin (5a)[c]	–	653+600	–	

[a] Methanol
[b] 6 m guanidinium chloride
[c] Ethyl acetate

by derivatization (e.g., cation or zinc complex formation) is characteristic for the chromophore type. The data of Table 1 show that phytochromobilin fits into the series of fully conjugated bilins (formerly called bilatrienes).

Biliproteins cannot directly be compared with free bile pigments in this respect because, in the native state, spectral properties of bilin chromophores are drastically modified by the protein (see Sect. 2.1.6). But after unfolding of the peptide chain, biliproteins behave similarly to free bilins (KÖST et al. 1975, GROMBEIN et al. 1975). Phytochrome (P_r) and PC unfolded with guanidinium chloride are included in Table 1.

The data of Table 1 are consistent with structure (5a) for phytochromobilin. A small red shift compared with the data of phycocyanobilin (6a) can be explained by the increment of the vinyl group at ring D (see Fig. 2). This increment (vinyl versus ethyl) can also be observed in other bile pigments. Differing λ_{max} values reported for the cation of (5a) (SIEGELMAN et al. 1966, WELLER and GOSSAUER 1980) are probably due to slightly different conditions of measurement which could lead to different populations of bilin conformers in solution. This is a basic problem in bile pigment spectroscopy because it was shown that solutions of bile pigments mostly consist of mixtures of conformers with different spectral properties (BRASLAVSKY et al. 1980a, LEHNER et al. 1978a, 1979, HOLZWARTH et al. 1978, 1980, SCHEER et al. 1977, PÉTRIER et al. 1979; see also Sect. 2.1.6). These discrepancies are especially pronounced with the free bases (Table 1). Solutions of free bases contain sometimes two peaks in varying intensity or one peak with pronounced shoulders which can best be resolved by derivative spectroscopy (RÜDIGER et al. 1980).

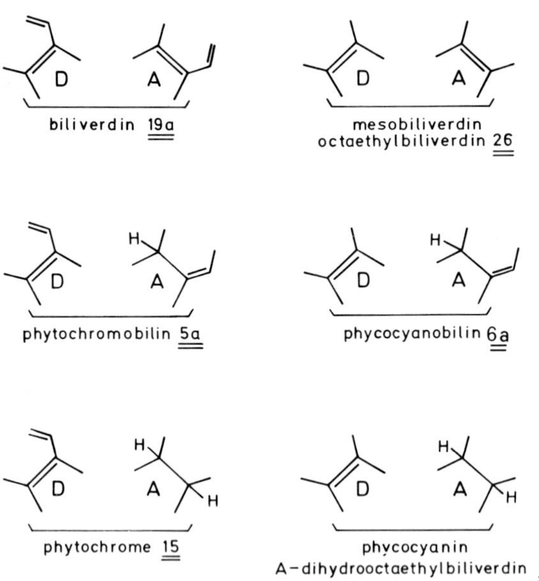

Fig. 2. Structural features of some bile pigments and biliprotein chromophores related to phytochrome P_r. Only substituents of rings A and D are given as relevant for spectral properties, all saturated substituents are indicated by a single line. Rings B/C and connection between all 4 rings are identical [see formula (5)] except for the octaethyl-derivatives bearing ethyl groups at all eight β-pyrrolic positions

A comparison of phytochromobilin and phytochrome (Table 1) reveals a spectral shift which is due to the ethylidene group at ring A in the former pigment. This group is absent in phytochrome (see Fig. 2). The same spectral shift is also observed with phycocyanobilin and PC (Table 1). Apparently, the ethylidene groups of the free bile pigments are absent as long as the pigments are covalently linked to the protein. Therefore the ethylidene side chain of ring A has been deduced as the site of linkage with the protein.

2.1.3 Cleavage from the Protein

The successful cleavage of the covalent linkage between bile pigments and proteins in plant biliproteins was a precondition for the elucidation of the structures of the free bile pigments. The first method applied to PC and PE, namely treatment with cold concentrated HCl (O'HEOCHA 1963, O'CARRA et al. 1964) was abandoned later because it can yield artifactual bile pigments (BEUHLER et al. 1976). The second method, cleavage with boiling methanol (O'CARRA and O'HEOCHA 1966) and higher alcohols (FU et al. 1979) led to isolation and structural elucidation of phycocyanobilin and phycoerythrobilin (CRESPI et al. 1967, COLE et al. 1967, RÜDIGER et al. 1967). However, the yield is low and possibly mixtures of isomeric bile pigments are obtained (FU et al. 1979). The best cleavage method so far which gives 100% yield of phycocyanobilin from PC is the cleavage with HBr in trifluoroacetic acid (KROES 1970, SCHRAM and KROES 1971). This method also cleaves phycoerythrobilin from PE (BRANDLMEIER, BLOS and RÜDIGER unpublished).

Whereas the treatment with concentrated HCl did not cleave the free chromophore from phytochrome, the method with boiling methanol was successful (SIEGELMAN et al. 1966). However, the yield was so poor that only an incomplete characterization was possible (Table 1). Also, the cleavage method with HBr in trifluoroacetic acid did not work with phytochrome (KROES 1970). This was later explained by secondary reactions of the vinyl group first with HBr and then with functional groups of the protein (BRANDLMEIER et al. 1980). The application of the HBr method to chromopeptides obtained from phytochrome yielded free phytochromobilin [(5a), see Fig. 1] and the methanol adduct (5c). Both were characterized (BRANDLMEIER et al. 1980, RÜDIGER et al. 1980; cf. Table 2) by comparison with authentic samples obtained by total synthesis (see Sect. 2.1.4).

Table 2. R_F values of bile pigments related to phytochromobilin (RÜDIGER et al. 1980). HPLC-plates (Merck, Darmstadt) coated with silica gel G, solvent A: carbon tetrachloride/ethyl acetate 1:1 (v:v), solvent B: carbon tetrachloride/acetic acid 1:1 (v:v)

	A	B
E-phytochromobilin (5a)	0.40	0.41
Z-phytochromobilin (5b)	0.45	0.48
E-phycocyanobilin (6a)	0.35	0.37
Z-phycocyanobilin (6b)	0.41	0.43
E-methanol adduct (5c)	0.27	0.35
Z-methanol adduct (5d)	0.33	0.43

2.1.4 Total Synthesis

The chemical structure of natural phytochromobilin was unequivocally confirmed by total synthesis of the racemic compound (5a) (WELLER and GOSSAUER 1980). The synthetic material furthermore allowed the investigation of the reactivity which was relevant to the cleavage reaction (RÜDIGER et al. 1980).

Important steps of the total synthesis were the connection of rings A and B, the introduction of the vinyl group at ring D and the condensation of the 2-pyrromethenone compounds (9) and (11) (rings A+B and C+D, respectively) to the final tetrapyrrole (see Fig. 3). The reaction of the monothioimide (7) (ring A) and the phosphorus ylide (8) (ring B), a general method for the synthesis of alkylidene lactams (GOSSAUER et al. 1977), had been applied before to the synthesis of phycocyanobilin (GOSSAUER and HINZE 1978) and phycoerythrobilin (GOSSAUER and WELLER 1978, GOSSAUER and KLAHR 1979). The introduction of the vinyl group starting from a primary hydroxyl function had also been applied to phycoerythrobilin (GOSSAUER and WELLER 1978). The final condensation reaction had also been applied before to a number of bile pigments. Interestingly, a photoisomerization at the ethylidene double bond of 5a was achieved (WELLER and GOSSAUER 1980). The thermodynamically more stable E-phytochromobilin (5a) was transformed into the Z-isomer (5b), which could thermally be reconverted to (5a). The analogous photoisomerization was also observed with phycocyanobilin (6a, 6b; formulas see Fig. 1).

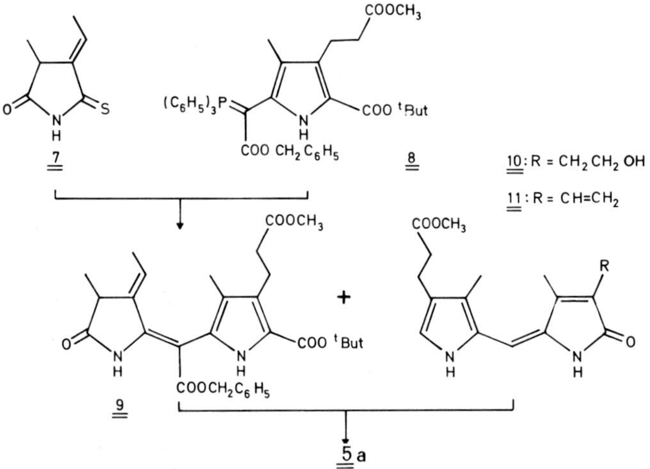

Fig. 3. Total synthesis of phytochromobilin (WELLER and GOSSAUER 1980). $tBu = C(CH_3)_3$

Treatment of E-phytochromobilin (5a) with HBr yielded a highly reactive bromo derivative which was not isolated as such. Addition of methanol led to quantitative formation of the methanol adduct (5c) (BRANDLMEIER et al. 1980). With Z-phytochromobilin (5b), the same reaction sequence yielded a mixture of (5a) and (5c) (RÜDIGER et al. 1980). Apparently, at least two reactions compete with each other, one of which finally leads to isomerization at the ethylidene group. Because some (5a) was obtained besides (5c) from the native phytochromobilinpeptide (RÜDIGER et al. 1980), a mixture of (5a) and (5b) is considered to be the primary product of the cleavage reaction.

2.1.5 Protein Linkage and Stereochemistry

Information about the covalent linkage between phytochromobilin and the peptide moiety in phytochrome came from analysis of phytochromobilinpeptides (FRY and MUMFORD 1971, LAGARIAS and RAPOPORT 1980; see Table 3). According to this analysis, the sequence of the main product (an undecapeptide) is Leu-Arg-Ala-Pro-His-Cys-Ser-His-Leu-Gln-Tyr. Minor chromopeptides were an octapeptide and presumably a hepta- and a decapeptide derived from the same region of the peptide chain. Because the blue color was extracted at that Edman degradation step which also removed cysteine, the thiol group was considered as the chromophore-binding function of the protein (LAGARIAS and RAPOPORT 1980). This situation is the same as in PC (FRANK et al. 1978, LAGARIAS et al. 1979, WILLIAMS and GLAZER 1978, BRYANT et al. 1978) and PE (KÖST-REYES et al. 1975, MUCKLE et al. 1978, KÖST-REYES and KÖST 1979).

Table 3. Amino acid sequence analysis of a phytochromobilinpeptide. (LAGARIAS and RAPOPORT 1980)

Amino acid	Original analysis	PTH derivative recovered after each step of the Edman degradation										
		1	2	3	4	5	6	7	8	9	10	11
His	2.0					+			+			
Arg	0.9		+									
Cys	0.9						+					
Ser	0.7							+				
Gln	0.9										+	
Pro	1.0				+							
Ala	1.0			+								
Leu	2.1	+								+		
Tyr	0.8											+

The site of linkage of the thiol group was elucidated by two independent approaches.

1. It was demonstrated that synthetic thioethers, in an elimination reaction, yield different alkene compounds for different positions of the sulfur substituent (Fig. 4). The C-3 thioether (12) yields the maleimide (13) whereas the C-3^1-thioether yields the ethylidene succinimide (1a) (SCHOCH et al. 1974). Because (1a) was the only product obtained from phytochrome in this reaction (KLEIN et al. 1977; see also Sect. 2.1.1) it was concluded that the sulfur linkage is localized at C-3^1 (see partial Structure 15).

2. The same conclusion was drawn from high resolution proton NMR spectroscopy of the phytochromobilin undecapeptide (LAGARIAS and RAPOPORT 1980). This investigation was based on a previous extensive investigation of a phycocyanobilinpeptide, a synthetic reference peptide lacking the chromo-

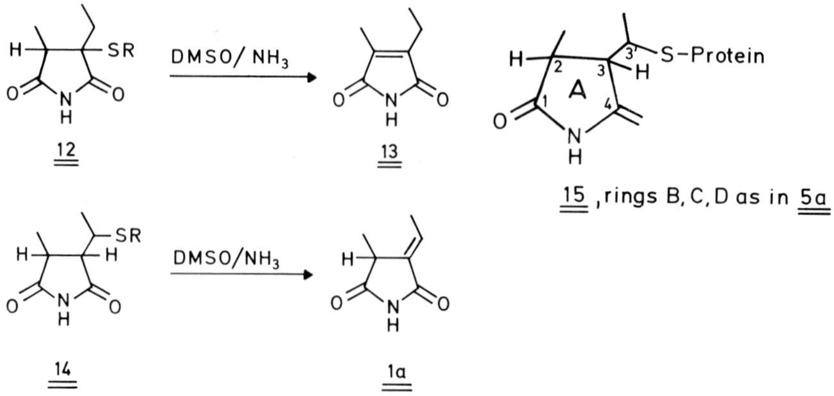

Fig. 4. Elimination reaction with synthetic thioether compounds as models for ring A of phytochromobilin. (SCHOCH et al. 1974)

phore and free phycocyanobilin (LAGARIAS et al. 1979). Double resonance experiments with the chromopeptide confirmed the hydrogenated ring A and the substitution at C-3[1]. The signals due to the chromophore in the phytochromobilinpeptide agreed well with those of the phycocyanobilinpeptide, including double resonance experiments. Therefore the structure of ring A and the thioether linkage are identical in PC and phytochrome. The only difference were the signals for the vinyl group of ring D (phytochromobilinpeptide) versus the signals for the ethyl group of ring D (phycocyanobilinpeptide).

Present knowledge on the stereochemistry of phytochromobilin and its protein linkage is only based on indirect evidence. It has been assumed that the absolute configuration at C-2 which is R in phycoerythrobilin (GOSSAUER and WELLER 1978) and probably in phycocyanobilin (BROCKMANN and KNOBLOCH 1973) is also R in phytochromobilin, but direct evidence is still lacking (KLEIN et al. 1977, LAGARIAS and RAPOPORT 1980). The assumption of the trans-configuration at ring A (i.e, 2R, 3R, or alternatively, 2S, 3S) was supported by the exclusive formation of trans-configurated products by addition of methanol or thiols to the ethylidene group of model imides and phycocyanobilin (KLEIN and RÜDIGER 1978, 1979, GOSSAUER et al. 1980). The observed coupling constant $^3J_{2H\text{-}3H}$ in the ^{1}H-NMR spectrum of both the phycocyanobilin-peptide and the phytochromobilin-peptide agrees with the trans-configuration at ring A (LAGARIAS et al. 1979, LAGARIAS and RAPOPORT 1980).

Evidence for the configuration at C-3[1] came from elimination experiments (chromic acid-ammonia-degradation) in which phytochrome behaved like the model compound (16a) and differently from model compound (16b) (KLEIN et al. 1977). The behavior of (16a) was also found with PC and PE (KLEIN and RÜDIGER 1978, MUCKLE et al. 1978) (Fig. 5). The stereochemistry of the model compounds (16a) and (16b) was unequivocally confirmed by X-ray analysis (LOTTER et al. 1977, LOTTER, KLEIN, RÜDIGER unpublished). Independent X-ray analysis was performed for the corresponding imides (17a) and (17b) obtained from phycocyanobilin methanol adducts and by total synthesis (GOSSAUER et al. 1980).

Fig. 5. Model imides for elimination (16) and addition (17) reactions at C-3[1]. The elimination was carried out with racemates, of which only one enantiomer has been drawn here. Configuration: (16a) 2R, 3R, 3^1R/2S, 3S, 3^1S; (16b) 2R, 3R, 3^1S/2S, 3S/3^1R. Phytochrome behaves like (16a)

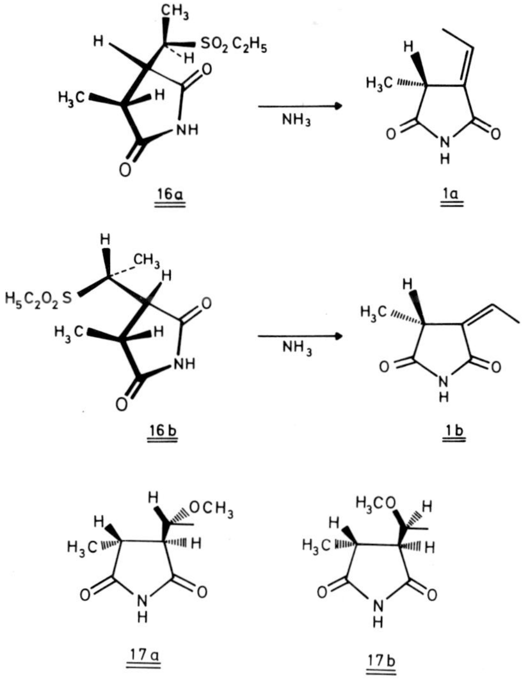

Thus the most probable configuration of the phytochrome chromophore and its protein linkage is 2R, 3R, 3^1R, but the alternative possibility 2S, 3S, 3^1S cannot yet be ruled out.

The question of a second linkage between chromophore and protein will be treated in Sect. 2.2.1.

2.1.6 The Native State

Native phytochrome (P_r) differs from denatured phytochrome, phytochromobilinpeptide and model compound (20) in its absorption spectrum (see Fig. 6) and many properties of the chromophore. These differences are not due to changes in the chemical structure of the chromophore and – if at all – only partly to protonation or deprotonation. They must rather be due to modification of the chromophore properties by non-covalent interactions with the native protein (see SCHEER 1980 for a discussion).

Because bile pigments are flexible molecules, the influence of chromophore conformation on spectral properties has been studied theoretically by several authors (BURKE et al. 1972, BLAUER and WAGNIÈRE 1975, SHAE and SONG 1975, SUGIMOTO et al. 1976, 1977, PASTERNAK and WAGNIÈRE 1979, SCHEER, FORMANEK and SCHNEIDER 1982). The main conclusion was that the oscillator strength of the long-wavelength band (f_1) is small compared to that of the short-wavelength band (f_2) in cyclic conformations. The predicted spectral properties were verified for biliverdin-type bile pigments with fixed extended [(25) BOIS-CHOUSSY

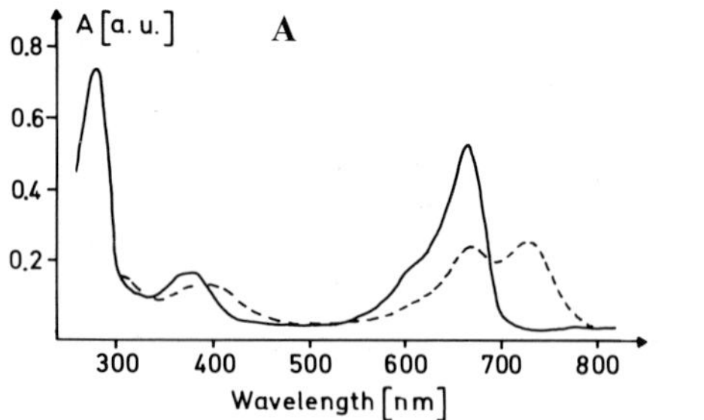

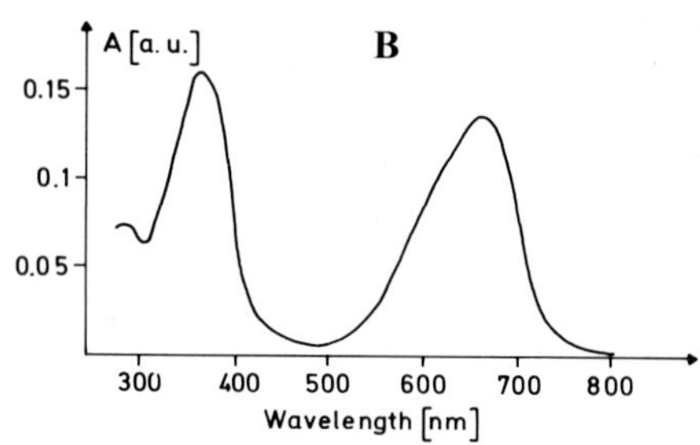

Fig. 6 A, B. Electronic spectra of phytochrome and its chromopeptide. **A** native P_r (——), native P_{fr} (------) both in 0.1 M sodium phosphate buffer, pH 7.8. **B** phytochromobilin-peptide in 10% acetic acid. (RÜDIGER, BRANDLMEIER, THÜMMLER unpublished data)

and BARBIER 1978)], and cyclic-helical (24) (formulas in Fig. 7) topologies (FALK and THIRRING 1981), and cyclic conformations have been determined for the conformationally unrestricted pigments both in solution and in the crystal (LEHNER et al. 1978a, b, FALK et al. 1978, FALK and HÖLLBACHER 1978, SHELDRICK 1976). Based on this criterion, the P_r chromophore should have a more extended conformation in its native state but a cyclic conformation in the chromopeptide and in unfolded P_r (BURKE et al. 1972, BRANDLMEIER et al. 1981 a).

Unfortunately, unfolding of P_r is irreversible. The process is fully reversible, however, with PC and PE and was investigated in more detail with these biliproteins by absorption, fluorescence, circular dichroism and chemical methods (SCHEER and KUFER 1977, LANGER et al. 1980, ZICKENDRAHT-WENDELSTADT et al. 1980). Especially PC exhibits differences between the native and the denatured state which are very similar to those in $P_r \cdot A_2/A_1$, which is roughly proportional to f_2/f_1, increases from 0.24 to 2.32 in PC and from 0.35 to 2.27 in P_r. This indicates rather similar non-covalent interactions of the two similar chromophores (15) and (28) with the two different peptide chains of P_r and PC, respectively. The precise conformation is still unsettled. Theoretical calculations (see above) predict semi-extended to fully extended conformations. The tentative structure (15) has been proposed for the P_r chromophore in its native state because A_2/A_1 is similar to that in the fully extended isophorcabilin (25, $A_2/A_1 = 0.25$).

Ramachandran-type calculations revealed the enantioselective preference of a twisted topology for cyclic and extended conformations of A-dihydrobilindiones (SCHEER et al. 1979). These calculations agree with strong CD bands in native *and* denatured P_r (Table 5, Sect. 2.3). Interestingly, the signs of both chromophoric bands of native P_r (positive at 660 nm, negative at 365 and 377 nm) are reverted by denaturation or proteolytic digestion (negative at 665–670 nm, positive at 373–375 nm). This is in contrast to PC, in which the chromophore CD bands do not change their signs upon denaturation (SCHEER et al. unpublished). The CD spectrum of denatured P_r is, however, no mirror image of that of native P_r.

Fig. 7. Proposal for the structure of native P_r chromophore, and structures for model compounds

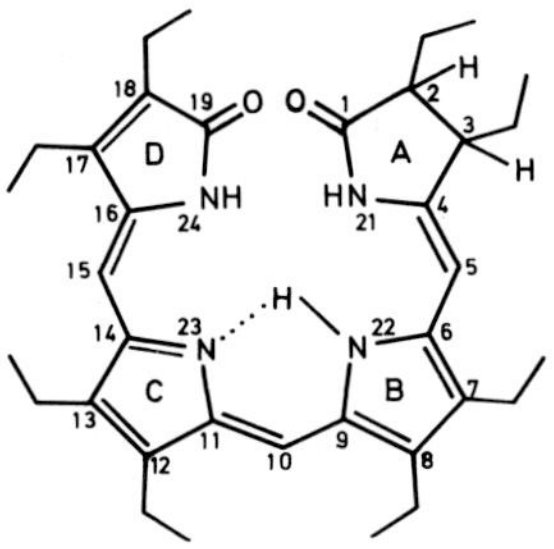

Thus the chromophores in native and denatured P_r do not represent merely an enrichment of different enantiomers, which is also likely from the different absorption spectra. Whereas a cyclic-helical conformation for the uncoupled chromophore is likely, a direct correlation of distinct conformations of the native chromophore species with its CD spectrum is not possible, but two factors relevant for the actual chromophore conformation and dissymmetry can be inferred from these experiments: The essential factors which govern the preferential chromophore helicity in P_r-peptide and denatured P_r are asymmetric centers of the chromophore itself. This influence is counteracted and overcome by the influence of the protein in native P_r.

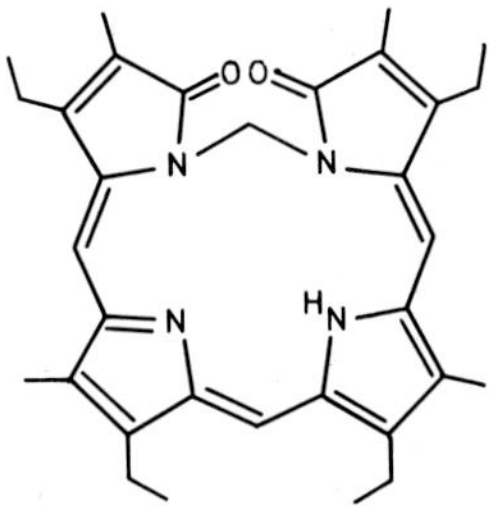

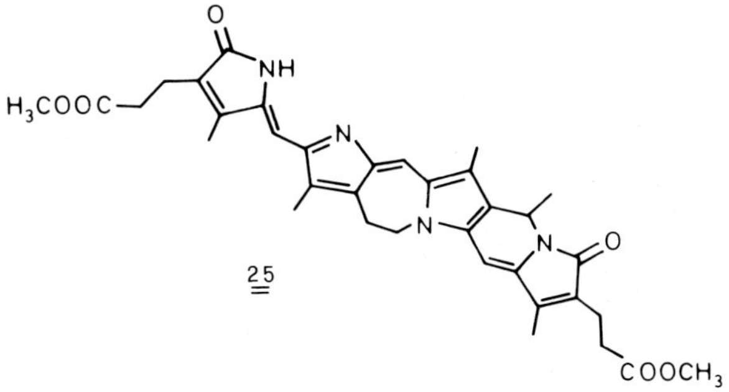

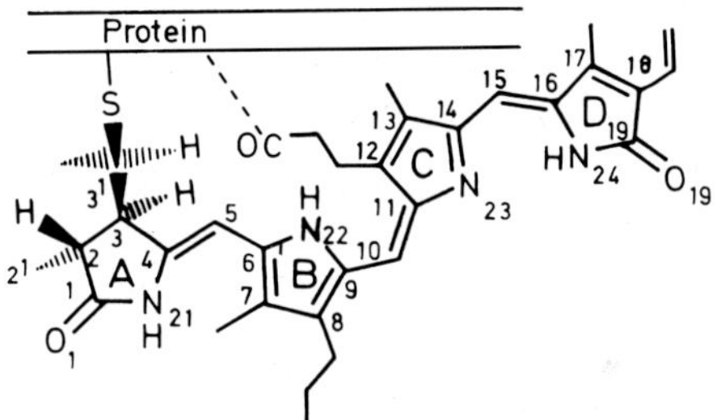

2.2 P_{fr} Structure

2.2.1 Degradation Studies

Chromic acid degradation yields essentially the same products as obtained by degradation of P_r, irrespective of the procedure applied. Methylvinylmaleimide (4) and hematinic acid imide (2) are obtained under non-hydrolytic conditions, and additional (2) as well as methylethylidene-succinimide (1a) are liberated by subsequent hydrolysis (see Fig. 1). (1a) is also obtained by the chromic acid-ammonia method (KLEIN et al. 1977). Since the chromophore of denatured P_{fr} is stable under acidic conditions (GROMBEIN et al. 1975), a rearrangement to P_r is unlikely during degradation. It is thus concluded, that not only the β-pyrrolic substituents are the same in P_r and P_{fr}, but that also the 3^1-thioether linkage and possibly an additional ester linkage of the P_{fr} chromophore to the protein are present in both forms. Whereas any reaction of the lactam carbonyl groups (CRESPI et al. 1968, LAGARIAS and RAPOPORT 1980), Z,E-isomerizations (FALK et al. 1978), and any reaction at the α-pyrrolic and methine positions may remain unnoticed by the chromic acid degradation, the release of the SH-group with formation of an endocyclic $\Delta 2$-double bond as present in biliverdin (SIEGELMAN et al. 1968, RÜDIGER and CORRELL 1969, SONG et al. 1979, LAGARIAS and RAPOPORT 1980) or of an additional double bond at the C-3 substituent (CRESPI et al. 1968) can be excluded.

As in P_r the presence of a bond to the protein via one of the propionic acid side chains is still unsettled. Such a bond has been implicated by the release of additional (2) after hydrolysis of the chromic acid degradation products (KLEIN et al. 1977) but later been questioned as a safe criterion for such a bond in biliproteins (TROXLER et al. 1978). However, quantitative studies with radiolabelled PE from *Porphyridium cruentum* indicate a second bond in this pigment (KÖST and TROXLER unpublished). A bond of this type is absent in two chromopeptides isolated from P_r (LAGARIAS and RAPOPORT 1980), but these peptides contain a serine residue next to the binding cystein, and ester bonds are susceptible to cleavage during proteolytic digestion. A chromopeptide containing a serine-propionate bond has been isolated from PE from *Pseudanabaena* sp. W1173, but again an artifact could not be excluded (MUCKLE et al. 1978). A definite proof may require milder degradation methods. Two such methods have been developed with PC and may be useful for phytochrome as well. The first method splits the tetrapyrrole skeleton selectively between rings A and B (SCHEER et al. 1977), the second between rings B and C. In a first application of these new degradation methods, a second protein bond at ring B in PC form *Spirulina platensis* has been suggested (KUFER et al. 1982a).

2.2.2 Spectral Studies

The spectrum of native phytochrome is shifted by approximately 70 nm (1,450 cm^{-1}) to the red upon conversion of P_r to P_{fr}. This has been taken as an indication of an increased length of the conjugated double bond system in P_{fr}, and consequently several proposals for the structure of the P_{fr} chromophore are based on this interpretation (CRESPI et al. 1968, STRUCKMEIER et al. 1976, SIEGELMAN et al. 1968, RÜDIGER and CORRELL 1969, SONG et al. 1979). As has been pointed out in Section 2.1.6., however, the spectra of native biliproteins are strongly influenced by non-covalent chromophore protein interactions, which render structural assignments on the basis of the spectra of the native chromophores ambiogous (see SCHEER 1980, for references). If these interactions are uncoupled by denaturation (GROMBEIN et al. 1975) or proteolysis (BRANDLMEIER et al. 1980, 1981a) at low pH, the 730 nm absorption of native P_{fr} is shifted to 620 nm. Denatured P_r absorbs under the same conditions at 660 nm (cation form). The P_{fr} chromophore uncoupled from the protein is stable only in its protonated form, and reverts to the P_r chromophore above pH 5 (GROMBEIN et al. 1975). From analogy with a series of free bile pigments (KÖST et al. 1975), the free base P_{fr} chromophore can be estimated to absorb around 570 nm, corresponding to a "purpurin" (SCHEER et al. 1977) or "violin" chromophore (SCHEER and KRAUSS 1977, KRAUS and SCHEER 1979). Two conclusions can be drawn from these results: (1) The chromophores of P_r and P_{fr} are chemically different, and do not only differ by their states of protonation, or conformation (STRUCKMEIER et al. 1976; see also the theoretical studies of BURKE et al. 1972, CHAE and SONG 1975, SUGIMOTO et al. 1977, PASTERNAK and WAGNIÈRE 1979). (2) The conjugation system of the P_{fr} chromophore is one double-bond *shorter* than that of the P_r chromophore, in contrast to conclusions derived from studies of the native pigments. A chromophore structure like (18a) (Fig. 8) in which the Δ4-double-bond is (at least spectroscopically, KRAUSS et al. 1979) abolished would best agree with these data (GROMBEIN et al. 1975).

2.2.3 Chemical Model Studies

The data from chromic acid degradation and denaturation are yet insufficient to establish a complete structure for the P_{fr} chromophore. Chemical studies starting from P_r model compounds as well as MO calculations have, therefore, been carried out to give additional information on the reactivity of P_r and spectroscopic properties of chemically reasonable structures derived thereof. Possible structures for the P_{fr} chromophore obtained on this basis can then be further scrutinized by the criteria summarized below, which are derived from the known differences of P_r and P_{fr}.

1. The cation of any model compound must have an absorption around 620 nm according to the denaturation studies described in Sect. 2.2.2.
2. Denatured (GROMBEIN et al. 1975) and pepsin-digested P_{fr} (THÜMMLER et al. 1981) are convertible back to P_r both thermally and photochemically, hence any model bilin for the P_{fr} chromophore should be thermodynamically less stable than and convertible back to its original form corresponding to P_r.
3. The chromophore of native P_{fr} (see below) is probably present in its deprotonated form (PASTERNAK and WAGNIÈRE 1979, RÜDIGER 1980). In this case, the pK_B value of any model should be within the physiological pH range.
4. Both P_r and P_{fr} carry probably one single chromophore in small as well as in large phytochrome, and the molecular weight remains within the same range upon photoconversion (see PRATT 1978, RÜDIGER 1980). Any dimeriza-

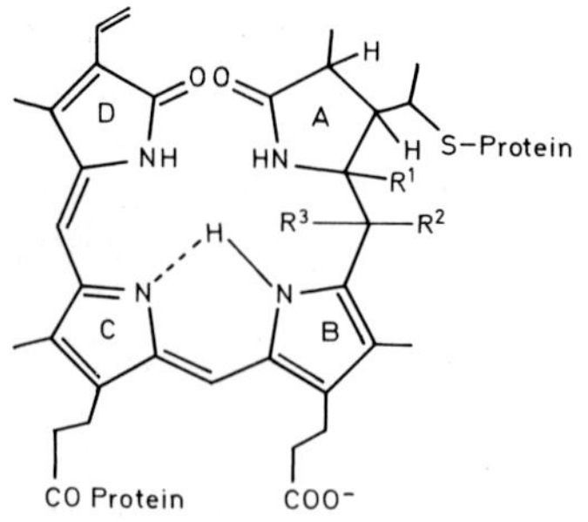

18a

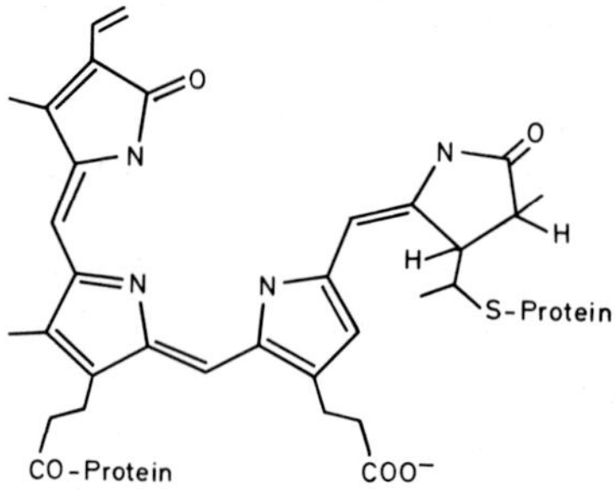

18b

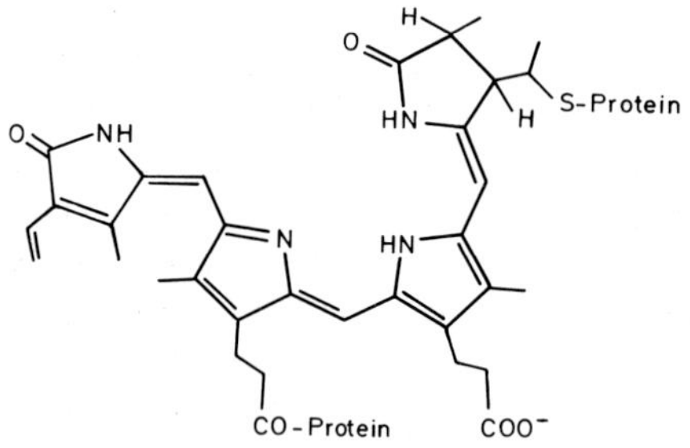

18c

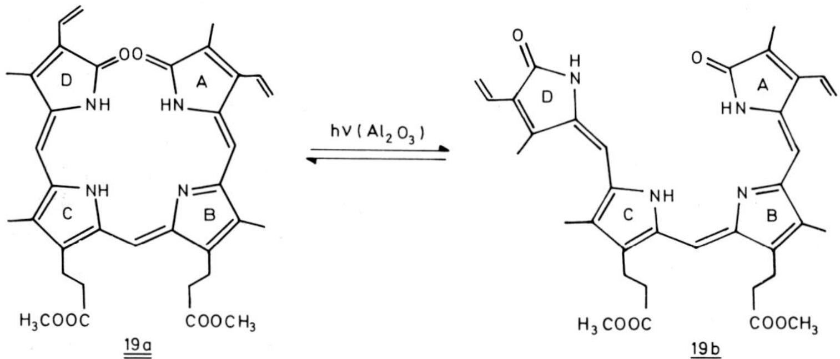

19a

hν (Al_2O_3)

19b

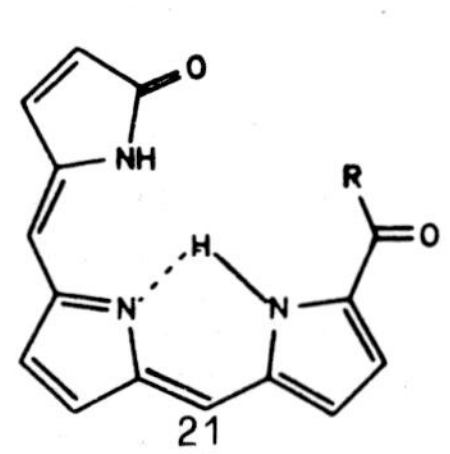

21

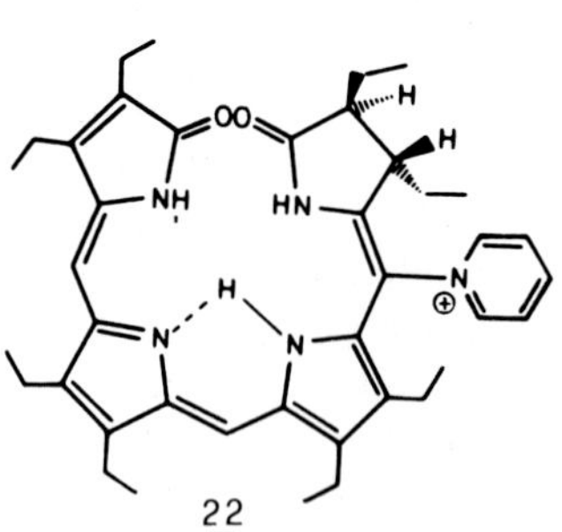

22

Fig. 8. Possible structures of the P_{fr} chromophore, and structures for model compounds

tion reactions of the chromophore (SCHEER and KRAUSS 1977) can, therefore, be excluded.

5. The reaction does not require any cofactors besides the protein since the phytochrome phototransformation occurs in solution of the purified pigment. Further possible criteria (e.g., stability of the P_{fr} chromophore towards reduction, oxidation, acids and bases) are discussed below.

Currently, there are two different models which meet most of the criteria summarized above.

The first model reaction is the sequence of oxidation, nucleophilic addition and tautomerization shown in Fig. 9. It is based on reactivity studies of the A-dihydro-bilindione (20), which has been taken as a model for the P_r-chromophore (SCHEER 1976). In neutral and alkaline media, (20) undergoes an easy and regioselective photooxidation at the C-5 methine bridge to a variety of products. In the presence of oxygen, "purpurins" are formed which share the oxotripyrrinone chromophore (21) (see Fig. 8) as a common substructure (SCHEER et al. 1977). In the absence of oxygen, "violins" are produced, which may all arise from one or two one-electron oxidation steps (KRAUSS and SCHEER 1979, EIVAZI et al. 1977) and a subsequent addition at C-5 (SCHEER and KRAUSS 1977, KRAUSS et al. 1979). Of particular interest is the pyridinium adduct (22) (KRAUSS et al. 1979). The spectral shift to 570 nm of the free base [criterion (1)] is – as in the E,Z,Z-biliverdin (19b) discussed below – not brought about by abolishing the Δ4-double bond, but rather by uncoupling it due to steric hindrance. The formation of (22) is thermo- but not photo-reversible [critn. (2)] and its pK for deprotonation [critn. (3)] is in the physiological region (KRAUSS et al. 1979). Criterium (4) is met as well. A reaction of this type would require very specific functions of the apoprotein as both the oxidant and the nucleophile (Fig. 9), since phytochrome is not known to contain any suitable cofactors (HUNT and PRATT 1980, QUAIL et al. 1978, ROUX et al. 1975). Cystine residues could serve as the oxidant. An involvement of cystine may be indicated by the finding of one more easily accessible SH group in P_{fr} than in P_r (HUNT and PRATT 1981). However, modification of either cysteine (HUNT and PRATT 1981) or cystine (HAHN and SONG 1981) with water-soluble reagents does not affect the phototransformation reactions. Tryptophan, tyrosin, serine, cysteine and others may serve as the nucleophile. (20) reacts readily with tryptophane derivatives to give products UV-vis spectroscopically similar to (22) (KRAUS and SCHEER 1981). In Fig. 9, tryptophane has been formulated arbitrarily as the nucleophile.

The second model is a geometric isomerization of a double bond between rings A and B or C and D. It has first been invoked by KROES (1970) and MUMFORD and JENNER (1971), but only recently gained further support. Based on earlier studies on the Z-E isomerization of dipyrroles (GOSSAUER and INHOFFEN 1970), FALK and coworkers conducted a systematic study of geometric isomerization reactions of bile pigments and partial structures derived therefrom (see FALK et al. 1978, 1980).

Biliverdin is most stable in solution in the all-syn, Z geometry (19a) (LEHNER et al. 1978, FALK and HÖLLBACHER 1978). The geometric isomers with anti-E, syn-Z, syn-Z geometry like (19b) are accessible by photoisomerization of all

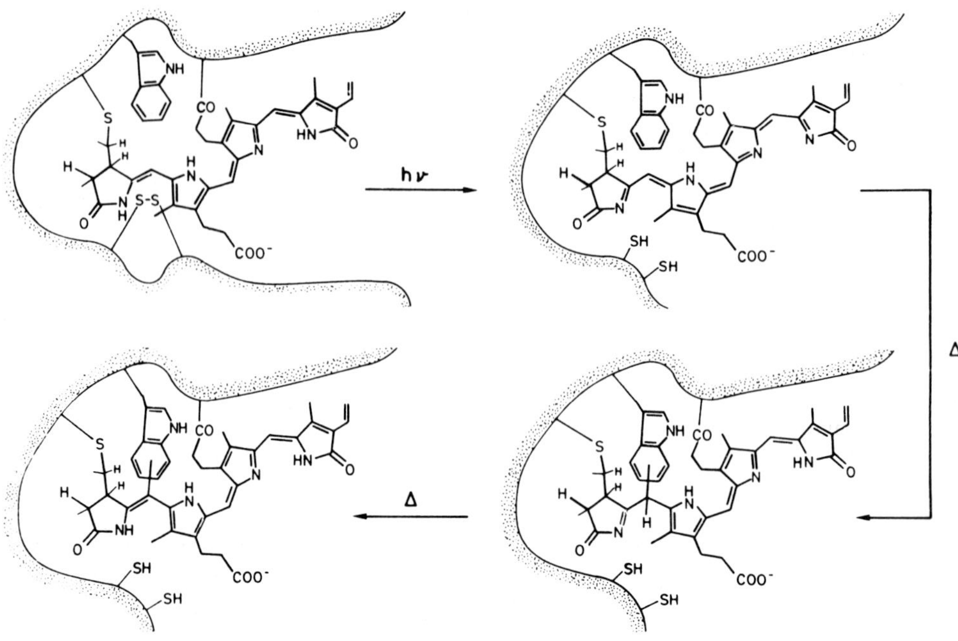

Fig. 9. Tentative model for the $P_r \rightleftharpoons P_{fr}$ interconversion as derived from photooxidations of the A-dihydrobilindion (20) as a model for the P_r-chromophore. (KRAUSS and SCHEER 1981)

syn-Z) isomers (FALK et al. 1978) or by direct synthesis (GOSSAUER et al. 1981). The isomer 19b absorbs at shorter wavelengths than (19a) which is not due to the Z,E-isomerization per se but rather to a twist of the Δ15-double bond which partially uncouples ring D from the remaining π-system (FALK and HÖLLBACHER 1978). The isomer (19b) is thermodynamically less stable ($\Delta H^+ =$ 105 kJ mol^{-1}) and reverts to (19a) both thermally ($\Delta H^+ = 20$ kJ mol^{-1}) and photochemically (FALK and GRUBMAYR 1979). In biliverdins, the Δ4 and Δ15 bonds between rings A and B, and rings C and D, respectively, are very similar (SHELDRICK 1976, LEHNER et al. 1978b), but this is no longer true in the A-dihydropigments, of which phytochrome is a member. Ramachandran-type calculations indicate a preferential isomerization at the Δ4 bond adjacent to the reduced ring which carries the bulky thioether substituent (SCHEER et al. 1979). Model compounds without this bulky substituent [e.g., (20)] yield predominantly the more stable 15E isomer (BLACHA-PULLER 1979, KUFER et al. 1982b). Because the UV-vis spectral properties of this 4Z, 10Z, 15E isomer are not much different from expected properties of the 4E, 10Z, 15Z compound it cannot yet be predicted whether the P_{fr}-peptide (and photoisomerized bilipeptides obtained from PC and P_r, THÜMMLER and RÜDIGER unpublished) contains a 4E, 10Z, 15Z (18b) or a 4Z, 10Z, 15E chromophore (18c). Recent results of high-resolution NMR spectroscopy demonstrated that the P_{fr} chromophore is the 4Z, 10Z, 15E isomer 18c (RÜDIGER, THÜMMLER, CMIEL, SCHNEIDER unpublished).

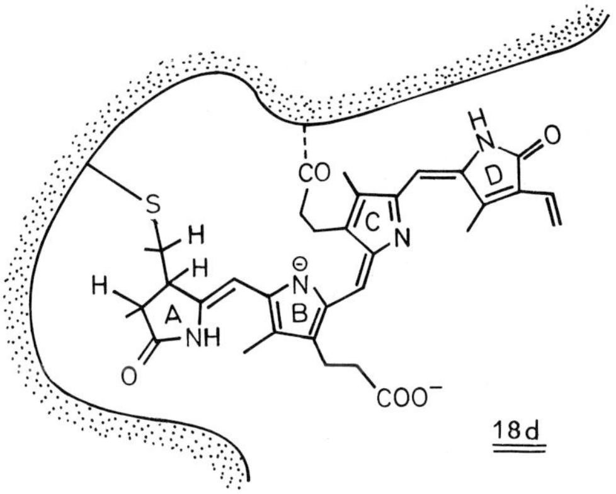

Fig. 10. Tentative structure for native P_{fr} chromophore on the basis of the Z,E-isomerization/deprotonation model

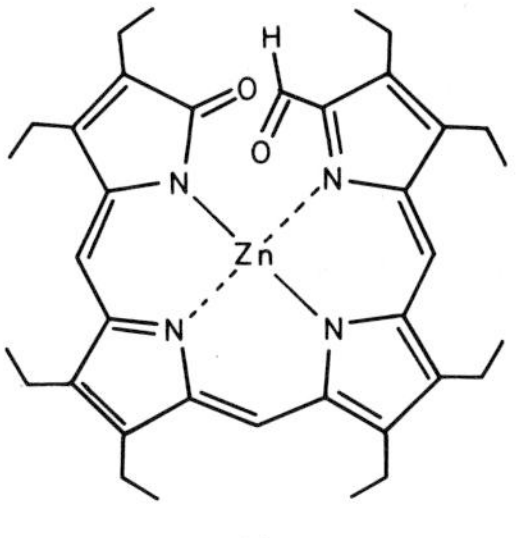

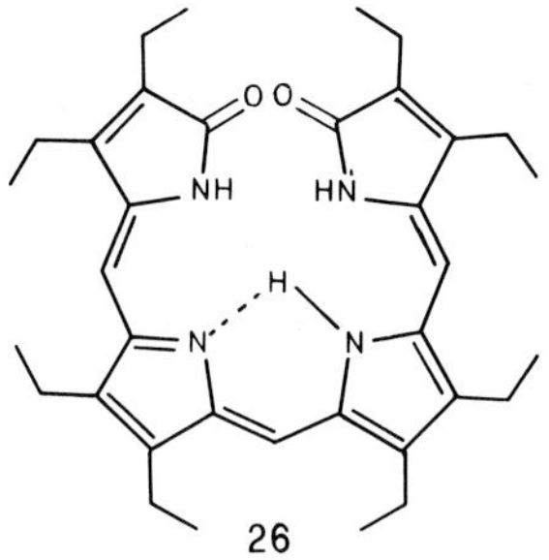

The Z,E-photoisomerization of (19) proceeds via a rubinoid pigment, which is the substrate proper for the photoisomerization (FALK et al. 1980, GOSSAUER and BLACHA-PULLER 1981). However, derivatives in which the tetrapyrrolic skeleton is strained or distorted from all-syn, Z geometry, can undergo a direct photoisomerization in solution as well (FALK and THIRRING 1979, 1980). In phytochrome, such distortions may arise both from the A-dihydrostructure and the influence of the native protein. In the P_r-peptide and in PC peptides in which a closed chromophore conformation predominates, photoisomerization so far was possible via a rubinoid intermediate. The photoisomerization product obtained in this way from the P_r-peptide is spectrally and chromatographically identically with the P_{fr}-peptide (THÜMMLER and RÜDIGER unpublished). Photoisomerization of the chromophore is, therefore, the currently most likely process during P_r–P_{fr} interconversion. It would meet the criteria (1) (2), (4), (5). The catalytic effect of redox-reagents on the P_{fr}–P_r-conversions would parallel the redox-reagent catalyzed isomerization of stilbenes (MUMFORD and JENNER 1971). A peculiar property of denatured P_{fr} and its peptides is their instability above pH 5. They are stable for hours in dilute acids, whereas the E,Z,Z-bilindiones are most stable as free bases around neutral pH. The destabilization of P_{fr} may be due to a catalytic effect of the two histidines situated next to the binding cysteine in the peptide chain (MUMFORD and JENNER 1971, LAGARIAS and RAPOPORT 1980).

2.2.4 The Native State of the Chromophore

The chromophore-protein interactions are even more pronounced in P_{fr} than in P_r (Sect. 2.1.6). The tentative structure (18d) (Scheme 9) is based on a geometric isomerization of the chromophore at the Δ4-bond discussed in the previous section, and the still rather indirect evidence presented below.

The long-wavelength absorption of native P_{fr} has its maximum at about 730 nm. Denatured P_{fr} is unstable at neutral pH, but from the absorption of the cation ($\lambda_{max} = 615$ nm), that of the free base can be estimated to $\lambda_{max} = 570$ nm (Sect. 2.2.2). This would correspond to a spectral shift of 160 nm (3,845 cm^{-1})

Table 4. Absorption maxima (λ_{max}) of and wavelength shifts ($\Delta\tilde{\nu}$) of different geometries and protonation states of bilindiones. The cations and anions are produced from the free bases dissolved in methanol by the addition of HCl and sodium methoxide, respectively

Pigment	λ_{max} (nm) (Form A)	λ_{max} (nm) (Form B)	$\Delta\tilde{\nu}$ (cm^{-1})	References
P_r	660 (native)	615 (denatured,	1,109	GROMBEIN et al. (1975)
P_{fr}	730	570 (free base)	3,845	GROMBEIN et al. (1975)
APC-I	655 (native)	600 (denatured)	1,376	CANAANI and GANTT (1980), GYSI and ZUBER (1976), ZILINSKAS et al. (1978)
APC-B	670 (native)	600 (denatured)	1,741	GLAZER and BRYANT (1975)
PC	620 (native)	600 (denatured)	537	SCHEER (1976)
PE	560 (native)	525 (denatured)	1,190	KÖST et al. (1975)
Formylbilindione zinc complex (23)[a]	830 (monomer)	750 (dimer)	1,285	STRUCKMEIER et al. (1976)
Bilindione	710 [Cyclic free base of (24)]	605 [extended, free base of (25)]	2,444	FALK and THIRRING (1981), BOIS-CHOUSSY and BARBIER (1978), BRANDLMEIER et al. (1981a)
Bilindione	740 [Cyclic, cation of (24)]	605 [extended, free base of (25)]	3,015	FALK and THIRRING (1980), BOIS-CHOUSSY and BARBIER (1978), BRANDLMEIER et al. (1981a)
Bilindione (26)[a]	657 (free base)	770 (anion)	2,234	SCHEER (1976)
A-Dihydro-bilindione (20)	594 (free base)	766 (anion)	3,780	SCHEER (1976)

[a] Formulas in Fig. 9

between the native and the denatured form. From the data in Table 4 it can be seen that this shift is much larger than the shifts induced in free bile pigments by conformational changes or protonation. One known process which leads to shifts of the same magnitude is the combination of a severe conformational change with a protonation of the chromophore [cation of the cyclic-helical (24) vs. free base of the extended (25)]. This would require the chromophore of denatured P_{fr} in an extended conformation, in contrast to all known free bilindions of the violin and verdin spectral type. It would also require the chromophore of native P_{fr} in a cyclic conformation, which is unlikely from a comparison of native P_r and P_{fr}. Linear dichroic data indicate at most moderate geometrical differences between the two forms (SONG et al. 1979). The CD bands of P_r and P_{fr} differ in sign and magnitude (BRANDLMEIER et al. 1981b, and references cited therein), however, and a direct comparison of P_r and P_{fr} may be ambiguous as long as the P_{fr} structure is unknown. The other process known to produce extreme shifts is the deprotonation of bilindiones, and especially of the 2,3-dihydrobilindiones typical for biliproteins (SCHEER 1976). The long-wavelength band of (20), a model for the P_r chromophore, is shifted by 3,780 cm^{-1} to the red upon deprotonation. It has been suggested on this basis that the chromophore of native P_{fr} is present in its deprotonated form (GROMBEIN et al. 1975), which has been supported by molecular orbital calculations (PASTERNAK and WAGNIÈRE 1979). This model would require a pK_B of the P_{fr} chromophore within the physiological range leading to criterion (3) which has been used to discriminate between P_{fr} models (Sect. 2.2.3). Protonation–deprotonation reactions have been suggested as primary processes in the low temperature photochemistry of biliproteins (FRIEDRICH et al. 1981a, b). Two recent observations may also be related to a deprotonation of the P_{fr} chromophore. The first is a pH-dependent proton uptake or release upon the irradiation of phytochrome in solution (TOKUTOMI et al. 1982). The second is the exposure of a hydrophobic protein surface in P_{fr} (TOKUTOMI et al. 1981, HAHN et al. 1980), which could be induced by the increased hydrophily of the chromophore.

The intensity changes of the long-wavelength absorption upon denaturation are less pronounced than in the case of P_r (GROMBEIN et al. 1975). However, molecular orbital calculations indicate less dramatic conformation dependencies in the spectra of violins (PASTERNAK and WAGNIÈRE 1979). Both the red and the blue DC bands of P_{fr} change sign upon denaturation, and the signs in both native and denatured P_{fr} are opposite to the CD bands of P_{fr} in the respective states (BRANDLMEIER et al. 1981b; see Table 5, Sect. 2.3). This is an independent proof of the different structures of the P_r and P_{fr} chromophores in the denatured states. As in P_r, it is again difficult to assign a precise conformation to P_{fr} in the native state, whereas the overriding influence of the native protein on the chromophore is again apparent.

2.3 Phytochrome Intermediates and Modifications of the Chromophore

The phototransformations $P_r \rightarrow P_{fr}$ and $P_{fr} \rightarrow P_r$ are multistep reactions. Several intermediates were detected by either one of the following methods: rapid kinet-

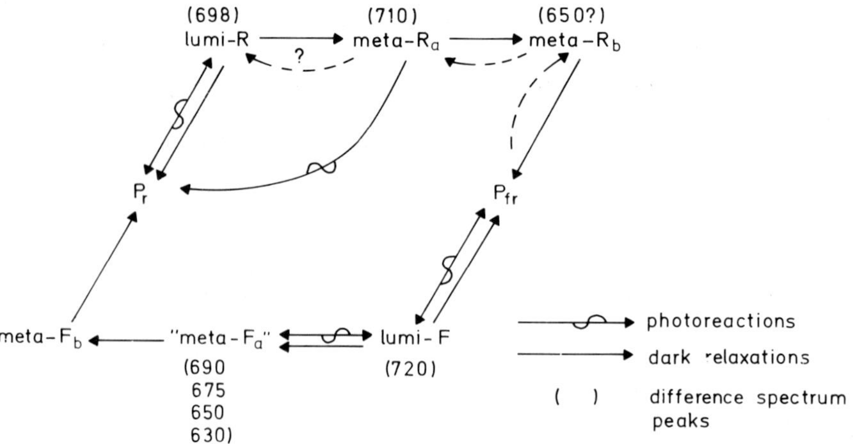

Fig. 11. Intermediates in phytochrome photoconversions including dark relaxations (*solid lines* according to KENDRICK and SPRUIT 1977; *broken lines* additional dark reactions according to Rüdiger 1980)

ics of absorption changes after flasch photolysis (LINSCHITZ et al. 1966, LINSCHITZ and KASCHE 1967, PRATT and BUTLER 1970, BRASLAVSKY et al. 1980b), low temperature spectral studies in vivo (KENDRICK and SPRUIT 1973a, b, SPRUIT and KENDRICK 1973, 1977) and in vitro (BURKE et al. 1972, CROSS et al. 1968, KROES 1970, PRATT and BUTLER 1970), dehydration in vivo (KENDRICK 1974, KENDRICK and SPRUIT 1974, SPRUIT et al. 1975) and in vitro (BALANGÉ 1974, TOBIN et al. 1973), absorption changes after continuous or during quasi-continuous irradiation ("pigment cycling", BRIGGS and FORK 1969a, b, KENDRICK and SPRUIT 1972, 1973a). Each type of study reveals the same sets of intermediates, which seem different for the forward ($P_r \rightarrow P_{fr}$) and the back reaction ($P_{fr} \rightarrow P_r$), respectively. The subject has been reviewed by KENDRICK and SPRUIT (1977), who also suggested a nomenclature similar to the one used for the rhodopsin transformations. An alternative nomenclature based on difference maxima is included in Fig. 11.

KENDRICK and SPRUIT (1977) distinguish photoreactions and dark relaxations; the latter are further divided into those which occur in non-aqueous medium and those which require liquid water (Fig. 11).

According to KENDRICK and SPRUIT, the photoreactions (formation of lumi-R and lumi-F) and the relaxations to meta-R_a and meta-F_a are chemical events essentially restricted to the chromophore, with only minor changes of the apoprotein. These events are rapid; they occur also at low temperature and in non-aqueous medium (e.g., in dehydrated tissue in vivo, in glycerol solution in vitro). The subsequent reactions via meta-R_b and meta-F_b are believed to involve conformational interaction of the apoprotein and chromophore, since they occur only in a less rigid matrix. They are slowest in the whole reaction sequence, they require liquid water and are strongly influenced by the molecular environment.

If this view is correct, the genuine chemical differences between the P_r and the P_{fr} chromophore should already exist between lumi R or meta R_a and P_r on the one hand and between lumi F or meta F_a and P_{fr} on the other; the subsequent reaction steps (to meta R_b and P_{fr}, to meta F_b and P_r, respectively) should only serve to stabilize these differences, e.g., by conformational rearrangements. The intermediates are certainly not as well stabilized as the final products; this follows from easy back reactions (either photochemically, or chemically in the dark), of lumi R, meta R_a and possibly meta R_b to P_r, and lumi F to P_{fr} (Fig. 11). The "inverse dark reversion" of phytochrome in dehydrated tissues has been related to such a back reaction of an intermediate (KENDRICK and SPRUIT 1974) (see Chap. 17, this Vol.). The molecular basis of the presumed stabilization reaction is not clear, however. This question can be answered only after a detailed structural investigation of the intermediates, including their geometries.

A sensitive tool for conformational changes of bilin chromophores are CD spectra. The CD spectra of native and denatured forms of P_r and P_{fr} are different from each other (Table 5). Whereas the CD spectra cannot be correlated directly with defined conformations of the different forms because these are chemically different species, it is obvious that some conformational changes of the chromophore occur during $P_r \rightarrow P_{fr}$ transformation in addition to (or as a consequence of) chemical reactions. This view agrees with the observation of BURKE et al. (1972) that meta R_b (then called P_{bl}) has a low absorption but – in relation to the absorption – a strong Cotton effect of the long-wavelength band. This has been interpreted as a cyclic, possibly helical conformation for the meta R_b chromophore whereas P_r and P_{fr} should contain more extended chromophores (BURKE et al. 1972).

Table 5. Circular dichroism data of phytochrome forms. (BRANDLMEIER et al. 1981b)

Samples	λ_{max} (nm)	$\Delta\varepsilon_{max}$ (M^{-1} cm^{-1})	Θ_{max} $\times 10^{-3}$
P_r native	660	−33.5	−110.5
	377	+35.9	+118.5
	365	+35.9	+118.5
P_r denatured[a]	670	+19.4	+64.6
	375	−14.6	−48.2
P_r-peptide[a]	665	+26.4	+87.1
	373	−15.8	−52.1
P_{fr} native (corrected)[b]	705	+8.1	
	375	−3.0	
P_{fr} denatured[a]	625	+4.75	+15.7
	380	−7.2	−23.8

[a] In 8 M urea, pH 2.5
[b] Corrected for the presence of 20% P_r

Whereas P_r is stable in the dark, P_{fr} can – at least in part – undergo a "dark reversion" to P_r. It is not yet known whether the dark reversion proceeds via similar intermediates as the photoreaction. The dark reversion depends strongly on the molecular environment. Its rate is increased by divalent metals (NEGBI et al. 1975), by pH changes and by reducing agents like sodium dithionite (ANDERSON et al. 1969; MUMFORD and JENNER 1971), as is the dark relaxation from meta R_b to P_{fr} (KENDRICK and SPRUIT 1973a, b). Interestingly, treatment of phytochrome with dithionite leads to an addition of sulfoxylate to the chromophore (KUFER and SCHEER 1979). The equilibria between the native chromophore and the rubinoid addition products favor the native form of P_r, but the rubinoid form of P_{fr}. Thus a reaction sequence $P_{fr} \rightarrow P_{add.} \rightarrow P_r$ is one possibiltiy to explain the catalytic effect of reducing agents during dark reversion (KUFER and SCHEER 1979). Such a sequence is not feasable, however, in the acceleration of the dark relaxation from meta R_b to P_{fr}.

3 Cryptochrome

Cryptochrome, the blue light photoreceptor, has been defined by GRESSEL (1980) "as that pigment system having an action spectrum somewhat characteristic of flavins and some carotenes. This name refers to its occurrence in cryptogams and its cryptic nature." It would, of course, be a paradox in itself to deal with the chromophore(s) of a compound of "cryptic nature". However, only two candidates for the cryptochrome chromophore are earnestly being discussed, namely flavins and carotenoids with a strong preference for the former. Arguments for and against each of these candidates are discussed in an excellent and comprehensive way in the book of SENGER (1980). Only some arguments and examples can be considered here.

3.1 Flavins

The best-known flavins are FMN (27a) and FAD (27b). Both have broad absorption maxima at 370 and 450 nm, the characteristic peaks of cryptochrome action spectra. However, most action spectra show a fine structure with two additional peaks or shoulders around the 450 nm peak (see Chap. 2 and 23, this Vol.). Although this is more similar to the absorption spectra of carotenoids in this spectral region, a fine structure of the FMN or FAD band can be observed in some flavoproteins (oxidized state, GHISLA 1980) or with protein-free flavin-derivatives in extreme environment (SONG et al. 1972).

A typical reaction of flavins is their photoreduction ("photobleaching") in the presence of a suitable electron donor (e.g., EDTA). This photoreduction is also possible with flavoproteins, e.g., nitrate reductase from *Neurospora crassa* (NINNEMANN and KLEMM-WOLFGRAMM 1980). Typical is the concomittant reduction of cytochrome b-557 of this enzyme. The action spectrum for photoreduction of cytochrome b in *Neurospora crassa* in vivo had previously been shown to be the typical cryptochrome spectrum including its fine structure; interestingly the absorbance measurements of *Neurospora* cells in these experiments (prolonged irradiation) point to photobleaching of flavins without any fine structure (MUÑOZ and BUTLER 1975). This example demonstrates a principal difficulty in identifying a flavin photoreceptor: because of the high photochemical reactivity of flavins, not only the specific photoreceptor but also all or most of the flavins present in abundant amounts in the cell wall react in the same manner. The classical argument of correlation between action spectrum and absorbance or absorbance difference spectrum (upon "bleaching") is not sufficient in this case. Additional correlation arguments are needed (see NINNEMANN and KLEMM-WOLFGRAMM 1980) until the true photoreceptor and then its chromophore can be identified.

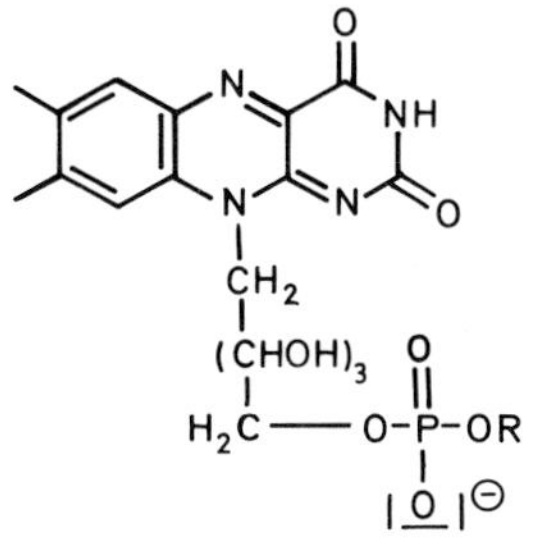

27a: R = H (FMN)

27b: R = AMP (FAD)

3.2 Carotenoids

A main argument for carotenoids as possible chromophores of cryptochrome has been the coincidence of the typical carotenoid absorption spectrum between 400 and 500 nm (including the fine structure) with the cryptochrome action spectrum. The UV absorption band which lacks in apolar solvents can be induced for carotenoids in polar solvents (HAGER 1970, SONG and MOORE 1974). However, photoreceptor chromophores are certainly not free in solution but probably bound to a membrane or protein. Protein binding can drastically change the absorption properties of chromophores, especially if they are conformationally flexible (e.g., retinal in rhodopsin, phytochromobilin in phytochrome, see above).

It can therefore be misleading to compare the absorption of chromophore candidates in solution with the action spectrum of photoreceptors of chromoprotein nature. No defined carotenoprotein has so far been described as candidate for cryptochrome. However, an interesting model has recently been suggested by SONG (1980). This is the peridinin-chlorophyll a-protein isolated from marine dinoflagellates (Fig. 12). Resonance (exciton) interactions between the carot-

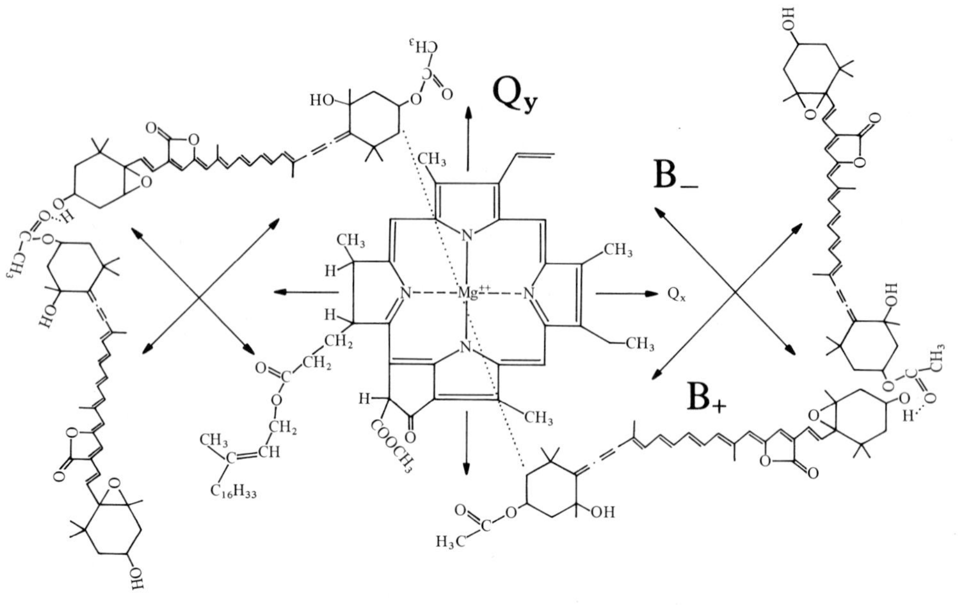

Fig. 12. The molecular topography of antenna photoreceptor complex, peridinin-chlorophyll a-protein from marine dinoflagellates. (SONG 1980)

enoid molecules and between the carotenoid and the chlorophyll as acceptor molecule allows efficient energy transfer in this system. By way of energy transfer, light absorbed by the carotenoid can be active not only for photosynthesis but also for phototaxis of these organisms. A similar situation has, however, not yet been detected for cryptochrome action.

4 Phycochrome, Phycomorphochromes and Adaptachromes

In many cyanobacteria and red algae, light-stimulated developmental responses (e.g., chromatic adaptation, induction of filamentous growth) have been observed which suggest the presence of photoreversibly photochromic pigments as photoreceptors (see BOGORAD 1975, BJÖRN and BJÖRN 1980) (Table 6). They have been termed phycochromes (in analogy to phytochrome), but as long as a correlation of distinct pigments with any of these responses is lacking the response oriented terms adaptachromes and phycomorphochromes (BOGORAD 1975) are recommended (see BJÖRN and BJÖRN 1980 for a discussion). The action spectra of the responses show a red-green photoreversibility (Table 6). It should be noted that red-green photoreversible morphogenic effects were also described in higher plants (KLEIN 1979).

From the shape and the maxima of the action spectra, biliproteins have been implicated as receptor pigments. SCHEIBE (1972) first isolated a biliprotein

Table 6. Action maxima for photoreversible photoresponses in blue green algae correlated to phycomorphochromes and adaptochromes, for (partially) purified phycochromes and for reversible absorption changes in partially denatured phycobiliproteins

	"Green" form		"Red" form		References
	λ_{max}	Action	λ_{max}	Action	
Tolypothrix tenuis, *Fremyella diplosiphon*	540–550 350–387	PE formation	641–660 360–463	PC formation	FUJITA and HATTORI (1962), DIAKOFF and SCHEIBE (1973), OHKI and FUJITA (1978), VOGELMANN and SCHEIBE (1978), HAURY and BOGORAD (1977)
Nostoc muscorum, *N. commune*	520–550	Reversion	640–650	Induction of filamentous growth	LAZAROFF and SCHIFF (1962), ROBINSON and MILLER (1970)
Synechocystis sp. 6701	540	PE formation	640	Reversion	TANDEAU DE MARSAC et al. (1980)
"Scheibe's pigment"	520	Formation P_{650}	650	Reversion	SCHEIBE (1972), OHKI and FUJITA (1979a)
Phycochrome a	630	Formation P_{580}	580	Reversion	BJÖRN and BJÖRN (1976), BJÖRN (1980a, b)
Phycochrome b	580	Formation P_{500}	500	Reversion	BJÖRN and BJÖRN (1976), BJÖRN (1979)
Phycochrome c[a]	630, 650[b]				BJÖRN and BJÖRN (1976), BJÖRN (1979)
Phycochrome d[a]	650	Formation P_{650}	610–620	Reversion	BJÖRN (1978, 1980, personal communication)
APC	645	Formation P_{555}	560	Reversion	OHAD et al. (1980)
PC, 0.5 M guanidinium chloride	570	Formation P_{630}	630	Reversion	OHKI and FUJITA (1979b)
APC, 0.5 M guanidinium chloride	600	Formation P_{650}	650	Reversion	OHKI and FUJITA (1979b)
APC, pH <4	645	Formation P_{555}	550–560	Reversion	OHAD et al. (1979, 1980)

[a] A correlation has recently been indicated between the occurence of phycochrome b and d and the presence of phycoerythrocyanin in the respective organism (BJÖRN 1979, 1980b)

[b] Probably two different forms

fraction from photobleached *Tolypothrix tenuis,* which gave photoreversible absorption difference spectra reminescent of the action spectra for chromatic adaptation in this species. The Björns have since characterized four different phycochromes (a, b, c, d) from various organisms, and purified to a different degree (Björn and Björn 1976, Björn 1978, 1979) (Table 6). All appear to be biliproteins, but the purified fractions generally exhibit only absorption differences which do not exceed a few percent. A notable exception is phycochrome b in which absorption differences of about one third of its maximum absorption allowed an in vivo detection (Björn 1979). To establish the identity of any such isolated fraction with a photoreceptor proper, it may be necessary to obtain further information besides the spectral data. As one such possibility, a differential temperature effect on the forward and back reaction of the photoreversion process has been suggested recently (Ohad et al. 1979).

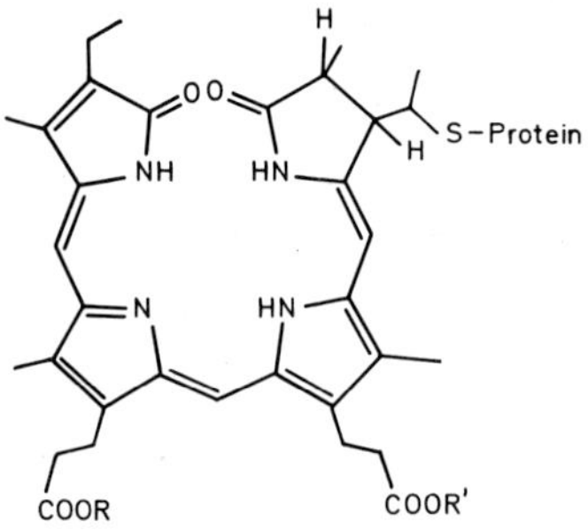

28

The need for a distinction becomes even more obvious from recent results on the photochemical properties of phycobiliproteins partially transformed in vivo (Ohki and Fujita 1979a) or partially denatured in vitro (Ohki and Fujita 1979b, Ohad et al. 1979, 1980) (Table 6). By this treatment, APC and PC containing the chromophore (28) obtain photoreversibly photochromic properties reminescent to the phycochromes. Especially the results with the isolated and purified biliproteins indicate that photochromicity is an inherent property of the bulk biliproteins and not related to a co-isolated impurity. The induced photochromicity of PC and APC decreases again at more severe denaturation conditions. Possibly, the "tickling" of the protein loosens the interactions with the protein sufficiently to open a photochemical channel, while internal conversion and destructive photochemistry of the pigments are still inhibited. More severe uncoupling (see Sect. 2.1.6 and 2.2.4) then favors the latter processes. Stepwise denaturation has been observed with PC from *Spirulina platensis,* suggesting a distinct intermediate in the unfolding process (Scheer and Kufer 1977). Similar intermediates may be present in the pigments isolated from *T. tenuis* (Ohki and Fujita 1979b) and *F. diplosiphon* (Ohad et al. 1979, 1980). In summary, the function of phycochromes as photoreceptors remains doubtful and needs further evaluation.

References

Anderson GR, Jenner EL, Mumford FE (1969) Temperature and pH studies on phytochrome in vitro. Biochemistry 8:1182–1187

Balangé AP (1974) Spectral changes of phytochrome in glycerol solution. Physiol Veg 12:95–105

Beuhler RJ, Pierce RC, Friedman L, Siegelman HW (1976) Cleavage of phycocyanobilin from C-phycocyanin separation and mass spectral identification of the products. J Biol Chem 251:2405–2411

Björn GS (1978) Phycochrome d, a new photochromic pigment from the blue-green alga, *Tolypothrix distorta.* Physiol Plant 42:321–323

Björn GS (1979) Action spectra for in vivo and in vitro conversions of phycochrome b, a reversibly photochromic pigment in blue-green algae, and its separation from other pigments. Physiol Plant 46:281–286

Björn GS (1980) Photoreversible pigments from blue-green algae (Cyanobacteria). Ph D Thesis, Univ Lund

Björn GS, Björn LO (1976) Photochromic pigments from blue-green algae phycochromes a, b and c. Physiol Plant 26:297–304

Björn LO, Björn GS (1980) Photochromic pigments and photoregulation in blue-green algae. Photochem Photobiol 32:849–852

Blacha-Puller M (1979) Chemische Synthese stereoisomerer Pyrromethenon-Derivate und davon abgeleiteter Gallenfarbstoffe. Dissertation Techn Univ Braunschweig

Blauer G, Wagnière G (1975) Conformation of bilirubin and biliverdin in their complexes with serum albumin. J Am Chem Soc 97:1949–1954

Bogorad L (1975) Phycobiliproteins and complementary chromatic adaptation. Annu Rev Plant Physiol 26:369–401

Bois-Choussy M, Barbier M (1978) Isomerizations and cyclizations in bile pigments. Heterocycles 9:677–690

Brandlmeier T, Blos I, Rüdiger W (1980) Structure and reactivity of the phytochrome chromophore. In: De Greef JA (ed) Photoreceptors and plant development. Antwerpen Univ Press, Antwerpen, pp 47–54

Brandlmeier T, Lehner H, Rüdiger W (1981b) Circular dichroism of the phytochrome chromophores in native and denatured state. Photochem Photobiol 34:69–73

Brandlmeier T, Scheer H, Rüdiger W (1981a) Chromophore content and molar absorptivity of phytochrome in the P_r form. Z Naturforsch 36c:431–439

Braslavsky SE, Holzwarth AR, Langer E, Lehner H, Matthews JI, Schaffner K (1980a) Conformational heterogeneity and photochemical changes of biliverdin dimethylesters in solution. Isr J Chem 20:196–202

Braslavsky SE, Matthews JI, de Kok HJ, Spruit CJP, Schaffner K (1980b) Characterization of a microsecond intermediate in the laser flash photolysis of small phytochrome from oat. Photochem Photobiol 31:417–420

Briggs WR, Fork DC (1969a) Long-lived intermediates in phy transformation. I. In vitro studies. Plant Physiol 44:1081–1088

Briggs WR, Fork DC (1969b) Long-lived intermediates in phy transformation. II. In vitro and in vivo studies. Plant Physiol 44:1089–1094

Briggs WR, Rice HV (1972) Phytochrome: Chemical and physical properties and mechanism of action. Annu Rev Plant Physiol 23:293–334

Brockmann H Jr, Knobloch G (1973) Substituierte Bernsteinsäuren. V. Die Absolute Konfiguration des 2E-Äthyliden-3-methylsuccinimids. Ein Beitrag zur Bestimmung der absoluten Konfiguration von Phycobilinen und Phytochrom. Chem Ber 106:803–811

Bryant DA, Hixson CS, Glazer AN (1978) Structural studies on phycobiliproteins. III. Comparison of bilin-containing peptides from the β subunits of C-phycocyanin, R-phycocyanin, and phycoerythrocyanin. J Biol Chem 253:220–225

Burke MJ, Pratt DC, Moscowitz A (1972) Low-temperature absorption and circular dichroism studies of phytochrome. Biochemistry 11:4025–4031

Canaani OD, Gantt E (1980) Circular dichroism and fluorescence polarization characteristics of blue-green algal allophycocyanins. Biochemistry 19:2950–2956

Chae Qu, Song PS (1975) Linear dichroic spectra and fluorescence polarization of biliverdin. J Am Chem Soc 97:4176–4179

Cole WJ, Chapman DJ, Siegelman HW (1967) The structure of phycocyanobilin. J Am Chem Soc 89:3643–3645

Correll DL, Edwards JL, Shropshire W Jr (1977) Phytochrome. Smithson Inst Press, Washington

Crespi HL, Boucher LJ, Norman GD, Katz JJ, Dougherty RC (1967) Structure of phycocyanobilin. J Am Chem Soc 89:3642–3643

Crespi HL, Smith U, Katz JJ (1968) Phycocyanobilin. Structure and exchange studies by nuclear magnetic resonance and its mode of attachment in phycocyanin. A model for phytochrome. Biochemistry 7:2232–2242

Cross DR, Linschitz H, Kasche V, Tenenbaum J (1968) Low temperature studies on phy: Light and dark reactions in the red-to-far red transformation and new intermediate forms. Proc Natl Acad Sci USA 61:1095–1101

Diakoff S, Scheibe J (1973) Action spectra for chromatic adaptation in *Tolypothrix tenuis*. Plant Physiol 51:382–385

Eivazi F, Lewis W, Smith KM (1977) Fragmentation reactions of bilindiones. Tetrahedron Lett 3083–3086

Falk H, Grubmayr K (1979) Die thermische Stabilität der geometrischen Isomeren von Bilatrienen-abc. Monatsh Chem 110:1237–1242

Falk H, Höllbacher G (1978) Über die Beziehung Lichtabsorption Struktur bei Bilatrienen-abc. Monatsh Chem 109:1429–1449

Falk H, Thirring K (1979) Darstellung, Struktur und Eigenschaften von isomeren N-Methyl-Bilatrienen-abc (N-Methyl-Etiobiliverdin IVγ). Z Naturforsch 34b:1448–1453

Falk H, Thirring K (1980) Zur anaeroben Photochemie von 21, 24-Dimethyl-etiobiliverdin-IVγ. Z Naturforsch 35b:376–380

Falk H, Thirring K (1981) Überbrückte Gallenpigmente: $N_{21}N_{24}$-Methylen-Aetiobiliverdin-IV- und N_{21}-N_{24}-Methylen-Aetiobilirubin-IV. Tetrahedron 37:761–766

Falk H, Grubmayr K, Haslinger E, Schlederer T, Thirring K (1978) Die diastereomeren (geometrisch Isomeren) Biliverdinmethylester-Struktur, Konfiguration und Konformation. Monatsh Chem 109:1451–1473

Falk H, Müller N, Schlederer T (1980) Eine regioselektive, reversible Addition an Bilatrienen a, b, c. Monatsh Chem 111:159–172

Frank G, Sidler W, Widmer H, Zuber H (1978) The complete amino acid sequence of both subunits of C-phycocyanin from the Cyanobacterium *Mastigocladus laminosus*. Hoppe-Seyler's Z Physiol Chem 359:1491–1507

Friedrich J, Scheer H, Zickendraht-Wendelstadt B, Haarer D (1981a) High-resolution optical studies on C-phycocyanin via photochemical hole burning. J Am Chem Soc 103:1030–1035

Friedrich J, Scheer H, Zickendraht-Wendelstadt B, Haarer D (1981b) Photochemical hole burning: A means to observe high resolution optical structures in phycoerythrin. J Chem Phys 74:2260–2266

Fry KT, Mumford FE (1971) Isolation and partial characterization of a chromophore-peptide fragment from pepsin digests of phytochrome. Biochem Biophys Res Commun 45:1466–1473

Fu E, Friedman L, Siegelman HW (1979) Mass-spectral identification and purification of phycoerythrobilin and phycocyanobilin. Biochem J 179:1–6

Fujita Y, Hattori A (1962) Photochemical interconversions between precursors of phycobilin chromoproteids in *Tolypothrix tenuis*. Plant Cell Physiol 3:209–220

Ghisla S (1980) Fluorescence and optical characteristics of reduced flavins and flavoproteins. Methods Enzymol 66:360–373

Glazer AN, Bryant DA (1975) Allophycocyanin B ($\lambda_{max} = 671$, 618 nm). A new cyanobacterial phycobiliprotein. Arch Microbiol 104:15–22

Gossauer A, Blacha-Puller M (1981) Darstellung von 3,4-Dihydro-5(1H)-pyrromethenonen aus 5(1H)-Pyrromethenonen sowie 5(2H)-Dipyrrylmethanonen. Liebigs Ann Chem: 1492–1504

Gossauer A, Blacha-Puller M, Zeisberg R, Wray V (1981) Notiz über die Synthese von E,Z,Z-konfigurierten Biliverdinen aus entsprechend konfigurierten 5(1H)-Pyrromethenonen. Liebigs Ann Chem 1981: 342–346

Gossauer A, Hinze RP (1978) An improved chemical synthesis of racemic phycocyanobilin dimethyl ester. J Org Chem 43: 283–285

Gossauer A, Hinze RP, Zilch H (1977) Umsetzung cyclischer Imide von Monothiodicarbonsäuren mit Phosphor-Yliden: Eine neue Methode zur Synthese von co-Alkylidenlactamen. Angew Chem 89: 429–430; Angew Chem Int Ed Engl 16: 418

Gossauer A, Hinze RP, Kutschan R (1980) Synthese von Gallenfarbstoffen XI. Totalsynthese und Zuordnung der relativen Konfigurationen zweier epimerer Methanol-Addukte des Phycocyanobilindimethylesters. Chem Ber 114: 132–146

Gossauer A, Inhoffen HH (1970) Darstellung von Gallenfarbstoff-Analoga mit endständigen Pyrrolidin-dion-(2.4)-Ringen. Liebigs Ann Chem 438: 18–30

Gossauer A, Klahr E (1979) Totalsynthese des racem. Phycoerythrobilindimethylesters. Chem Ber 112: 2243–2255

Gossauer A, Weller JP (1978) Chemical total synthesis of (+)-(2R, 16R)- and (+)-(2R, 16R)-phycoerythrobilin dimethyl ester. J Am Chem Soc 100: 5928–5933

Gressel J (1980) Blue light and transcription. In: Senger H (ed) The blue light syndrome. Springer, Berlin Heidelberg New York, pp 133–153

Grombein S, Rüdiger W, Zimmermann H (1975) The structures of the phytochrome chromophore in both photoreversible forms. Hoppe-Seyler's Z Physiol Chem 356: 1709–1714

Gysi J, Zuber H (1976) APC I-a second cyanobacterial APC? Isolation, characterization and comparison with APC II form the same alga. FEBS Lett 68: 49–54

Hagar A (1970) Ausbildung von Maxima im Absorptionsspektrum von Carotinoiden im Bereich um 370 nm: Folgen für die Interpretation bestimmter Wirkungsspektren. Planta 91: 38–53

Hahn T-R, Kang S-S, Song P-S (1980) Difference in the degree of exposure of chromophores in the P_r and P_{fr} forms of phytochrome. Biochem Biophys Res Commun 97: 1317–1323

Hahn T-R, Song P-S (1981) Hydrophobic properties of phytochrome as probed by 8-anilinonaphthalene-1-sulfonate fluorescence. Biochemistry 20: 2602–2609

Haury JF, Bogorad L (1977) Action spectra for phycobiliprotein synthesis in a chromatically adapting cyanophyte, fremyella diplosiphon. Plant Physiol 60: 835

Holzwarth AR, Langer E, Lehner H, Schaffner K (1980) Luminescence and solvent-induced circular dichroism of bilirubin dimethyl ester: Evidence for heterogeneous composition of the solute. J Mol Struct 60: 367–371

Holzwarth AR, Lehner H, Braslavsky SE, Schaffner K (1978) Phytochrome models. II. The fluorescence of biliverdin dimethyl ester. Liebigs Ann Chem 1978: 2002–2017

Hunt RE, Pratt LH (1980) Partial characterization of undegraded oat phytochrome. Biochemistry 19: 390–394

Hunt RE, Pratt LH (1981) Physico-chemical differences between the red and far-red-absorbing forms of phytochrome. Biochemistry 20: 941–945

Kendrick RE (1974) Phytochrome intermediates in freeze-dried tissue. Nature 250: 159

Kendrick RE, Spruit CJP (1972) Phytochrome properties and the molecular environment. Plant Physiol 52: 327–331

Kendrick RE, Spruit CJP (1973a) Phytochrome intermediates in vivo. I. Effects of temperature, light intensity, wavelength and oxygen on intermediate accumulation. Photochem Photobiol 18: 139–144

Kendrick RE, Spruit CJP (1973b) Phytochrome intermediates in vivo. III. Kinetic analysis of intermediate reactions of low temperature. Photochem Photobiol 18: 153–159

Kendrick RE, Spruit CJP (1974) Inverse dark reversion of phytochrome: An explanation. Planta 120: 265–272

Kendrick RE, Spruit CJP (1977) Phototransformations of phytochrome. Photochem Photobiol 26: 201–214

Klein G, Grombein S, Rüdiger W (1977) On the linkages between chromophore and protein in biliproteins. VI. Structure and protein linkage of the phytochrome chromophore. Hoppe-Seyler's Z Physiol Chem 358:1077–1079

Klein G, Rüdiger W (1978) Über die Bindungen zwischen Chromophor und Protein in Biliproteiden. V. Stereochemie von Modell-Imiden. Liebigs Ann Chem 1978:267–279

Klein G, Rüdiger W (1979) Thioether formation of phycocyanobilin: A model reaction of phycocyanin biosynthesis. Z. Naturforsch 34c:192–195

Klein RM (1979) Reversible effects of green and orangered radiation on plant cell elongation. Plant Physiol 63:114–116

Köst HP, Rüdiger W, Chapman DJ (1975) Über die Bindung zwischen Chromophor und Protein in Biliproteiden. I. Abbauversuche und Spektraluntersuchungen. Liebigs Ann Chem 1975:1582–1593

Köst-Reyes E, Köst HP (1979) The protein-chromophore bond in B phycoerythrin from porphyridium cruentum. Eur J Biochem 102:83–91

Köst-Reyes E, Köst HP, Rüdiger W (1975) Nachweis von Cystein als bindende Aminosäure in B-Phycoerythrin. Liebigs Ann Chem 1975:1494–1600

Kraus C, Bubenzer C, Scheer H (1979) Photochemically assisted reaction of A-dihydrobilindione with nucleophiles as a model for phytochrome interconversion. Photochem Photobiol 30:473–477

Krauss C, Scheer H (1979) Long-lived π-cation radicals of bilindionato zinc complexes. Tetrahedron Lett 3553–3556

Kraus C, Scheer H (1981) Studies on plant bile pigment −9. Photooxidation of A-dihydrobilindione as phytochrome model in acid medium. Photochem Photobiol 34:385–391

Kroes HH (1970) A study of phytochrome, its isolation, structure and photochemical transformation. Meded Landbouwhogesch Wageningen 70–18:1–112

Kufer W, Scheer H (1979) Reversible reductions of biliprotein chromophores and model pigments. Z Naturforsch 35c:776–781

Kufer W, Krauss C, Scheer H (1982a) Zwei milde, regioselektive Abbaumethoden von Biliprotein Chromophoren. Angew Chem 94:455–456; Angew Chem Suppl 1050–1060; Angew Chem Int Ed 21:446–447

Kufer W, Cmiel E, Thümmler F, Rüdiger W, Schneider S, Scheer H (1982b) Regioselective photochemical and acid catalyzed Z,E isomerization of dihydrobilindione as phytochrome model. Photochem Photobiol 36:603–608

Lagarias JC, Glazer AN, Rapoport H (1979) Chromopeptides from C-phycocyanin. Structure and linkage of a phycocyanobilin bound to the β subunit. J Am Chem Soc 101:5030–5037

Lagarias JC, Rapoport H (1980) Chromopeptides from phytochrome. The structure and linkage of the P_r form of the phytochrome chromophore. J Am Chem Soc 102:4821–4828

Langer E, Lehner H, Rüdiger W, Zickendraht-Wendelstadt B (1980) Circular dichroism of C-phycoerythrin: A conformational analysis. Z Naturforsch 35c:367–375

Lazaroff N, Schiff J (1962) Action spectrum for developmental photoinduction in the blue-green alga N. muscorum. Science 137:603–604

Lehner H, Braslavsky SE, Schaffner K (1978a) Phytochrome models I. Isolation, characterization, and solution conformation of biliverdin dimethylester and its XIIIα isomer. Liebigs Ann Chem 1978:1990–2001

Lehner H, Braslavsky SE, Schaffner K (1978b) Enantiomere und Diastereomere bei bishelicalen Bilatrien-Dimeren im Kristallgitter. Angew Chem 90:1012–1013

Lehner H, Riemer W, Schaffner K (1979) Die Interkonversionsbarriere der Bilatrien-Helix. Liebigs Ann Chem 1979:1758–1801

Linschitz H, Kasche V (1967) Kinetics of phytochrome conversion: Multiple pathways in the P_r to P_{fr} reaction, as studied by double-flash technique (flash spectroscopy). Proc Natl Acad Sci US 58:1059–1064

Linschitz H, Kasche V, Butler WL, Siegelman HW (1966) The kinetics of phytochrome conversion. J Biol Chem 241:3395–3403

Lotter H, Klein G, Rüdiger W, Scheer H (1977) Röntgenstrukturanalyse von (Äthylsulfonyl)-Methylsuccinimid. Tetrahedron Lett 1977:2317–2320

Mitrakos K, Shropshire W Jr (1972) Phytochrome Academic Press, London New York

Muckle G, Otto J, Rüdiger W (1978) Amino acid sequence in the chromophore regions of c-phycoerythrin from *Pseudanabaena* W 1173 and *Phormidium persicinum*. Hoppe-Seyler's Z Physiol Chem 359:345–355

Mumford FE, Jenner EL (1971) Catalysis of the phytochrome dark reaction by reducing agents. Biochemistry 10:98–101

Muñoz V, Butler WL (1975) Photoreceptor pigment for blue light in *Neurospora crassa*. Plant Physiol 55:421–426

Negbi M, Hopkins DW, Briggs WR (1975) Acceleration of dark reversion of phytochrome in vitro by Ca and Mg. Plant Physiol 56:157–159

Ninnemann H, Klemm-Wolfgramm E (1980) In: Senger H (ed), The blue light syndrome. Springer, Berlin Heidelberg New York, pp 238–243

O'Carra P, O'hEocha C (1966) Bilins released from algae and biliproteins by methanolic extraction. Phytochemistry 5:993–997

O'Carra P, O'hEocha C, Carroll DM (1964) Spectral properties of the phycobilins II phycoerythrobilin. Biochemistry 3:1343–1350

Ohad L, Clayton RK, Bogorad L (1979) Photoreversible absorbance changes in solutions of allophycocyanin purified from *F. diplosiphon*. Proc Natl Acad Sci USA 76:5655–5659

Ohad L, Schneider H-JAW, Gendel S, Bogorad L (1980) Light-induced changes in allophycocyanin. Plant Physiol 65:6–12

O'hEocha C (1963) Spectral properties of the phycobilins I phycocyanobilin. Biochemistry 2:375–382

Ohki K, Fujita Y (1978) Photocontrol of phycoerythrin formation in the blue-green alga *Tolipothrix tenuis* growing in the dark. Plant Cell Physiol 19:7–15

Ohki K, Fujita Y (1979a) In vivo transformations of phycobiliproteins during photobleaching of *T. tenuis* to forms active in photoreversible. Plant Cell Physiol 20:1341–1348

Ohki K, Fujita Y (1979b) Photoreversible absorptions changes of guanidine-HCl-treated phycocyanin and allophycocyanin isolated from the blue-green alga *Tolypothrix tenuis*. Plant Cell Physiol 20:483–490

Parker MW, Hendricks SB, Borthwick HA (1950) Action spectrum for the photoperiodic control of floral initiation of the long-day plant *Hyoscyamus niger*. Bot Gaz 111:242–252

Pasternak R, Wagnière G (1979) Possible interpretation of long-wavelength spectral shifts in P_r and P_{fr}. J Am Chem Soc 101:1662–1667

Pétrier C, Dupuy C, Jardon P, Gautron R (1979) Studies on tetrapyrrol pigments I. Absorption and fluorescence of biliverdin dimethyl esters of the IX series. Photochem Photobiol 29:389–392

Pratt LH (1978) Properties of phytochrome. Photochem Photobiol 27:81–105

Pratt LH, Butler WL (1970) The temperature dependence of phytochrome transformations. Photochem Photobiol 11:361–369

Quail PH, Briggs WR, Pratt LH (1978) In vivo phosphorylation of phytochrome. Carnegie Inst Wash Year Book 77:342–344

Robinson BL, Miller JH (1970) Photomorphogenesis in the blue-green alga, Nostoc commune. Physiol Plant 23:461–472

Roux SJ, Lisansky SG, Stoker BM (1975) Purification and partial carbohydrate analysis of phytochrome from *Avena sativa*. Physiol Plant 35:85–90

Rüdiger W (1969) Chromsäure- und Chromatabbau von Gallenfarbstoffen. Hoppe-Seyler's Z Physiol Chem 350:1291–1300

Rüdiger W (1970) Neues aus Chemie und Biochemie der Gallenfarbstoffe. Angew Chem 82:527–534, Angew Chem Int Ed Engl 9:473–480

Rüdiger W (1971) Gallenfarbstoffe und Biliproteide. Fortschr Chem Org Naturst 29:60–139

Rüdiger W (1972) Isolation and purification of phytochrome. Chemistry of the phytochrome chromophore. In: Mitrakos K, Shropshire W Jr (eds) Phytochrome, Academic Press, London New York, pp 105–141

Rüdiger W (1980) Phytochrome, a light receptor of plant photomorphogenesis. Struct Bonding 40:101–140

Rüdiger W, Brandlmeier T, Blos I, Gossauer A, Weller JP (1980) Isolation of the phytochrome chromophore. The cleavage reaction with hydrogen bromide. Z Naturforsch 35c:763–769

Rüdiger W, Correll DL (1969) Über die Struktur des Phytochrom-Chromophors und seine Protein-Bindung. Liebigs Ann Chem 723:208–212

Rüdiger W, Klose W, Vuillaume M, Barbier M (1968) On the structure of pterobilin, the blue pigment of *Pienis brassicae*. Experientia 24:1000

Rüdiger W, O'Carra P, O'hEocha C (1967) Structure of phycoerythrobilin and phycocyanobilin. Nature 215:1477–1478

Scheer H (1976) Studies on plant bile pigments. I. Characterization of a model for the phytochrome P_r chromophore. Z Naturforsch 31c:413–417

Scheer H (1980) Biliproteine. Angew Chem 93:230–250, Angew Chem Int Ed Engl 20:241–261

Scheer H, Formanek H, Rüdiger W (1979) The conformation of bilin chromophores in biliproteins: Ramachandrantype calculations. Z Naturforsch 34c:1085–1093

Scheer H, Krauss C (1977) Oxidative photodimerization of a phytochrome P_r model pigment and its thermal reversion. Photochem Photobiol 25:311–314

Scheer H, Kufer W (1977) Conformational studies on C-phycocyanin from *Spirulina platensis*. Z Naturforsch 32c:513–519

Scheer H, Linsenmeier U, Krauss C (1977) Chemical and photochemical oxygenation of a phytochrome P_r chromophore model pigment in purpurins. Hoppe-Seyler's Z Physiol Chem 358:185–196

Scheer H, Formanek H, Schneider S (1982) Theoretical studies of biliprotein chromophores and related pigments by molecular orbital and Ramachandran type calculations. Photochem Photobiol 36:259–272

Scheibe J (1972) Photoreversible pigment: Occurence in a blue-green alga. Science 176:1037–1039

Schoch S, Klein G, Linsenmeier U, Rüdiger W (1974) Synthese und Reaktionen von Äthyliden-methylsuccinimid. Tetrahedron Lett 2465–2468

Schram BL, Kroes HH (1971) Structure of phycocyanobilin. Eur J Biochem 19:581–594

Senger H (ed) (1980) The blue light syndrome. Springer, Berlin Heidelberg New York

Sheldrick WS (1976) Crystal and molecular structure of biliverdin. J Chem Soc Perkin Trans II 1976:1457–1465

Siegelman HW, Chapman DJ, Cole WJ (1968) In: Goodwin TW (ed) Prophyrins and related compounds biochem soc symp 28:107–120. The bile pigments of plants, Academic Press, London New York

Siegelman HW, Turner BC, Hendricks SB (1966) The chromophore of phytochrome. Plant Physiol 41:1289–1292

Smith H (1975) Phytochrome and photomorphogenesis. McGraw-Hill, London

Smith H, Kendrick RE (1976) In: Goodwin TW (ed) Chemistry and biochemistry of plant pigments. The structure and properties of phytochrome. Academic Press, London New York

Song P-S (1980) Spectroscopic and photochemical characterization of flavoproteins and carotenoproteins as blue light photoreceptors. In: Senger H (ed) The blue light syndrome. Springer, Berlin Heidelberg New York, pp 157–171

Song P-S, Chae A, Gardner JD (1979) Spectroscopic properties and chromophore conformations of the photomorphogenetic receptor: phytochrome. Biochim Biophys Acta 576:479–495

Song P-S, Moore TA (1974) On the photoreceptor pigment for phototropism and phototaxis: is a carotenoid the most likely candidate? Photochem Photobiol 19:435–441

Song P-S, Moore TA, Sun M (1972) Excited states of some plant pigments. In: Chichester CO (ed) The chemistry of plant pigments. Academic Press, London New York, pp 33–74

Spruit CJP, Kendrick RE (1973) Phytochrome intermediates in vivo. II. Characterization of intermediates by difference spectroscopy. Photochem Photobiol 18:145–152

Spruit CJP, Kendrick RE (1977) Phototransformations of phytochrome: The characterization of lumi-F and meta-F_a. Photochem Photobiol 26:133–138

Spruit CJP, Kendrick RE, Cooke RJ (1975) Phytochrome intermediates in freeze-dried tissue. Planta 127:121–132

Struckmeier G, Thewaldt U, Fuhrhop J-H (1976) Structures of zinc-octaethyl-formylbiliverdinate hydrate and its dehydroated bis-helical dimer. J Am Chem Soc 98:278–279

Sugimoto T, Ishikawa K, Suzuki H (1976) On the models for phytochrome chromophore III. J Physic Soc Jpn 40:258–266

Sugimoto T, Oishi M, Suzuki H (1977) On the models for phytochrome chromophore IV. J Physic Soc Jpn 43:619–626

Tandeau de Marsac N, Castets AM, Cohen-Bazire G (1980) Wavelength modulation of phycoerythrin synthesis in synechocystis sp. 6701. J Bacteriol 142:310–314

Thümmler F, Brandlmeier T, Rüdiger W (1981) Preparation and properties of chromopeptides from the P_{fr} form of phytochrome. Z Naturforsch 36c:440–449

Tobin E, Briggs WR, Brown PK (1973) The role of hydration in the phototransformation of phytochrome. Photochem Photobiol 18:497–503

Tokutomi S, Yamamoto KT, Furuya M (1981) Photoreversible changes in hydrophobicity of undegraded pea phytochrome determined by partition in a aqueous two-phase system. FEBS Lett 134:159–162

Tokutomi S, Yamamoto KT, Miyoshi Y, Furuya M (1982) Photoreversible changes in pH of pea phytochrome solutions. Photochem Photobiol 35:431–434

Troxler RF, Brown AS, Köst HP (1978) Quantitative degradation of radiolabeled phycobiliproteins: Chromic acid degradation of C-phycocyanin. Eur J Biochem 87:181–189

Vogelmann TC, Scheibe J (1978) Action spectra for chromatic adaptation in the blue green alga, F. diplosiphon. Planta 143:233–239

Weller JP, Gossauer A (1980) Synthese und Photoisomerisierung des racem. Phytochromobilin-dimethylesters. Chem Ber 113:1603–1611

Williams VP, Glazer AN (1978) Structural studies on phycobiliproteins. I. Bilin-containing peptides of C-phycocyanin. J Biol Chem 253:202–211

Zickendraht-Wendelstadt B, Friedrich J, Rüdiger W (1980) Spectral characterization of monomeric C-phycoerythrin from pseudanabaena W 1173 and its α and β subunits: Energy transfer in isolated subunits and C-phycoerythrin. Photochem Photobiol 31:367–376

Zilinskas BA, Zimmermann BK, Gantt E (1978) Allophycocyanin forms isolated from *Nostoc* sp. phycobilisomes. Photochem Photobiol 27:587–596

8 Assay of Photomorphogenic Photoreceptors

L.H. PRATT

1 Introduction

The scope of this chapter is confined to direct assays of photomorphogenic photoreceptors. Photoreceptors may also, at least in principle, be assayed indirectly by extrapolation from data obtained by study of photomorphogenesis of intact plant systems. Such indirect assays, however, would be of limited applicability because they would depend upon generally untested assumptions concerning the relationship between the photoreceptor and the photomorphogenic response being studied. They consequently will not be considered here.

A second limitation to the scope of this chapter is inherent to the subject. To devise an assay for a photoreceptor, that photoreceptor must first be identified and at least partially characterized. Since phytochrome is the only receptor thus far identified and characterized (BUTLER et al. 1959, Chap. 1, this Vol.), it is the only one that can be dealt with in any detail here. Other potential photoreceptors will be treated only briefly.

Direct photoreceptor assays may be divided into four categories. (1) An assay analogous to the bioassays commonly used with plant hormones would be suitable. No biological response to an exogenously supplied photoreceptor has yet been described, however, that could serve as the basis for such a photoreceptor bioassay. (2) An assay could derive from the expression of some molecular activity of the photoreceptor analogous to the catalytic activity of an enzyme. This assay approach is also not possible because no such molecular activity of a photoreceptor has yet been identified. (3) Assay may depend upon unique spectrophotometric characteristics of the photoreceptor. It is this approach that has received the greatest attention. (4) Assay may also depend upon the unique antigenic properties of a photoreceptor. These immunochemical assays are useful because they are not only independent of the spectral properties of the photoreceptor but are also highly specific. A third restriction to the scope of this chapter, then, will be to a consideration of only spectrophotometric and immunochemical assays because they are the only ones presently available.

2 Spectrophotometric Assay of Phytochrome

2.1 Background

A brief discussion of the events leading to the discovery and identification of phytochrome as a photomorphogenic photoreceptor (BORTHWICK 1972,

Chap. 2, this Vol.) is worthwhile in the present context for two reasons. First, they provide the fundamental information that serves as the basis for spectrophotometric assays of phytochrome. Second, they serve as a useful model that illustrates the logic that is being used in attempts to identify other photoreceptors and to develop assays for them (Sect. 3).

The discovery of phytochrome derives from the study of two very different photomorphogenic phenomena. One is the induction by light of light-sensitive lettuce seed germination (BORTHWICK et al. 1952b, Chap. 17, this Vol.) while the other is the interruption by light of the measurement of night length that serves as the basis for photoperiodic phenomena such as flowering (BORTHWICK et al. 1952a, Chap. 18, this Vol.). Action spectra for both light effects were similar (PARKER et al. 1946, BORTHWICK et al. 1952b), indicating that the photoreceptor should be the same for both and should have maximum absorbance near 660 nm (BORTHWICK 1972). From this information alone, however, identification of phytochrome, and thus its assay, would probably have been impossible. Two other observations were of vital significance. First, the effect of red light, both in inducing seed germination and in interrupting night length measurement, was reversed or nullified by subsequent exposure to far-red light (BORTHWICK et al. 1952a, 1952b). Second, the antagonistic effects of red and far-red light were repeatedly reversible. To explain these observations BORTHWICK et al. (1952b) postulated the existence of a pigment that must exist in two different, photointerconvertible forms:

1. Pigment + RX $\rightleftharpoons$ Pigment·X + R

or

2. Pigment·X + R $\rightleftharpoons$ Pigment + RX

where R is a hypothetical reaction partner and the form of the pigment on the left is the red-absorbing form, that on the right the far-red-absorbing form. Red light would drive the equilibrium to the right, far-red light to the left.

This postulate led directly to a further prediction. Namely, that plant tissue upon irradiation with red light should exhibit an absorbance decrease in the red spectral region coupled with an increase in the far-red spectral region. Conversely, far-red irradiation should lead to a decrease in far-red absorbance and an increase in red absorbance (BUTLER et al. 1959). Seven years after BORTHWICK et al. (1952b) postulated the existence of this photoreversible pigment, BUTLER et al. (1959) demonstrated its existence by confirming the occurrence in etiolated maize coleoptiles of the predicted absorbance changes, which have since served as the basis for spectrophotometric assays of phytochrome. Comparable measurements of light-induced absorbance changes in etiolated oat shoots are shown in Fig. 1. The initial red actinic irradiation photoconverts both protochlorophyll to chlorophyll and the red-absorbing form of the pigment (P_r) to the far-red-absorbing form (P_{fr}) (Fig. 1a) while the second red irradiation, given after an intervening far-red irradiation, converts only P_r to P_{fr} (Fig. 1b).

This photoreversible pigment was also extracted by BUTLER et al. (1959), shown to be a chromoprotein, and later given the name phytochrome (BORTH-

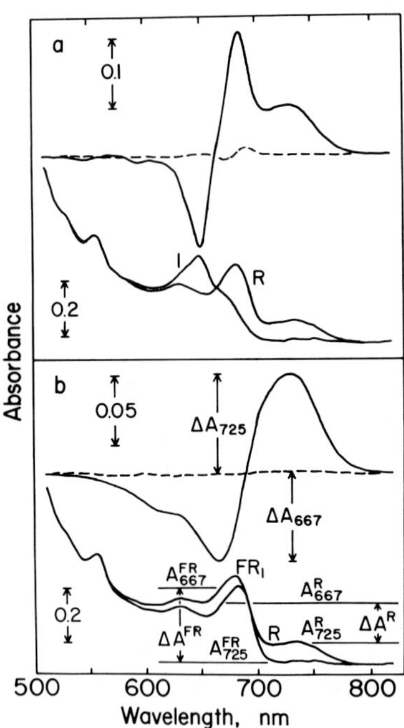

Fig. 1a, b. Absorbance spectra of 4-day-old, dark-grown (25 °C) oat shoots. Shoots were harvested under dim green light, chopped into approximately 5 mm segments and packed loosely into an ice-water-cooled cuvette with a nominal light path of 10 mm. Spectra were measured with reference to filter paper in an Hitachi 557 spectrophotometer interfaced with a Hewlett Packard Model 9845T computer. **a** The initial spectrum (*I*) was measured prior to any actinic irradiation. Spectrum *R* was measured immediately after 1 min actinic red light. A third spectrum (not shown) was measured after an additional 15 min of darkness. Difference spectra between spectrum *R* and spectrum I (*R*−*I*, *solid line*) and between the third spectrum and that measured after red light [(R+15 min D)−*R*, *dashed line*] are shown. **b** The same sample was irradiated with 1 min far-red light and then an additional 4 min red light to ensure saturation of the protochlorophyll-to-chlorophyll transformation. The sample was then irradiated again with 1 min far-red light and a spectrum was recorded (FR_1). Another spectrum was recorded after a subsequent 1 min red irradiation (*R*) and after a final 1 min far-red irradiation (not shown). Difference spectra are shown between spectrum *R* and spectrum FR_1 ($R-FR_1$, *solid line*) and between the spectrum taken after the final far-red irradiation and spectrum FR_1 (FR_2-FR_1, *dashed line*). Note the different absorbance scales used. (Spectra obtained with the assistance of Y. INOUE and Y. SHIMAZAKI)

WICK and HENDRICKS 1960). Subsequent work with highly purified phytochrome indicates that the initial postulate of BORTHWICK et al. (1952b) was in error only in that phytochrome can be photointerconverted without the participation of a reaction partner (Chap. 6, this Vol.). If it did exist, of course, then it may not have been possible to use these photoreversible absorbance changes as the basis for phytochrome assay in vitro, since in vitro phytochrome might have been separated from its hypothetical reaction partner. Absorbance spectra of highly purified oat phytochrome (Fig. 2) indicate that it exhibits absorbance characteristics similar to those of the pigment in vivo (Fig. 1), especially as seen in a difference spectrum.

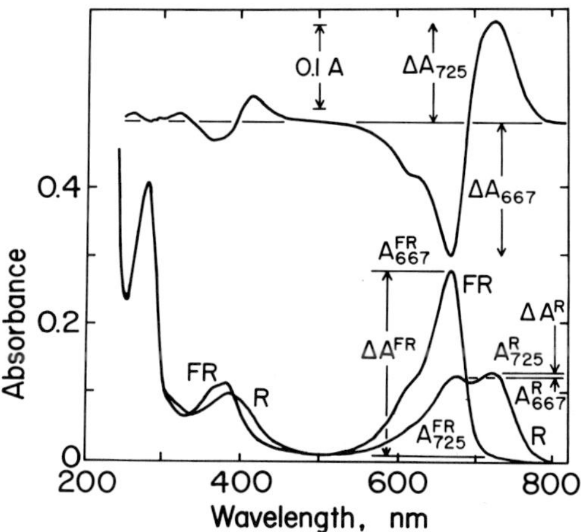

Fig. 2. Absorbance spectra of immunopurified oat phytochrome. Spectra were recorded with an Hitachi 557 spectrophotometer interfaced with a Hewlett Packard Model 9845T computer. Phytochrome was in 0.1 M Na-phosphate, 1 mM ethylenediaminetetraacetate (pH 7.8) in an ice-water-cooled, 5 mm pathlength cuvette. Spectra were recorded after saturating red (*R*) or saturating far-red (*FR*) irradiation. The difference spectrum between these two (*R*−*FR*) is shown with a slightly expanded absorbance scale. (Sprectra obtained with the assistance of Y. INOUE and Y. SHIMAZAKI)

2.2 Simple Assays

Direct absorbance measurements provide the simplest assay for phytochrome in optically clear solutions when other pigments absorbing in the red spectral region are absent. It is preferable to assay for phytochrome in its P_r form at about 667 nm (Fig. 2) because the extinction of P_{fr} in the far-red region (a) is lower than that of P_r in the red (TOBIN and BRIGGS 1973, PRATT 1978) and (b) is more strongly dependent upon its environment than is the extinction of P_r (BUTLER et al. 1964b, LISANSKY and GALSTON 1974, 1976, PRATT and CUNDIFF 1975, PRATT 1978). Absorbance is related to phytochrome concentration by the Lambert-Beer relationship, $A = \varepsilon lc$, where A is absorbance, ε is the extinction coefficient of phytochrome at the wavelength of assay, l is the optical pathlength of the measuring beam, and c is phytochrome concentration. One unit of phytochrome is defined conveniently as the quantity that after saturating, far-red irradiation gives an absorbance of 1.0 at 667 nm when dissolved in 1.0 ml and assayed in a 10-mm light path. Based upon extinction coefficients for rye (TOBIN and BRIGGS 1973) and oat (HUNT and PRATT 1980a) phytochrome, one unit is equal to 1.7 or 1.8 mg, respectively. More recently, however, ROUX et al. (1982) have reported a significantly higher extinction coefficient for oat phytochrome, derived by quantitative amino acid analysis, which indicates that one unit is equal to about 1.2 mg. Since this latter method is more absolute, it would be most prudent to use this newer value. In each case the phytochrome referred to here exhibits approximately 120,000 molecular weight monomers by sodium dodecyl sulfate, polyacrylamide gel electrophoresis.

2.3 Assays Based upon Light-Induced Absorbance Changes

A light-induced absorbance change (LIAC) measurement is useful whenever the pigment to be assayed undergoes a unique absorbance change in response to actinic irradiation. In the case of phytochrome, a LIAC measurement is relatively easy to make since the LIAC's induced by actinic red and far-red irradiations normally persist in darkness (Figs. 1 and 2). It is necessary to assay for phytochrome by its LIACs only when other pigments absorbing in the same spectral region are present and/or when the samples are light scattering. Dual-wavelength assays of LIACs are utilized most frequently in order to minimize the effects of time-dependent changes in the optical properties of the sample.

Three different wavelength pairs are convenient for phytochrome assay (Figs. 1 and 2): red (about 667 nm) and far-red (about 725 nm), red and near infra-red (about 800 nm), and far-red and near infra-red. The use of red and far-red measuring beams gives the greatest sensitivity since the absorbance changes in both regions are summed. Nevertheless, far-red and near infra-red wavelengths are often used in an attempt to minimize interference caused by chlorophylls (BUTLER and LANE 1965, Sect. 2.8), which absorb strongly in the red region. While this approach has obvious merit, too little attention is paid to a different problem, which is the dependence of the extinction of P_{fr} in the far-red upon a variety of factors such as metal ions (LISANSKY and GALSTON 1974, 1976, PRATT and CUNDIFF 1975), a naturally occurring substance referred to as P_{fr}-killer (FURUYA and HILLMAN 1966), and other unknown factors (PRATT and CUNDIFF 1975, YAMAMOTO 1980). The possibility that such extinction changes might occur in vivo as well as in vitro is generally not widely acknowledged in the interpretation of photoreversible absorbance changes monitored only in the far-red region. In the absence of a compelling reason to limit assay to the far-red region, it would therefore seem prudent to assay in the red region as well.

As an illustrative example of how the photoreversible absorbance changes of phytochrome are used for a dual-wavelength assay (BUTLER et al. 1959, 1963) we can consider a photoreversibility measurement utilizing 667 nm and 725 nm as the measuring wavelengths (refer to Figs. 1 and 2). The difference in absorbance at 667 nm and 725 nm is measured after actinic far-red ($\Delta A^{FR} = A_{667}^{FR} - A_{725}^{FR}$) and actinic red ($\Delta A^{R} = A_{667}^{R} - A_{725}^{R}$) irradiations. The quantity of phytochrome in the sample is linearly proportional to the absolute value of the difference between these two absorbance difference measurements: $\Delta\Delta A = |\Delta A^{FR} - \Delta A^{R}|$. Rearrangement of the terms in this final relationship indicates that $\Delta\Delta A = |(A_{667}^{R} - A_{667}^{FR})| + |(A_{725}^{R} - A_{725}^{FR})| = |\Delta A_{667}| + |\Delta A_{725}|$, an expression that is related more obviously to the concentration of phytochrome in the sample. This rearranged expression also makes it clear that any influence either of background attenuation of the measuring beam caused by light scattering or of the presence of other pigments is eliminated. Obviously, however, one $\Delta\Delta A$ measurement may be compared to another only when the effects of light scattering on the measuring pathlength are the same for both measurements (BUTLER 1962, SPRUIT 1972).

2.4 Interconversions Among Different Assay Units

Investigators in different laboratories have used many different units, including arbitrary units, to quantitate phytochrome. Arbitrary units, however, should be avoided because they do not permit either critical evaluation of data or careful comparison of data obtained in different laboratories.

It is a simple matter to convert units used by one research group to those used by another given the details of how the spectrophotometric assays were performed [wavelength(s) used, nominal measuring pathlength, physical composition of sample to indicate the extent of light scattering (Sect. 2.5), type of assay] (PRATT 1978). For example, it can be seen by reference to Fig. 2 that for large oat phytochrome, the $\Delta\Delta A$ for the wavelength pair of 667 nm and 725 nm is 102% of the absorbance at 667 nm following a saturating actinic far-red irradiation. Thus it is evident that one unit ml^{-1} of large oat phytochrome, as defined above, would give in a 1-cm pathlength a $\Delta\Delta A$ of 1.02, 0.55, or 0.47, for the wavelength pairs 667 nm and 725 nm, 667 nm and 800 nm, or 725 nm and 800 nm, respectively.

When interconverting among different units, one must ensure that the phytochrome spectra used for that purpose properly reflect the properties of the phytochrome being assayed. For example, units used to assay for phytochrome that exhibits as P_{fr} reduced extinction in the far-red region must be interconverted by reference to absorbance spectra of the same kind of phytochrome (PRATT and CUNDIFF 1975).

2.5 Assay in Light-Scattering Samples

Absorption spectroscopy of light-scattering samples is a complex subject (BUTLER 1962, 1964, SPRUIT 1972, BRITZ et al. 1977, FUKSHANSKY 1978, FUKSHANSKY and KAZARINOVA 1980) whose theoretical basis has not yet been defined fully (GROSS et al. 1983). Detailed discussion of this subject here is not appropriate. Only the two most obvious effects of light scattering will be mentioned. One effect is the increased attenuation of the measuring beam(s) caused by scatter as opposed to absorption. This effect may be eliminated by measuring a LIAC as discussed above (Sect. 2.3). The other effect is the increase in measuring light path that occurs as the consequence of multiple scattering within the sample (BUTLER 1962). This effect of scattering is often used to advantage because it leads to amplification of absorption which thereby increases assay sensitivity (BUTLER 1962, 1964). An empirical approach to account for this increase in the measuring light path is to modify the Lambert-Beer relationship by inclusion of a term (β) that expresses the amplification that results from light scattering: $A = \beta\varepsilon lc$ (BUTLER 1962).

In vivo assay of phytochrome is susceptible to a variety of artifacts that may lead to faulty interpretation of data (BUTLER 1964, SPRUIT 1972, PRATT 1978, GROSS et al. 1983, Sect. 2.8). In the present context it is important to add that it is virtually impossible to determine β with precision for an in vivo sample. It is thus important to describe precisely the tissue being assayed, the sample geometry, and the method of sample preparation. In addition it is important to ensure that the measuring beam or beams do not have a significant actinic effect on phytochrome phototransformation, that the sample does not have pinholes through which light could pass without interacting with the sample, that the sample is not moved during actinic irradiation, and that the sample is not so opaque that stray light becomes a problem (BUTLER 1964).

The sensitivity of in vitro assays of phytochrome, as already noted, is often enhanced by addition of a light-scattering agent, typically $CaCO_3$. Enhancement of sensitivity by a factor of 10 or more is achieved readily (BUTLER 1962). This enhancement may be quantified (i.e., β determined) simply by inclusion of a known quantity of phytochrome in a representative light-scattering sample (BUTLER 1962).

An artifact associated with the use of added scattering agent has been described (BOISARD et al. 1979). When red-irradiated plant extracts were assayed in cuvettes with horizontal measuring beams and with added $CaCO_3$, sample photoreversibility increased by 20% to 40% over that observed for dark control extracts as the $CaCO_3$ settled. The cause of this signal increase is unknown, but might result from trapping of phytochrome by the scattering agent as the scattering agent settles in the cuvette. Phytochrome would thus be concentrated in the lower part of the cuvette, which is where the measuring beams pass through. This artifact is not observed in cuvettes with vertical measuring beams when only just enough liquid to saturate the $CaCO_3$ is used (CORDONNIER and PRATT unpublished data).

2.6 Specialized Spectrophotometers

Phytochrome assay in optically clear solutions does not require a specially designed spectrophotometer. Even when phytochrome concentration is low, as in a crude plant extract, a conventional single-wavelength spectrophotometer may be used (JUNG and SONG 1979). Spectrophotometer design becomes important only when working with strongly light-scattering samples (BUTLER 1964, SPRUIT 1970).

A simple dual-wavelength spectrophotometer designed and constructed by the author in 1969, and since copied by others (PRATT and MARMÉ 1976, SCHÄFER 1978, GARDNER personal communication), serves as an example of an instrument intended for phytochrome assay in light-scattering samples (KIDD and PRATT 1973, see Fig. 3). A light chopper, which consists of a large wheel with three windows and which is driven by a high-speed (30 rev/s) synchronous motor, divides the measuring lamp output into two beams separated in time. A photomultiplier with a 5-cm-diameter photocathode, which is placed close to the sample, is used to detect as much of the light coming from the sample as possible (BUTLER 1964). The photomultiplier signal, which approximates a square wave of 90 Hz, is passed through a logarithmic amplifier so that the output of the spectrophotometer will be in absorbance units. A lock-in amplifier tuned to the frequency of the measuring signal is used to obtain an output that is linearly proportional to the absorbance difference (ΔA) between the two measuring wavelengths.

Within the past year (1982), the author has modified the instrument just described by interfacing the photomultiplier directly with a microcomputer.

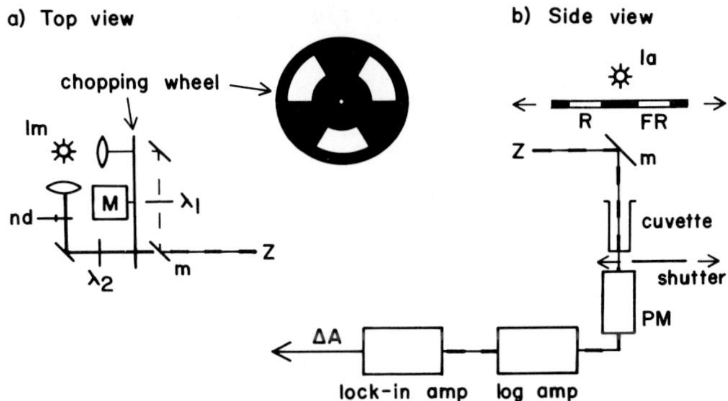

Fig. 3a, b. Schematic diagram of a dual-wavelength spectrophotometer designed for phytochrome assay. **a** Top view. **b** Side view. The measuring light is collected from two sides of a lamp (*lm*), focused, and passed through a chopping wheel driven by a 30 rev s^{-1} motor (*M*) to produce a 90 Hz time-resolved signal composed of two alternating wavelengths (λ_1 and λ_2 selected by interference filters). The two wavelengths are recombined by a partially silvered mirror (*m*), passed through the sample, and detected by an EMI 9658R photomultiplier (*PM*). The 90 Hz output of the photomultiplier is converted to a logarithmic signal by a Philbrick/Nexus logarithmic amplifier (No. 4351) and then fed into a Princeton Applied Research lock-in amplifier (Model 120) tuned to 90 Hz. The output, which is the absorbance difference between the two wavelengths (ΔA^{R}, ΔA^{FR}, or ΔA^{I}, see Sects. 2.3, 2.7.2), is displayed by a digital multimeter or recorder. A neutral density wheel (*nd*) is inserted into one optical path to permit balancing of the two measuring beams. The measuring beam is deflected downward by a second partially silvered mirror (*m* in **b**), which permits insertion of an actinic lamp (*la*) above the cuvette. During measurement a plate holding red (*R*) and far-red (*FR*) filters is positioned with an opaque region between the actinic lamp and the cuvette while a movable shutter is removed from a position between the cuvette and the photomultiplier. During actinic irradiation this shutter is inserted between the cuvette and the photomultiplier to protect the photomultiplier. Insertion of the logarithmic amplifier permits one to measure ΔA values without the necessity to balance precisely the two measuring beams. Phase of the lock-in amplifier is adjusted with a circuit built around a small lamp and photoresistor placed on opposite sides of the chopping wheel

Among other advantages, the new design permits one to account for photomultiplier dark current in calculating absorbance values. In addition, the logarithmic and lock-in amplifiers then are eliminated, making the computerized version not only more powerful but also less expensive. An extra advantage derived from use of a microcomputer is that it may also control the duration of actinic irradiations and measuring intervals, making it easier to automate the spectrophotometer.

Custom-built spectrophotometers intended for phytochrome assay have also been designed by others (BUTLER et al. 1959, 1963, SPRUIT 1970, LARCHER 1971). In addition commercial spectrophotometers that may be adapted to phytochrome assay in light-scattering samples are also becoming more common, although the commercial instrument originally designed for this application (known as the Ratiospect) is no longer available.

2.7 Applications Other than Quantitation

2.7.1 Phytochrome Distribution

Four approaches to the study of phytochrome distribution have utilized spectrophotometric assays. (1) Plants have been dissected and comparable pieces have been pooled and assayed (BRIGGS and SIEGELMAN 1965, CORRELL et al. 1968, MCARTHUR and BRIGGS 1970). Highest phytochrome content has been found in grass coleoptile tips and in meristematic and recently meristematic regions, such as the plumular hook. (2) The intracellular distribution of phytochrome has been studied by spectrophotometric assay of purified subcellular fractions. Phytochrome has been reported to be associated with the plasma membrane (YU 1975, MARMÉ et al. 1976), endoplasmic reticulum (WILLIAMSON et al. 1975, MARMÉ et al. 1976), etioplast envelope (EVANS and SMITH 1976), degraded ribonucleoprotein particles (QUAIL and GRESSEL 1976), nuclei (WAGLE and JAFFE 1980), and mitochondria (MANABE and FURUYA 1975). One must conclude either that phytochrome is associated with many subcellular fractions or that it adheres non-specifically to a wide variety of structures. (3) Microspectrophotometry has been used to detect phytochrome associated with nuclei in vivo (GALSTON 1968). It is highly unlikely for theoretical reasons, however, that phytochrome could be detected by this method (SPRUIT 1972). The significance of this single report of the use of microspectrophotometry is therefore uncertain at best. (4) An improved approach for determining the distribution of phytochrome along a plant axis has been described (KONDO et al. 1973) but has so far not found additional application. This approach is analogous to that used for scanning a cylindrical acrylamide gel in that phytochrome within a plant axis is assayed as the plant is moved through the measuring beam.

2.7.2 Phytochrome Photoequilibria and Separate Assay of P_r and P_{fr}

Absorbance spectra of P_r and P_{fr} overlap (Fig. 2). Actinic irradiation therefore produces a photostationary equilibrium in which the relative concentrations of P_r and P_{fr} are a function of wavelength (BUTLER et al. 1964a, PRATT 1978). Pr has almost no absorption in the far-red (Fig. 2) so that actinic far-red irradiation converts virtually all P_{fr} to P_r. At shorter wavelengths, however, the proportion of phytochrome present as P_{fr} at photostationary equilibrium ($P_{fr\infty}$) must be determined experimentally. $P_{fr\infty}$ in red light ($P^R_{fr\infty}$) is typically used as a reference value. $P^R_{fr\infty}$ (or any other reference value) may be determined only with an optically clear sample in which phytochrome is the only absorber in the red and far-red regions (BUTLER et al. 1964a, BUTLER 1972). It is theoretically impossible, therefore, to determine independently a $P_{fr\infty}$ value for phytochrome in vivo. Any $P_{fr\infty}$ determination in vivo must of necessity be made with reference to a value obtained in vitro if it is to be expressed in other than relative units.

BUTLER et al. (1964a) reported a $P^R_{fr\infty}$ value of 0.81 for oat phytochrome that was later shown to have been proteolytically degraded to about one-half its original size (GARDNER et al. 1971, PRATT 1975). Remeasurement of $P^R_{fr\infty}$ for large oat phytochrome, which possesses 120,000 molecular weight monomers and if degraded has lost no more than about 6,000 molecular weight in size (VIERSTRA and QUAIL 1982), gave a significantly lower value of 0.75 (PRATT 1975). The same value was also reported for large rye phytochrome (PRATT 1975). YAMAMOTO and SMITH (1981) have remeasured $P^{R}_{fr\infty}$ with large rye phytochrome, obtaining a value of 0.84. This discrepancy in reported values indicates a need for additional investigation to determine whether one of the measurements is in error or whether $P^{R}_{fr\infty}$ varies for as yet unknown reasons.

Given a value for $P^{R}_{fr\infty}$, it is a relatively simple matter to assay separately for P_r and P_{fr}. The following procedure is widely used (BUTLER and LANE 1965). The ΔA of the sample (Sect. 2.3) is measured prior to any actinic irradiation (ΔA^{I}). The sample then is irradiated with far-red light to convert all P_{fr} to P_r and the ΔA is measured again (ΔA^{FR}). Red light is typically not used for this initial actinic irradiation because it converts not only P_r to P_{fr} but also protochlorophyll to chlorophyll (Fig. 1a). The first red-induced absorbance change may therefore reflect both transformations. Once all protochlorophyll has been converted to chlorophyll, however, virtually no interference by chlorophyll in subsequent measurements is to be anticipated if the sample is kept near 0 °C [Fig. 1a, (R+15 min D)−R]. Total phytochrome (P_{tot}) then is assayed by determining the $\Delta\Delta A$ as outlined above (Sect. 2.3). $[P_{tot}]$, $[P_{fr}]$, and $[P_r]$ may then be expressed as follows (BUTLER and LANE 1965):

$$[P_{tot}] = k \cdot |\Delta\Delta A| / P^{R}_{fr\infty}$$

$$[P_{fr}] = k \cdot |\Delta A^{I} - \Delta A^{FR}|$$

$$[P_r] = [P_{tot}] - [P_{fr}],$$

where k is a proportionality constant. The validity of this method depends, of course, upon the validity of the value chosen for $P^{R}_{fr\infty}$. For measurements in vivo, it is most prudent to use a value for $P^{R}_{fr\infty}$ that was determined with a phytochrome sample most similar in properties to phytochrome in vivo. The value of 0.81, which was determined with proteolytically degraded phytochrome, is no longer appropriate.

2.7.3 Spectrophotometric Assay of Purity

The purity of a phytochrome sample is, of course, best expressed as the ratio of phytochrome to total protein. The simplest such expression is the specific absorbance ratio (A_{667}/A_{280} with phytochrome as P_r) (BRIGGS and RICE 1972, PRATT 1973). As originally discussed by BRIGGS and RICE (1972), however, there are two limitations inherent to this expression of purity that bear repeating. (1) A specific absorbance ratio is at best a minimum indication of purity because the visible extinction of phytochrome may be selectively reduced or lost under a wide variety of conditions (e.g. BUTLER et al. 1964b). (2) Proteolytically degraded, 60,000 molecular weight phytochrome has a higher specific absorbance ratio than large, 120,000 molecular weight monomer phytochrome of equal purity (BRIGGS and RICE 1972, RICE and BRIGGS 1973b). A specific absorbance ratio may be misleading in the absence of evidence concerning whether or not the phytochrome in question is proteolytically degraded and, if so, by how much (PRATT 1978, 1979).

2.8 Limitations Inherent to Spectrophotometric Assays

Limitations inherent to spectrophotometric assay of phytochrome have been reviewed several times (e.g. BUTLER 1964, SPRUIT 1972, PRATT 1978). These limitations include the following. (1) Light scattering, especially with in vivo samples, may vary from one sample to the next, leading to consequent changes in β (Sect. 2.5). (2) Screening of phytochrome by other pigments, especially chlorophyll, can yield artifactual data. (3) Reversible or irreversible changes in phytochrome extinction may lead to changes in signal levels that do not reflect changes in phytochrome content (Sects. 2.2, 2.3). (4) Spectral assays for phytochrome obviously cannot detect the phytochrome apoprotein, bleached forms of phytochrome, or non-chromophore-containing fragments derived from phytochrome by proteolysis. (5) Spectral assays are relatively insensitive, limited by the extinction coefficient of phytochrome and by spectrophotometric sensitivity to about 0.1 to 1.0 μg (PRATT 1978). (6) Non-homogeneous pigment distribution may produce misleading data. (7) Fluorescence induced by the spectrophotometer measuring beam(s) may result in artifactual measurements. These last two problems will be reviewed below briefly (Sects. 2.8.1, 2.8.2) because they are neither widely appreciated nor well understood.

Some of the limitations just summarized (2, 5, 7) make it impossible to assay for phytochrome spectrophotometrically in normal light-grown plants. Even though chlorophyll absorbs only weakly in the far-red, GRILL (1977) has nevertheless shown that chlorophyll can interfere with phytochrome assay at 730 nm even in only partially greened plants. Two approaches have been used to quantitate phytochrome in light-grown plants. One has utilized a radioimmunoassay (Sect. 4.1.2) that is insensitive to chlorophyll and is many times more sensitive than spectrophotometric assays (HUNT and PRATT 1979, 1980b). The other has utilized light-grown but achlorophyllous plants obtained by treatment with the herbicide Norflurazon (JABBEN and DEITZER 1978a, 1978b). Both approaches, however, have limitations. The radioimmunoassay at present will detect only phytochrome that can be made soluble while the herbicide-treated plants lack photosynthetic capacity and are therefore energy-deficient as compared to untreated plants. Even though the herbicide does not appear to alter some phytochrome-mediated responses (JABBEN and DEITZER 1979), there are important differences between the data obtained by these two approaches (HUNT and PRATT 1980b, Sect. 4.1.2) that indicate that the herbicide treatment might markedly alter rates of phytochrome synthesis and/or destruction, thus limiting the value of data obtained with achlorophyllous, herbicide-treated plants.

2.8.1 Nonhomogeneous Pigment Distribution and the Sieve Effect

The potential effects of nonhomogeneous pigment distributions and the associated sieve effect on phytochrome quantitation by spectrophotometric assay have been considered several times in recent years (SPRUIT 1972, BRITZ et al. 1977, MACKENZIE et al. 1978a, FUKSHANSKY 1978, FUKSHANSKY and KAZARINOVA 1980, GROSS et al. 1983). SPRUIT (1972) argued, based upon a theoretical treatment, that non-homogeneous pigment distributions might be responsible for many phytochrome spectra that exhibit distortion in the red spectral region. As shown below (Sect. 2.8.2), however, a trivial alternative explanation for these distorted spectra is available. In addition, SPRUIT (1972) argued that a rearrangement of phytochrome from a homogeneous to a non-homogeneous distribution could account for the loss of phytochrome photoreversibility by the process known as phytochrome destruction (FRANKLAND 1972). In contrast, BRITZ et al. (1977) concluded, based upon an alternative theoretical treatment, that changes in phytochrome photoreversibility levels would not result from an intracellular redistribution of phytochrome. MACKENZIE et al. (1978a) demonstrated experimentally that a massive, immunocytochemically detected intracellular redistribution of phytochrome had no effect on the level of phytochrome photoreversibility. Finally, a more complete theoretical approach to the question of how non-homogeneous pigment distributions should influence phytochrome assays, as well as other spectrophotometric measurements, is being developed by Fukshansky and colleagues (FUKSHANSKY 1978, FUKSHANSKY and KAZARINOVA 1980, GROSS et al. 1983). While their analysis is not yet complete, it does indicate that phytochrome redistributions should have a negligible effect on photoreversibility levels. In contrast, changes in the extent of vacuolarisation, for example as cells mature, would be expected to have a marked effect on photoreversibility levels via the sieve effect (GROSS et al. 1983).

2.8.2 Fluorescence Induced by the Spectrophotometer Measuring Beam(s)

Perturbation of phytochrome assays by fluorescence induced by the spectrophotometer measuring beam(s) was recognized as a problem very early (LANE et al. 1963, BUTLER 1964). This problem is by no means obvious. One is tempted to argue, intuitively, that such a fluorescence artifact is eliminated by subtraction of a reference signal, assuming that the reference exhibits the same level of fluorescence as the sample. A simple treatment of this problem (PRATT 1978) and a more thorough theoretical discussion (FUKSHANSKY and KAZARINOVA 1980) have appeared recently. Here I shall present only a simple demonstration that, if it exists, there is no escape from fluorescence induced by the measuring beam(s) when the assay sample is strongly light scattering.

Sample fluorescence is a potentially serious problem with a spectrophotometer designed for light-scattering samples, regardless of whether the sample is light scattering, because the large solid angle of light collection detects sample fluorescence with high efficiency. If the sample is optically clear this potential artifact may be reduced by moving the sample away from the light detector. Detection of the transmitted measuring beam is not impaired because it is collimated, while detection of fluorescence, which is not collimated, is greatly decreased because the angle of light collection is greatly reduced. With a light-scattering sample, however, there is no practical way to detect selectively only transmitted measuring light, regardless of the spectrophotometric optics used, because the transmitted light is now as uncollimated as the fluoresced light. The following argument indicates why fluorescence induced by the measuring beam is a serious problem in phytochrome assay (PRATT 1978).

The absorbance difference (ΔA) between a sample and reference (they may be the same but at two time points such as after red and far-red actinic irradiations) is by definition

$$\Delta A = A_s - A_r = \log(I_s^\circ/I_s) - \log(I_r^\circ/I_r)$$

where A_s (A_r) is the sample (reference) absorbance, I_s° (I_r°) is the fluence rate of the measuring beam incident upon the sample (reference), and I_s (I_r) is the measuring beam fluence rate transmitted by the sample (reference). By assuming that $I_s^\circ = I_r^\circ$ (this simplification does not change the outcome) and by rearranging the above relationship one obtains

$$\Delta A = \log(I_r/I_s).$$

To account for fluorescence, one must add a parameter (I_s^f for the sample, I_r^f for the reference) to this expression to obtain

$$\Delta A^* = \log[(I_r + I_r^f)/(I_s + I_s^f)],$$

where ΔA^*, which must be less than or equal to ΔA, is the value actually measured. When $I_s = 0.1\ I_r$, $\Delta A = 1$ by definition. This value is obtained of course when $I_s^f = I_r^f = 0$. When, however, $I_s^f = I_r^f = I_s$, $\Delta A^* = 0.74$, a distortion of 26%. When $I_s^f = I_r^f = I_r$, $\Delta A^* = 0.26$, a distortion of 74%. Clearly, the effect of fluorescence induced by the measuring beam increases as I^f increases (even though $I_s^f = I_r^f$) and as I_r and I_s decrease relative to I° as they do when absorbance increases and/or when light attenuation by scattering increases (BUTLER 1964). As shown elsewhere (PRATT 1978, FUKSHANSKY and KAZARINOVA 1980), these relative values for I^f are to be anticipated under a variety of conditions, some of which are summarized below.

An experimental demonstration of this artifact is most easily made by measuring the apparent absorbance of serially diluted protein samples with a spectrophotometer designed for light-scattering samples (Fig. 4). With the cuvette close to the photomultiplier, deviation from linearity begins at less than 1*A*. That this non-linearity is caused by measuring-beam-induced fluorescence is verified by moving the cuvette away from the photomultiplier, in which case linearity is increased to 2*A*. These data further indicate that fluorescence-induced distortion is most evident at wavelengths that the fluorescing chromophore absorbs most strongly (note that non-linearity at 280 nm occurs at lower protein concentration that at 250 nm). This simple demonstration of how easily fluorescence induced by the measuring beam can lead to spectral artifacts should not only serve as a warning when assaying light-scattering samples but should also indicate the need for caution in all spectrophotometric assays that utilize optics intended for application to light-scattering samples.

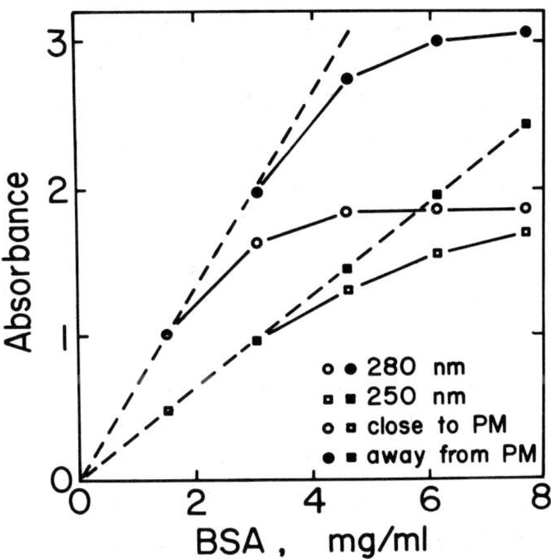

Fig. 4. Absorbance at 280 nm (○, ●) and at 250 nm (□, ■) of bovine serum albumin solutions. Measurements were made with an Hitachi 557 spectrophotometer with the 1-cm pathlength cuvette close to the photomultiplier (about 2 cm from the photocathode; ○, □) or in the position away from the photomultiplier (about 7 cm from the photocathode; ●, ■). *Dashed lines* indicate the expected outcome assuming no deviation from the Lambert-Beer relationship

What are the practical effects of these considerations on phytochrome assays? Three conditions that are often met favor the introduction of fluorescence artifacts. (1) As chlorophyll concentration is increased, for example during greening, distortions in the region of chlorophyll absorption should be anticipated (PRATT 1978, FUKSHANSKY and KAZARINOVA 1980). (2) As I_s becomes much smaller than I_s°, which it does for a strongly absorbing and/or light-scattering sample, the fluorescence artifact increases in a greater than linear fashion. For example, when $A=2$, $I_s=0.01\ I_s^\circ$ and $I_s^f \propto 0.99\ I_s^\circ$, but when $A=3$, $I_s=0.001\ I_s^\circ$ while $I_s^f \propto 0.999\ I_s^\circ$. If $I_s^f=0.01\ (I_s^\circ-I_s)$, then a 50% increase in A (from 2 to 3) would result in an approximately 230% increase in the magnitude of the fluorescence-induced artifact (from 15% to 35%). Furthermore, a ΔA of 0.01 at 667 nm in a sample with a chlorophyll-derived background absorbance of 2 or 3 and with $I^f=0.01\ (I^\circ-I)$ would be observed as a ΔA^* of only 0.005 or 0.0009, respectively. Only a modest amount of fluorescence obviously can have a disastrous effect on photoreversibility assays or difference spectrum measurements. (3) As sample temperature is lowered, with a concomitant increase in fluorescence quantum yields both for chlorophyll (GOVINDJEE and GOVINDJEE 1975) and for phytochrome (SONG et al. 1975), distortions should become more evident.

These fluorescence-induced artifacts include a flattening of absorption bands, especially in the red, blue and ultraviolet regions (Fig. 4; see also Fig. 2 in PRATT 1978), or a shift in a difference spectrum toward the baseline, and a shift toward the blue in the apparent wavelength of maximum absorption by P_r (PRATT 1978). Hence, measurements that exhibit distortions of this type (SPRUIT 1966a, 1966b, 1966c, 1967a, 1967b, SPRUIT and SPRUIT 1972, GRILL 1972, 1977, SPRUIT and KENDRICK 1973, 1977, SPRUIT et al. 1975, KENDRICK and SPRUIT 1976, JOSE et al. 1977), especially as a function of increasing chlorophyll concentration or decreasing sample temperature, might more sensibly be attributed to fluorescence rather than to other causes such as screening or nonhomogeneous pigment distributions (SPRUIT 1972, SPRUIT and SPRUIT 1972).

The interpretation (KENDRICK and SPRUIT 1973, 1977) of data quite likely distorted by fluorescence also must be viewed with caution. As FUKSHANSKY and KAZARINOVA (1980) have shown theoretically, and GRILL (1977) has demonstrated experimentally, artifacts can be obtained even when one is using far-red and near infra-red measuring wavelengths.

One other artifact arising from fluorescence induced by the measuring beam is the obvious underestimate of protein concentration that may occur when an in vitro sample is assayed in a spectrophotometer that has optics designed for light-scattering samples. Whenever $A_{280} > 1$ (Fig. 4) such optics would be expected to result in an artifactual overestimate of the specific absorbance ratio (Sect. 2.7.3) of the sample.

3 Spectrophotometric Assay of Other Photoreceptors

3.1 Phycochromes

Action spectra for chromatic adaptation in blue-green algae (FUJITA and HATTORI 1962, DIAKOFF and SCHEIBE 1973) led to a search for a photoreversible pigment analogous to phytochrome that would explain the antagonistic effects of green and red light. Photochromic phycobiliproteins that have absorption spectra indicating that they may serve as photoreversible morphogenically active photoreceptors have been isolated from blue-green algae (SCHEIBE 1972, BJÖRN and BJÖRN 1976, see BJÖRN and BJÖRN 1980 for review). These photoreversible phycobiliproteins have been named phycochromes. Recent work, however, demonstrates that these phycobiliproteins become photochromic for artifactual reasons (OHAD et al. 1979, OHKI and FUJITA 1979), leading to the suggestion that the biological role, if any, of photochromic phycobiliproteins must be reconsidered (OHKI and FUJITA 1979). It is therefore inappropriate here to discuss assays specific for phycochromes since it is not yet evident that they are photomorphogenic photoreceptors.

3.2 Mycochrome

Conidial development in some fungi is modulated by blue and near ultraviolet light, with the response being a function of the last irradiation given (see KUMAGAI 1978 for review, also Chap 23, this Vol.). A blue, near ultraviolet photoreaction has been observed in a light-minus-dark difference spectrum that indicates the presence of two pigments that might function as photomorphogenic photoreceptors (KUMAGAI 1979). This pigment complex has been called mycochrome. It is not yet firmly established that mycochrome functions in fact as the photomorphogenic photoreceptor in this system and, as for the phycochromes, it is at present inappropriate here to discuss mycochrome assays in any detail.

3.3 Blue-Light Photoreceptor

A search for the photoreceptor(s) responsible for blue-light-induced photomorphogenesis was initiated a little over 10 years ago, based upon the premise that the blue-light photoreceptor(s) might produce a light-induced absorbance change that could serve as the basis for an assay (BERNS and VAUGHN 1970). POFF and BUTLER (1974) later identified absorbance changes that were induced by blue light with an action spectrum similar to those obtained for blue-light-mediated photomorphogenesis. They suggested that these absorbance changes might result from a flavin-mediated photoreduction of a b-type cytochrome. Subsequent work led to the isolation and characterization of membrane systems containing the flavin-cytochrome pigment complex, which has been proposed as a candidate for the blue-light photoreceptor (see BRIGGS 1980 for review). Nevertheless, it is not yet evident that a photomorphogenically active blue-light photoreceptor has been unequivocally identified. Discussion of assays specific for this photoreceptor is thus premature in the present context.

4 Immunochemical Assay of Phytochrome

Immunochemical assays do not suffer from the limitations inherent to spectrophotometric assays (Sect. 2.8) because they are based upon totally different properties of phytochrome. Immunochemical assays, however, do have a few serious limitations of their own. They do not discriminate between P_r and P_{fr} nor can they always discriminate between spectrophotometrically active and inactive forms of phytochrome. Nevertheless, immunochemical assays can provide information unavailable by dependence solely upon spectrophotometric assays, thus providing an invaluable alternative assay capability. Only a few of the wide variety of immunochemical assays that exist (e.g. see WEIR 1978) have so far been applied to phytochrome (PRATT 1983). The present discussion will include only those few.

4.1 Quantitative Assays

4.1.1 Radial Immunodiffusion

Radial immunodiffusion assay of phytochrome is sensitive to about 5 $\mu g\ ml^{-1}$ or, since only about 3 µl is needed for assay, about 15 ng (PRATT et al. 1974). The assay utilizes an agar plate containing antiphytochrome immunoglobulins at a low, uniform concentration. When phytochrome is put into a well formed in this agar plate, it diffuses outward until its concentration is diluted enough that precipitation by the antiphytochrome immunoglobulins will take place. A precipitation ring, which encompasses an area linearly proportional to the amount of phytochrome placed into the well, is thus formed (PRATT et al. 1974).

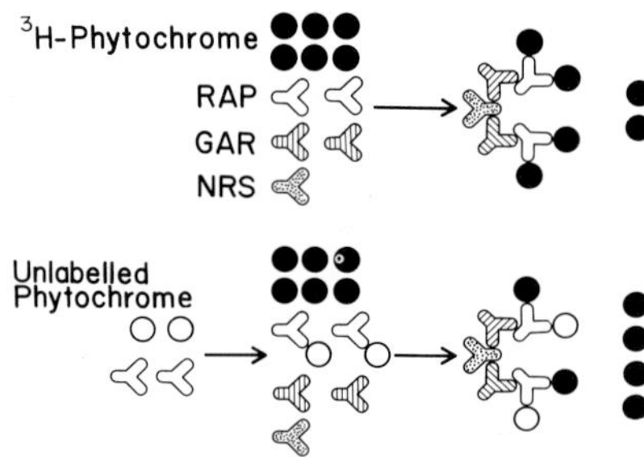

Fig. 5. Protocol for phytochrome radioimmunoassay. *RAP* rabbit antiphytochrome immunoglobulin; *GAR* goat antirabbit-immunoglobulin immunoglobulin; *NRS* nonimmune rabbit serum immunoglobulin. (After PRATT 1983. Drawing courtesy of R. HUNT)

Radial immunodiffusion assay has been used to follow phytochrome destruction both as P_{fr} (PRATT et al. 1974) and as P_r following a red, far-red irradiation sequence (STONE and PRATT 1979). In both cases, antigenically detectable phytochrome was lost in parallel with spectrophotometrically detectable phytochrome indicating that destruction of both P_r and P_{fr} resulted from a general proteolytic degradation of phytochrome.

4.1.2 Radioimmunoassay

The radioimmunoassay developed for use with phytochrome (HUNT and PRATT 1979) has a present sensitivity of about 500 pg. While this sensitivity may be improved, it is already enough to permit quantitation of phytochrome in crude extracts of green, light-grown plants (HUNT and PRATT 1980b).

A so-called second-antibody method was chosen for use with phytochrome (Fig. 5). In this assay, a second antibody (goat antirabbit-immunoglobulin immunoglobulins) is used to form an immunoprecipitate large enough to separate bound from unbound phytochrome by centrifugation. The assay depends upon the replacement in the ultimate immunoprecipitate of ^{3}H-phytochrome by non-radioactive phytochrome. The amount of radioactivity lost from the immunoprecipitate is proportional to the amount of non-radioactive phytochrome added. The parameters of the assay were designed for maximum sensitivity, so that the first increment of non-radioactive phytochrome will replace ^{3}H-phytochrome. A standard curve with a sensitivity of about 500 pg may thus be obtained (Fig. 6). This sensitivity permits quantitation of phytochrome in only 100 nl of a crude extract of etiolated oats or 10 μl of a comparable extract from green, light-grown oats (HUNT and PRATT 1979, 1980b).

The phytochrome radioimmunoassay has not yet been widely utilized. Nevertheless, the recent preparation of antisera against dicotyledonous phytochrome (CORDONNIER and PRATT 1980, 1982a, b) means that this method may now be applied to virtually any angiosperm. So far the radioimmunoassay has been used to demonstrate that the phytochrome content of light-grown oats oscillates by a factor of three on a diurnal cycle (HUNT and PRATT 1980b). In addition, an extended 48-h dark period was found to lead to a 50-fold increase in phytochrome level (HUNT and PRATT 1980b). These observations are in sharp contrast to those reported by JABBEN and DEITZER (1978b) who observed by spectrophotometric assay of achlorophyllous, light-grown oats (Sect. 2.8) only a two fold

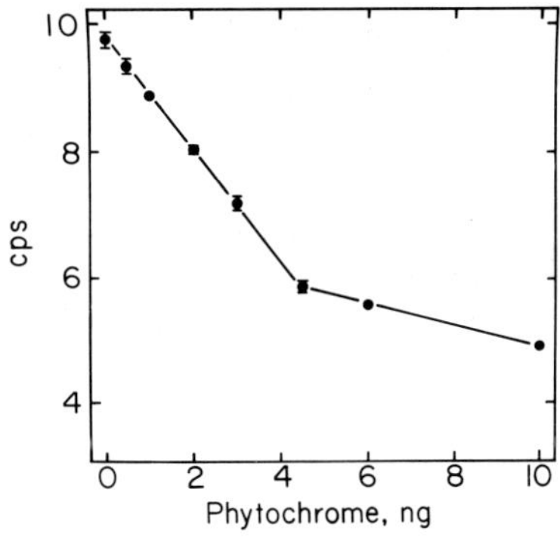

Fig. 6. Phytochrome radioimmunoassay standard curve showing the immunoprecipitate radioactivity in counts per second (cps) as a function of the amount of non-radioactive large oat phytochrome added. (After PRATT 1983. Data courtesy of R. HUNT)

increase in phytochrome content during a 48-h dark period. The reason for the apparent discrepancy remains to be resolved although several possible explanations exist (HUNT and PRATT 1980b, Sect. 2.8).

4.2 Qualitative Assays

4.2.1 Immunoelectrophoresis and Ouchterlony Double Immunodiffusion

Immunoelectrophoresis and Ouchterlony double immunodiffusion have been used extensively for phytochrome assay (HOPKINS and BUTLER 1970, PRATT 1973, RICE and BRIGGS 1973a, CUNDIFF and PRATT 1973, 1975a, 1975b, PRATT et al. 1974, CORDONNIER and PRATT 1982a, b) because of their simplicity. Even though simple, these assays have provided important information about phytochrome. The observation that 60,000 molecular weight degraded phytochrome and large phytochrome, which has 120,000 molecular weight monomers, are related by proteolysis (GARDNER et al. 1971) was confirmed by the detection of antigenic non-identity between the two sizes of phytochrome both by Ouchterlony double diffusion assay (CUNDIFF and PRATT 1975b, PRATT 1978) and by immunoelectrophoresis (CUNDIFF and PRATT 1973). Furthermore, antiserum against large phytochrome was used to demonstrate that the phytochrome monomer is asymmetric with respect to its primary structure rather than being composed of two similar photoreversible chromopeptides linked by a covalent bond (CUNDIFF and PRATT 1975a). Phytochrome preparations obtained from different plant species have also been compared by Ouchterlony double diffusion. Phytochrome from monocotyledonous plants is immunologically distinct from that obtained from dicotyledonous plants as shown by the poor reactivity between dicotyledonous phytochrome and antimonocotyledonous-phytochrome serum (PRATT 1973, RICE and BRIGGS 1973a) and between monocotyledonous phytochrome and antidicotyledonous-phytochrome serum (CORDONNIER and PRATT 1980, 1982a). Nevertheless, there do seem to be a few antigenic determinants that might be common to all phytochrome molecules, leading to the suggestion that the immunoglobulins that recognize these common determinants might serve as an assay for a putative "active site" of phytochrome (CORDONNIER and PRATT 1982a).

4.2.2 Micro Complement Fixation

Micro complement fixation is a very sensitive indicator of antigenic activity (LEVINE and VAN VUNAKIS 1967). HOPKINS and BUTLER (1970) first applied this assay to phytochrome and demonstrated a difference between P_r and P_{fr}. Subsequent application of this technique, however, has failed to repeat their observation (PRATT 1973, CUNDIFF and PRATT 1975b). Micro complement fixation assay has also been used to compare phytochrome preparations obtained from different plants. This assay not only confirmed the marked immunological dissimilarity between monocotyledonous and dicotyledonous phytochrome but also detected differences among phytochrome preparations obtained from different grasses (PRATT 1973).

4.2.3 Immunocytochemistry

Because immunocytochemical assay is in principle sensitive to the presence of only a single antigen molecule (STERNBERGER 1979), it has been an important tool for the study of phytochrome distribution. The sensitivity and resolution provided by immunocytochemistry are many-fold greater than that provided by spectrophotometric assay (Sect. 2.7.1).

The immunocytochemical assay most extensively used for the visualization of phytochrome is the double-indirect, peroxidase-antiperoxidase method (STERNBERGER et al. 1970). Tissue first is fixed histologically to immobilize phytochrome. The fixed tissue is sectioned and peroxidase is cross-linked to phytochrome in the sections by a bridge of immunoglobulins. Immunospecifically adsorbed peroxidase then is visualized by reaction with diaminobenzidine and H_2O_2, which produces a reaction product that is visible by light microscopy. The reaction product may be made electron dense for electron microscopy by subsequent incubation with OsO_4 (see PRATT et al. 1976, for detailed discussion of the method as applied to phytochrome). More recently a single-indirect immunofluorescence method has been applied to the problem of phytochrome visualization (EPEL et al. 1980). This simpler method appears to offer both higher sensitivity and better structural preservation than that obtained with the original peroxidase-antiperoxidase method.

Observations concerning phytochrome distribution have been consistent with those obtained by spectrophotometric assay (Sect. 2.7.1), although information obtained by immunocytochemistry is of much greater spatial resolution. Within a grass seedling, phytochrome has been found in relatively high concentration in the root cap, the apex of emerging adventitious roots, the tip of the coleoptile, the coleoptilar node, and, in some instances, the basal region of the young leaves (PRATT and COLEMAN 1971, 1974). As a general rule phytochrome has been found in young, rapidly expanding cells recently derived from meristematic cells but not in the meristematic cells themselves.

Within a single cell, phytochrome as P_r has been observed to be distributed generally throughout the cytosol as though it were a soluble chromoprotein (COLEMAN and PRATT 1974a, 1974b). In at least some plants, however, phytochrome following its photoconversion to P_{fr} becomes associated with discrete but as yet unidentified subcellular regions about 1 μm in size (MACKENZIE et al. 1975, 1978b, EPEL et al. 1980). While by no means established, this sequestering of phytochrome in its active, P_{fr} form may represent an interaction between P_{fr} and its putative receptor or reaction partner.

5 Effects of Proteolysis and Denaturation on Phytochrome Assays

Phytochrome assay in vitro is influenced strongly by both proteolytic degradation and denaturation. Many such influences have already been mentioned above. The intent here is to focus briefly on these effects.

5.1 Effects of Proteolysis

Proteolytic degradation of phytochrome to the 60,000 molecular weight photoreversible chromopeptide (GARDNER et al. 1971) is accompanied by changes in phytochrome properties that influence both spectrophotometric and immunochemical assays. Of greatest importance are (a) the increased molar extinction in the visible as compared to the ultraviolet spectral regions, which leads to an increase in specific absorbance ratio without a concomitant increase in purity (Sect. 2.7.3), and (b) the decreased electrophoretic mobility under nondenaturing conditions (PRATT 1973), which can serve as the basis of an assay to detect the extent of proteolytic degradation of a phytochrome preparation (PRATT 1978, 1979). In addition, although the extinction spectra of P_r and P_{fr} are virtually the same whether or not the protein moiety is degraded (PRATT and CUNDIFF 1975, PRATT 1978), $P^R_{fr\infty}$ may be greater for degraded than for undegraded phytochrome (Sect. 2.7.2), which means that for a given molar concentration of phytochrome, degraded phytochrome would give a significantly greater photoreversibility signal (Sect. 2.7.2). Thus, if no phytochrome is lost during proteolysis, photoreversibility measurements could indicate a presumptive increase in phytochrome content by 8% [(0.81/0.75)−1)].

5.2 Effects of Denaturation

Limited denaturation of phytochrome leads both to reduced extinction in the visible region (BUTLER et al. 1964b, Sect. 2.2) and to a small blue shift in the wavelength of maximum absorption by P_r (BUTLER et al. 1964b). More extensive denaturation leads to a virtually complete loss of visible extinction (BUTLER et al. 1964b). Denaturation thus will lead obviously to decreases in spectrophotometrically detectable phytochrome levels even though no change in phytochrome concentration occurs (Sects. 2.2, 2.3, 2.4). Since in at least one case denaturation is caused by a naturally occurring substance (FURUYA and HILLMAN 1966), the possibility that changes in phytochrome photoreversibility detected in vivo result from denaturation-induced extinction changes must be considered. Also obvious is the fact that spectral assays become almost useless when working with phytochrome in vitro under conditions that might lead to denaturation, such as when studying the interaction between phytochrome and lipoidal materials (e.g. GEORGEVICH et al. 1976). In contrast, antiphytochrome immunoglobulins appear to recognize spectrally denatured phytochrome at least as well, if not better, than undenatured phytochrome (PRATT et al. 1974, PRATT unpublished observations). Immunochemical methods therefore appear to be the best approach to assaying for phytochrome when denaturation is probable.

6 Future

The vast majority of investigators have utilized only spectrophotometric assays because (a) they were the first ones available, (b) they are relatively simple, and (c) they do not require any involvement with biochemistry. Nevertheless, the importance of immunochemical assays should be apparent from the preceeding discussion and it is anticipated that they will be of increasing importance in the future. In particular, extensive application of radioimmunoassay methods to phytochrome should yield our first directly obtained information about phytochrome in green plants grown under natural environmental conditions. Recent improvements in immunocytochemical methodology (EPEL et al. 1980) should also make it possible to begin examining phytochrome distribution in green, light-grown plants. Future application of the monoclonal antibody technique (MÖLLER 1979) to the production of antiphytochrome immunoglobulins might permit among many other applications the production of an immunoglobulin specific for P_r or P_{fr}. The possibility would then exist for developing a radioimmunoassay and/or immunocytochemical assay that would be selective for only one form of phytochrome. The advantages of such a development should be obvious.

Although my emphasis on the future is placed on immunochemical assays, a recent focus on theoretical treatments of spectrophotometric assays by FUKSHANSKY and his colleagues (Sects. 2.8.1, 2.8.2) should provide a better basis for the interpretation of spectrophotometric data, especially those obtained by in vivo measurements. It is disconcerting that it is still impossible to predict with certainty the practical consequences of pigment non-homogeneities and light scattering on optical measurements of photomorphogenic photoreceptors.

Acknowledgment. My research and the preparation of this contribution were supported by grants from the National Science Foundation.

References

Berns DS, Vaughn JR (1970) Studies on the photopigment system in Phycomyces. Biochem Biophys Res Commun 39:1094–1103

Björn GS, Björn LO (1976) Photochromic pigments from blue-green algae: phycochromes a, b, and c. Physiol Plant 36:297–304

Björn LO, Björn GS (1980) Photochromic pigments and photoregulation in blue-green algae. Photochem Photobiol 32:849–852

Boisard J, Cordonnier M-M, Gabriac B, Marmé D, Pratt LH (1979) Apparent increase in photodetectable phytochrome in the presence of calcium carbonate. Photochem Photobiol 30:513–517

Borthwick H (1972) History of phytochrome. In: Mitrakos K, Shropshire W Jr (eds) Phytochrome. Academic Press, London New York, pp 3–23

Borthwick HA, Hendricks SB (1960) Photoperiodism in plants. Science 132:1223–1228

Borthwick HA, Hendricks SB, Parker MW (1952a) The reaction controlling floral initiation. Proc Natl Acad Sci USA 38:929–934

Borthwick HA, Hendricks SB, Parker MW, Toole EH, Toole VK (1952b) A reversible photoreaction controlling seed germination. Proc Natl Acad Sci USA 38:662–666

Briggs WR (1980) A blue light photoreceptor system in higher plants and fungi. In: DeGreef J (ed) Photoreceptors and plant development. Antwerpen Univ Press, Antwerpen, pp 17–28

Briggs WR, Rice HV (1972) Phytochrome: chemical and physical properties and mechanism of action. Annu Rev Plant Physiol 23:293–334

Briggs WR, Siegelman HW (1965) Distribution of phytochrome in etiolated seedlings. Plant Physiol 40:934–941

Britz SJ, Mackenzie JM Jr, Briggs WR (1977) Non-homogeneous pigment distribution, multiple cell layers, and the determination of phytochrome by in vivo spectrophotometry. Photochem Photobiol 25:137–140

Butler WL (1962) Absorption of light by turbid materials. J Opt Soc Am 52:292–299

Butler WL (1964) Absorption spectroscopy in vivo: theory and application. Annu Rev Plant Physiol 15:451–470

Butler WL (1972) Photochemical properties of phytochrome in vitro. In: Mitrakos K, Shropshire W Jr (eds) Phytochrome. Academic Press, London New York, pp 185–192

Butler WL, Lane HC (1965) Dark transformations of phytochrome in vivo. II. Plant Physiol 40:13–17

Butler WL, Norris KH, Siegelman HW, Hendricks SB (1959) Detection, assay, and preliminary purification of the pigment controlling photoresponsive development of plants. Proc Natl Acad Sci USA 45:1703–1708

Butler WL, Lane HC, Siegelman HW (1963) Nonphotochemical transformations of phytochrome in vivo. Plant Physiol 38:514–519

Butler WL, Hendricks SB, Siegelman HW (1964a) Action spectra of phytochrome in vitro. Photochem Photobiol 3:521–528

Butler WL, Siegelman HW, Miller CO (1964b) Denaturation of phytochrome. Biochemistry 3:851–857

Coleman RA, Pratt LH (1974a) Electron microscopic localization of phytochrome in plants using an indirect antibody-labeling method. J Histochem Cytochem 22:1039–1047

Coleman RA, Pratt LH (1974b) Subcellular localization of the red-absorbing form of phytochrome by immunocytochemistry. Planta 121:119–131

Cordonnier M-M, Pratt LH (1980) Preparation, characterization, and utilization of antiserum against zucchini phytochrome. In: DeGreef J (ed) Photoreceptors and plant development. Antwerpen Univ Press, Antwerpen, pp 69–78

Cordonnier M-M, Pratt LH (1982a) Comparative phytochrome immunochemistry as assayed by antisera against both monocotyledonous and dicotyledonous phytochrome. Plant Physiol 70:912–916

Cordonnier M-M, Pratt LH (1982b) Immunopurification and initial characterization of dicotyledonous phytochrome. Plant Physiol 69:360–365

Correll DL, Edwards JL, Medina VJ (1968) Phytochrome in etiolated annual rye. II. Distribution of photoreversible phytochrome in the coleoptile and primary leaf. Planta 79:284–291

Cundiff SC, Pratt LH (1973) Immunological determination of the relationship between large and small sizes of phytochrome. Plant Physiol 51:210–213

Cundiff SC, Pratt LH (1975a) Immunological and physical characterization of the products of phytochrome proteolysis. Plant Physiol 55:212–217

Cundiff SC, Pratt LH (1975b) Phytochrome characterization by rabbit antiserum against high molecular weight phytochrome. Plant Physiol 55:207–211

Diakoff S, Scheibe J (1973) Action spectra for chromatic adaptation in *Tolypothrix tenuis*. Plant Physiol 51:382–385

Epel BL, Butler WL, Pratt LH, Tokuyasu KT (1980) Immunofluorescence localization studies of the P_r and P_{fr} forms of phytochrome in the coleoptile tips of oats, corn and wheat. In: DeGreef J (ed) Photoreceptors and plant development. Antwerpen Univ Press, Antwerpen, pp 121–133

Evans A, Smith H (1976) Spectrophotometric evidence for the presence of phytochrome in the envelope membranes of barley etioplasts. Nature 259:323–325

Frankland B (1972) Biosynthesis and dark transformations of phytochrome. In: Mitrakos K, Shropshire W Jr (eds) Phytochrome. Academic Press, London New York, pp 195–225

Fujita Y, Hattori A (1962) Photochemical interconversion between precursors of phycobilin chromoproteids in *Tolypothrix tenuis*. Plant Cell Physiol 3:209–220

Fukshansky L (1978) On the theory of light absorption in non-homogeneous objects. J Math Biol 6:177–196

Fukshansky L, Kazarinova N (1980) Extension of the Kubelka-Munk theory of light propagation in intensely scattering materials to the fluorescent media. J Opt Soc Am 70:1101–1111

Furuya M, Hillman WS (1966) Rapid destruction of the P_{fr} form of phytochrome by a substance in extracts of *Pisum* tissue. Plant Physiol 41:1242–1244

Galston AW (1968) Microspectrophotometric evidence for phytochrome in plant nuclei. Proc Natl Acad Sci USA 61:454–460

Gardner G, Pike CS, Rice HV, Briggs WR (1971) "Disaggregation" of phytochrome in vitro – a consequence of proteolysis. Plant Physiol 48:686–693

Georgevich G, Krauland J, Roux S (1976) Phytochrome interaction with lipid bilayers. Plant Physiol Suppl 57:20

Govindjee, Govindjee R (1975) Introduction to photosynthesis. In: Govindjee (ed) Bioenergetics of photosynthesis. Academic Press, London New York, pp 1–50

Grill R (1972) The influence of chlorophyll on in vivo difference spectra of phytochrome. Planta 108:185–202

Grill R (1977) Influence of chlorophyll content on phytochrome measurements in turnip cotyledons. Planta 134:11–16

Gross I, Fukshansky L, Schäfer E (1983) In vivo spectroscopy. In: Smith H (ed) Techniques in photomorphogenesis. Academic Press, London New York (in press)

Hopkins DW, Butler WL (1970) Immunochemical and spectroscopic evidence for protein conformational changes in phytochrome transformations. Plant Physiol 45:567–570

Hunt RE, Pratt LH (1979) Phytochrome radioimmunoassay. Plant Physiol 64:327–331

Hunt RE, Pratt LH (1980a) Partial characterization of undegraded oat phytochrome. Biochemistry 19:390–394

Hunt RE, Pratt LH (1980b) Radioimmunoassay of phytochrome content in green, light-grown oats. Plant Cell Environ 3:91–95

Jabben M, Deitzer GF (1978a) A method for measuring phytochrome in plants grown in white light. Photochem Photobiol 27:799–802

Jabben M, Deitzer GF (1978b) Spectrophotometric phytochrome measurements in light-grown *Avena sativa* L. Planta 143:309–313

Jabben M, Deitzer GF (1979) Effects of the herbicide San 9789 on photomorphogenic responses. Plant Physiol 63:481–485

Jose AM, Vince-Prue D, Hilton JR (1977) Chlorophyll interference with phytochrome measurement. Planta 135:119–123

Jung J, Song P-S (1979) A simple modification of the spectrophotometer for rapid phytochrome assay. Photochem Photobiol 29:419–421

Kendrick RE, Spruit CJP (1973) Phytochrome intermediates in vivo. III. Kinetic analysis of intermediate reactions at low temperature. Photochem Photobiol 18:153–159

Kendrick RE, Spruit CJP (1976) Intermediates in the photoconversion of phytochrome. In: Smith H (ed) Light and plant development. Butterworth, London, pp 31–43

Kendrick RE, Spruit CJP (1977) Phototransformations of phytochrome. Photochem Photobiol 26:201–214

Kidd GH, Pratt LH (1973) Phytochrome destruction: an apparent requirement for protein synthesis in the induction of the destruction mechanism. Plant Physiol 52:309–311

Kondo N, Inoue Y, Shibata K (1973) Phytochrome distribution in *Avena* seedlings measured by scanning a single seedling. Plant Sci Lett 1:165–168

Kumagai T (1978) Mycochrome system and conidial development in certain fungi imperfecti. Photochem Photobiol 27:371–379

Kumagai T (1979) Blue and near ultraviolet reversible photoreaction in conidial development of certain fungi. In: Senger H (ed) The blue light syndrome. Springer, Berlin Heidelberg New York

Lane HC, Siegelman HW, Butler WL, Firer EM (1963) Detection of phytochrome in green plants. Plant Physiol 38:414–416

Larcher G (1971) Étude théorique et mise au point d'un spectrophotomètre différentiel à deux longueurs d'onde de très grande sensibilité. Nouv Rev d'Optique 2:331–336

Levine L, van Vunakis H (1967) Micro complement fixation. Methods Enzymol 11:928–936

Lisansky SG, Galston AW (1974) Phytochrome stability in vitro. I. Effect of metal ions. Plant Physiol 53:352–359

Lisansky SG, Galston AW (1976) Phytochrome stability in vitro. II. A low molecular weight protective factor. Plant Physiol 57:188–191

Mackenzie JM Jr, Coleman RA, Briggs WR, Pratt LH (1975) Reversible redistribution of phytochrome within the cell upon conversion to its physiologically active form. Proc Natl Acad Sci USA 72:799–803

Mackenzie JM Jr, Briggs WR, Pratt LH (1978a) Phytochrome photoreversibility: empirical test of the hypothesis that it varies as a consequence of pigment compartmentation. Planta 141:129–134

Mackenzie JM Jr, Briggs WR, Pratt LH (1978b) Intracellular phytochrome distribution as a function of its molecular form and of its destruction. Am J Bot 65:671–676

Manabe K, Furuya M (1975) Distribution and non-photochemical transformation of phytochrome in subcellular fractions from *Pisum* epicotyls. Plant Physiol 56:772–775

Marmé D, Bianco J, Gross J (1976) Evidence for phytochrome binding to plasma membrane and endoplasmic reticulum. In: Smith H (ed) Light and plant development. Butterworth, London, pp 95–110

McArthur JA, Briggs WR (1970) Phytochrome appearance and distribution in the embryonic axis and seedling of Alaska peas. Planta 91:146–154

Möller G (ed) (1979) Hybrid myeloma monoclonal antibodies against MHC products. Immunol Rev 47

Ohad I, Clayton RK, Bogorad L (1979) Photoreversible absorbance changes in solutions of allophycocyanin purified from *Fremyella diplosiphon*: temperature dependence and quantum efficiency. Proc Natl Acad Sci USA 76:5655–5659

Ohki K, Fujita Y (1979) In vivo transformation of phycobiliproteins during photobleaching of *Tolypothrix tenuis* to forms active in photoreversible absorption changes. Plant Cell Physiol 20:1341–1347

Parker MW, Hendricks SB, Borthwick HA, Scully NJ (1946) Action spectrum for the photoperiodic control of floral initiation of short-day plants. Bot Gaz 108:1–26

Poff KL, Butler WL (1974) Absorbance changes induced by blue light in *Phycomyces blakesleeanus* and *Dictyostelium discoideum*. Nature 248:799–801

Pratt LH (1973) Comparative immunochemistry of phytochrome. Plant Physiol 51:203–209

Pratt LH (1975) Photochemistry of high molecular weight phytochrome in vitro. Photochem Photobiol 22:33–36

Pratt LH (1978) Molecular properties of phytochrome. Photochem Photobiol 27:81–105

Pratt LH (1979) Phytochrome: function and properties. Photochem Photobiol Rev 4:59–124

Pratt LH (1983) Phytochrome immunochemistry. In: Smith H (ed) Techniques in photomorphogenesis. Academic Press, London New York (in press)

Pratt LH, Coleman RA (1971) Immunocytochemical localization of phytochrome. Proc Natl Acad Sci USA 68:2431–2435

Pratt LH, Coleman RA (1974) Phytochrome distribution in etiolated grass seedlings as assayed by an indirect antibody-labelling method. Am J Bot 61:195–202

Pratt LH, Cundiff SC (1975) Spectral characterization of high-molecular-weight phytochrome. Photochem Photobiol 21:91–97

Pratt LH, Marmé D (1976) Red light-enhanced phytochrome pelletability: a re-examination and further characterization. Plant Physiol 58:686–692

Pratt LH, Kidd GH, Coleman RA (1974) An immunochemical characterization of the phytochrome destruction reaction. Biochim Biophys Acta 365:93–107

Pratt LH, Coleman RA, Mackenzie JM Jr (1976) Immunological visualisation of phytochrome. In: Smith H (ed) Light and plant development. Butterworth, London, pp 75–94

Quail PH, Gressel J (1976) Particle-bound phytochrome: interaction of the pigment with ribonucleoprotein material from *Cucurbita pepo* L. In: Smith H (ed) Light and Plant Development. Butterworth, London, pp 111–128

Rice HV, Briggs WR (1973a) Immunochemistry of phytochrome. Plant Physiol 51:939–945

Rice HV, Briggs WR (1973b) Partial characterization of oat and rye phytochrome. Plant Physiol 51:927–938

Roux SJ, McEntire K, Brown WE (1982) Determination of extinction coefficients of oat phytochrome by quantitative amino acid analyses. Photochem Photobiol 35:537–543

Schäfer E (1978) Variation in the rates of synthesis and degradation of phytochrome in cotyledons of *Cucurbita pepo* L. during seedling development. Photochem Photobiol 27:775–780

Scheibe J (1972) Photoreversible pigment: occurrence in a blue-green alga. Science 176:1037–1039

Song P-S, Chae Q, Briggs WR (1975) Temperature dependence of the fluorescence quantum yield of phytochrome. Photochem Photobiol 22:75–76

Spruit CJP (1966a) Low-temperature action spectra for transformations of photoperiodic pigments. Biochim Biophys Acta 120:454–456

Spruit CJP (1966b) Thermal reactions following illumination of phytochrome. Meded Landbouwhogesch Wageningen 66–15:1–7

Spruit CJP (1966c) Photoreversible pigment transformations in etiolated plants. Biochim Biophys Acta 112:186–188

Spruit CJP (1967a) Photoreactions in phytochrome-containing extracts from etiolated pea seedlings. Meded Landbouwhogesch Wageningen 67–15:1–9

Spruit CJP (1967b) Phytochrome decay and reversal in leaves and stem sections of etiolated pea seedlings. Meded Landbouwhogesch Wageningen 67–14:1–6

Spruit CJP (1970) Spectrophotometers for the study of phytochrome in vivo. Meded Landbouwhogesch Wageningen 70–14:1–18

Spruit CJP (1972) Estimation of phytochrome by spectrophotometry in vivo: instrumentation and interpretation. In: Mitrakos K, Shropshire W Jr (eds) Phytochrome. Academic Press, London New York, pp 77–104

Spruit CJP, Kendrick RE (1973) Phytochrome intermediates in vivo. II. Characterisation of intermediates by difference spectrophotometry. Photochem Photobiol 18:145–152

Spruit CJP, Kendrick RE (1977) Phototransformations of phytochrome: the characterization of lumi-F and meta-Fa. Photochem Photobiol 26:133–138

Spruit CJP, Spruit HC (1972) Difference spectrum distortion in non-homogeneous pigment associations: abnormal phytochrome spectra in vivo. Biochim Biophys Acta 275:401–413

Spruit CJP, Kendrick RE, Cooke RJ (1975) Phytochrome intermediates in freeze-dried tissue. Planta 127:121–132

Sternberger L (1979) Immunocytochemistry, 2nd edn. Wiley and Sons, New York

Sternberger LA, Hardy PH Jr, Cuculis JJ, Meyer HG (1970) The unlabeled antibody enzyme method of immunohistochemistry. Preparation and properties of soluble antigen-antibody complex (horseradish peroxidase-antihorseradish peroxidase) and its use in identification of spirochetes. J Histochem Cytochem 18:315–333

Stone HJ, Pratt LH (1979) Characterization of the destruction of phytochrome in the red-absorbing form. Plant Physiol 63:680–682

Tobin EM, Briggs WR (1973) Studies on the protein conformation of phytochrome. Photochem Photobiol 18:487–495

Vierstra RD, Quail PH (1982) Native phytochrome: inhibition of proteolysis yields a homogeneous monomer of 124 kilodaltons from *Avena*. Proc Natl Acad Sci USA 79:5272–5276

Wagle J, Jaffe MJ (1980) The association and function of phytochrome in pea nuclei. Plant Physiol Suppl 65:3

Weir DM (ed) (1978) Handbook of experimental immunology, 3rd edn. Blackwell, Oxford

Williamson FA, Morré DJ, Jaffe MJ (1975) Association of phytochrome with rough-surfaced endoplasmic reticulum fractions from soybean hypocotyls. Plant Physiol 56:738–743

Yamamoto K (1980) Molecular and spectroscopic properties of phytochrome purified from etiolated pea shoots. PhD Dissertation, Univ Tokyo

Yamamoto KT, Smith WO Jr (1981) A re-evaluation of the mole fraction of P_{fr} at the red-light-induced photostationary state of undegraded rye phytochrome. Plant Cell Physiol 22:1159–1164

Yu R (1975) Characterization of the phytochrome-containing particles obtained by glutaraldehyde pre-fixation of maize coleoptiles. J Exp Bot 26:808–822

9 Rapid Action of Phytochrome in Photomorphogenesis

P.H. QUAIL

1 Introduction

The kinetics of phytochrome-mediated responses can be a powerful analytical tool in the quest to elucidate this photoreceptor's molecular mechanism of action. Clearly those responses most rapidly detectable upon photoconversion are the ones most likely to be closest in sequence to the primary action of the pigment. The observation that several rapid phytochrome-mediated responses appeared to involve changes in membrane properties led HENDRICKS and BORTHWICK (1967) to propose the "membrane hypothesis" of phytochrome action. In its most explicit formulation, this hypothesis proposes that phytochrome modifies the functional properties of one or more cellular membranes as its primary action upon photoconversion, and that this modification results from the direct physical interaction of the pigment with the components of those membranes. All other observed alterations in cellular and molecular function are then postulated to ensue in cascade fashion from this single molecular mechanism of action.

Since it was first proposed there have been two main categories of investigation directed at testing the validity of this hypothesis: (a) Further kinetic studies aimed at examining the nature of the most rapid phytochrome-induced responses; and (b) localization studies aimed at determining the subcellular location of the phytochrome molecule itself to establish whether there is direct evidence of physical association with membranes. A detailed review of the localization data will not be attempted here. However, as was concluded recently (QUAIL 1982), there is currently no compelling evidence from immunocytochemical or cell fractionation studies for a biologically meaningful association of phytochrome with cellular membranes.

This chapter focuses instead on the kinetic aspects of phytochrome-induced responses. There are now a considerable number of responses that have been interpreted to indicate that phytochrome action has occurred within the first 5 min of photoconversion, the most rapid within seconds. The nature of these responses is examined here and the rigor with which the assembled data support the membrane hypothesis of phytochrome action is evaluated. The data from the various studies are assessed, first for evidence that the criteria for phytochrome involvement per se in the response have been satisfactorily fulfilled, and second for evidence that the changes observed represent changes in membrane properties. Those responses that satisfy these two criteria are then examined for kinetic evidence that the induced changes result from the direct, physical interaction of the phytochrome molecule with the membranes in question. As will be apparent, the critical question ultimately becomes: are those rapid responses that do indeed involve membranes rapid *enough* to constitute evidence of direct phytochrome-membrane interaction? A definitive answer to this question requires a detailed understanding of each of the potentially rate-limiting steps in the catenary sequence between photon absorption and response expression. To this end, therefore, the photoconversion kinetics and potential intracellular motion of the phytochrome molecule are examined and compared with the kinetics of the most rapid responses thus far observed.

2 Kinetic Categories of Phytochrome-Mediated Responses

The period from photon absorption to the appearance of a measurable response can be divided into three phases:

photon absorption	→ P_{fr} formation	→ P_{fr} action	→ response detection (expression)

These three phases correspond to the formal framework common to other sensory perception systems (HAUPT and FEINLEIB 1979, SHROPSHIRE 1979): signal perception (P_{fr} formation), signal transduction (P_{fr} action, referring here specifically to the *first* molecular change induced in the responding system by the active form of the photoreceptor) and response expression (the appearance of a measurable change in the particular response parameter under investigation).

All phytochrome-induced responses exhibit a period of latency between the onset of irradiation and the appearance of a detectable change in the parameter being monitored. This latency period varies dramatically from a few seconds for changes in cellular electrical properties to days or weeks for gross morphogenetic changes such as flowering. In addition, the period over which P_{fr} action is still required during the latent period is variable. For the purposes of the present discussion it is conventient to define three kinetic classes of response:

a) rapid action/rapid expression responses
b) rapid action/delayed expression responses
c) delayed action/delayed expression responses

Rapid is arbitrarily defined here as 5 min or less; delayed as more than 5 min.

In the first category the rapid action of P_{fr} is implied from the rapid appearance of the response under study. In the second category rapid action is deduced from observations such as rapid escape from far-red (FR) reversibility, where the response itself may not be detectable for hours or even days. The third category encompasses the majority of "classical" phytochrome responses where partial or complete FR reversibility is demonstrable over relatively long periods following P_{fr} formation. This type of response indicates a requirement for P_{fr} action either continuously or at critical points in the transduction chain over significant portions of the latency period for ultimate response expression to occur. The present discussion is restricted to responses in the first two categories.

3 Rapid Action/Rapid Expression Responses

3.1 In Vivo

3.1.1 Pelletability and Sequestering

The P_{fr}-dependent intracellular dark reaction that leads to enhanced phytochrome pelletability in oats and maize has a half-time of about 2 s at 25 °C (Fig. 1), the most rapid P_{fr}-mediated event thus far recorded (QUAIL and BRIGGS 1978). Immunocytochemically detectable phytochrome, diffusely distributed throughout the cytoplasm of unirradiated cells, likewise rapidly redistributes to small (1 µm), discrete loci in the cytoplasm upon P_{fr} formation (Fig. 2) (MACKENZIE et al. 1978, KASS and PRATT 1978, PRATT 1979, EPEL et al. 1980). This process has been termed "sequestering." The similarity between the kinetics of sequestering and of in vivo pelletability induction suggests that these two phenomena may reflect the same cellular process. However, neither the pelletable structures nor the 1 µm loci with which the phytochrome becomes associated have been identified. The loci do not appear to be mitochondria, plastids, nuclei, plasma membrane or tonoplast (MACKENZIE et al. 1978, EPEL et al. 1980). There is thus currently no unequivocal evidence that either pelletability or sequestering results from the binding of P_{fr} to membrane-bound receptors in the cell (see MARMÉ 1977, PRATT 1978, QUAIL 1982, for reviews) although this clearly remains an attractive hypothesis. The possibility of spontaneous, intracellular self-aggregation of P_{fr} is one suggested alternative (MACKENZIE et al. 1978, YAMAMOTO et al. 1980, QUAIL 1982). Regardless of the ultimate mechanistic explanation for these phenomena the rapid rate of the P_{fr}-dependent reaction in each case

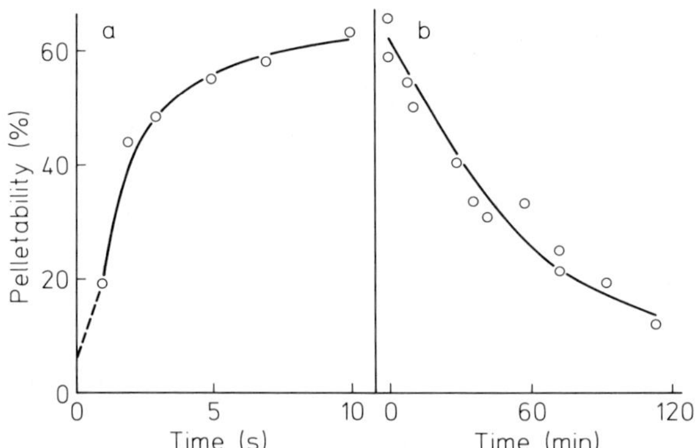

Fig. 1 a, b. Kinetics of the induction and reversal (="relaxation") of pelletability in *Avena* shoot tissue at 25 °C. The percentage of phytochrome that is pelletable in extracts of irradiated tissue is plotted as a function of the duration of a dark period interposed **a** between 1 s R and 10 s FR (followed immediately by extraction), and **b** between a 5 min R/10 min FR sequence and subsequent extraction. The half-time for induction = 2 s; for relaxation = 25 min. (Data from QUAIL and BRIGGS 1978 and PRATT and MARMÉ 1976, adapted by PRATT 1979)

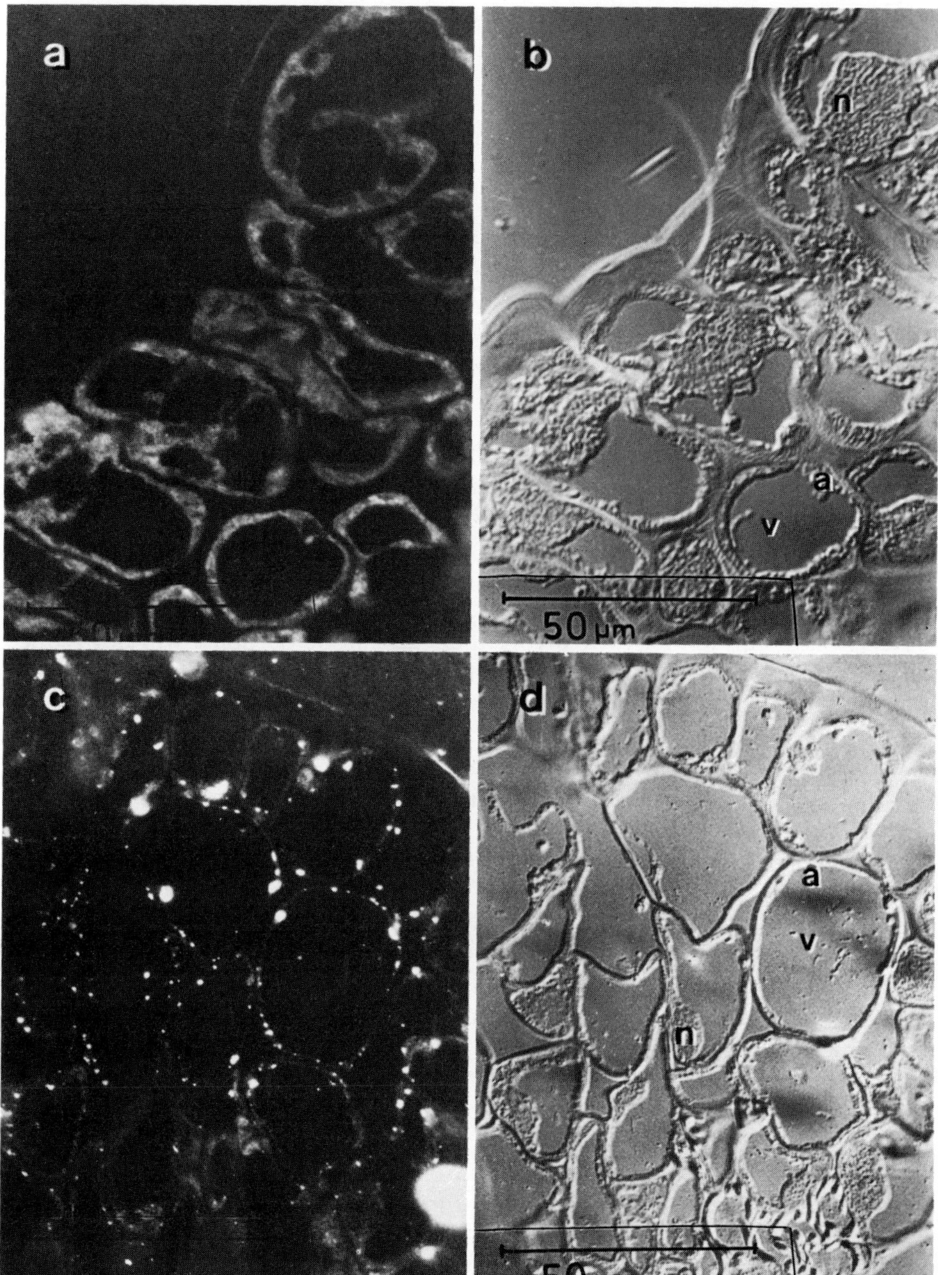

Fig. 2a–d. Phytochrome sequestering in *Avena* coleoptile cells. Phytochrome has been localized by immunofluorescence (*white areas*) in sections taken from **a** unirradiated tissue and **c** tissue irradiated with 3 min R at 25 °C before fixation. **b** and **d** are Nomarski optics micrographs of the corresponding fields shown in **a** and **c**. *n* nucleus; *a* amyloplast; *v* vacuole. (Micrographs from Epel et al. 1980)

makes it a candidate for involvement in the primary action of the pigment and therefore worthy of further investigation.

SPRUIT (1980) has recently reported a physiological response with some intriguing qualitative similarities to the sequestering and pelletability responses. The enhanced capacity for chlorophyll production in bean seedlings induced by a 380 ms pulse of red (R) light is not reversed by an immediately subsequent 3 s of FR but is reversed increasingly if a dark period is interposed between the R and FR pulses. The half-time of the dark reaction leading to maximal FR effectiveness is estimated to be 3.2 s. The possibility that this reaction is simply the rate-limiting step on the $P_r \rightarrow P_{fr}$ phototransformation pathway is considered improbable given current kinetic measurements of this process (see Sect. 5.1 below). Instead, the interpretation of the data suggested by Spruit is the same as that proposed by Hendricks (see VANDERHOEF et al. 1979) to explain the phytochrome "paradoxes." It is proposed that initially cytoplasmically distributed phytochrome will bind upon P_{fr} formation to sites of limited number in the cell and that this association induces the response. The number of sites is so small as to be saturated by the number of P_{fr} molecules formed by a saturating FR pulse alone. Once a P_{fr} molecule occupies a site, it continues to do so, at least in the short term, even after reconversion to P_r, thereby blocking the binding of other free P_{fr} molecules. Thus a pulse of FR that rapidly follows an inductive R before the sites can be occupied will not reduce the ultimate level of site occupation nor the response. FR given *after* full occupation of the sites by R-transformed P_{fr} molecules, on the other hand, will reduce the number of molecules in the P_{fr} form that finally occupy the sites by back transforming to P_r the bulk of the subpopulation already bound to the sites. SPRUIT'S (1980) data would suggest that binding to the hypothetical sites proceeds in the dark with a half-time of 3 s at 20 °C upon P_{fr} formation (cf. Fig. 1). There is no basis for arguing that these sites need be membrane localized. It is worth noting here that, although interpreted differently by the authors, the data of BLAAUW-JENSEN and BLAAUW (1976) are also fascinatingly consistent with two other predictions of the Hendrick's hypothesis: (a) the optimum fluence for FR induction of the response, and (b) the shift to higher fluences of the R fluence response curve after an initial cycle of sequential R and FR irradiations.

3.1.2 Double-Flash Experiments

Phytochrome-induced chloroplast rotation in *Mougeotia* is first detectable within 60 s of the onset of irradiation (HAUPT and ÜBEL 1975) and shows 50% escape from FR reversal within 5 min (HAUPT 1959). Initial indications from double-flash experiments of the existence of an extremely rapid P_{fr}-mediated event in the signal chain of this already rapid response (HAUPT et al. 1980) are now considered instead to be a trivial consequence of the inherent kinetic properties of the photoconversion process. HAUPT and BRETZ (1976) irradiated *Mougeotia* with two, 3 ms flashes of R light separated by a dark interval ranging from zero (simultaneous) to 10 s. No response was induced by a single flash nor by the two flashes given simultaneously. Potentiation of the response began

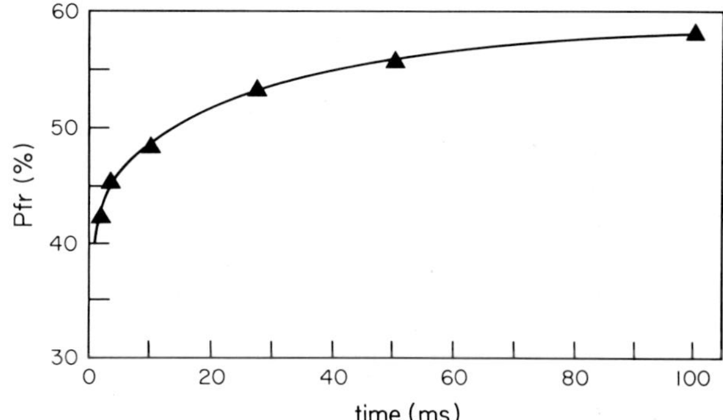

Fig. 3. Percentage of maize coleoptile phytochrome converted to P_{fr} as a function of the dark interval interposed between two flashlamp pulses (emission max 650 nm) of ~1 ms each. The phytochrome measurements were made in vivo following irradiation of the intact tissue. (Data from FUAD 1979)

as the dark interval between the two consecutive flashes exceeded about 3 ms and reached a maximum for dark intervals of 30 ms or more. The sensitivity of the system to the second flash was reversed by FR. Moreover, further investigations (HAUPT et al. 1980) suggested that, whereas the effectiveness of the first flash was apparently independent of the plane of polarization, that of the second flash was strongly dependent on this plane in the manner well established for longer irradiations (HAUPT and WEISENSEEL 1976). Thus, this apparent induction of "dichroic sensitivity" exhibited the operational criteria for a P_{fr}-mediated process that is completed within 30 ms.

More recent data, however, do not support the notion of second flash-induced "dichroic sensitivity", and the induced second flash sensitivity per se now appears to be the product of the intrinsic kinetic properties of phytochrome photoconversion (KRAML and HAUPT 1981). It is suggested that one or more intermediates on the path to P_{fr} formation may revert in the dark to P_r with a given probability and a half-time ensuring completion by ~30 ms (see KENDRICK and SPRUIT 1973). The net level of P_{fr} formed would thereby be greater for R flashes separated by longer than 30 ms than for shorter dark intervals. Such an explanation also appears to account for the effects of analogous double-flash irradiation on lettuce seed germination (HAUPT and SCHEUERLEIN 1977, SCHEUERLEIN 1980) and *Avena* mesocotyl growth inhibition (KRAML 1980). Direct evidence supporting this proposal has been obtained from spectrophotometric measurements of percent phototransformation of phytochrome following double-pulse irradiation of corn coleoptiles (Fig. 3) (FUAD 1979; M. KRAML and E. SCHÄFER personal communication).

3.1.3 Bioelectric Potentials

There are several reports attributing rapid light-induced changes in electrical potential to phytochrome mediation. Both surface (external) electrodes and microelectrodes have been used. RACUSEN (1976) used microelectrodes to monitor the transmembrane potential of *Avena* coleoptile cells and reported a R-induced *de*polarization (Fig. 4). NEWMAN and SULLIVAN (1976), in direct contrast, recorded a *hyper*polarization with the same system. Both sets of authors report lags of only 5–10 s, but Racusen indicates that the depolarization persists subse-

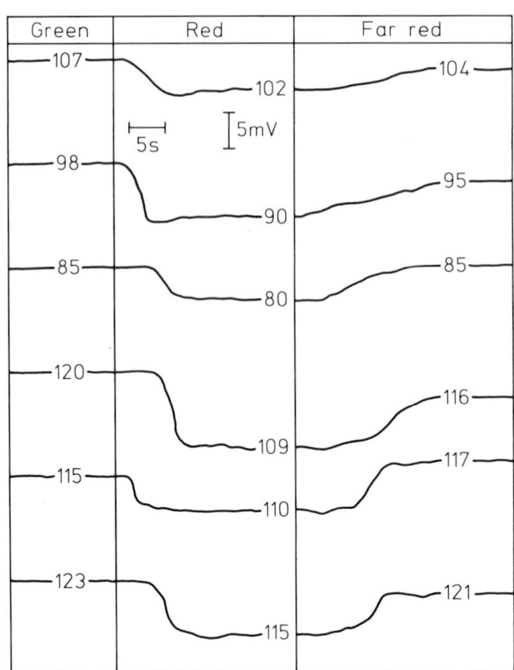

Fig. 4. Transmembrane potential changes recorded with implanted microelectrodes in six *Avena* coleoptile parenchyma cells during successive exposures to continuous green, R and FR light. The numerical values associated with sections of each trace are the mean absolute potentials in negative millivolts for that section of the trace. (Data from RACUSEN 1976)

quently in the dark whereas Newman and Sullivan recorded only transient changes following a 2 s pulse, with return to the starting potential within minutes. Both groups report that FR induces changes in potential that are opposite to those of R: Racusen observed a repolarization to the original pre-R resting potential (Fig. 4); NEWMAN and SULLIVAN observed a transient FR-induced depolarization with return to the starting potential in the dark over minutes following the 2 s pulse. NEWMAN and BRIGGS (1972) using external surface electrodes earlier had reported R-induced depolarizations in *Avena* coleoptiles with lags of 10–15 s consistent with the opposite signs of the effects obtained using microelectrodes (NEWMAN and SULLIVAN 1976). More recently NEWMAN (1981) has demonstrated that the lag to the initiation of detectable changes following 1-s pulses of R and FR can be accurately measured to be 4.5 s in both cases (Fig. 5).

RACUSEN and SATTER (1975) report dark-stable, R (hyperpolarization) FR (repolarization) reversible changes in *Samanea* flexor cell transmembrane potentials detectable within ~90 s of the start of 3 min irradiations. Apparent R/FR reversible changes in transmembrane potentials in *Nitella* have been reported (WEISENSEEL and RUPPERT 1977). However, the small size of the effect superimposed on a large photosynthetically driven background component renders questionable the postulated phytochrome involvement. Finally, a R/FR reversible enhancement of blue-light-induced changes in the transmembrane potential of corn colepotiles has been observed as early as 5 min from the onset of the R irradiation (RACUSEN and GALSTON 1980).

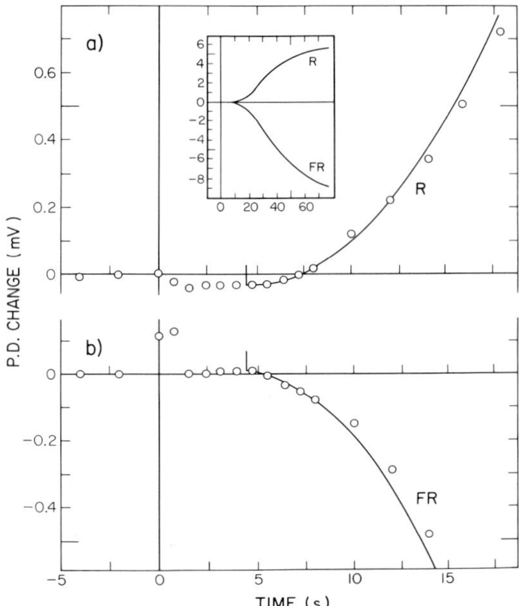

Fig. 5a, b. R- and FR-induced potential difference changes in *Avena* coleoptiles recorded with surface contact electrodes. Pulses of 1 s **a** R and **b** FR light were administered starting at time zero in each case. The FR pulse was given 10 min after the initial R. The surface potential before each light pulse has been designated "zero" and deviations from this value recorded. *Open circles* are data points, *solid lines* are half-Gaussian curves fitted using an assumed lag period of 4.5 s. *Inset* Longer-term response on lower sensitivity scale. (Data from NEWMAN 1981)

The conflicting data obtained by different authors in the same system and by the same authors in different systems in the above group of studies, underscore the difficulty of obtaining reliable measurements with microelectrodes in plant cells in general (GOLDSMITH and GOLDSMITH 1978). Thus, despite the fact that changes in transmembrane potential represent the most rapid R/FR-reversible response in a recognized cellular function yet recorded, caution is needed in the interpretation of these effects.

External electrodes have been used to quantitate the rapid ($\leqq$30 s lag), R/FR reversible changes in surface potential of the root cap cells of barley and mung bean (JAFFE 1968, RACUSEN and MILLER 1972, RACUSEN and ETHERTON 1975) originally observed by TANADA (1968) as root tip adhesion to negatively charged glass surfaces. RACUSEN and ETHERTON (1975) report further than when the root cap cells are plasmolized the protoplast moves within the cell wall "cage" in response to an imposed, oscillating field following R irradiation. This observation is taken to indicate that phytochrome induces an alteration in the distribution of fixed charges on the surface of the plasma membrane of these cells. The presence of immunocytochemically detectable phytochrome in root cap cells is consistent with its involvement in this response (PRATT and COLEMAN 1974). However, despite the long-standing acceptance that this response is phytochrome-mediated, close inspection of published data reveals that the criteria for phytochrome involvement have not been rigorously fulfilled. Attachment of the root tips is induced and observed under continuous R light. Detachment occurs and is observed under subsequent continuous FR light that is activated immediately upon termination of the R irradiation. An essential control is lacking. It has not been demonstrated that upon termination of the inductive R the root tips remain attached in the dark for a period equivalent to the normal subsequent FR irradiation. The possibility has not been excluded that detachment occurs automatically upon R termination and is indifferent to the imposed FR.

3.1.4 Ion and Water Flux

Several rapid, phytochrome-controlled changes in ion flux have been reported although none have been shown convincingly to be correlated with any membrane potential change. The best characterized system is the redistribution of K^+ ions observed during leaf movement in certain legumes. FONDEVILLE et al. (1966) showed that leaflet closure in *Mimosa* is under phytochrome control and is detectable within 5 min. SATTER and GALSTON (1973), who have examined this response in *Albizzia* and *Samanea* in great detail, have shown directly that K^+ moves out of the ventral motor cells during leaflet closure and that this movement is detectable within 5 min. P_{fr} promotes K^+ efflux and leaflet closure with time-courses that are parallel, while P_r inhibits K^+ efflux and leaflet closure concomitantly. BROWNLEE and KENDRICK (1977) report that the rate of uptake of K^+ by mung bean subhypocotyl hook sections is inhibited within 5 min of a 30 s R light pulse and that the R effect is completely reversed by an immediately subsequent 30 s of FR. Changes in passive permeability of the plasma membrane have been suggested as one possible interpretation of the data (BROWNLEE and KENDRICK 1979) but the validity of the kinetic analysis used to support this notion can be questioned.

DREYER and WEISENSEEL (1979) have documented a rapid phytochrome-controlled alteration in the rate of exogenous Ca^{2+} entry into *Mougeotia* cells. A two- to ten-fold enhancement of the rate of $^{45}Ca^{2+}$ accumulation is detectable when tested at 3 min from the start of 30 s R. This effect is completely reversed by an immediately subsequent 30 s FR. The true rate at which the effect is induced by P_{fr} is yet to be determined, as 3 min is the earliest time point thus far reported. These data raise the possibility that the signal chain from photoreception to physiological response in *Mougeotia* is very short: Phytochrome → elevated cytoplasmic $[Ca^{2+}]$ → calmodulin activation → myosin light-chain kinase activation → actomyosin contractile movement → chloroplast rotation. All components of the chain are present or potentially so. Apart from phytochrome and Ca^{2+}, there is ultrastructural evidence for actomyosin cables between plastid and plasma membrane (WAGNER and ROSSBACHER 1980) and calmodulin, known to activate myosin light-chain kinase (KLEE et al. 1980), has recently been demonstrated in plants (VAN ELDIK et al. 1980).

An attempt has also been made to monitor the kinetics of Ca^{2+} transport in oat coleoptile protoplasts with murexide, a dye sensitive to free Ca^{2+} (HALE and ROUX 1980). A change in absorbance attributed to a change in murexide extinction was detectable upon monitoring immediately following termination of 2 min R. Reversal of this effect by a subsequent 2 min FR was also already complete when monitoring was resumed at the end of the irradiation. The data were interpreted to indicate a R-induced efflux of free Ca^{2+} into the medium (opposite in direction to that of *Mougeotia*), and a curiously complete reaccumulation by the protoplasts of the exported Ca^{2+} in response to FR. However, the lack of a control with protoplasts in the absence of murexide to check for light scattering artifacts renders the report equivocal. Proton extrusion by mung bean roots has been claimed to be under phytochrome control (YUNGHANS and JAFFE 1972, RACUSEN and ETHERTON 1975) but the effect is small and unconvincing.

The earliest time-point reported in any of the above ion flux studies was 2 min from the onset of irradiation, with others being 3–5 min. Detection is thereby obviously precluded of any potentially more rapid changes that might have been initiated as P_{fr} first appeared in the cell. Thus, with the possible exception of *Mougeotia,* these studies are of limited value in assessing whether the changes in ion flux are a consequence of direct or indirect effects of phytochrome on membrane properties. With *Mougeotia* additional evidence from separate action dichroism studies has been interpreted to indicate that phytochrome might be plasma membrane localized in this alga (HAUPT and WEISENSEEL 1976).

Apparent phytochrome control of rates of plasmolysis and deplasmolysis in *Mougeotia,* evident after 5-min irradiations, has been interpreted to indicate regulation of the water-permeability of the plasma membrane by the pigment (WEISENSEEL and SMEIBIDL 1973). However, phytochrome was also found to determine the volume of the protoplasts at full plasmolysis, indicating that this interpretation is too simplistic. A previous claim of phytochrome-induced changes in water permeability in *Taraxacum* (CARCELLER and SÁNCHEZ 1972) can be disregarded as no FR reversal of the reported R effect was demonstrated.

3.1.5 ATP Levels

Possible phytochrome control of ATP levels is the subject of a considerable number of contradictory reports. SANDMEIER and IVART (1972) first reported an increase in ATP in etiolated *Avena* mesocotyls within 15 s of the onset of R irradiation but no FR reversal was demonstrated. A recent careful attempt to repeat these observations with oat shoot tissue has been unsuccessful (H. GORTON personal communication). YUNGHANS and JAFFE (1972) recorded a tenfold decrease in ATP in mung bean root tips after 4 min R irradiation with restoration to the dark control level when 4 min FR were administered after the R. However, because the R samples were extracted 4 min, and the R/FR samples 8 min from the onset of the R irradiation, the asserted reversal of the effect by FR cannot be distinguished from a simple R-induced transient change in ATP that had disappeared by 8 min. Moreover, BÜRCKY and KAUSS (1974) using the same mung bean root tip system were unable to repeat even the R effect on ATP levels. PIKE and co-workers (WHITE and PIKE 1974, KIRSHNER et al. 1975) have documented transient increases in ATP in etiolated bean buds in response to 5 min R. Two peaks of activity were recorded: a five-fold increase at 4 min from the start of the irradiation, followed by a decrease to the control level at 5 min and a second transient increase to about double the control at 1 min after termination of the 5 min R. Once again, however, the reported attempt to demonstrate FR reversal must be considered inconclusive. This is because the protocol used for the R/FR sample resulted in extraction at 11 min from the onset of the initial R irradiation. At this time the transiently elevated ATP levels would have already returned almost to the control level in the absence of any subsequent FR. FRIEDERICH and MOHR (1975) report an absence of any effect of continuous FR on ATP in *Sinapis*. It should be noted that in no case is there evidence that the changes in ATP levels that were observed result from changes in membrane properties.

3.1.6 Enzyme Activities

Two enzymes show sufficiently rapid light-induced changes in detectable activity for potential inclusion in the rapid response category. Lipoxygenase (LOG) accumulation in *Sinapis* is inhibited completely upon formation of more than a few percent P_{fr} (OELZE-KAROW and MOHR 1973). The standard errors reported indicate that no distinction can be made between instantaneous cessation of accumulation and cessation with a lag of about 5 min. A longer lag would be detectable, however, so that phytochrome action in this system appears to occur in 5 min or less. Nitrate reductase in *Sinapis* shows a rapid transient increase in activity during a 5 min R irradiation (JOHNSON 1976). Although the long-term rate of accumulation of this enzyme is clearly under phytochrome control, FR reversal of the rapid R-induced transient per se has not been demonstrated. Moreover, the possibility that the in vivo assay used by JOHNSON (1976) might detect rapid changes in the leakiness of the tissue for NO_3 or NO_2 rather than bona fide changes in enzyme activity, has been raised (STARR et al. 1980).

3.1.7 Growth Responses

The majority of rapid light-regulated growth responses are mediated via the blue-light photoreceptor (MEIJER 1968, GABA and BLACK 1979, COSGROVE 1981). There are, however, also a few reports of rapid phytochrome-mediated growth responses. WEINTRAUB and LAWSON (1972) observed a R/FR reversible stimulation of *Avena* coleoptile growth rate within 60 s of the start of the R irradiation. NAUNOVIĆ and NESKOVIĆ (1979) have documented a R/FR reversible decrease in growth rate in etiolated *Pisum* detectable within about 6 min of the onset of 3 min R. PILET and NEY (1978) report that white light reduces the elongation rate of maize roots within 5 min of the start of irradiation of the root cap but not if the elongation zone itself is irradiated. A variety of circumstantial evidence suggests that phytochrome in the root cap may be responsible for this effect (QUAIL 1980). The localization of the bulk of the immunocytochemically detectable root phytochrome in the cap is consistent with this proposal (PRATT and COLEMAN 1974). MORGAN and SMITH (1978) have documented reversible changes in the elongation rate of light-grown *Chenopodium* within 5 min of altering the photoequilibrium level of P_{fr}. None of these growth responses provides direct evidence of phytochrome-mediated alterations in membrane properties.

3.2 In Vitro

Theoretically, at least, the most direct approach to determining whether phytochrome modifies membrane properties by direct interaction with those membranes is to monitor for rapid R/FR reversible responses in isolated subcellular fractions. The R/FR reversible modulation in vitro of some molecular function intrinsic to the isolated fraction is interpreted to indicate that, not only is phytochrome physically present, but it is in addition regulating the activity of a second molecule or function in a potentially biologically meaningful way.

There are numerous reports claiming to demonstrate rapid phytochrome-mediated changes in vitro but most either have proven to be irreproducible or do not withstand critical examination. In brief, these include: enhanced rate of NADP reduction in mitochondria-rich fractions (MANABE and FURUYA 1973, CEDEL and ROUX 1980, FURUYA personal communication); altered NADH dehydrogenase activity in mitochondria (CEDEL and ROUX 1980, BILLET and SMITH 1978; PRATT 1982); modulation of peroxidase activity in crude particulate fractions (PENEL et al. 1976, PRATT 1978, QUAIL unpublished); control of NAD kinase activity (TEZUKA and YAMAMOTO 1974, HOPKINS and BRIGGS 1973); and a change in choline acetyl transferase activity (JAFFE 1976).

3.2.1 Gibberellins from Etioplasts

An apparent exception to this catalog of equivocal reports is the R/FR reversible alteration in vitro of the properties of GA-like substances extractable from etioplast-rich fractions (COOKE and SAUNDERS 1975, COOKE et al. 1975, EVANS and SMITH 1976, HILTON and SMITH 1980). This alteration is first detectable 5 min from the onset of R irradiation and involves the apparent conversion of nonacidic GAs to acidic GAs (COOKE et al. 1975, COOKE and KENDRICK 1976). HILTON and SMITH (1980) have provided convincing evidence that the phenomenon is localized in intact etioplasts. There is no definitive evidence, however, (a) that the phytochrome controlling the response is membrane-localized as opposed to residing in the lumen of the organelle nor (b) that the observed change in the partitioning properties of extractable GA-activity involves any alteration in the functional properties of the organellar membranes. On the contrary, both the long lag and the apparent precursor-product nature of the response constitute evidence against the direct modulation of membrane properties and are indicative instead that the response involves enzymatic interconversion of the GA forms. A claim by COOKE and KENDRICK (1976) based on subfractionation studies that the complete phytochrome–GA response system is located in the plastid envelope has been strongly challenged. GRAEBE and ROPERS (1978) point out that the purported differences in GA-like activity reported are too small to be considered significant in the bioassay system used.

3.2.2 Enzyme Activities in Crude Particulate Fractions

JOSE and colleagues have described experiments in which the effects of both in vivo and in vitro irradiations in various combinations have been tested for their effects on certain enzymatic activities in an undefined particulate fraction from mung bean (JOSE 1977, JOSE and SCHÄFER 1979). ATPase, hexokinase and adenylate kinase were reported to exhibit rapid R/FR reversible modulation in vitro within the 10 s to 1 min irradiation periods used. There is no evidence, however, that the observed changes represent a meaningful modulation of membrane properties. Even the purported involvement of phytochrome remains in question given the observed artifactual photodynamic action of protochlorophyll(ide) on enzymatic activities that include adenylate kinase (GROSS et al. 1979). Moreover, in a recent attempt to repeat the effects of in vitro irradiations on ATPase activity, THOMAS and TULL (1981) report a change in activity that is precisely opposite in direction to that observed by JOSE (1977) and an inconsistent reversibility of the FR effect by subsequent R.

3.2.3 Ca^{2+} Flux in Mitochondria

A recent study presents evidence for R/FR modulation in vitro of the rate of substrate-dependent Ca^{2+} accumulation by isolated mitochondria (ROUX et al. 1981). The data indicate that P_{fr} reversibly enhances the rate of passive Ca^{2+} efflux thereby reducing net active uptake within the 1 min actinic irradiation period used. Superficially the results suggest the direct modulation of a membrane property by phytochrome in vitro. However, whereas the rate of Ca^{2+} flux is controlled by the inner mitochondrial membrane, the phytochrome responsible for the observed effect is reported to reside on the outer membrane (ROUX et al. 1981). Caution is needed in the interpretation of these results, as R light inhibition of mitochondrial Ca^{2+} uptake was found in an earlier study to result artifactually from the destructive photodynamic action of contaminating protochlorophyll(ide) (GROSS et al. 1979). Indeed, DIETER and MARMÉ (1981) were unable to repeat the results of ROUX et al. (1981) when care was taken to eliminate protochlorophyllide from the preparations.

3.2.4 Artificial Membranes

ROUX and YGUERABIDE (1973) have shown R/FR reversible changes in black lipid membrane resistance in the presence of purified phytochrome within 60 s of the onset of irradiation. The length of the lag raises questions about the relevance of the observation to the molecule's biological function, however, as direct effects on conductance would be expected to be much more rapid (HUBBELL and BOWNDS 1979, HERRMANN and RAYFIELD 1978). More recently GEORGEVICH and ROUX (1979) have reported immediate enhancement of K^+ efflux from preloaded liposomes in response to phytochrome addition, the rate being two to three times higher with added P_{fr} than with added P_r. Post-addition, light-induced reversal of the effects is yet to be reported, however, and the possibility of differential rates of nonspecific liposome lysis as opposed to differential permeability induction remains to be eliminated.

It is clear that none of the in vitro studies thus far reported provides unequivocal evidence of phytochrome-induced changes in membrane properties.

4 Rapid Action/Delayed Expression Responses

The rapid action of phytochrome in eliciting responses that are themselves not detected for hours or days after P_{fr} formation has been inferred from the three types of observations outlined below.

4.1 Rapid Escape from FR Reversal

FREDERICQ (1964) first demonstrated that the effect of a brief R night-break on flowering in *Pharbitis* could only be reversed by FR if given within the subsequent 2 min and then only with rapidly declining effectiveness. Flowering was scored several days later. WHITELAM et al. (1979) report that nitrate reductase accumulation induced by 2 min white light in cauliflower curd is reversed by 10 min FR given immediately, but not when given after 15 min R. The enzyme is assayed 8 h after the light treatments. The phytochrome-mediated enhancement of nitrate reductase accumulation in *Sinapis* escapes from FR reversibility with a half-time of 7 min (JOHNSON and WHITELAM 1982). JABBEN and MOHR (1975) have demonstrated that R pre-irradiations of *Sinapis* accelerate the Shibata shift when it is assayed 2 h or more later. The effect of 5 or 15 s R pre-irradiation can be reversed by FR given immediately but 5 min of R is no longer reversible. Complete escape from P_{fr} control is apparently effected between 2 and 5 min. The rate of stroma thylakoid accumulation in *Sinapis* exhibits an analogous effect (GIRNTH et al. 1978). Pre-irradiation with R accelerates this accumulation when seedlings are transferred to white light several hours later. FR given immediately after 15 s R completely reverses this effect but not when given after 5 min R. OELZE-KAROW and MOHR (1980b) have observed that 3% P_{fr} established by a 5 min FR pretreatment of *Sorghum* seedlings enhances the rate of chlorophyll accumulation upon transfer to white light 12 h later. The FR effect is reversed by subsequent 756 nm light (establishing $<1\%$ P_{fr}) but begins to escape from reversal as early as 6 min from the start of the FR. In an analogous study KASEMIR and MOHR (1981) have recently reported that FR reversal is lost even more rapidly in *Pinus sylvestris*, being complete within 2 min of a 15 s R pulse. TOBIN (1981) has demonstrated that increased levels of translatable mRNAs for the small subunit of ribulose 1,5-bisphosphate carboxylase and the light-harvesting chlorophyll a/b-protein induced by 1 min R can be reversed by immediately subsequent FR, whereas the same R dose given over 10 min is not reversible. WARNER et al. (1981) report that FR reversal of the stimulatory effect of 10 s R on the elongation of excised maize coleoptile sections is lost when the interval between R and FR treatments exceeds 45 s. The most rapid escape from FR reversibility thus far recorded is that observed by LEUNG and BEWLEY (1981). These authors found that an increase in α-galactosidase activity induced in lettuce seeds by multiple 2 s pulses of R light could be reversed by FR given immediately after each R pulse but that 5 s R pulses could not be reversed by immediately subsequent FR pulses. These data indicate that for the induction of this response P_{fr} action is complete somewhere between 2 and 10 s of the start of the R pulse, although the rise in enzyme activity does not occur until 2 h later. These various escape data obviously imply that only the very first steps in the signal chain leading to each particular response are directly controlled by P_{fr} and that these steps can be completed within as little as a few seconds.

4.2 Intra- and Interorgan Signal Transmission

Apparently rapid intra- or interorgan transmission of inductive signals following P_{fr} formation has been implied from a variety of surgical experiments. WAGNÉ (1965) reported that when part of a cereal leaf was irradiated with R light the unirradiated part exhibited the phytochrome-controlled unrolling response (measured 24 h later) even when severed from the irradiated section as early as 20 s from the start of irradiation. This observation was interpreted to indicate the rapid intraorgan transmission of a P_{fr}-induced signal for a delayed response. KANG and ZEEVAART (1968) have demonstrated, however, that the effect is an artifact of light scattering from the irradiated to the unirradiated region (see also MANDOLI and BRIGGS 1982). CAUBERGS and DEGREEF (1975) report that selective 5 min R irradiation of the leaf of a bean seedling induced opening of the unirradiated hook even when the leaf was removed immediately at the end of the irradiation. The data were interpreted as indicating a rapid interorgan transmission of a phytochrome-mediated signal. However, as no FR reversal of the response could be detected even when the leaf was left attached, the postulated involvement of phytochrome in this particular response has not been established. While rapid escape from FR reversibility remains a possible explanation for the results, this is yet to be demonstrated.

OELZE-KAROW and MOHR (1974) have reported that phytochrome control of lipoxygenase (LOG) accumulation in *Sinapis* cotyledons depends on the presence of an attached hypocotyl hook. Since the LOG response in the intact system occurs with a lag of $\leqq 5$ min, it was concluded that the rapid interorgan transmission of a signal from hook-localized phytochrome was necessary to account for the results. The data presented, however, do not preclude the possibility that hook removal indirectly destroys the control of the response exerted by cotyledon-localized phytochrome, by, for example, terminating the supply of some vital hook produced factor(s). Thus, the speculative suggestion that signal transmission may occur via a "highly ordered biophysical system" lacks direct supportive evidence. Even should interorgan signal transmission exist, the possibility that the signal is a diffusible agent is by no means ruled out by the rapidity of the response. Light-induced inhibition of root growth results from transmission of a diffusible substance at least as rapidly as for the LOG response (PILET and NEY 1978). More recently OELZE-KAROW and MOHR (1980a) have reported that phytochrome-enhanced chlorophyll synthesis occurs normally in excised cotyledons in white light when they are preirradiated with a pulse of R light 12–24 h before excision. At first these data might be taken as evidence against a general uncoupling of cotyledon responses from phytochrome control upon excision and therefore in favor of interorgan control of LOG accumulation (QUAIL 1980). However, since the R effect on chlorophyll synthesis escapes from FR reversal within 3 h (OELZE-KAROW and MOHR 1980b), the response has clearly passed beyond phytochrome control at the time of excission 12 h later. Thus while *expression* of the response does not require the presence of the hook, there is no evidence that phytochrome *action* does not require its presence just as for LOG.

4.3 Permissive Temperature Transient

The induction of high temperature germination competence in *Amaranthus* seeds, during brief exposure to permissive temparatures provides a unique demonstration of the rapid action of P_{fr} in the early events of a delayed response (HENDRICKS and TAYLORSON 1978). *Amaranthus* seeds held at 40 °C germinate poorly. Transferring the seeds briefly to a temperature <32 °C before returning them again to 40 °C greatly enhances germination after 3 days if phytochrome is in the P_{fr} form during the low temperature window but not if it is in the P_r form. The half-time for P_{fr} action during a permissive temperature transient of 15 °C is 8 min. The authors speculate that membrane phase transitions may account for the permissive and non-permissive configurations for P_{fr} action.

It should be noted that while the rapid action/delayed expression responses considered above reaffirm that P_{fr} action can indeed be rapid, they provide little definitive information on the *nature* of the initial events.

5 Are Cellular Membranes the Locus of the Primary Action of Phytochrome?

There is little dispute that, of the variety of phytochrome-mediated events demonstrable within the first 5 min from P_{fr} formation, several involve changes in membrane properties. Even more significantly, the *most* rapid R/FR reversible change in a recognized cellular function yet recorded is membrane-localized – the modulation of the transmembrane potential of *Avena* coleoptile cells (Figs. 4 and 5). The point at issue is, however, do these changes result from the *direct* interaction of phytochrome with the relevant membranes as proposed by the "membrane hypothesis" or are they instead secondary consequences of a yet more rapid primary molecular event?

To determine whether kinetic analysis of phytochrome-induced responses can provide an answer to this question it is necessary to examine the kinetics of each of the potentially rate-limiting steps between photon absorption and P_{fr} action. These steps are photoconversion and the diffusion to, association with and induced conformational change of any presumptive reaction partner.

5.1 Photoconversion Kinetics

How rapidly can P_{fr} be expected to appear in the cell after the onset of irradiation? It is necessary to distinguish between the time required for an individual molecule to complete the transformation upon photon absorption and the time required to convert a large population of such molecules from P_r to P_{fr}. The rate limiting factor is different in each case.

CORDONNIER et al. (1981) have measured the kinetics of P_{fr} appearance in purified, 120,000 molecular weight *Avena* phytochrome at 2 °C following a 15 ns laser pulse (Fig. 6). The complication of two apparent populations aside, it is clear that the first P_{fr} molecules appear within milliseconds and that 96% of the molecules complete the transformation within 1 s of excitation. The half-times for P_{fr} appearance in 120,000 molecular weight pea phytochrome under the same conditions are markedly slower, being 172 ms and 773 ms for dual kinetic populations in similar proportion to those of oats (CORDONNIER et al. 1981). Nevertheless, 92% of the molecules are in the P_{fr} form 2 s from photoexcitation. Comparable flash photolysis data, but for the entire spectrum from 350 to 750 nm, have been obtained with a newly designed spectrophotometer (SHIMAZAKI et al. 1980, PRATT et al. 1982).

Direct evidence that the $P_r \rightarrow P_{fr}$ transition is complete within 1 s in the living cell comes from two recent studies. SPRUIT (1982) measured the rate of P_{fr} appearance directly in several species by in vivo spectrophotometry. Half-

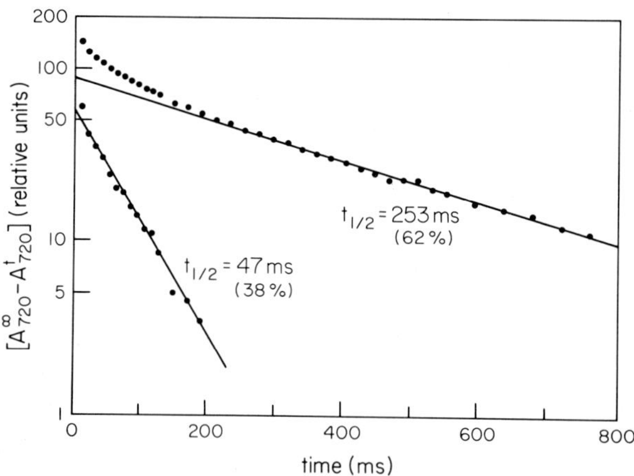

Fig. 6. Phototransformation kinetics of $P_r \rightarrow P_{fr}$ at 2 °C following pulse excitation. Purified, 120,000 molecular weight *Avena* phytochrome in the P_r form was subjected to an 8 ns laser pulse (emission maximum 630 nm, power output 30 mJ) and the increase in P_{fr} absorbance was monitored at 720 nm. A^{t}_{720} = absorbance at 720 nm at any time t following the pulse; A^{∞}_{720} = final absorbance at 720 nm. The data are plotted semilogarithmically. The closed circles in the upper curve are the original data. The lower curve is computed by "peeling" the upper curve on the assumption of two parallel first order reactions. The half-time value for each of these reactions is indicated together with the percentage in parentheses of the total population exhibiting that half-time. (Data from CORDONNIER et al. 1982)

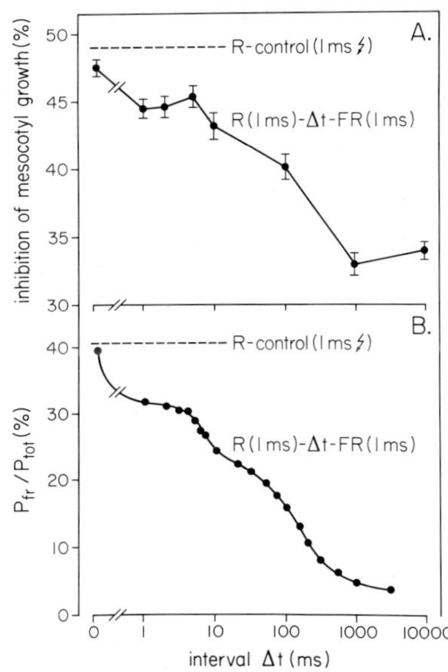

Fig. 7. A Reversal by FR of R-induced mesocotyl growth inhibition as a function of the dark interval between sequential pulses of R and FR. Etiolated *Avena* seedlings were exposed to a 1 ms pulse of R followed by a dark period of variable duration (Δt, ms) and a FR pulse of 1 ms. Data are expressed as the percent inhibition of growth relative to a non-irradiated control (which=0% inhibition). The interval Δt is plotted on a log scale. Full reversal becomes possible 1 s or longer after the start of the R pulse indicating that transformation of excited molecules is complete within 1 s in the cell at 25 °C. (Data from M. KRAML, H. DAUT and M. JOHN personal communication). **B** Reversal by FR of R-induced P_{fr} formation as a function of the dark interval between sequential R and FR pulses. Etiolated *Avena* seedlings were irradiated as in **A** and the percent phytochrome present as P_{fr} determined directly in the tissue by in vivo spectrophotometry. Maximum reverse photoconversion to P_r occurs for FR pulses given 1 s or longer after the start of the R pulse confirming that $P_r \rightarrow P_{fr}$ conversion occurs within 1 s in the cell. (Data from M. KRAML and E. SCHÄFER personal communication)

times of $\leqq$200 ms can be estimated for tissue at 25 °C. In a study of phytochrome-regulated mesocotyl elongation, M. KRAML (personal communication) irradiated *Avena* seedlings sequentially with 1 ms of R and 1 ms of FR separated by a dark interval of increasing duration. The effectiveness of the FR pulse in reversing the R effect increased with the duration of the dark interval up to 1 s where reversal was maximized (Fig. 7a). This result indicates that essentially all molecules destined to form P_{fr} had completed the transition and were capable of reverse photoconversion when the FR pulse was administered at 1 s. In vivo spectroscopy of phytochrome in identically treated tissue provides direct verification that this is so (Fig. 7b) (M. KRAML and E. SCHÄFER personal communication).

Recent data suggest that the multiple kinetic populations observed in the flash photolysis studies on purified phytochrome referred to above (Fig. 6) might be an artifact of purification. The immunoaffinity-purified phytochrome used in these studies consists of 2 to 3 polypeptides that are derived from a single, larger (124,000 m.w.), and presumptively native molecule by limited proteolysis during purification (VIERSTRA and QUAIL 1982). As the proteolytically modified preparations have altered absorption spectra, the multiple species present might generate correspondingly multiple photoconversion kinetics.

In contrast to the above flash photolysis studies, the rate-limiting factor in the photoconversion of a large phytochrome population from P_r to P_{fr} under commonly encountered experimental conditions, is not the rate of internal rearrangement of individual molecules but rather the rate of photon absorption. The latter is in turn a function of the photon fluence rate of the actinic irradiation used (generally considerably less than 300 μmol m^{-2} s^{-1}). The absolute rate of photoconversion of a population is given by the equation (BUTLER 1972):

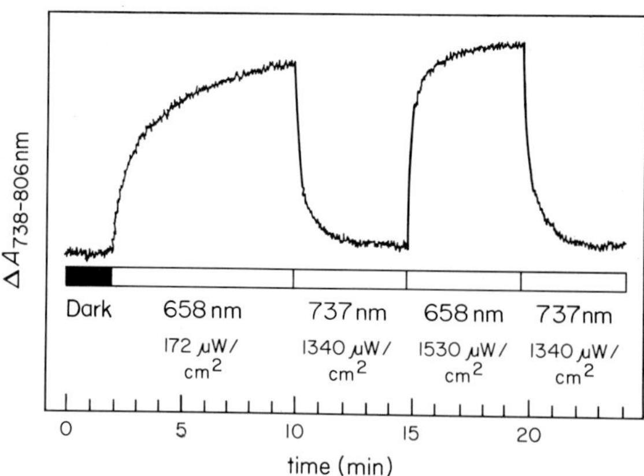

Fig. 8. Time-course of in vivo phytochrome phototransformation in *Amaranthus* seedlings during irradiation with R (658 nm) and FR (737 nm) light. The change in absorbance of P_{fr} at 738 nm is monitored continuously during the actinic irradiations, the wavelength and fluence rate of each being indicated below the appropriate portion of the curve. Note that two different fluence rates of R are compared. The $\Delta(\Delta A)$ value at full saturation is 2×10^{-2}. (SPRUIT and KENDRICK 1972)

$$\frac{d[Pr]}{dt} = I_\lambda \, \xi_{fr\lambda} \, \phi_{fr} [P_{fr}] - I_\lambda \, \xi_{r\lambda} \, \phi_r [P_r] \quad (1)$$

where I_λ = photon fluence rate at wavelength λ; $\xi_{r\lambda}$ and $\xi_{fr\lambda}$ = molar extinction coefficients to the base 10 for P_r and P_{fr} respectively at λ; ϕ_r and ϕ_{fr} = quantum yields for P_r and P_{fr} photoconversion respectively; and $[P_r]$ and $[P_{fr}]$ = concentrations of P_r and P_{fr}.

That the overall photoconversion process is exponential with respect to time of irradiation has been verified experimentally both in vivo and in vitro (PRATT and BRIGGS 1966, SCHMIDT et al. 1973, PRATT 1975). Departures from first order are observed, however, with optically thick samples (Fig. 8) as a result of the steep light intensity gradient across the sample (SCHMIDT et al. 1973, SPRUIT and KENDRICK 1972).

In practice, irradiations of several minutes are often necessary to saturate photoconversion with commonly used actinic fluence rates (Fig. 8). Under these circumstances the rate-limiting step for phytochrome action may not be that intrinsic to the specific molecular process involved (i.e., signal transduction), but rather the absolute rate of P_{fr} formation (i.e., signal perception). Since unambiguous analysis of the kinetics of the phytochrome action step obviously requires minimizing the probability of overlap of the perception and action phases of the signal chain, short, high intensity pulses of actinic light are a necessity in such studies.

5.2 Kinetics of Intracellular Molecular Motion and Interaction

Given that the first P_{fr} molecules can begin to appear in the cell within milliseconds of the onset of irradiation (Fig. 6), how rapidly can initiation of P_{fr} action be expected? Lack of unequivocal evidence on the intracellular location of the active pigment presently precludes a definitive answer to this question.

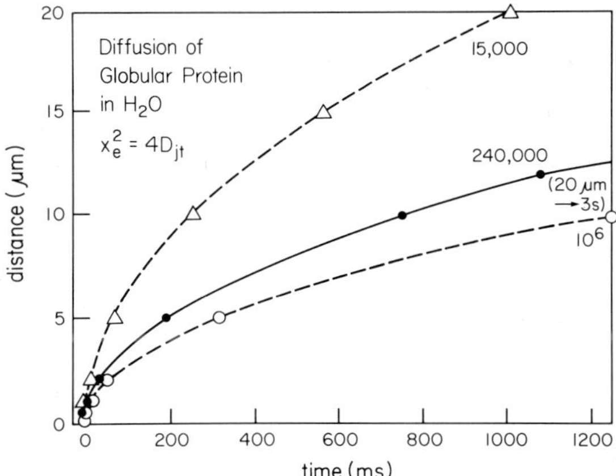

Fig. 9. Diffusion of globular proteins in water calculated from the equation $x_e^2 = 4\,D_j t$, where x_e = distance diffused; D_j diffusion coefficient; t time (NOBEL 1974). The rate of diffusion of a globular protein having the molecular weight of the phytochrome dimer (240,000, PRATT 1979) is compared with those of proteins of 15 and 10^6 molecular weights

Crude upper limits can be estimated, however, for certain of the formal, hypothetical possibilities.

For those possibilities involving a bimolecular reaction (i.e., interaction of phytochrome with a specific reaction partner), the next step that is potentially rate-limiting after photoconversion is diffusional motion. Four such possibilities are considered:

a) Soluble phytochrome, soluble reaction partner. The half-time for a diffusion-limited reaction between the two proposed molecules is estimated to be of the order of 1 ms at 25 °C based on the following assumptions: (1) The reaction is second-order with a rate constant of 10^9 mol l^{-1} s^{-1} (MORRIS 1968, FERSHT 1977); (2) The concentration of the phytochrome dimer (240 kdaltons) in the cytosol is 10^{-6} mol l^{-1} [calculated from published yields (PRATT 1978) and assuming the cytoplasm is 10% of the cell volume]; (3) The concentration and molecular weight of the reaction partner are equal to those of phytochrome; (4) The cytosol is a dilute aqueous medium. Some of these assumptions are clearly tenuous. For example, the diffusion rate of fluorescently labeled albumin in the cytoplasm of human fibroblasts following microinjection has been found to be 1/70 the rate in aqueous buffer (WIJCIESZYN et al. 1981). Nevertheless, it is instructive to note that, even if underestimated by an order of magnitude or more, the half-time for the bimolecular reaction calculated above remains less than that for the intrinsic rate-limiting step for photoconversion of P_r to P_{fr} (Fig. 6). For this case then P_{fr} action appears unlikely to be diffusion limited.

b) Soluble phytochrome, membrane-localized reaction partner. A globular protein the size of the phytochrome dimer will diffuse 1 µm in 7 ms, 2 µm in 30 ms and 5 µm in 190 ms (Fig. 9). Since the shell of cytoplasm in many cells is frequently less than 1–2 µm thick it can be argued that phytochrome is capable of traversing the likely maximum diffusional distance between itself and a membrane-bound reaction partner at a rate faster than or equal to the half-time for $P_r \rightarrow P_{fr}$ photoconversion. The rate of P_{fr} action once again would not appear to be diffusion limited.

c) Membrane-bound phytochrome, soluble raction partner. Arguments analogous to those for (b) apply here also.

d) Both phytochrome and reaction partner membrane-bound. Bimolecular interaction in this instance occurs as a result of lateral diffusion in the plane of the membrane (SINGER 1974, CHERRY 1975). The rate of this diffusion is largely determined by the fluidity of the lipid bilayer and the molecular radius of the proteins (SAFFMAN and DELBRÜCK 1975). Several direct measurements have been made of the rates of lateral mobility of surface membrane proteins in mammalian cells (CHERRY 1975, POO et al. 1979) but none are available for plant cells. Extrapolation of the mammalian cell data indicates that a half-time of the order of 50 ms might be expected for a diffusion-limited interaction between phytochrome and a reaction partner (SCHLESSINGER et al. 1978, DELISI 1980). It should be strongly emphasized, however, that this estimate is potentially subject to considerable error. Apart from the need for assumptions regarding local phytochrome and reaction partner concentrations, the lateral diffusion constants determined for mammalian cells range over three orders of magnitude (CHERRY 1975, SCHLESSINGER et al. 1978, POO et al. 1979). Thus, although the fluidity of plant membranes (BOROCHOV et al. 1976) appears to lie in the same range as for mammalian cells (SHINITZKY and BARENHOLZ 1978), the lateral mobility of any bilayer localized phytochrome could vary an order of magnitude either way from that suggested above. Despite these qualifications, it is nevertheless not unreasonable to expect that the rate of P_{fr} action may also not be diffusion-limited in this case.

After diffusional motion, binding and induction of a conformational change are the next two potentially rate-limiting steps in sequence in any postulated bimolecular reaction between P_{fr} and a reaction partner. Association rate constants for protein-protein interactions (the binding step) lie mostly in the range 10^6 to $10^8\ s^{-1}\ M^{-1}$ (FERSHT 1977). These values correspond to half-times of 10 ms to 1 s for 10^{-6} M phytochrome and reaction partner. Rate constants for induced protein conformational changes commonly range from 10 to $10^4\ s^{-1}$ (FERSHT 1977) corresponding to half-times between 70 μs and 70 ms. Thus, potentially the slowest step in the chain from photon absorption to P_{fr} action in a bimolecular reaction involving a P_{fr}-induced conformational change in a macromolecular reaction partner is the binding step itself rather than phototransformation, diffusion or induced conformational change in the reaction partner. The half-times, for each of these individual steps are summarized in Table 1. On the basis of these values, P_{fr} action might be anticipated to begin as early as a few ms and to be effectively completed at maximum within 5 to 6 s from the start of a pulse irradiation.

Alternative to the bimolecular reactions discussed above, is the possibility that phytochrome exerts its action without inducing a change in a specific, single macromolecular reaction partner. In this case the pigment would need to be, or to become, a transbilayer, integral membrane protein (SINGER 1974). Two mechanistically different possibilities exist. The molecule might function in its own right either as a photoreversible gated channel (BROWNLEE et al. 1979) or as a light-driven permease (SMITH 1970, JOHNSON and TASKER 1979)

Table 1. Rate constants and half-times ($t_{1/2}$) for discrete steps between photon absorption and P_{fr}-induced conformational change for a bimolecular reaction between phytochrome and a hypothetical macromolecular reaction partner

Step	Rate constant	$t_{1/2}$ (ms)
Photoconversion (P_{fr} formation)[a]	$3.46\ s^{-1}$	200
Diffusion[b]	10^9 mol $l^{-1}\ s^{-1}$	1–50
Association (binding)[c]	10^6–$10^8\ s^{-1}\ M^{-1}$	10–1000
Induced conformational change[d]	10–$10^4\ s^{-1}$	0.07–70

[a] From Fig. 6

[b] Rate constant (FERSHT 1977) is for interaction between soluble phytochrome and soluble reaction partner. Range of $t_{1/2}$ s is for various topographical combinations of phytochrome and reaction partner discussed in the text. Concentration of each assumed to be 10^{-6} M in the cytoplasm

[c] Rate constant range from FERSHT (1977). Cytoplasmic concentration assumed to be 10^{-6} M

[d] Rate constant range from FERSHT (1977)

or photocoupler (QUAIL 1976). In either case phytochrome action would constitute the direct regulation of the vectorial transport of small, ionic species across the bilayer. Precedent for the kinetic properties of such systems exists in the cases of rhodopsin and bacteriorhodopsin, both integral membrane protein photoreceptors that span the bilayer. Ion movements across membranes containing these photoreceptors are detectable within tens of ms or less of photoexcitation (HUBBELL and BOWNDS 1979, GOLD and KORENBROT 1980, HAMDORF and KIRSCHFELD 1980, HERRMANN and RAYFIELD 1978). Given this precedent and the fact that diffusion of small ions across the required transcytoplasmic and transmembrane distances can occur in the order of μs (NOBEL 1974), it would appear certain that any phytochrome regulated system of this nature would be rate-limited by the photoconversion process only (Figs. 6, 7).

5.3 Kinetic Analysis of Phytochrome-Induced Responses

Are the rapid responses that have been shown definitively to involve membrane changes rapid enough to constitute evidence that they result from direct modulation of membrane properties by phytochrome? As indicated above, the most rapidly the initiation of a P_{fr}-induced membrane change can be expected is in the range of 10–100 ms, rising to a maximum within ~1 s under flash photolysis conditions. Any response slower than this could in principle be rate-limited by:

a) the rate of photon absorption and thereby of P_{fr} formation (signal perception);

b) the rate of P_{fr} action (first transduction step);

c) subsequent transduction or amplification steps;

d) the sensitivity and/or response-time of the assay itself.

No phytochrome response yet recorded exhibits the theoretical maximum rate of appearance. It is thus necessary in the first instance to assess whether the observed latencies of 5 s or more are a function of the rate of photon absorption in each case or of some subsequent step. A definitive answer to this question is possible only where short, inductive pulse-irradiations that do not overlap significantly with the initiation of the response proper are used. Studies from only a single laboratory have satisfied this requirement.

NEWMAN and SULLIVAN (1976) and more recently NEWMAN (1981) have shown that 1 to 2 s pulses of high intensity R and FR light induce transient changes in transmembrane potential that develop in the subsequent dark period with a minimum latent period of 4.5 s (Fig. 5). In addition, a FR pulse given within the latent period immediately after a R pulse partially negates the R-induced change in membrane potential (NEWMAN 1974), indicating that back-conversion circumvents P_{fr} action before it has been completed. In these studies there is a clear separation of signal perception and the initiation of the membrane response. The rate-limiting step is not the rate of photon absorption set by the actinic source nor the rate of internal rearrangement of the phytochrome molecule (Figs. 6, 7), but some subsequent step in the signal chain. The observation that the response develops in the dark rather than being light-driven argues strongly against a direct photocoupling or photopermease function for phytochrome (SMITH 1970, QUAIL 1976, JOHNSON and TASKER 1979). Likewise, the length of the latent period and the transient nature of the response are evidence against the pigments functioning directly as a transmembrane, gated-ion channel. As discussed in Section 5.2, P_{fr} formation would be the rate-limiting step under these circumstances with latencies of only tens of ms expected.

The possibility that the response (Fig. 5) results from a bimolecular reaction in which P_{fr} directly induces a conformational change in a transmembrane, gated-channel reaction partner, also would not appear to be supported by the data in Table 1. None of the discrete events in the sequence from photon absorption to P_{fr}-induced conformational change has a half-time that would result in a 4.5 s latent period. A more rapid response is expected. This observation raises the possibility that the change in potential is indirect and is mediated by one or more second messengers that transmit the signal from phytochrome to the membrane. It should be emphasized that this possibility is by no means unequivocally established by the data as several of the assumptions made in estimating the half-times in Table 1 are presently unverifiable. It can be concluded, however, that the latent period for the most rapid phytochrome-induced change in a membrane function yet recorded is too long to provide unambiguous support for the notion that this change results from direct interaction of the photoreceptor with that membrane.

It should be noted that although the pelletability and sequestering phenomena like NEWMAN's data also exhibit a clear separation of the perception and response phases, and are even more rapid than the transmembrane potential changes (Figs. 1 and 5), both these "responses" describe the behavior of the phytochrome molecule itself rather than the induction of a second process *by* phytochrome. There is no evidence that induced changes in membrane properties are involved.

Table 2. Percent P_{fr} estimated to be present at the onset of response detection in various studies

Response	Lag to response detection (s)	% P_{fr} at onset of response detection[a]	References
Surface contact potential (*Avena*)	4.5	30.0	NEWMAN (1981)
Transmembrane potential (*Avena*)	10.0	75.0	NEWMAN and SULLIVAN (1976)
Transmembrane potential (*Avena*)	5.0	0.6	RACUSEN (1976)
Transmembrane potential (*Samanea*)	90.0	23.0	RACUSEN and SATTER (1975)
Root tip adhesion (to glass)	30.0	0.4	TANADA (1968)
Root tip adhesion (to platinum electrode)	50.0	13.0	RACUSEN and MILLER (1972)

[a] Calculated using the photoconversion rate constant of PRATT (1975) and the total photon fluence estimated to have been delivered during the lag period. The latter was estimated from the photon fluence rates reported in the individual studies. The effects of light scattering by the tissue have been ignored and the assumption made that the rate of photoconversion of phytochrome in the cell is the same as that of the purified pigment. Where reported, the experimental temperature was in the range 23°–25 °C

The experimental design of the remaining studies of documented membrane changes precludes definitive analysis of the likely nature of the rate-determining step in each case. Two types of problem are evident. First, in the case of the ion flux studies, Ca^{2+} in *Mougeotia* (DREYER and WEISENSEEL 1979) and K^+ in *Samanea* (SATTER and GALSTON 1973), the earliest time-point measured was 3 and 5 min respectively from the start of irradiation. In both cases the onset of the response had clearly occurred at some undetermined time prior to the first measurement thereby eliminating possible detection and quantitation of any lag. No conclusions regarding potential direct phytochrome-membrane interactions are possible.

The second problem arises from the use of actinic irradiations that continue during and after the initiation of the response. The induction of surface fixed-charge alterations in root cap cells (TANADA 1968, RACUSEN and ETHERTON 1975) and of transmembrane potential changes in oat coleoptiles (RACUSEN 1976) and *Samanea* (RACUSEN and SATTER 1975) comprise this category. It is not possible from the data presented in these studies unambiguously to determine whether the latencies observed are a function of the rate of P_{fr} formation or a subsequent step. Direct measurements of P_{fr} formation under the conditions used are invariably lacking from these studies. Rough estimates of the likely degree of photoconversion achieved during the various latent periods are nevertheless possible using published rate constants for phytochrome photoconversion (PRATT 1975) and the photon fluence rates reported in these studies. Table 2 compares these values with those estimated for the pulse-irradiation studies of NEWMAN and coworkers. No definite pattern is apparent. Thus, although the low level of presumptive photoconversion occurring during the pre-response irradiation period in some cases might indicate that photon absorption is rate-limiting, the data are not conclusive on this point.

From the above discussion it must be concluded (a) that the experimental design of most existing kinetic studies precludes definitive identification of the rate-determining parameter in the system; and (b) that even in the single instance where signal transduction, as opposed to perception, has been shown to be rate-limiting, the kinetic analysis does not permit the conclusion that the membrane change observed is the direct manifestation of P_{fr} action.

6 Summary Evaluation of Rapid Action Phytochrome Responses

Table 3 summarizes the evaluation of rapid action phytochrome responses detailed in the preceding sections. Each study has been subjected to three sequential questions:

1. Are the criteria for phytochrome involvement per se in the response satisfactorily fulfilled? If so,
2. Is there conclusive evidence that the changes observed are, in fact, changes in membrane properties? If so,
3. Is there evidence that the induced changes are the result of direct, physical interaction of the phytochrome molecule with the membranes in question?

Compliance with all three criteria is clearly essential for a given response to be considered to provide definitive support for the membrane hypothesis of phytochrome action. It will be clear from perusal of Table 3 that no single existing response satisfies all three criteria. Those studies lacking the necessary supportive data are progressively eliminated from contention by the three question sequence. A significant number of rapid action responses claimed to be under phytochrome control fail to provide the minimum necessary evidence to that effect. In a further number of cases the response itself has proven to be irreproducible or questionable. Of the remainder where phytochrome involvement has been definitely established, only four responses have been shown to represent alterations in membrane properties. However, there is no conclusive evidence from either localization or kinetic studies that any of these alterations result from direct phytochrome–membrane interactions. The action dichroism for chloroplast rotation in *Mougeotia* (HAUPT and WEISENSEEL 1976) remains the strongest evidence to date for a possible direct association of phytochrome with a cellular membrane. Paradoxically, however, despite strongly suggestive data from Ca^{2+} studies (DREYER and WEISENSEEL 1979, WAGNER and ROSSBACHER 1980), there is as yet no *direct* evidence that the plastid rotation itself does, in fact, result from the direct alteration by phytochrome of the membrane with which it appears to be associated.

Table 3. Summary evaluation of rapid action phytochrome responses

Response	Criteria satisfied			Reference
	Phytochrome involvement	Membrane changes	Direct phytochrome-membrane interaction	
Pelletability	+	−	+	Quail (1981)
Sequestering	+	−	−	Epel et al. (1980)
Chloroplast rotation	+	−	+	Haupt and Übel (1975)
Bioelectric potentials				
Avena coleoptiles	+	+	−	Racusen (1976), Newman and Sullivan (1976), Newman (1981)
Samanea	+	+	−	Racusen and Satter (1975)
Nitella	±	+	−	Weisenseel and Rupert (1977)
Tanada effect	−	+	−	Tanada (1968), Racusen and Etherton (1975)
Ion fluxes				
K^+ *Samanea*	+	+	−	Satter and Galston (1973)
K^+ soybean	+	−	−	Brownlee and Kendrick (1977)
Ca^{2+} *Mougeotia*	+	+	−	Dreyer and Weisenseel (1979)
Ca^{2+} oat protoplasts	+	−	−	Hale and Roux (1980)
H^+ mung bean	±	−	−	Yunghans and Jaffe (1972)
H_2O flux				
Mougeotia	±	−	−	Weisenseel and Smeibidl (1973)
Taraxacum	−	−	−	Carceller and Sánchez (1972)
ATP levels				
Avena shoots	−	−	−	Sandmeier and Ivart (1972)
Mung bean root	−	−	−	Yunghans and Jaffe (1972), Bürcky and Kauss (1974)
Bean bud	−	−	−	Kirshner et al. (1975)
Sinapis	−	−	−	Friederich and Mohr (1975)

Table 3 (continued)

Response	Criteria satisfied			Reference
	Phytochrome involvement	Membrane changes	Direct phytochrome-membrane interaction	
Enzyme activities				
Lipoxygenase	+	–	–	OELZE-KAROW and MOHR (1973)
Nitrate reductase	–	–	–	JOHNSON (1976)
Growth responses				
Avena coleoptile	+	–	–	WEINTRAUB and LAWSON (1972)
Pisum	+	–	–	NAUNOVIĆ and NESKOVIĆ (1979)
Maize roots	+	–	–	PILET and NEY (1978), QUAIL (1980)
Chenopodium	+	–	–	MORGAN and SMITH (1978)
In vitro responses				
NADP reduction	±	–	–	MANABA and FURUYA (1973), CEDEL and ROUX (1980)
NADH dehydrogenase	–	–	–	BILLET and SMITH (1978), CEDEL and ROUX (1980)
Peroxidase	±	–	–	PENEL et al. (1976), PRATT (1978)
NAD kinase	±	–	–	TEZUKA and YAMAMOTO (1974), HOPKINS and BRIGGS (1973)
Choline acetyl transferase	–	–	–	JAFFE (1976)
GA in etioplasts	+	–	–	COOKE et al. (1975), EVANS and SMITH (1976), HILTON and SMITH (1980)
GA in etioplast envelopes	±	–	–	Cooke and KENDRICK (1976), GRAEBE and ROPERS (1978)
ATPase, hexokinase, adenylate kinase crude pellet	±	–	–	JOSE (1977), JOSE and SCHÄFER (1979), GROSS et al. (1979), THOMAS and TULL (1981)
Ca^{2+} flux in mitochondria	±	–	–	ROUX et al. (1981), GROSS et al. (1979), DIETER and MARMÉ (1981)
BLM resistance	+	(+)	(+)	ROUX and YGUERABIDE (1973)
K^+ flux liposomes	+	(+)	(+)	GEORGEVICH and ROUX (1979)

Rapid escape from FR reversal				
Flowering *Pharbitis*	+	−	−	Fredericq (1964)
Nitrate reductase	+	−	−	Whitelam et al. (1979), Johnson and Whitelam (1981)
Shibata shift	+	(−)	−	Jabben and Mohr (1975)
Thylakoid accumulation	+	(−)	−	Girnth et al. (1978)
Chlorophyll accumulation	+	−	−	Oelze-Karow and Mohr (1980b), Kasemir and Mohr (1981)
mRNA levels	+	−	−	Tobin (1981)
Coleoptile elongation	+	−	−	Warner et al. (1981)
α-galactosidase activity	+	−	−	Leung and Bewley (1981)
Signal transmission				
Cereal leaf unrolling	±	−	−	Wagné (1965), Kang and Zeevaart (1968)
Hook opening	−	−	−	Caubergs and DeGreef (1975)
Lipoxygenase	±	−	−	Oelze-Karow and Mohr (1974)
Chlorophyll synthesis	+	(−)	−	Oelze-Karow and Mohr (1980a, b)
Root growth	+	−	−	Pilet and Ney (1978), Quail (1980)
Permissive temperature transient				
Amaranthus germination	+	±	−	Hendricks and Taylorson (1978)

+ = criteria satisfied i.e., positive answer to questions posed in Section 6. − = criteria not satisfied i.e, negative answer to questions posed in Section 6. ± = equivocal data or interpretation. (+) = indications are that effects on artificial membranes may be unrelated to the normal biological function of phytochrome. (−) = delayed response itself involves changes in thylakoid membrane components but no information on nature of rapid escape reaction

7 Conclusions

There is a steadily increasing array of phytochrome–mediated events that are detectable within the first 5 min of the initial appearance of P_{fr} in the cell. Of these events, several among the most rapid involve alterations in the functional properties of cellular membranes. Critical examination of localization and kinetic studies indicates, however, that there is currently no rigorous evidence that the alterations result from direct, physical interaction of the phytochrome molecule with these membranes. The possibility that the changes are an indirect consequence of a more rapid primary event elsewhere in the cell is not excluded by existing data. Acceptance of the membrane hypothesis of phytochrome action in its most explicit molecular formulation is clearly premature at the present time.

Other current hypotheses of the mechanism of phytochrome action, the "gene regulation" (MOHR 1966, 1972) and "multiple primary action" (MOHR 1974, OELZE-KARROW and MOHR 1973, DRUMM and MOHR 1974) hypotheses, also lack definitive supportive evidence. For the "gene regulation" hypothesis to be meaningful at the molecular level, it must be postulated that the photoreceptor interacts *directly* with the genome to alter transcription qualitatively and/or quantitatively. H. MOHR (personal communication) has suggested that P_{fr} might interact with the genome in a manner analogous to that postulated to occur with steroid hormone-receptor complexes in mammalian cells (MILGROM 1981, PALMITER et al. 1981). However, as has been repeatedly stressed (SMITH 1970, SATTER and GALSTON 1973, MARMÉ 1977), rapid responses such as changes in transmembrane potential, detectable within 5 s, clearly cannot result from a genome-localized primary action of phytochrome. Thus, while the long-anticipated phytochrome-mediated changes in the levels of rRNA (THIEN and SCHOPFER 1982) and certain translatable mRNAs (APEL 1979, TOBIN 1981) have now been documented, these exhibit lags of at least 1 h or more, obviously too long to preclude the possibility that the changes are secondary consequences of a more rapid regulatory action.

The "multiple primary action" hypothesis (DRUMM and MOHR 1974, OELZE-KAROW and MOHR 1973) was conceived to circumvent the inability of the gene regulation hypothesis to account for the rapid responses. The multiple action hypothesis permits retention of the notion of direct phytochrome–genome interaction while accommodating the existence of responses too rapid to be explained in this way. To be meaningful at the molecular level this hypothesis must propose that P_{fr} interacts with more than one molecular species of reaction partner, physically located at sites in the cell as distinctly different as the cellular membrane and the genome, and that the direct consequence of this interaction is different at each site. In support of this idea it has been argued that the existence of both "graded" and "all or none" changes in enzyme activities as a function of P_{fr} level can only be understood if P_{fr} has such multiple primary actions (DRUMM and MOHR 1974, OELZE-KAROW and MOHR 1973). This argument is specious, however, since any graded signal is convertible to a cooperative one at any step in a signal cascade. The available data are entirely consistent with the operation of a single molecular action of P_{fr}. Thus while multiple primary actions might be hypothetically feasible, there is presently no experimentally sound reason to abandon the simpler concept of a phytochrome-induced change in a single type of reaction partner, with a single immediate consequence in molecular terms and with amplification occurring at subsequent steps. There is no compelling reason to suggest that the primary molecular event induced by P_{fr} in the 5 s leading to the onset of a change in transmembrane potential (NEWMAN 1981) is not mechanistically identical to the molecular event that ultimately culminates, via a more elaborate signal cascade, in changes in gene expression.

Acknowledgement. Supported by the National Science Foundation Grant PCM8003792.

References

Apel K (1979) Phytochrome-induced appearance of mRNA activity for the apoprotein of the light-harvesting chlorophyll a/b protein of barley (*Hordeum vulgare*). Eur J Biochem 97:183–188

Billet EE, Smith H (1978) Phytochrome and mung bean mitochondria. Annu Eur Symp Photomorphogenesis, Abstr p 5

Blaauw-Jensen G, Blaauw OH (1976) Further evidence for the existence of two phytochrome systems from two distinct effects of far-red light on lettuce seed germination. Acta Bot Neerland 25:213–219

Borochov A, Halevy AH, Shinitzky M (1976) Increase in microviscosity with ageing in protoplast plasmalemma of rose petals. Nature 263:158–159

Brownlee C, Kendrick RE (1977) Phytochrome and potassium uptake by mung bean hypocotyl sections. Planta 137:61–64

Brownlee C, Kendrick RE (1979) Ion fluxes and phytochrome protons in mung bean hypocotyl segments. I. Fluxes of potassium. Plant Physiol 64:206–210

Brownlee C, Roth-Bejerano N, Kendrick RE (1979) The molecular mode of phytochrome action. Sci Prog Oxf 66:217–229

Bürcky K, Kauss H (1974) Veränderung im Gehalt an ATP and ADP in Wurzelspitzen der Mungobohne nach Hellrotbelichtung. Z Pflanzenphysiol 73:184–186

Butler WL (1972) Photochemical properties of phytochrome in vitro. In: Mitrakos K, Shropshire W Jr (eds) Phytochrome. Academic Press, London New York, pp 185–192

Carceller MS, Sánchez RA (1972) The influence of phytochrome in the water exchange of epidermal cells of *Taraxacum officinale*. Experientia 28:364

Caubergs R, DeGreef JA (1975) Studies on hook-opening in *Phaseolus vulgaris* L. by selective R/FR pretreatments of embryonic axis and primary leaves. Photochem Photobiol 22:139–144

Cedel TE, Roux SJ (1980) Modulation of a mitochondrial function by oat phytochrome in vitro. Plant Physiol 66:704–709

Cherry RJ (1975) Protein mobility in membranes. FEBS Lett 55:1–7

Cooke RJ, Kendrick RE (1976) Phytochrome-controlled gibberellin metabolism in etioplast envelopes. Planta 131:303–307

Cooke RJ, Saunders PF (1975) Phytochrome mediated changes in extractable gibberellin activity in a cell-free system from etiolated wheat leaves. Planta 123:299–302

Cooke RJ, Saunders PF, Kendrick RE (1975) Red-light-induced production of gibberellin-like substances in homogenates of etiolated wheat leaves and in suspensions of intact etioplasts. Planta 124:319–328

Cordonnier MM, Mathis P, Pratt LH (1981) Phototransformation kinetics of undegraded oat and pea phytochrome initiated by laser flash excitation of the red-absorbing form. Photochem Photobiol 34:733–740

Cosgrove DJ (1981) Rapid suppression of growth by blue light. Occurrence, time-course and general characteristics. Plant Physiol 67:584–590

DeLisi C (1980) The biophysics of ligand-receptor interactions. Q Rev Biophys 13:201–230

Dieter P, Marmé D (1981) Far-red light irradiation of intact corn seedlings affects mitochondrial and calmodulin-dependent microsomal Ca^{2+} transport. Biochem Biophys Res Commun 101:749–755

Dreyer EM, Weisenseel MH (1979) Phytochrome-mediated uptake of calcium in *Mougeotia* cells. Planta 146:31–39

Drumm H, Mohr H (1974) The dose-response curve in phytochrome-mediated anthocyanin synthesis in the mustard seedling. Photochem Photobiol 20:151–157

Epel BL, Butler WL, Pratt LH, Tokuyasu KT (1980) Immunofluorescence localization studies of the P_r and P_{fr} forms of phytochrome in the coleoptile tips of oats, corn and wheat. In: DeGreef JA (ed) Photoreceptors and plant development. Antwerpen Univ Press, Antwerpen, pp 121–133

Evans A, Smith H (1976) Localization of phytochrome in etioplasts and its regulation in vitro of gibberellin levels. Proc Natl Acad Sci USA 73:138–142

Fersht A (1977) Enzyme structure and mechanism. Freeman, San Francisco

Fondeville JA, Borthwick HA, Hendricks SB (1966) Leaflet movement in *Mimosa pudica* L. indicative of phytochrome action. Planta 69:357–364

Fredericq H (1964) Conditions determining effects of far-red and red irradiations on flowering response of *Pharbitis nil*. Plant Physiol 39:812–816

Friederich KE, Mohr H (1975) Adenosine 5′-triphosphate content and energy charge during photomorphogenesis of the mustard seedling *Sinapis alba* L. Photochem Photobiol 22:49–53

Fuad N (1979) Phytochrome pelletability, phototransformation and destruction in maize coleoptiles. Ph D thesis, Australian National Univ, Canberra

Gaba V, Black M (1979) Two separate photoreceptors control hypocotyl growth in green seedlings. Nature 278:51–54

Georgevich G, Roux SJ (1979) Ion fluxes in liposomes induced by phytochrome. Plant Physiol Suppl 63:155

Girnth C, Bergfeld R, Kasemir H (1978) Phytochrome-mediated control of grana and stroma thylakoid formation in plastids of mustard cotyledons. Planta 141:191–198

Gold HG, Korenbrot JI (1980) Light-induced calcium release by intact retinal rods. Proc Natl Acad Sci USA 77:5557–5561

Goldsmith TH, Goldsmith MHM (1978) The interpretation of intracellular measurements of membrane potential, resistance and coupling in cells of higher plants. Planta 143:267–274

Graebe JE, Ropers HJ (1978) Gibberellins. In: Letham SL, Goodwin PB, Higgins TJ (eds) Phytohormones and related compounds: A comprehensive treatise Vol 1. Elsevier, Amsterdam, pp 107–203

Gross J, Ayadi A, Marmé D (1979) Protochlorophyll(ide)-630 photosensitizes active Ca^{2+} accumulation in microsomal and mitochondrial fractions isolated from plants. Photochem Photobiol 30:615–621

Hale CC, Roux SJ (1980) Photoreversible calcium fluxes induced by phytochrome in oat coleoptile cells. Plant Physiol 65:658–662

Hamdorf K, Kirschfeld K (1980) Reversible events in the transduction process of photoreceptors. Nature 283:859–860

Haupt W (1959) Die Chloroplastendrehung bei *Mougeotia*. I. Über den quantitativen und qualitativen Lichtbedarf der Schwachlichtbewegung. Planta 53:484–501

Haupt W, Bretz N (1976) Short-term reactions of phytochrome in *Mougeotia*. Planta 128:1–3

Haupt W, Feinleib ME (1979) Introduction. In: Haupt W, Feinleib ME (eds) Physiology of movements. Encyclopedia of plant physiology new ser Vol 7. Springer, Berlin Heidelberg New York, pp 1–9

Haupt W, Scheuerlein W (1977) Short-term phytochrome response in seed germination. Z Pflanzenphysiol 85:445–450

Haupt W, Übel H (1975) Zum Mechanismus der Phytochromwirkung bei der Chloroplastbewegung von *Mougeotia*. Z Pflanzenphysiol 75:165–171

Haupt W, Weisenseel MH (1976) Physiological evidence and some thoughts on localized responses, intracellular localization and action of phytochrome. In: Smith H (ed) Light and plant development. Butterworth, London

Haupt W, Hupfer B, Kraml M (1980) Induction of chloroplast movement in *Mougeotia* with light flashes: spectral sensitivity and action dichroism. Z Pflanzenphysiol 96:331–342

Hendricks SB, Borthwick HA (1967) The function of phytochrome in regulation of plant growth. Proc Natl Acad Sci USA 58:2125–2130

Hendricks SB, Taylorson RB (1978) Dependence of phytochrome action in seeds on membrane organization. Plant Physiol 61:17–19

Herrmann TR, Rayfield GW (1978) The electrical response to light of bacteriorhodopsin in planar membranes. Biophys J 21:111–125

Hilton JR, Smith H (1980) The presence of phytochrome in purified barley etioplasts and its in vitro regulation of biologically active gibberellin levels in etioplasts. Planta 148:312–318

Hopkins DW, Briggs WR (1973) Phytochrome and NAD kinase: A re-examination. Plant Physiol Suppl 51:52

Hubbell WL, Bownds MD (1979) Visual transduction in vertebrate photoreceptors. Annu Rev Neurosci 2:17–34

Jabben M, Mohr H (1975) Stimulation of the Shibata shift by phytochrome in cotyledons of the mustard seedling *Sinapis alba* L. Photochem Photobiol 22:55–58

Jaffe MJ (1968) Phytochrome-mediated bioelectric potentials in mung bean seedlings. Science 162:1016

Jaffe MJ (1976) Phytochrome-controlled acetylcholine synthesis at the endoplasmic reticulum. In: Smith H (ed) Light and plant development. Butterworth, London

Johnson CB (1976) Rapid activation by phytochrome of nitrate reductase in the cotyledons of *Sinapis alba* L. Planta 128:127–131

Johnson CB, Tasker R (1979) A scheme to account quantitatively for the action of phytochrome in etiolated and light-grown plants. Plant Cell Environ 2:259–265

Johnson CB, Whitelam GC (1982) Phytochrome action in light-grown plants: The control of nitrate reductase as a model response. Phtotochem Photobiol 35:251–254

Jose AM (1977) Phytochrome modulation of ATPase activity in a membrane fraction from *Phaseolus*. Planta 137:203–206

Jose AM, Schäfer E (1979) Red/far-red modulation in vitro of enzyme activity in a membrane fraction from *Phaseolus aureus*. Planta 146:75–81

Kang BG, Zeevaart JAD (1968) An alternative explanation for the transmissible light effect in wheat leaf sections. Annu Rep MSU/AEC Plant Res Lab Mich State Univ

Kasemir H, Mohr H (1981) The involvement of phytochrome in controlling chlorophyll and 5-aminolevulinate formation in a gymnosperm seedling (*Pinus sylvestris*). Planta 152:369–373

Kass LB, Pratt LH (1978) Immunocytochemical assay of the time-course of phytochrome sequestering. Plant Physiol Suppl 61:13

Kendrick RE, Spruit CJP (1973) Phytochrome intermediates in vivo. III. Kinetic analysis of intermediate reactions at low temperature. Photochem Photobiol 18:153–159

Kirshner RL, White JM, Pike CS (1975) Control of bean bud ATP levels by regulatory molecules and phytochrome. Physiol Plant 34:373–377

Klee CB, Crouch TH, Richman PG (1980) Calmodulin. Annu Rev Biochem 49:489–515

Kraml M (1980) Photoconversion of phytochrome by short red flashes in *Mougeotia* and *Avena*. In: DeGreef JA (ed), Photoreceptors and plant development. Antwerpen Univ Press, Antwerpen

Kraml M, Haupt W (1981) Phytochrome-controlled chloroplast orientation in *Mougeotia*: Action dichroism in double-flash experiments. Plant Sci Lett 21:145–150

Leung DWM, Bewley JD (1981) Immediate phytochrome action in inducing α-galactosidase in lettuce seeds. Nature 289:587–588

MacKenzie JM, Briggs WR, Pratt LH (1978) Intracellular phytochrome distribution as a function of its molecular form and of its destruction. Am J Bot 65:671–676

Manabe K, Furuya M (1973) A rapid phytochrome-dependent reduction of nicotinamide adenine dinucleotide phosphate in particle fraction from etiolated bean hypocotyl. Plant Physiol 51:982–983

Mandoli DF, Briggs WR (1982) The photoreceptive sites and the function of tissue light piping in photomorphogenesis of etiolated oat seedlings. Plant Cell Environ 5:137–145

Marmé D (1977) Phytochrome: Membranes as possible sites of primary action. Annu Rev Plant Physiol 28:173–198

Meijer G (1968) Rapid growth inhibition of gherkin hypocotyls in blue light. Acta Bot Neerl 17:9–14

Milgrom E (1981) Activation of steroid-receptor complexes. In: Litwack G (ed) Biochemical actions of hormones Vol 8. Academic Press, London New York

Mohr H (1966) Differential gene activation as a mode of action of phytochrome 730. Photochem Photobiol 5:469–483

Mohr H (1972) Lectures on photomorphogenesis. Springer, Berlin Heidelberg New York

Mohr H (1974) Advances in phytochrome research. Photochem Photobiol 20:542–546

Morgan DC, Smith H (1978) Simulated sunflecks have large rapid effects on plant stem extension. Nature 273:534–536

Morris JG (1968) A biologist's physical chemistry. Arnold, London

Naunović G, Nesković M (1979) Rapid responses to light and gibberellic acid in etiolated pea stems. Photochem Photobiol 29:1173–1175

Newman I (1974) Electric responses of oats to phytochrome transformation. In: Bieleski RL, Ferguson AR, Cresswell MM (eds) Mechanisms of regulation of plant growth. Bull 12, Soc NZ, Wellington

Newman I (1981) Rapid electric responses of oats to phytochrome show membrane processes unrelated to pelletability. Plant Physiol 68:1494–1499

Newman IA, Briggs WR (1972) Phytochrome-mediated electric potential changes in oat seedings. Plant Physiol 50:687–693

Newman IA, Sullivan JK (1976) Auxin transport in oats: A model for the electric changes. In: Wardlaw IF, Passioura JB (eds) Transport and transfer processes in plants. Academic Press, London New York

Nobel PS (1974) Biophysical plant physiology. Freeman, San Francisco

Oelze-Karow H, Mohr H (1973) Quantitative correlation between spectrophotometric phytochrome assay and physiological response. Photochem Photobiol 18:319–330

Oelze-Karow H, Mohr H (1974) Interorgan correlation in a phytochrome-mediated response in the mustard seedling. Photochem Photobiol 20:127–131

Oelze-Karow H, Mohr H (1980a) Studies about the effect of phytochrome on the development of the capacity of photophosphorylation and for chlorophyll synthesis in mustard seedling cotyledons. In: DeGreef JA (ed) Photoreceptors and plant development. Antwerpen Univ Press, Antwerpen, pp 253–255

Oelze-Karow H, Mohr H (1980b) Two steps in initial phytochrome action on chlorophyll synthesis. In: Akoyunoglou G (ed) Photosynthesis Vol 5. Proc 5th Int Congr photosynthesis

Palmiter RD, Mulvihill ER, Shepherd JH, McKnight GS (1981) Steroid hormone regulation of ovalbumin and conalbumin gene transcription. A model based upon multiple regulatory sites and intermediary proteins. J Biol Chem 256:7910–7916

Penel C, Greppin H, Boisard J (1976) In vitro photomodulation of a peroxidase activity through membrane-bound phytochrome. Plant Sci Lett 6:117–121

Pilet PE, Ney D (1978) Rapid, localized light effect on rooth growth in maize. Planta 144:109–110

Poo MM, Lam JW, Orida N (1979) Electrophoresis and diffusion in the plane of the cell membrane. Biophys J 26:1–22

Pratt LH (1975) Photochemistry of high molecular weight phytochrome in vitro. Photochem Photobiol 22:33–36

Pratt LH (1978) Molecular properties of phytochrome. Photochem Photobiol 27:81–105

Pratt LH (1979) Phytochrome: function and properties. In: Smith KC (ed) Photochemical and photobiological reviews, Vol 4, Plenum, New York London, pp 59–124

Pratt LH (1982) Phytochrome: The protein moeity. Annu Rev Plant Physiol 33:557–582

Pratt LH, Briggs WR (1966) Photochemical and nonphotochemical reactions of phytochrome in vivo. Plant Physiol 41:467–474

Pratt LH, Coleman RA (1974) Phytochrome distribution in etiolated grass seedlings as assayed by an indirect antibody-labeling method. Am J Bot 61:195–202

Pratt LH, Marmé D (1976) Red light enhanced phytochrome pelletability. Plant Physiol 58:686–692
Pratt LH, Shimazaki Y, Inoue Y, Furuya M (1982) Spectral analysis of phototransformation intermediates in the pathway from the red-absorbing to the far-red-absorbing form of phytochrome. Photochem Photobiol 36:471–478
Quail PH (1976) Phytochrome. In: Bonner J, Varner JE (eds) Plant biochemistry. Academic Press, London New York
Quail PH (1980) Phytochrome: The first five minutes from P_{fr} formation. In: Degreef JA (ed) Photoreceptors and plant development. Antwerpen Univ Press, Antwerpen
Quail PH (1982) Intracellular location of phytochrome. In: Helen C, Charlier M, Montenay-Garestier TL (eds) Trends in photobiology, Plenum, New York London
Quail PH, Briggs WR (1978) Irradiation-enhanced phytochrome pelletability: Requirement for phosphorylative energy in vivo. Plant Physiol 62:773–778
Racusen RH (1976) Phytochrome control of electrical potentials and intercellular coupling in oat coleoptile tissue. Planta 132:25–29
Racusen RH, Etherton B (1975) Role of membrane-bound fixed-charge changes in phytochrome-mediated mung bean root tip adherence phenomenon. Plant Physiol 55:491–495
Racusen RH, Galston AW (1980) Phytochrome modified blue-light-induced electrical changes in corn coleoptiles. Plant Physiol 66:534–535
Racusen RH, Miller K (1972) Phytochrome-induced adhesion of mung bean root tips to platinum electrodes in a direct current field. Plant Physiol 49:654–655
Racusen RH, Satter RL (1975) Rhythmic and phytochrome-regulated changes in transmembrane potential in *Samanea* pulvini. Nature 255:408–410
Roux SJ, Yguerabide J (1973) Photoreversible conductance changes induced by phytochrome in model lipid membranes. Proc Natl Acad Sci USA 70:762–764
Roux SJ, McEntire K, Slocum RD, Cedel TE, Hale CC (1981) Phytochrome induces photoreversible calcium fluxes in purified mitochondrial fractions from oats. Proc Natl Acad Sci USA 78:283–287
Saffman PG, Delbrück M (1975) Brownian motion in biological membranes. Proc Natl Acad Sci USA 72:3111–3113
Sandmeier M, Ivart J (1972) Modification du taux des nucléotides adenyliques (ATP, ADP et AMP) par un éclairement de lumière rouge-clair (660 nm). Photochem Photobiol 16:51–59
Satter RL, Galston AW (1973) Leaf movements: Rosetta stone of plant behvior? Bioscience 23:407–416
Scheuerlein R (1980) Short-term reactions of phytochrome: Flash-induction of seed germination in *Lactuca sativa*. In: Degreef JA (ed) Photoreceptors in plant development, Antwerpen Univ Press, Antwerpen, pp 375–380
Schlessinger J, Schechter Y, Cuatrecasas P, Willingham MC, Pastan I (1978) Quantitative determination of the lateral diffusion coefficients of the hormone-receptor complexes of insulin and epidermal growth factor on the plasma membrane of cultured fibroblasts. Proc Natl Acad Sci USA 75:5353–5357
Schmidt W, Marmé D, Quail PH, Schäfer E (1973) Phytochrome: First-order phototransformation kinetics in vivo. Planta 111:329–336
Shimazaki Y, Inoue Y, Yamamoto KT, Furuya M (1980) Phototransformation of the red-light-absorbing form of undegraded pea phytochrome by laser flash excitation. Plant Cell Physiol 21:1619–1625
Shinitzky M, Barenholz Y (1978) Fluidity parameters of lipid regions determined by fluorescence polarization. Biochim Biophys Acta 515:367–394
Shropshire W Jr (1979) Stimulus perception. In: Haupt W, Feinleib ME (eds) Physiology of movements. Encyclopedia of plant physiology new ser Vol 7. Springer, Berlin Heidelberg New York, pp 10–41
Singer SJ (1974) The molecular organization of membranes. Annu Rev Biochem 43:805–833
Smith H (1970) Phytochrome and photomorphogenesis in plants. Nature 227:665–668
Spruit CJP (1980) Short term far-red reversibility of red potentiated chlorophyll accumulation in bean. In: DeGreef JA (ed) Photoreceptors and plant development. Antwerpen Univ Press, Antwerpen, pp 179–183

Spruit CJP (1982) Phytochrome intermediates in vivo. IV. Kinetics of Pfr emergence. Photochem Photobiol 35:117–121

Spruit CJP, Kendrick RE (1972) On the kinetics of phytochrome photoconversion in vivo. Planta 103:319–326

Starr R, Gupta S, Acton J (1980) Rapid phytochrome activation of nitrate reductase in mustard cotyledons. Is it an artefact? In: DeGreef JA (ed) Photoreceptors and plant development. Antwerpen Univ Press, Antwerpen, pp 293–295

Tanada T (1968) A rapid photoreversible response of barley root tips in the presence of 3-indoleacetic acid. Proc Natl Acad Sci USA 59:376–380

Tezuka T, Yamamoto Y (1974) Kinetics of activation of nicotinamide adenine dinucleotide kinase by phytochrome, far-red absorbing form. Plant Physiol 53:717–722

Thien W, Schopfer P (1982) Control by phytochrome of cytoplasmic rRNA synthesis in the cotyledons of mustard seedlings. Plant Physiol 69:1156–1160

Thomas B, Tull SE (1981) Photoregulation of K^+-ATPase in vitro by red and far-red light in extracts from cucumber hypocotyls. Z Pflanzenphysiol 102:283–292

Tobin E (1981) Phytochrome-mediated regulation of messenger RNAs for the small subunit of ribulose 1,5-bisphosphate carboxylase and the light-harvesting chlorophyll a/b-protein in *Lemna gibba*. Plant Mol Biol 1:35–51

Vanderhoef LN, Quail PH, Briggs WR (1979) Red light-inhibited mesocotyl elongation in maize seedlings. II. Kinetic and spectral studies. Plant Physiol 63:1062–1067

Van Eldik LJ, Grossman AR, Iverson DB, Watterson DM (1980) Isolation and characterization of calmodulin from spinach leaves and in vitro translation mixtures. Proc Natl Acad Sci USA 77:1912–1916

Vierstra RD, Quail PH (1982) Native phytochrome: inhibition of proteolysis yields a homogeneous monomer of 124 kilodaltons from *Avena*. Proc Natl Acad Sci USA 79:5272–5279

Wagné C (1965) The distribution of the light effect from irradiated to nonirradiated parts of grass leaves. Physiol Plant 18:1001–1006

Wagner G, Rossbacher R (1980) X-ray microanalysis and chlorotetracycline staining of calcium vesicles in the green alga *Mougeotia*. Planta 149:298–305

Warner TJ, Ross JD, Coombs J (1981) Phytochrome control of maize coleoptile section elongation. Plant Physiol 67:355–357

Weintraub RL, Lawson VR (1972) Mechanism of phytochrome-mediated effects of light on cell growth. In: Book of Abstracts, VI Int Cong Photobiol Bochum, No 161

Weisenseel MH, Ruppert HK (1977) Phytochrome and calcium ions are involved in light induced membrane depolarization in *Nitella*. Planta 137:225–229

Weisenseel MH, Smeibidl E (1973) Phytochrome controls the water permeability in *Mougeotia*. Z Pflanzenphysiol 70:420–431

White JM, Pike CS (1974) Rapid phytochrome-mediated changes in adenosine 5′triphosphate content of etiolated bean buds. Plant Physiol 53:76–79

Whitelam GC, Johnson CB, Smith H (1979) The control by phytochrome of nitrate reductase in the curd of light-grown cauliflower. Photochem Photobiol 30:589–594

Wojcieszyn JW, Schlegel RA, Wu ES, Jacobson KA (1981) Diffusion of injected macromolecules within the cytoplasm of living cells. Proc Natl Acad Sci USA 78:4407–4410

Yamamoto KT, Smith WO, Furuya M (1980) Photoreversible Ca^{2+}-dependent aggregation of purified phytochrome from etiolated pea and rye seedlings. Photochem Photobiol 32:233–240

Yunghans H, Jaffe MJ (1972) Rapid respiratory changes due to red light or acetylcholine during early events of phytochrome-mediated photomorphogenesis. Plant Physiol 49:1–7

10 Photocontrol of Gene Expression

C.J. LAMB and M.A. LAWTON

1 Introduction

Development involves the spatial and temporal, selective regulation of gene expression and thus information about the photocontrol of gene expression is of central importance to our understanding of the cellular transduction of environmental stimuli leading to the initiation of developmental change during photomorphogenesis. The aim of this chapter is to review from a biochemical point of view our current understanding of the photocontrol of gene expression, paying particular attention to those systems where the underlying molecular mechanisms have been elucidated. Hence the treatment is not comprehensive but rather represents a critical consideration of selected experimental results and theoretical concepts currently under intense investigation and debate. In the last few years there have been a number of reviews related to the present topic and the reader is referred to these for historical details (MOHR 1974, SCHOPFER 1977, ZUCKER 1972).

2 Conceptual and Technical Background

A structural gene cannot be considered fully expressed until the final direct gene product is formed and active at its physiological site of action. This definition emphasises the functional relationship between selective gene expression and developmental change. It is also convenient because the starting point for many studies of the molecular events underlying photocontrol of gene expression is the observation of photomodulation of the cellular level of an enzyme. Clearly therefore, measurement of the in vivo activity of an enzyme is of crucial importance but the immense technical problems of a true in vivo assay have rarely been overcome, the method of AMRHEIN and GODEKE (1976), AMRHEIN et al. (1976) for measurement of phenylalanine ammonia-lyase being an elegant exception. Hence in general it is necessary to extract the enzyme and assay its activity in vitro. Conceptual and technical problems associated with this approach have been critically considered (SCHOPFER 1977). If appropriate checks are made, precautions taken and the enzyme is assayed under defined, optimised conditions, then it is possible to obtain from extraction and in vitro assay procedures a useful measure of enzyme activity level which can be taken as a relative estimate of the amount of active enzyme molecules in the tissue (SCHOPFER 1977). The term enzyme level is used throughout this chapter in purely this operational sense. Similarly following the suggestion of SCHOPFER (1977) the terms induction and repression refer only to increases and decreases respectively in the level of some specified parameter without implying any particular mechanism.

On this basis it is clear that the activity levels of a wide range of enzymes are under photocontrol (Table 1). Hence many metabolic activities, including photosynthesis, photorespiration, chlorophyll synthesis, fat degradation, starch degradation, nitrate assimilation, nucleic acid and protein synthesis and degradation and natural product synthesis, may be under photocontrol exerted at the level of selective gene expression. It is clear from Table 1 that with a few exceptions the effect of light is to increase the activity levels of specific enzymes, but that the apparent lag periods for these responses vary widely. Another generalisation which is not clear from Table 1 is that most light-mediated changes in enzyme activity are transient even on transfer from darkness to continuous illumination. This point is of considerable importance since the timing, magnitude and duration of the increase in specific enzyme levels will be crucial in determining the precise overall photomorphogenic response. Therefore attention should be given not only to the molecular mechanisms underlying the initial light-mediated change in gene expression, but also to the interaction of environmental signals with internal control mechanisms such as developmental control of the competence to respond to the environment (MOHR 1972) and pathway specific feedforward and feedback control systems.

A change in enzyme activity level might reflect modulation of one or more of the following processes: (a) de novo synthesis, (b) degradation, (c) activation and (d) inactivation. The post-translational processes (b)–(d) are not molecular mechanisms as such but refer to broad categories of control.

Generally an enzyme is active immediately following termination of translation and release of the complete polypeptide, but transformation of an inactive pro-enzyme to the mature active enzyme by limited proteolysis may occur. In a number of cases excision of an N-terminal peptide or leader sequence is required to produce the mature polypeptide, the leader sequence being involved in the injection of the polypeptide across a cellular membrane, either co-translationally in the case of the endoplasmic reticulum (BLOBEL and DOBBERSTEIN 1975) or independently of translation in the cases of mitochondria and chloroplasts (HIGHFIELD and ELLIS 1978). In some cases (LETO and MILES 1980) it has been shown that the processing procedure is under genetic control and clearly the existence of sophisticated molecular mechanisms associated with the transport of a newly synthesised polypeptide from its site of synthesis to its site of action introduces the possibility of specific photocontrol of these processes. Other possible mechanisms for activation of newly made polypeptides include attachment of a prosthetic group, the accumulation of which may itself be under photocontrol, glycosylation or assembly of an oligomeric enzyme from inactive subunits. Where the oligomer consists of two or more different types of subunit, then expression of the gene for one particular subunit may regulate the assembly or stability of the oligomer and hence overall expression of the genes for the remaining subunit(s). These considerations are particularly relevant to the expression of activity of multimeric proteins in chloroplasts and mitochon-

Table 1. Enzymes under photocontrol

Enzyme	Tissue	+/−	Lag	Mechanism	References
Adenylate kinase	*Chenopodium rubrum*: seedling	+			Frosch and Wagner (1973a, b)
	Zea mays: leaf	+			Butler and Bennett (1969)
Aminoacyl tRNA synthetases	*Pisum sativum*: leaf	+			Henshall and Goodwin (1964)
	Euglena gracilis: greening cells	+		Increased de novo synthesis	Nover (1976)
Aminolevulinate dehydratase	*Phaseolus vulgaris*: leaves	+	24 h		Steer and Gibbs (1969b)
	Avena sativa: isolated etioplasts	+			Hampp and Ziegler (1975)
	Sinapis alba: cotyledon, hypocotyl	+			Kasemir and Masoner (1975)
Amylase	*Raphanus sativus*: cotyledon	+			Huff and Ross (1975)
	Sinapis alba: cotyledon, rest of seedling	+			Drumm et al. (1971)
Ascorbate oxidase	*Sinapis alba*: cotyledon, hypocotyl taproot	+			Drumm et al. (1972)
	cotyledon	+	2 h	Increased de novo synthesis No change in immunoprecipitable protein	Acton et al. (1974), Attridge et al. (1974) Newbury and Smith (1981)
Aspartate aminotransferase	*Kalanchoe blossfeldiana*: leaf	+			Brulfert et al. (1973)
Catalase	*Helianthus annuus*			De novo synthesis	Betsche and Gerhardt (1978)
	Sinapis alba: cotyledon, hypocotyl	+		New isoenzymes	Drumm and Schopfer (1974)
	Triticum aestivum: leaf	+			Feierabend (1975)
Chalcone isomerase	*Petroselinum hortense*: cell culture	+	2 h		Hahlbrock et al. (1971b), Wellmann and Baron (1974)
	Sinapis alba: cotyledon	+			Wellmann (1974)

Table 1 (continued)

Enzyme	Tissue	+/−	Lag	Mechanism	References
Chalcone synthase	*Petroselinum hortense*: cell culture	+	2 h	Increased mRNA activity	SCHRÖDER et al. (1979b)
Chlorophyll a/b binding protein	*Hordeum vulgare*: leaf	+		Increased mRNA activity	APEL (1979)
				Stabilisation by chlorophylls	BELLEMARE et al. (1981)
	Lemna gibba: cultures	+		Increased mRNA activity	TOBIN (1981)
	Petroselinum hortense: cell culture	+		Increased mRNA activity	MULLER et al. (1980b)
Cinnamate 4-hydroxylase	*Petroselinum hortense*: cell culture	+	1 h		HAHLBROCK et al. (1971b)
	Pisum sativum: terminal leaf buds	+	4 h		RUSSELL (1971)
				Increase in cytochrome P-450	BENEVENISTE et al. (1978)
Cytochrome c oxidase	*Sinapis alba*: cotyledon	+			BAJRACHARAYA et al. (1976)
Enolase	*Pisum sativum*: stem apex	+			GRAHAM et al. (1971)
Ferredoxin	*Phaseolus vulgaris*: leaf	+			HASLETT et al. (1973)
Ferredoxin-$NADP^+$ reductase	*Phaseolus vulgaris*: leaf	+			HASLETT and CAMMACK (1976)
Fructose bisphosphate aldolase	*Phaseolus vulgaris*: leaf	+			BRADBEER (1971)
Fructose bisphosphatase	*Pisum sativum*: leaf	+			GRAHAM et al. (1968)
	Zea mays: leaf	+			GRAHAM et al. (1970)
Fructosidase	*Raphanus sativus*: cotyledon, hypocotyl, root	+	24 h	Increased activity and transfer from ER to cell wall	ZOUAGHI and ROLLIN (1976) ZOUAGHI (1976), ZOUAGHI et al. (1979)
Fumarate hydratase	*Sinapis alba*: cotyledon	+			BAJRACHARAYA et al. (1976)
α-Galactosidase	*Lactuca sativa*: seed	+	Nil	Increased activity	LEUNG and BEWLEY (1981)

Glucose 6-phosphate dehydrogenase	*Chenopodium rubrum*: seedling	+			WAGNER (unpublished)
	Pinus roxburghii: germinating pollen	+			DHARVAN and MALIK (1979)
Glutamine synthetase	*Chlorella sorokiniana*: synchronous cultures	+		Activation	TISCHNER and HÜTTERMANN (1980)
Glutamate synthetase	*Chlorella sorokiniana*: synchronous cultures	+			TISCHNER and HÜTTERMANN (1980)
	Oryza sativa: leaves	+		Change of isoenzyme pattern	GUIZ et al. (1979)
Glutathione reductase	*Sinapis alba*: cotyledon	+			DRUMM and MOHR (1973)
Glyceraldehyde phosphate dehydrogenase (NADP$^+$) phosphorylating	*Chenopodium rubrum*: seedling	+	12 h		FROSCH and WAGNER (1973a, b)
	Phaseolus vulgaris: leaf	+			FILNER and KLEIN (1968), KLEIN (1969)
	Pisum sativum: stem apex	+			GRAHAM et al. (1971)
	Secale cereale: shoot	+			FEIERABEND (1969)
	Sinapis alba: cotyledon	+			BRUNING et al. (1975), CERFF (1973, 1974)
	Zea mays: leaf	+			GRAHAM et al. (1970)
Glyceraldehyde phosphate dehydrogenase (NADP$^+$)	*Pisum sativum*: leaf	+			KELLY and GIBBS (1973)
Glyceraldehyde phosphate dehydrogenase (NAD$^+$)	*Chenopodium rubrum*: seedling	+			FROSCH and WAGNER (1973a, b)
	Phaseolus vulgaris: leaf	+	12 h		BRADBEER (1971), FILNER and KLEIN (1968)
	Sinapis alba: cotyledon	+			CERFF 1973, 1974)
Glycollate oxidase	*Phaseolus vulgaris*: leaf	+			KLEIN (1969)
	Sinapis alba: seedling, cotyledon	+	6 h		GERHARDT (1974), VAN POUCKE and BARTHE (1970)
	Triticum aestivum: leaf	+			FEIERABEND (1975)
Glyoxylate reductase	*Sinapis alba*: seedling, cotyledon	+			CERFF (1973, 1974), VAN POUCKE et al. (1970)
	Triticum aestivum: leaf	+			FEIERABEND (1975)

Table 1 (continued)

Enzyme	Tissue	+/−	Lag	Mechanism	References
Hydroxycinnamate: CoA ligase	*Petroselinum hortense*: cell culture	+	2 h		HAHLBROCK et al. (1976)
				Increased mRNA activity	RAGG et al. (1981)
Hydroxymethylglutaryl-CoA reductase	*Pisum sativum*: terminal bud	+/−	5 min	Induction of plastid enzyme Inactivation of microsomal enzyme	BROOKER and RUSSELL (1974, 1979)
Indole acetic acid oxidase	*Campsis grandiflora*	−			SHARMA and MALIK (1978)
Isocitrate dehydrogenase	*Pisum sativum*: stem apex	+			GRAHAM et al. (1971)
Lipoxygenase	*Cucurbita moscata*: cotyledon	−			SURREY (1976)
	Sinapsis alba: cotyledon	−	0 h		MOHR and OELZE-KAROW (1976), OELZE-KAROW and MOHR (1970, 1973), OELZE-KAROW et al. (1970)
Malate dehydrogenase	*Kalanchoe blossfeldiana*: leaf	+			QUEIROZ (1969)
	Pisum sativum: stem apex	+			GRAHAM et al. (1971)
Malic enzyme	*Kalanchoe blossfeldiana*: leaf	+			BRULFERT et al. (1973), QUEIROZ (1969)
NAD^+ kinase	*Avena sativa*: coleoptile	+	0 min		TEZUKA and YAMAMOTO (1975)
	Pharbitis nil: cotyledon			In vitro activation, but see also	
	Pisum sativum: stem apex				HOPKINS and BRIGGS (1973)
$NAD(P)^+$ Transhydrogenase	*Phaseolus vulgaris*: leaf	+			KEISTER et al. (1962)
NADPH: Protochlorophyllide oxidoreductase	*Hordeum vulgare*	−	5 min	Increased removal	MAPPLESTON and GRIFFITHS (1980), SANTEL and APEL (1981)
				Decreased mRNA activity	APEL (1981)

Nitrate reductase	*Chlorella vulgaris*: cultures	+		Activation	TISCHNER and HÜTTERMANN (1978)
	Oryza sativa: seedlings	+			SASAKAWA and YAMAMOTO (1979)
	Phaseolus vulgaris: leaves	+			SLUITERS-SCHOLTEN (1973)
	Pisum sativum: terminal bud	+			JONES and SHEARD (1975)
	Raphanus sativus: cotyledons	+			BEEVERS et al. (1965)
	Sinapis alba: cotyledons	+	5 min	Activation	JOHNSON (1976)
	Triticum aestivum	+			VIJAYARAGHAVAN et al. (1979)
	Zea mays: leaves, seedlings	+			RÁO et al. (1980)
	Saccharum hybrid	+	24 h	Not increased de novo synthesis: possibly inter-conversion of forms	GOATLY and SMITH (1974), GOATLY et al. (1975)
Peroxidase	*Cucurbita pepo*: hypocotyl hook	−	1 min	In vitro inhibition	PENEL et al. (1976)
	Sinapis alba: cotyledon	+			SCHOPFER and PLACHY (1973)
	Spinacia oleracea: leaf	+			PENEL and GREPPIN (1973)
	Zea mays: leaf	+			SHARMA et al. (1976, 1977)
Phenylalanine ammonia-lyase	*Brassica oleracea*: shoot, hypocotyl	+			PECKET and BASSIM (1974), ENGELSMA (1970)
	Cucumis sativus: hypocotyl, cotyledon	+			ENGELSMA (1967, 1968)
	hypocotyl	+		Activation	ATTRIDGE and SMITH (1973, 1974)
	Fagopyrum esculentum: hypocotyl	+			SCHERF and ZENK (1967)
				Application of in vivo assay	AMRHEIN et al. (1976)
	Helianthus tuberosus: tuber	+			DURST and DURANTON (1970)
	Hordeum vulgare: shoot	+			MCCLURE (1974)
	Oenethera hybrids: seedling	+			HACHTEL (1972)
	Petroselinum hortense: cell culture	+	1 h		WELLMANN and BARON (1974)

Table 1 (continued)

Enzyme	Tissue	+/−	Lag	Mechanism	References
				Increased mRNA activity	SCHRÖDER et al. (1979b)
	Pisum sativum: terminal bud	+			ATTRIDGE and SMITH (1967), SMITH and ATTRIDGE (1970)
	Raphanus sativis: cotyledon, hypocotyl, taproot				HUAULT (1974), TOME et al. (1975)
	Sinapis alba: cotyledon, hypocotyl	+	2 h	Increased de novo synthesis/activation	ATTRIDGE et al. (1974), JOHNSON and SMITH (1978), TONG and SCHOPFER (1976)
	Solanum tuberosum: tuber	+	2 h	Increased de novo synthesis	LAMB and MERRITT (1979), LAMB et al. (1979)
Phenylalanine transaminase	*Rhaphanus sativis*: cotyledon	+			TOME et al. (1975)
Phosphoenolpyruvate carboxylase	*Kalanchoe blossfeldiana*: leaf	+			BRULFERT et al. (1973), QUEIROZ (1969)
	Zea mays: leaf	+			GRAHAM et al. (1970)
				Increased mRNA activity	HAGUE and SIMS (1980, 1981), HAYAKAWA et al. (1981)
Phosphoglycerate kinase	*Phaseolus vulgaris*: leaf	+			BRADBEER (1971)
Photogene 32	*Zea mays*: leaf	+		Increased mRNA level	BEDBROOK et al. (1978)
	Sinapis alba: cotyledons	+			LINK (1981, 1982)
Phytochrome	*Avena sativa*: whole seedling	−		Decreased levels in radioimmunoassay	HUNT and PRATT (1980)
				Decreased mRNA activity	GOTTMANN and SCHÄFER (1982)
	Cucurbita pepo: hook, cotyledon	−		Decreased stability	QUAIL et al. (1973a, b)

Plastocyanin	*Phaseolus vulgaris*: leaf	+		Haslett and Cammack (1974)
Pyrophosphatase	*Zea mays*: leaf	+		Butler and Bennett (1969)
Pyruvate: orthophosphate dikinase	*Zea mays*: leaf	+		Graham et al. (1970)
Ribosephosphate isomerase	*Secale cereale*: shoot	+		Feierabend and Pirson (1966)
Ribonuclease II	*Lupinus albus*: hypocotyls	+	Increased de novo synthesis	Acton (1972, 1974), Acton and Schopfer (1974)
Ribulose bisphosphate carboxylase	*Hordeum vulgare*: leaf	+	Increased de novo synthesis	Kleinhopf et al. (1970), Tobin (1978)
	Lemna gibba: cultures	+	Increased mRNA activity	Tobin and Suttie (1980)
	Nicotiana sylvestris	+	Increased mRNA activity	Lett et al. (1980)
	Phaseolus vulgaris: leaf	+		Bradbeer (1971), Filner and Klein (1968)
			Increased de novo synthesis	Gray and Kekwick (1974)
	Pisum sativum: stem apex	+		Graham et al. (1968, 1971)
			Increased mRNA level	Bedbrook et al. (1980)
	Secale cereale: shoot	+		Feierabend (1969), Feierabend and Pirson (1966)
	Sinapis alba: cotyledon	+		Bruning et al. (1975), Frosch et al. (1976)
	Zea mays: leaf	+		Graham et al. (1970)
			Increased mRNA level	Link et al. (1978)
RNA nucleotidyl transferase	*Pisum sativum*: terminal buds	+		Bottomley (1970)
Stilbene synthase	*Cissus antartica*: leaves	+		Fritzemeier and Kindl (1981)

Table 1 (continued)

Enzyme	Tissue	+/−	Lag	Mechanism	References
Succinate dehydrogenase	*Pinus roxburghii*: germinating pollen	+			DHARVAN and MALIK (1979)
	Sinapis alba: cotyledons	+			BAJRACHARYA et al. (1976)
Succinyl-CoA synthase	*Phaseolus vulgaris*: leaf	+			STEER and GIBBS (1969a)
Transketolase	*Phaseolus vulgaris*: leaf	+			BRADBEER (1971)
	Secale cereale: shoot	+			FEIERABEND and PIRSON (1966)
UDP-apiose: flavone apiosyl transferase	*Petroselium hortense*: cell culture	+			WELLMANN and BARON (1974), HAHLBROCK et al. (1971a, b)
			2 h	Increased mRNA activity	GARDINER et al. (1980)
UDP-galactose: 1,2-diacylglycerol galactosyl transferase	*Sinapis alba*: cotyledon	+			UNSER and MASONER (1972)
Urate oxidase	*Sinapis alba*: cotyledon, rest of seedling	+			HONG and SCHOPFER (unpublished)

dria where in all cases so far studied the gene for at least one of the subunits is present in the nuclear genome and the gene for at least one of the other subunits is present in the appropriate organellar genome (ELLIS 1977).

Binding of specific proteins can lead to marked changes in the properties and activities of specific enzymes. For example, specific proteinaceous anti-enzymes have been discovered of which the best characterised is the anti-enzyme in mammalian cells specific for ornithine decarboxylase (HELLER et al. 1976).

It is well established that protein synthesis in plants occurs against a background of considerable protein degradation (BOUDET et al. 1975, HUFFAKER and PETERSON 1974). Furthermore, different proteins exhibit markedly different rates of turnover, implying some specificity of the turnover pathway(s). Studies of a limited number of plant enzymes suggest that turnover rate is correlated with subunit molecular weight (ACTON and GUPTA 1979). Proteinaceous inactivators that act by limited proteolysis have been identified for specific plant enzymes (e.g. WALLACE 1975), and group-specific proteases have been identified and characterised in a number of cases (KATUNUMA et al. 1972, KOMINAMI and KATUNUMA 1976). Furthermore, whilst in general binding of low molecular weight effectors such as substrates, products and allosteric effectors will be of significance only to fine metabolic control, it is possible that changes in the levels of such effectors will alter the conformation of the enzyme in vivo, thereby specifically altering the susceptibility of that enzyme to inactivation or degradation by otherwise non-specific processes.

Thus there are a number of specific post-translational molecular processes that represent potential sites for the photocontrol of enzyme levels in plants. Therefore the observation of a light-mediated change in enzyme level immediately raises the problem of the molecular mechanisms underlying the effect and some consideration is now given to possible experimental approaches to this question.

In principle, the simplest way to demonstrate the involvement of protein synthesis is by application of inhibitors such as cycloheximide or chloramphenicol. However, there are a number of pitfalls in this approach (ELLIS and McDONALD 1970, SMITH et al. 1977), and use of inhibitors of protein synthesis cannot provide rigorous proof of photocontrol of the synthesis of specific proteins and inhibition of a response should be considered only as a preliminary indication of the involvement of protein synthesis, or as an operational criterion for comparative purposes (LAMB 1977a, 1979).

Modulation of the rates of synthesis or degradation leads to changes in enzyme activity level accompanied by changes in enzyme protein level, whereas modulation of activation/inactivation processes leads to changes in the specific activity of the enzyme molecule. The most powerful techniques for measurement of changes in the levels of specific proteins are those based on immunological procedures including: immunotitration of enzyme activity, rocket immunoelectrophoresis and radioimmunoassay (DAUSSANT et al. 1977, SCHIMKE 1975, SCHRÖDER and SCHÄFER 1980).

Observation of a change in enzyme protein level does not necessarily establish control over protein synthesis, and direct positive evidence for the regulation of enzyme de novo synthesis can only be obtained by a method based on in vivo labelling of newly synthesised proteins by pursuing the incorporation of labelled amino acids into enzyme protein. In general enzymes undergo continuous turnover (ACTON and GUPTA 1979) and in these cases clear labelling evidence for a change of de novo synthesis can be obtained only by a careful comparative investigation of the kinetics of labelling in the presence and absence of the inducing agents (SCHOPFER 1977).

Two in vivo labelling techniques have been extensively employed: (a) density labelling with a stable isotope such as ^{13}C, ^{2}H, ^{15}N or ^{18}O followed by analysis of the equilibrium distribution of enzyme activity in appropriate density gradients, (b) labelling with radioactive isotopes, usually ^{35}S in the form of ^{35}S-methionine, followed by specific immunoprecipitation of radiolabelled subunits of the enzyme under study. The advantages, disadvantages and limitations of these techniques have been extensively reviewed (ARIAS et al. 1969, BETZ et al. 1978, BURGESS et al. 1978, HÜTTERMANN and WENDLBERGER 1976, JOHNSON 1977, PASKIN and MAYER 1978, SCHIMKE 1975). A major problem with both techniques as applied to photocontrol of gene expression is possible photoeffects on the specific activity of label in the amino acid pool from which enzyme is synthesised (JOHNSON 1977, JOHNSON and SMITH 1978, LAWTON et al. 1980).

It is generally held that specific changes in the rate of protein synthesis reflect changes in the mRNA population present in polysomes (LODISH 1976). This can be checked by in vitro translation of polysomal mRNA populations from the tissue of interest. Widely used in vitro translation systems are those obtained from wheat germ (IYNEDJIAN 1979) and rabbit reticulocytes (PELHAM and JACKSON 1976, SCHMIKE et al. 1974). Degradation of mRNA by endogenous ribonuclease during extraction is possible and methods are available to detect and inhibit this (TAYLOR 1979). Fidelity of translation can be checked by comparison of the molecular weight distribution of polypeptides synthesised in vitro and in vivo.

Such heterologous translation systems may exhibit a differential preference for particular mRNA's compared to the protein-synthesising machinery in the tissue from which the mRNA was derived. Furthermore, the composition of the final translation products may partially reflect differential in vitro stability of particular mRNAs and for these reasons only relative changes in the activity level of a specific mRNA can be determined by these procedures. However, measurement of light-induced changes in protein synthesis by analysis of in vitro translation of polysomal mRNA preparations has the advantage that the pitfalls associated with light effects on the in vivo specific activity of label in the tissue during in vivo labelling experiments are avoided. In a heterologous translation system processing of a high molecular weight precursor polypeptide, or glycosylation may not occur and therefore a comparison of the molecular weights of polypeptides synthesised in vitro and in vivo may indicate the involvement of such events in production of specific mature polypeptides.

Changes in the activity level of a polysomal mRNA might reflect modulation of one or more of the following processes: (a) de novo production of the mature mRNA, (b) stability of the mRNA and (c) selective recruitment of the mRNA into the polysomes. For example there is evidence for considerable amounts of translatable mRNA in seeds which may be mobilised during germination

(PAYNE 1976). Recent advances in eukaryote molecular biology have shown that de novo production of mature mRNA may involve a number of processing steps including: (a) excision of non-coding intervening sequences from the primary transcripts of nuclear-encoded genes (DUGAICZYK et al. 1978, JEFFREYS and FLAVELL 1977, SUN et al. 1981, TSAI et al. 1980), (b) capping at the 5′-terminus (MINTY and BIRNIE 1981) and (c) polyadenylation at the 3′-terminus (MINTY and BIRNIE 1981).

Selective recruitment into polysomes can be conveniently checked by comparison of in vitro translation products from polysomal and total cellular RNA preparations. However, for more rigorous demonstration of change in mRNA level as opposed to activity (cf. methods for enzyme protein and enzyme activity), or for study of synthesis, processing and degradation of specific mRNA molecules a hybridisation probe, which is a nucleic acid sequence complementary to the mRNA, is required. Generally an excess of mRNA over probe is used, hybridisation proceeding with apparent first-order kinetics and the rate of hybridisation being proportional to the concentration of the individual mRNA within the population (EFSTRATIADIS and KAFATOS 1976, DAVIDSON 1976, WILLIAMSON 1976). The most common hybridisation probe is that obtained by synthesis of radio-labelled DNA complimentary to the mRNA (cDNA) using the mRNA as a template for avian myeloblastosis virus reverse transcriptase (EFSTRATIADIS and KAFATOS 1976). Preparation of specific cDNA therefore requires either the purification of mRNA before reverse transcription, or the purification of cDNA transcribed from a mixed population of mRNAs. The physicochemical properties of mRNAs are sufficiently similar that direct purification may not be completely satisfactory even for superabundant mRNAs. In favourable cases it may be possible to isolate the mRNA of interest by immunoprecipitation of the polysomes whose nascent polypeptide chains have the antigenicity of the mature polypeptide (SHAPIRO et al. 1974, SCHUTZ et al. 1977).

In general, however, recombinant DNA technology will be used to purify cDNA transcribed from a mixed population of (partially purified) mRNA by insertion of the cDNA into a suitable plasmid followed by molecular cloning (WU 1979). This procedure generates many clones containing cDNA sequences corresponding to the heterologous mRNA preparation. Clones containing relevant cDNA sequences can be identified by their ability to influence translation of the mRNA of interest in hybridisation-arrest (PATTERSON et al. 1977) and hybridisation-release translation procedures (SOBEL et al. 1978). Appropriate cDNA hybridisation probes can be used to measure changes in the level of specific mRNAs and coupled with pulse and pulse-chase labelling techniques can be used to study synthesis and degradation of the mature mRNA.

Hybridisation probes can also be used to study gene structure and indeed cloned genomic fragments can themselves be used as hybridisation probes. For example in studies of the expression of the ovalbumin gene, cloned genomic fragments containing the sequence of particular introns have been used to examine the loss of these sequences as message precursors are processed to mature messages (ROOP et al. 1978, TSAI et al. 1980).

Application of methods based on recombinant DNA technology will reveal the molecular mechanisms underlying mRNA production in plants and provide the basis for elucidation of the sites of photocontrol of gene expression at the nucleic acid level. In a few cases, in vitro transcription systems are being developed (JOLLY et al. 1981) and together with developments in photoreceptor biochemistry and reconstitution techniques we can anticipate in vitro study of the effects of photoreceptors and other putative regulatory components upon the transcription of isolated genes.

3 General Control by Light

The protein complement of light-treated plants is different to that of etiolated plants (COBB and WELLBURN 1973, 1974, GREBANIER et al. 1979, MULLETT et al. 1980), reflecting particularly the transition from etioplast to chloroplast. Illumination of plants leads to an increase in the overall rate of protein synthesis (FOURCROY et al. 1979, ZUCKER 1963, 1972) and application of in vivo labelling coupled with autoradiographic or fluorographic analysis of labelled proteins following 1D or 2D gel electrophoresis has allowed the demonstration of light-induced changes in the pattern of protein synthesis (KAVEH and HAREL 1973, MULLETT et al. 1980). The molecular mechanisms underlying the changes in pattern of protein synthesis are a key to an understanding of the photocontrol of gene expression.

During germination the total number of ribosomes increases both in darkness and under continuous illumination (FOURCROY et al. 1979). However, light increases the overall proportion of ribosomes present as polysomes (FOURCROY et al. 1979, KLEIN and PINE 1977, MALCOLM and RUSSELL 1974, PINE and KLEIN 1972, SMITH 1976, YAMAMOTO et al. 1974, 1975), and also the average size of the polysomes (SMITH 1976). In some cases at least, this effect seems to be mediated by phytochrome. Pre-treatment of tissue with cordycepin at concentrations sufficient to inhibit both total and poly(A)-rich mRNA synthesis 85%–95% within 3 h does not inhibit the effect of brief red light treatment on polysome levels and only partially inhibits the effect of continuous far-red light, suggesting that continuous mRNA synthesis is not necessary for the light-mediated increase in polysome proportions (SMITH 1976). There is also evidence for post-transcriptional control during the light-induced accumulation of thylakoid membrane proteins with molecular weights 24,000 and 28,000 during the greening of *Chlamydomonas reinhardtii* y-1 cells (HOOBER and STEGEMAN 1976).

Both white light and brief red light treatment of etiolated seedlings of *Zea mays* increase the subsequent capacity for protein synthesis in vitro of isolated monosomes (TRAVIS et al. 1974). This increased in vitro protein synthetic capacity was associated with changes in ribosomal properties including increased association of peptidyl-tRNA with ribosomes, which might suggest that light regulates the availability of initiation factors. These studies show that light can regulate gene expression at the translational level but do not indicate a mechanism for selective change.

Recently, light-induced changes of polysomal poly(A)-rich mRNA during greening of etiolated plants of *Hordeum vulgare* have been studied by the technique of cDNA-mRNA hybridisation (HEINZE et al. 1980). Hybridisation data of the homologous reactions reveal that in etiolated as well as in greened shoots a complexity of 5×10^7 nucleotides, or about 33,000 different average-sized mRNA's are present. These are organised in different abundancy classes with 94% of the total complexity present in each of the slowest reacting classes, presumably representing rare messages. Heterologous hybridisations indicate that 92% of all polysomal poly(A)-rich mRNA's in etiolated shoots are complementary with those of greened and 82% of green poly(A)-rich mRNAs are complementary to etiolated ones. The abundant mRNA classes are essentially responsible for these differences. The prevalent classes, making up 15% (etiolated) and 31% (green) of the poly(A-)rich mRNA mass but comprising only a complexity of 1.8×10^4 and 2.1×10^4 nucleotides respectively, are identical only to 50% with each other. Hybridisation of isolated prevalent green cDNA with whole etiolated poly(A)-rich mRNA indicates that the additionally appearing 50% prevalent green messengers must be regarded as green-specific, only present in polysomal poly(A)-rich mRNA after illumination. Similarly there is a proportion of the prevalent etiolated messengers which are not found in polysomal poly(A)-rich mRNA after illumination. The appearance of the prevalent, green-specific messengers in functional polysomes is not caused by a light-induced shift from poly(A)-free mRNA to poly(A)-rich mRNA and the results suggest that light induces greening, at least in part, by increasing the level of specific mRNA species in polysomes.

This might represent light-mediated selective recruitment of particular mRNAs into polysomes. In vitro translation in a wheat germ system of poly(A)-rich mRNA from polysomes of *Phaseolus vulgaris* seedlings that had been pre-irradiated with continuous white light results in the synthesis of polypeptides with molecular weights of 34, 32 and 25×10^3 (GILES et al. 1977). The mRNAs for these species were present at much lower levels in polysomes from dark-grown leaves, but were present in roughly equal amounts in total poly(A)-rich mRNA preparations from dark-grown and illuminated leaves. This result suggests that the mRNA molecules responsible for the changed pattern of protein synthesis observed after illumination were already present in dark-grown cells and that light merely mobilised the mRNAs to the polysomes. Taken with previous findings it seems that light regulates the initiation of translation of specific mRNAs. However, even a general effect on initiation or elongation rates can have seemingly specific effects on protein synthesis if particular mRNAs are preferentially translated and in situations where the overall rate of initiation is changing with time, the translation of different mRNAs will be variously affected. For example, if the mRNA for a given polypeptide having a below-average rate of attachment to ribosomes is a poorly translated messenger and if light stimulates the overall rate of polypeptide initiation, one would expect to see a light-mediated effect on the translation of this mRNA relative to other mRNAs that was apparently specific but reflected operation of a general mechanism (LODISH 1976).

The absolute amounts of poly(A)-rich mRNA have been measured in developing leaves of *Phaseolus aureus* (GRIERSON and COVEY 1975). Continuous white light causes rapid increases in plastid and cytoplasmic rRNA. Illumination also leads to a significantly larger proportion of poly(A)-rich mRNA but the response is relatively slow, no effect of light being observed within 6 h. However, the lack of gross quantitative changes does not preclude marked qualitative changes and indeed light regulation of the level of specific mRNA's have recently been demonstrated (Sect. 4.6, 4.9).

Light stimulates the synthesis of RNA (COHEN and SCHIFF 1976, HAREL and BOGORAD 1973, POULSON and BEEVERS 1970, SCHROTT and RAU 1977), but often the earliest effects reflect increased rRNA synthesis rather than changes in the rate of synthesis of putative mRNA molecules (COHEN and SCHIFF 1976, HAREL and BOGORAD 1973, POULSON and BEEVERS 1970). A general mechanism for increased RNA synthesis might involve stimulation of DNA-dependent RNA polymerases, and light-mediated increases of both nuclear and plastid enzymes have been observed (APEL and BOGORAD 1976, BOTTOMLEY 1970, POULSON and BEEVERS 1970, STOUT et al. 1967). However, detailed studies suggest that such effects are not related to either a quantitative rise in the amount of RNA polymerase or qualitative changes in the enzyme (APEL and BOGORAD 1976, SASIKI et al. 1979).

4 Control of Specific Gene Products

Although the photocontrol of a number of specific gene products has now been examined by inhibitor treatments, in vivo labelling, immunological and molecular biological techniques, in this section only those examples can be considered in which progress has been made in the elucidation of the mechanism of photocontrol at the molecular level. Information about other systems of interest can be obtained from Table 1 and references therein.

4.1 Chlorophyll a/b Binding Protein

The final stage of light-induced plastid membrane differentiation during the transformation of etioplasts into chloroplasts involves the insertion of light-harvesting structures into the chloroplast membrane (ARMOND et al. 1977). A major part of the light-harvesting chlorophyll appears to be associated with the light-harvesting chlorophyll a/b protein (THORNBER 1975), and together with the insertion of chlorophylls a and b, a massive incorporation of the apoprotein of the light-harvesting chlorophyll a/b protein into the membrane occurs (ARMOND et al. 1977, BAR-NUN et al. 1977). There is evidence from interspecific hybridisation experiments with *Nicotiana* species that the apoprotein is encoded in the nuclear genome and inhibitor studies suggest that it is synthesised outside the chloroplast on 80 S ribosomes (MACHOLD and AURICH 1972). It is well established that light stimulates chlorophyll synthesis (BOARDMAN et al. 1978 and Chap. 11, this Vol) and it has recently been shown in etiolated leaves of *Hordeum vulgare* that light also specifically induces the appearance of a prominent mRNA species which codes for a polypeptide of molecular weight 29,500 (APEL and KLOPPSTECH 1978). This component was identified by immunoprecipitation and peptide mapping as a precursor of the apoprotein of the light-harvesting chlorophyll a/b protein, molecular weight 25,500, which is the major light-induced chloroplast membrane component in vivo. This change in mRNA activity can also be induced by a 15 s red light pulse followed by 4 h in darkness and the red light effect is reversed by a subsequent far-red light treatment, indicating that the change in mRNA activity is controlled by phytochrome (APEL 1979). The light-induced increases in mRNA activity in total poly(A)-rich fractions and in the polysome fractions follow identical time courses, indicating that initiation of mRNA into polysomal complexes is not a rate-limiting step (MULLER et al. 1980a). Light-induced increases in mRNA activity for the chlorophyll a/b binding protein have also been observed in *Petroselinum hortense* (MULLER et al. 1980b) and *Lemna gibba* (TOBIN 1981).

In continuous white light the protein appears within the membrane after a lag period of 4–5 h which can be partially eliminated by a red light pulse given prior to the onset of continuous illumination (APEL 1979). However, the light-harvesting chlorophyll a/b binding protein cannot be detected amongst the plastid membrane proteins of *Hordeum vulgare* plants treated with a red light pulse alone and accumulation of the protein only takes place under continuous illumination which allows chlorophyll synthesis. In illuminated plants of the chlorophyll b-deficient mutant *chlorina f2,* as in the etiolated wild-type plants treated with a red light pulse, the appearance of mRNA activity for the apoprotein is induced and the mRNA is incorporated into the polysomes and presumably actively translated (APEL 1979, APEL and KLOPPSTECH 1978, 1980). Immunoprecipitation of the in vivo-labelled apoprotein from an extract of total protein of mutant plants indicates that, at least during the initial phase of the greening period, the protein is not only synthesised but also accumulated to a certain degree in the mutant (APEL and KLOPPSTECH 1980). However, in etiolated wild-type plants given only a red light pulse and which therefore remain deficient in both chlorophyll a and chlorophyll b, accumulation and insertion of the newly synthesised protein into the membrane is not observed (APEL and KLOPPSTECH 1980). Thus even though the apoprotein can accumulate in the membrane to a certain degree without chlorophyll b, accumulation in the thylakoid membrane is not possible without both chlorophylls a and b. Recently it has been directly demonstrated that in vitro-synthesised apoprotein can be integrated into thylakoid membranes of isolated chloroplasts in both wild-type plants and mutants lacking chlorophyll b (BELLEMARE et al. 1981). Therefore uptake, processing and integration into the membrane are not dependent on chlorophyll b. However, in vivo labelling experiments show that in the mutant, although integrated into the membrane, the apoprotein turns over more rapidly than in the wild-type plants and hence chlorophyll b appears to stabilise the inserted apoprotein.

The picture that emerges from these studies is that the synthesis and assembly of the chlorophyll a/b binding protein depends on at least two different light reactions (a) the phytochrome-mediated increase in the activity level of the mRNA for the apoprotein and (b) light-induced chlorophyll accumulation.

Light regulates chlorophyll accumulation both directly in the photoconversion of protochlorphyll(ide) to chlorophyll(ide) and by phytochrome-mediated stimulation of the appearance of δ-aminolevulinic acid formation (KLEIN et al. 1977, MASONER and KASEMIR 1975). The latter step has been found to be rate-limiting in chlorophyll biosynthesis (BEALE 1971, NADLER and GRANICK 1970, SISLER and KLEIN 1963) and has been held responsible for the lag period during the greening process (BEALE 1971, NADLER and GRANICK 1970). However, a large proportion of chlorophyll a and all of chlorophyll b is attached to the apoprotein of the chlorophyll a/b-binding protein and hence incorporation of chlorophyll into the membrane as part of the complex may be regulated not only by the concentration of chlorophyll, but also by the concentration of the apoprotein. Thus the greening process is not necessarily solely determined by the rate of chlorophyll synthesis but probably also depends on the rate at which the protein moiety of the chlorophyll a/b-binding protein is synthesised (APEL 1979). Hence induction of mRNA activity for the apoprotein of the chlorophyll a/b-binding protein may be a rate-limiting step in the assembly of the functional chloroplast. Furthermore, this system seems particularly well suited for study of phytochrome-mediated changes in the synthesis and accumulation of specific mRNA molecules using hybridisation probes obtained by molecular cloning techniques.

4.2 NADPH:Protochlorophyllide Oxidoreductase

NADPH:protochlorophyllide oxidoreductase is a major component of plastid membranes which undergoes rapid decrease in activity following illumination of dark-grown plants (APEL and KLOPPSTECH 1978, 1980, MAPLESTON and GRIFFITHS 1980, SANTEL and APEL 1981). Within the first 5 min of continuous light 90% of enzyme activity and 60% of enzyme protein are lost from etiolated seedlings of *Hordeum vulgare* and no stable polypeptide degradation fragments can be detected (SANTEL and APEL 1981). In this phase the rate of enzyme synthesis is not greatly affected. However, subsequently there is a marked inhibition of in vivo synthesis as measured by labelling with ^{35}S-methionine followed by specific immunoprecipitation, which is concomitant with a continuous light-dependent degradation of enzyme protein. Thus NADPH:protochlorophyllide oxidoreductase functions for only a short period after the onset of light. Following illumination there is a rapid decrease in the translatable activity of a mRNA which encodes for a polypeptide of molecular weight 44,000 (APEL 1981). This change can be induced by a 15 s red light pulse followed by 5 h in darkness and can be reversed by a subsequent far-red light treatment, indicating phytochrome involvement. The polypeptide product has been identified as a precursor of the NADPH:protochlorophyllide oxidoreductase, molecular weight 36,000. This represents a model system for study of phytochrome-mediated photorepression of gene expression.

4.3 Nitrate Reductase

Nitrate reductase is a key enzyme in the biological fixation of nitrogen and protein yields have been correlated to the levels of nitrate reductase during plant development (JOHNSON et al. 1976). The enzyme is induced by its substrate and by light. It has been suggested that the effect of light, at least in part, reflects light-mediated changes in nitrate pool levels (BEEVERS et al. 1965) or accumulation of photosynthetic products (ASLAM et al. 1973), but it is now established that light has a direct effect on the levels of nitrate reductase in a number of systems (see Table 1). Phytochrome control has been clearly demonstrated in, for example, buds of *Pisum sativum* (JONES and SHEARD 1975), cotyledons of *Sinapis alba* (JOHNSON 1976), etiolated leaves of *Zea mays* (RAO et al. 1980), etiolated seedlings of *Oryza sativa* (SASAKAWA and YAMAMOTO 1979) and excised leaves of *Triticum aestivum* (VIJAYARAGHAVAN et al. 1979).

In cotyledons of *Sinapis alba* seedlings that have been raised on a nitrate medium, light acting through phytochrome causes a very rapid, cycloheximide-insensitive increase in enzyme activity as measured in situ, whereas the increase in enzyme levels in response to nitrate treatment of water-grown material is inhibited by cycloheximide (JOHNSON 1976). After due allowance for problems associated with the interpretation of data from in situ assays, these results indicate a post-translational control by phytochrome at least over the short term.

In a number of systems, light stimulation of enzyme activity is prevented by inhibitors of RNA and protein synthesis (BEEVERS et al. 1965, HEWITT 1975, RAO et al. 1980, SAWNHEY and NAIK 1972, SLUITERS-SCHOLTEN 1973, TISCHNER and HÜTTERMAN 1978). However, it would be premature to conclude that in these cases light stimulates de novo synthesis of nitrate reductase. Thus in *Chlorella vulgaris,* although cycloheximide completely inhibits the light-mediated increase in enzyme activity, density labelling studies revealed relatively little incorporation of label, which was taken to indicate the operation of an activation mechanism (TISCHNER and HÜTTERMAN 1978), in addition to possible stimulation of de novo synthesis of the enzyme (JOHNSON 1979). Similar conclusions were drawn from nitrate-mediated induction of the enzyme in *Chlorella vulgaris* (JOHNSON 1979). However, in these experiments no estimates were made of the dead time between administration of label and the availability of labelled amino acids for enzyme synthesis. In the present cases where the increase in enzyme activity is rapid, it is possible to envisage the synthesis of significant amounts of enzyme from unlabelled amino acids before labelled amino acids are available. Hence the quantitative contributions of activation and de novo synthesis remain to be established. More recently, radio-isotope labelling (FUNKHOUSER et al. 1980) and immunological analysis (FUNKHOUSER and RAMADOSS 1980) indicate the presence of an inactive precursor protein. Nitrate stimulates and ammonia inhibits the activation of the precursor, but the effect of light on this process has not been reported. The activation is inhibited by cycloheximide, but whether this blocks the synthesis of a regulatory protein needed for activation of nitrate reductase is not yet clear (FUNKHOUSER and RAMADOSS 1980).

Following transfer to darkness there is a rapid loss of nitrate reductase activity in a number of systems (NICHOLAS et al. 1976, TRAVIS et al. 1969). In leaves of *Hordeum vulgare* the decay is inhibited by cycloheximide and actinomycin D (TRAVIS et al. 1969). It is clear from isotope labelling studies that under certain conditions nitrate reductase undergoes rapid turnover (JOHNSON 1979, ZIELKE and FILNER 1971). Specific inactivators of nitrate reductase have been isolated from *Glycine max* (JOLLY and TOLBERT 1978), *Oryza sativa* (KADAM et al. 1974, YAMAYA and OHIRA 1978) and *Zea mays* (WALLACE 1974, 1975). The inactivator from *Zea mays* roots is irreversible (WALLACE 1974, 1975), whereas those from cell cultures of *Oryza sativa* and *Glycine max* act reversibly (YAMAYA and OHIRA 1978, YAMAYA et al. 1980). The inactivator from *Glycine max* is a heat-labile protein molecular weight 31,000 (JOLLY and TOLBERT 1978), which is itself inhibited by light, probably acting through a flavin moiety, although the inactivator is present in similar amounts in dark- and light-grown leaves. The physiological role of the various inactivators is not known.

4.4 Phenylpropanoid Biosynthetic Enzymes

One of the earliest biochemical observations relating to photomorphogenesis was that light treatment often caused the accumulation of large amounts of anthocyanin (MOHR 1957, SIEGELMAN and HENDRICKS 1958). In other cases light causes the accumulation of different phenylpropanoid derivatives such as other flavonoids (FURUYA et al. 1962) or esters of hydroxycinnamic acids (ZUCKER 1963, ENGELSMA and MEIJER 1965).

The pathway of phenylpropanoid biosynthesis is well established and involves a central sequence from phenylalanine through cinnamic and p-coumaric acids to p-coumaroyl-CoA which is a key intermediate from which various branch pathways lead to the production of particular phenylpropanoid end-products (HAHLBROCK and GRISEBACH 1979).

In general, onset of light-induced accumulation of specific phenylpropanoid end-products is accompanied by marked but transient increases in the activity levels of the appropriate biosynthetic enzymes (CAMM and TOWERS 1973, HAHLBROCK and GRISEBACH 1979), and the pathway is being increasingly examined as a model system for the study of co-ordinate control of gene expression by light (BILLETT and SMITH 1980, HAHLBROCK and GRISEBACH 1979, LAMB 1979). For example, illumination of dark-grown cell cultures of *Petroselinum hortense* causes marked but transient increases in the activity levels of the enzymes of flavonoid glycoside biosynthesis (HAHLBROCK et al. 1976). These enzymes have been classified into two groups according to various operational criteria. Group I comprises the three enzymes of the central sequence: phenylalanine ammonia-lyase, cinnamic acid 4-hydroxylase and p-coumarate: CoA ligase; Group II consists of about 10 to 12 enzymes of the branch pathway from p-coumaroyl-CoA associated with flavonoid glycoside biosynthesis. Particularly useful criteria for the distinction of the two groups are the differential effects of treatment with fungal cell surface "elicitor" preparations or dilution of the cell cultures in the absence of light. Only the enzymes of Group I are induced by these treatments (HAHLBROCK et al. 1981, SCHRÖDER et al. 1977), whereas both groups are induced concomitantly by irradiation (EBEL and HAHLBROCK 1977, HAHLBROCK et al. 1976). The time after illumination at which maximum activity is reached is different for each enzyme and is correlated with the subsequent rate of decay of enzyme activity (HAHLBROCK et al. 1976). Detailed studies showed that the kinetics of induction of phenylalanine ammonia-lyase were most closely correlated with changes in the rate of accumulation of flavonoid glycosides (HAHLBROCK et al. 1976), suggesting that this enzyme is the primary site of photocontrol of phenylpropanoid biosynthesis in this system. However, the induction of the other enzymes of the pathway is probably part of the overall control system, since in general their basal activities are insufficient to support the subsequent light-induced flux through the pathway and it has been suggested that chalcone synthase, the first enzyme of the flavonoid glycoside branch pathway, is a second key site of photocontrol (SCHRÖDER et al. 1979a).

Following the preparation of discs of *Solanum tuberosum* tuber tissue and incubation in darkness, the level of chlorogenic acid (caffeoylquinic acid) increases after a lag of several hours (LAMB and RUBERY 1976b). The central pathway enzymes increase in activity concomitantly in dark-incubated discs but in contrast, hydroxycinnamoyl-CoA:quinate hydroxycinnamoyl transferase, the first enzyme of the branch pathway leading from p-coumaroyl-CoA to chlorogenic acid, is present at relatively high levels in dormant tissue and does not initially increase in dark-incubated discs (LAMB 1977b). However, chlorogenic acid accumulation is further stimulated by continuous illumination with white light; of the biosynthetic enzymes, only phenylalanine ammonia-lyase and hydroxycinnamoyl transferase are under photocontrol (LAMB 1977b, LAMB and RUBERY 1976b). The data show that chlorogenic acid accumulation is regulated by the transient, inter-related changes in activity of these biosynthetic enzymes with phenylalanine ammonia-lyase and hydroxycinnamoyl transferase as the sites of primary and secondary photocontrol respectively (LAMB 1977b, LAMB and RUBERY 1976b). The situation with regard to photocontrol of chlorogenic acid accumulation in etiolated seedlings of *Fagopyrum esculentum* is somewhat different in that the activity level of cinnamic acid 4-hydroxylase is also under photocontrol but, as in *Solanum tuberosum,* the level of p-coumarate: CoA ligase is insensitive to light (AMRHEIN and ZENK 1970, MCCLURE and GROSS 1975).

It is clear from these examples that photocontrol of phenylpropanoid accumulation is achieved by transient, inter-related changes in activity of appropriate biosynthetic enzymes and that within this framework there is considerable variation in the degrees of co-ordination of enzyme changes. This flexibility presumably reflects the various biological requirements of specific metabolic programmes in different systems. Current research focuses on two related problems: (a) the molecular mechanisms underlying the transient increase and subsequent decay in activity levels of the various biosynthetic enzymes and (b) the molecular mechanisms providing for flexible co-ordination of these changes.

In a number of systems the initial increase in enzyme level in response to illumination involves stimulation of the rate of de novo enzyme synthesis. This has been demonstrated most powerfully by HAHLBROCK and co-workers in studies of *Petroselinum hortense* cell cultures. Early work demonstrated that the increase in phenylalanine ammonia-lyase activity in response to illumination was inhibited by actinomycin-D and cycloheximide (HAHLBROCK and RAGG 1975) and density-labelling experiments using ^{15}N from $^{15}NH_4$ showed photocontrol over the rate of phenylalanine ammonia-lyase de novo production (WELLMANN and SCHOPFER 1975). These observations have been confirmed and extended to show that light causes large, concomitant increases in mRNA activities for phenylalanine ammonia-lyase and chalcone synthase (SCHRÖDER et al. 1979b). The rates of enzyme synthesis both in vitro in a reticulocyte lysate and in vivo were quantitated by direct immunoprecipitation of the labelled enzyme subunits. The two mRNA activities increased rapidly in irradiated cells for several hours. Phenylalanine ammonia-lyase mRNA reached a peak in activity a few hours earlier than chalcone synthase mRNA. The apparent half-lives of the enzyme activities were about 7 to 10 h for phenylalanine ammonia-lyase and 5 to 7 h for chalcone synthase. Recently it has been shown that changes in mRNA activity for UDP-apiose synthetase follow a similar time course to chalcone synthase mRNA activity (GARDINER et al. 1980). These data were used to calculate the expected light-induced changes in enzyme activity from the measured changes in mRNA activity. The results for all three enzymes were in agreement with experimental data indicating that the light-induced changes in enzyme activity can largely be explained by changes in the rates of enzyme synthesis which are in turn governed by changes in the respective mRNA activity (GARDINER et al. 1980, SCHRÖDER et al. 1979b). However, relatively small changes (2- to 3-fold) in the rate of enzyme inactivation or degradation might not be detected by this approach. The mature mRNA for phenylalanine ammonia-lyase is too small to encode for other enzymes of the pathway and hence the co-ordination does not reflect translation of a polycistronic mRNA (RAGG and HALBROCK 1980). Recently co-ordinate induction by light of the mRNA's encoding phenylalanine ammonia-lyase and p-coumarate: CoA ligase have been observed (RAGG et al. 1981).

During the phase of decay in enzyme activity the exponential rates of decrease of immunoprecipitable radioactivity in phenylalanine ammonia-lyase subunits and enzyme activity were about the same (BETZ et al. 1978, SCHRÖDER et al. 1979b). In contrast, with chalcone synthase, the exponential rate of decrease was much greater for enzyme activity than for radioactivity in immunoprecipitable subunits, indicating a more rapid loss of catalytic activity than actual degradation of the enzyme molecule (SCHRÖDER et al. 1979b). Comparison of the kinetics of light-induced changes in chalcone synthase catalytic activity and chalcone synthase protein as determined by a sensitive and specific radioimmunoassay reveals the accumulation of a pool of inactive molecules during only the later stages of enzyme induction (SCHRÖDER and SCHÄFER 1980). These observations raise the possibility that modulation of the rate of enzyme inactivation contributes to the overall regulation of the activity level of chalcone synthase in illuminated *Petroselinum hortense* cells.

Dual control over both the rate of de novo enzyme production and the rate of removal of enzyme activity occurs during the transient increase in phenylalanine ammonia-lyase activity in illuminated tuber discs of *Solanum tuberosum* (LAMB and MERRITT 1979, LAMB et al. 1979). Comparative in vivo density labelling with ^2H from ^2H_2O shows that light stimulates the rate of de novo production of the enzyme (LAMB and MERRITT 1979). Quantitative analysis of pulse ^2H-labelling experiments indicates that the initial increase in enzyme activity in illuminated discs reflects a transient increase in the rate of de novo enzyme production against a low background rate of enzyme removal (LAMB et al. 1979). The abrupt transition to a phase of enzyme decay is caused by not only a reduction in the rate of de novo enzyme production but also by a marked increase in the rate of removal of active enzyme. The ^2H-labelling technique does not allow discrimination between enzyme inactivation and enzyme degradation (LAMB and RUBERY 1976a). Inhibitor studies that the co-ordinate increase in phenylalanine and ammonia-lyase and cinnamic acid 4-hydroxylase activities originate from concomitant increases in the mRNA activity levels in a similar manner to phenylpropanoid enzymes in *Petroselinum hortense* (HAHLBROCK and RAGG 1975, LAMB 1979, SCHRÖDER et al. 1979b). Cinnamic acid 4-hydroxylase is a cytochrome P-450-dependent enzyme and accounts for most of the cytochrome P-450 present in *Solanum tuberosum* tuber discs (RICH and LAMB 1977). The increase in enzyme activity is concomitant with an increase in cytochrome P-450 protein determined spectrophotometrically, further suggesting de novo synthesis of the enzyme (RICH and LAMB 1977). Similar co-induction of enzyme activity and cytochrome P-450 by light acting through phytochrome has been demonstrated in *Pisum sativum* seedlings (BENEVENISTE et al. 1978).

The role of light modulation of enzyme de novo production in the photocontrol of phenylalanine ammonia-lyase levels in intact, etiolated seedlings is much less clear. In cotyledons of *Sinapis alba* phenylalanine ammonia-lyase activity can be dramatically induced by light operating through the phytochrome system (SCHOPFER and MOHR 1972). Contradictory conclusions on the role of light modulation of (a) de novo enzyme production and (b) enzyme activation from inactive or less active forms in this system have been reached by two laboratories using density labelling with ^2H from ^2H_2O. The controversy centres on possible effects of light on the specific activity of label in the amino acid pools from which the enzyme was synthesised (ACTON and SCHOPFER 1975, ATTRIDGE et al. 1974, JOHNSON and SMITH 1978, TONG and SCHOPFER 1976). Recently light-induced increase in phenylalanine ammonia-lyase synthesis has been observed in seedlings of *Pisum sativum* using ^{35}S-methionine labelling coupled with specific immunoprecipitation of enzyme subunits (LOSCHKE et al. 1981).

The best evidence for post-translational control of phenylalanine ammonia-lyase by light in intact seedlings comes from studies, using in vivo density labelling and cycloheximide treatments, on the blue-light-mediated induction of enzyme activity in *Cucumis sativus* seedlings (ATTRIDGE and SMITH 1973, 1974): see SCHOPFER (1977) for a detailed discussion.

Interestingly an inactivator for phenylalanine ammonia-lyase has been detected in extracts of *Cucumis sativus* seedlings. Originally it was thought to be a small molecule (FRENCH and SMITH 1975) but subsequent investigation suggests it is a macromolecule, probably a protein which is found in both soluble and membraneous fractions (BILLETT et al. 1978). The inactivator is a reversible time-independent inhibitor that competes with the substrate phenylalanine. The inactivator also inhibits cinnamic acid 4-hydroxylase, the second enzyme of the phenylpropanoid pathway, but has no effect on a wide variety of other, metabolically unrelated enzymes. The physiological role of this inactivator in relation to light modulation of phenylalanine ammonia-lyase and cinnamic acid 4-hydroxylase has not been established (BILLETT and SMITH 1980) and recently it has been suggested that the inactivator is simply phenylpropanoid intermediates non-specifically bound to cellular proteins (GUPTA and ACTON 1979). This proposal would explain the localisation of the inactivator in both soluble and membrane-bound fractions and might account for the conflicting reports with regard to molecular weight.

Time-dependent inactivators of phenylalanine ammonia-lyase have been identified in extracts of *Helianthus annuus* and *Ipomoea batatas* (CREASY 1976, TANAKA et al. 1977). In the latter case the level of inactivator increases markedly in the later stages of enzyme induction, concomitant with the increase in rate of removal of the enzyme detected by in vivo labelling experiments (TANAKA and URITANI 1977).

4.5 Phosphoenolpyruvate Carboxylase

Treatment of dark-grown *Zea mays* seedlings with 80 h of white light leads to a fivefold increase in the extractable activity of phosphoenolpyruvate carboxylase, which is prevented by inhibitors of protein synthesis (GRAHAM et al. 1970). This increase in enzyme activity is accompanied by an increase in the amount of enzyme protein as detected by SDS-PAGE and identified by peptide mapping (HAGUE and SIMS 1980). During the light period there was an increase in the rate of incorporation of ^{35}S-methionine into the subunit band tentatively identified as phosphoenolpyruvate carboxylase (HAGUE and SIMS 1980), suggesting a light-mediated increase in de novo synthesis of the enzyme. This has been confirmed by specific immunoprecipitation of the enzyme during greening (HAYAKAWA et al. 1981). There is a specific increase in the amount of enzyme protein as detected by rocket immunoelectrophoresis and increased incorporation of radiolabel into immunoprecipitable subunits. More recently it has been shown that an increase in translatable mRNA for the enzyme accompanies the incresed rate of synthesis in vivo (HAGUE and SIMS 1981).

Phosphoenolpyruvate carboxylase exists in multiple forms associated with different metabolic pathways (TING and OSMOND 1973). In *Saccharum* hybrids the enzyme from dark-grown leaves has the characteristics of the dark CO_2-fixation enzyme commonly found in C-3 plants, whereas the isoenzyme from light-grown leaves has the characteristics of the photosynthetic enzyme present in C-4 plants (GOATLY and SMITH 1974). In vivo density labelling and biochemical studies suggest that the long term increase in activity following illumination might in this case reflect interconversion of the two forms leading to a change in properties and increased catalytic activity (GOATLY et al. 1975).

4.6 Photogene 32

A major light-induced component of thylakoid membranes of *Zea mays* is a polypeptide of molecular weight 32,000: photogene 32 (Bedbrook et al. 1978) which is the binding site for s-triazine type inhibitors of photosystem II (McIntosh et al. 1981). The onset of the accumulation of photogene 32 product is correlated with the appearance of (a) mRNA activity coding for a polypeptide molecular weight 34,500 and (b) a major light-induced, non-ribosomal RNA species which hybridises to a Bam fragment of *Zea mays* ctDNA (Bedbrook et al. 1978). Although the fragment is sufficiently large to contain more than one gene, it is probable that these three co-induced parameters relate to the expression of the same gene. Recently it was shown by hybridisation studies that light operating through phytochrome increases the level of the plastid mRNA encoding a polypeptide molecular weight 35,000 in *Sinapis alba* seedlings (Link 1981, 1982). The gene for this polypeptide is homologous to the *Zea mays* plastid gene encoding the precursor of the photogene 32 polypeptide (Bogorad et al. 1980). Photogene 32 may also be under photocontrol by a similar mechanism in *Spirodela oligorrhiza* (Weinbaum et al. 1979) and *Nicotiana sylvestris* (Lett et al. 1980). Although photogene 32 is a major product during greening, isolated chloroplasts of *Zea mays* accumulate the 34,500 polypeptide instead, and hence isolated chloroplasts seem unable to process this precursor to give the mature photogene 32 product (Grebanier et al. 1978, 1979). Interestingly, nuclear mutants of *Zea mays* blocking photosystem II have been obtained which lack the mature chloroplast-encoded photogene 32 product (Leto and Miles 1980). However, the relationship between these nuclear mutants and the processing of the precursor polypeptide has not been established.

4.7 Phytochrome

The level of spectrally detectable phytochrome increases during the growth of etiolated seedlings. Following irradiation there is a rapid decrease in phytochrome level followed by a subsequent increase upon return of the irradiated seedlings to darkness. Recently, the effect of light on phytochrome levels has been confirmed using a radio-immunoassay (Hunt and Pratt 1980) and also by spectrophotometric measurements of photoreversibility in Norflurazon-treated seedlings (Jabben and Deitzer 1978). Density labelling with ^{2}H from ^2H_2O followed by analysis of the equilibrium distribution of spectrally detectable phytochrome in CsCl density gradients showed that de novo synthesis of the protein moiety of the photoreceptor occurred in the hook (Quail et al. 1973a) and cotyledons (Quail et al. 1973b) during the developmental increase of phytochrome level in *Cucurbita pepo* seedlings. Similarly, de novo synthesis occurred in the recovery of phytochrome levels following return of irradiated seedlings to darkness. A preliminary report has demonstrated in vitro synthesis of immunoprecipitable phytochrome directed by poly(A)-rich mRNA from *Avena sativis* but the light regime was not specified (Bolton and Quail 1981). More recently a light-mediated reduction of the activity level of a mRNA encoding for a protein apparent molecular weight 125,000 has been observed (GOTTMANN and SCHÄFER 1982) which was identified by immunochemical means as the phytochrome apoprotein. Hence light appears to inhibit synthesis of the photoreceptor but may also modulate the rate of phytochrome destruction (HEIM et al. 1981). The molecular weight of the in vitro translation product was somewhat greater than phytochrome synthesised in vivo which might reflect post-transcriptional modification or slight proteolytic degradation during purification of phytochrome synthesised in vivo.

4.8 rRNA and tRNA

Light acting by the phytochrome system stimulates the accumulation of both cytoplasmic rRNA (molecular weights 1.3 and 0.7×10^6) and plastid rRNA (molecular weights 1.1 and 0.56×10^6) in cotyledons of *Sinapis alba* (THIEN and SCHOPFER 1975a). Pulse-labelling of cotyledons with ^{3}H-uridine showed that light stimulated the synthesis of the respective high molecular weight precursors of both cytoplasmic rRNA and plastid rRNA (2.9 and 2.4×10^6 m.w. respectively) (THIEN and SCHOPFER 1975b). Large and small ribosomal subunit RNAs are maintained in a 1:1 molar ratio in both cytoplasm and plastids, irrespective of seedling age and light treatment. Continuous white fluorescent light, which saturates chlorophyll synthesis is less effective than continuous far red light, which gives essentially no chlorophyll synthesis, in producing the response, indicating that the accumulation of plastid rRNA is independent of the build-up of a functional photosynthetic apparatus. The data are consistent with phytochrome-mediated photocontrol of rRNA synthesis.

Light also increases the level of plastid rRNA in *Euglena gracilis* (COHEN and SCHIFF 1976) but in this system the action spectrum suggests that the photoreceptor is protochlorophyll(ide) and furthermore, the level of cytoplasmic rRNA is not affected by light. It was previously claimed on the basis of size fractionation studies that plastid rRNA did not occur in dark-grown cells (BROWN and HASELKORN 1971) but more sensitive methods of analysis using hybridisation probes indicate that plastid rRNA accounts for 2% of total cellular RNA in dark-grown cells (CHELM et al. 1977). Light increases the proportion of plastid rRNA to about 25%, whereas the amount of plastid rDNA increases only two to three fold. Hence light-stimulated increase in the gene dosage of plastid rRNA is not the major factor in the accumulation of plastid rRNA implying photocontrol of the rate of synthesis of plastid rRNA or possibly increased rRNA stability (CHELM et al. 1977).

Light-induced increase in the level of plastid tRNA has been detected in greening *Zea mays* by ctDNA: ^{125}I-tRNA hybridisation probes (HAFF and BOGORAD 1976).

4.9 Ribulose Bisphosphate Carboxylase

Ribulose bisphosphate carboxylase is localised in the chloroplast and catalyses the only reaction in higher plants that leads to net fixation of carbon dioxide.

The enzyme also exhibits an oxygenase activity, which is the origin of phosphoglycollate associated with photorespiratory release of CO_2. The enzyme can account for up to 50% of total leaf protein and is therefore probably the most abundant protein in nature (ELLIS 1979, KUNG 1976). In higher plants it is an aggregate composed of 16 subunits: 8 large subunits with molecular weight = 50–55,000 and 8 small subunits with molecular weight = 14–16,000 (KAWASHIMA and WILDMAN 1970). The large subunit contains the active sites for both carboxylase and oxygenase functions, and the small subunit appears to have regulatory functions.

Studies with interspecific hybrids of *Nicotiana* sp. show that the small subunit is coded for in the nuclear genome and there are one or very few copies of the gene per haploid genome, whereas the large subunit is coded for in the chloroplast genome with one copy per molecule of circular ctDNA (BEDBROOK et al. 1979, CHAN and WILDMAN 1972, CASHMORE 1979, KAWASHIMA and WILDMAN 1972, KUNG 1976). The gene for the large subunit has been physically mapped in the chloroplast genome (BEDBROOK et al. 1979, COEN et al. 1977, LINK and BOGORAD 1980, MALNOE et al. 1979) and sequenced (MCINTOSH et al. 1980). No "classical" prokaryotic promoter sequences or putative eukaryotic promoter sequences were found, although there is a typical prokaryotic ribosome binding site close to where translation is initiated (MCINTOSH et al. 1980). The mRNA for the large subunit is found in chloroplasts (HARTLEY et al. 1975, WHEELER and HARTLEY 1975) and the large subunit is the major soluble product of light-driven protein synthesis in isolated chloroplasts (BLAIR and ELLIS 1973, MORGENTHALER and MENDIOLA-MORGENTHALER 1976), indicating that the protein is synthesised within the organelle. The mRNA is not polyadenylated although the data do not rule out a sequence of less than 20 residues (WHEELER and HARTLEY 1975). The small subunit is synthesised as a precursor polypeptide with a molecular weight of 4–8,000 greater than the mature small subunit (depending on the species) when poly(A)-rich mRNA from cytoplasmic polysomes is translated in vitro (CHUA and SCHMIDT 1978, DOBBERSTEIN et al. 1977, HIGHFIELD and ELLIS 1978, TOBIN and SUTTIE 1980). Recently, polyadenylated RNA from leaves of *Pisum sativum* have been copied into DNA and cloned in the *Escherichia coli* plasmid pBR322 (BEDBROOK et al. 1980). From these clones the DNA encoding the mRNA for the precursor polypeptide has been identified and sequenced. The precursor enters isolated chloroplasts with cleavage to its final size by a mechanism which is independent of concomitant translation (CHUA and SCHMIDT 1978, HIGHFIELD and ELLIS 1978), and assembles in the stroma with the large subunit molecules into the holoenzyme (CHUA and SCHMIDT 1978, SMITH and ELLIS 1979).

During greening there is a long-term increase in the activity level of ribulose bisphosphate carboxylase, and this response is under phytochrome control in many but not all cases (see Table 1 and SCHOPFER 1977, TOBIN and SUTTIE 1980). In vivo labelling followed by immunoprecipitation of the enzyme has demonstrated that light stimulates the rate of synthesis of the enzyme (GRAY and KEKWICK 1974, KLEINHOPF et al. 1970, TOBIN and SUTTIE 1980). Light causes an increase in the activity level of mRNA for the small subunit precursor in preparations of polyribosomal RNA (HIGHFIED and ELLIS 1978, LETT et al.

1980, TOBIN and SUTTIE 1980). This effect is not associated with selective recruitment of mRNA into polysomes (unpublished results of TEPFER and TOBIN in TOBIN and SUTTIE 1980). Using cloned cDNA encoding for the mRNA for small subunit precursor as a hybridisation probe, it has recently been shown that light causes an increase in the amount of mRNA for the precursor in *Pisum sativum* (BEDBROOK et al. 1980). Using similar techniques it has been shown that in leaves of *Zea mays* mRNA for the large subunit is present in bundle-sheath cells, but is almost or entirely absent from mesophyll cells reflecting the distribution of the holoenzyme in these two types of photosynthetic cells of C_4 plants (LINK et al. 1978). Although it has not been directly demonstrated, it is very probable that light causes an increase in the amount of mRNA for the large subunit in bundle-sheath cells. However, in cotyledons of *Sinapis alba,* hybridisation of in vitro ^{32}P-labelled cellular RNA to a cloned cDNA fragment encoding the large subunit failed to reveal red- or far-red-light-mediated changes in the level of the mRNA encoding for the large subunit (LINK 1982).

Early work suggested that synthesis of the two subunits of ribulose bisphosphate carboxylase was tightly coupled and it was postulated that accumulation of the small subunit in the chloroplast stroma might stimulate expression of the gene encoding the large subunit (ELLIS 1975). There is strong evidence for such coupling in algal cells (IWANIJ et al. 1975), but in higher plants there is evidence that synthesis of the two subunits is not tightly coupled at least over short periods. Thus treatment of isolated cells of *Glycine max* with cycloheximide and 2-(4-methyl-2,6-dinitroaniline)-N-methylpropionamide, which are inhibitors of protein synthesis on 80S ribosomes inhibits small subunit synthesis but large subunit synthesis continues for up to 4 h. Furthermore, under these conditions the newly synthesised large subunit enters active ribulose bisphosphate carboxylase molecules, implying the existence of a pool of small subunits in cells (BARRACLOUGH and ELLIS 1979). Earlier work had suggested the presence of a pool of large subunits in *Phaseolus vulgaris* during greening and hence it is possible that assembly of the holoenzyme is a rate-limiting step and a potential site for photomodulation (GRAY and KEKWICK 1974). Sucrose density gradient centrifugation of extracts of ^{35}S-methionine-labelled pea leaves reveals the presence of a slowly sedimenting form of small subunit which may reflect assembly intermediates (ROY et al. 1978). Furthermore, in greening *Hordeum vulgare* plants a major portion of newly synthesised immunoprecipitable ribulose bisphosphate carboxylase protein preceded the light-induced increase in enzyme activity, consistent with the idea that both subunits increase before associating to give the holoenzyme (SMITH et al. 1974). The same study revealed that there was zero turnover of the holoenzyme during the greening period. However, it is well established that during senescence there is marked degradation of the enzyme when leaves are put in darkness and there is some evidence that this is preferential when compared with other proteins (PETERSON and HUFFAKER 1975, WILLIAMS and KENNEDY 1978, WITTENBACH 1978). Hence it is possible that regulation of the turnover of the enzyme may be important in determining enzyme levels, especially in the later stages of induction of the enzyme.

5 Endogeneous Regulation of the Photocontrol of Gene Expression

In the preceding sections the molecular mechanisms underlying photocontrol of the expression of specific genes have been considered. The question now arises as to how these disparate responses are integrated in photomorphogenesis.

A number of observations illustrate the specificity of phytochrome action with respect to gene expression. First, some enzymes are induced, some are repressed and some are unresponsive to phytochrome action (MOHR 1972, SCHOPFER 1977). Second, there are characteristic differences in the kinetics and dose-response behaviour with respect to phytochrome of different enzymes within the same tissue and even the same organelle. Detailed studies of the response of phenylalanine ammonia-lyase and ribulose bisphosphate carboxylase show that there are multiple switches for phytochrome control of gene expression rather than a single master switch (FROSCH et al. 1977). Even with two functionally linked enzymes of the Calvin cycle, although the induction kinetics are broadly similar there are significant differences in detail and the data support the idea that the multiple switches concept may apply within a metabolic pathway (BRUNING et al. 1975). Clearly it will be of interest to apply this concept to the co-ordinate control of the enzymes of phenylpropanoid biosynthesis. Third, differences with respect to the quality of the primary actions of phytochrome can be postulated. Thus in graded responses such as the induction of phenylalanine ammonia-lyase, the degree of induction is a continuous function of the amount of the active form of phytochrome established by light pulses and similarly a smooth irradiance dependence is observed in the high irradiance response (MOHR 1972, SCHOPFER and MOHR 1972). The amount of phytochrome needed to obtain a response may be very small. In contrast, in threshold responses such as the repression of lipoxygenase accumulation in seedlings of *Sinapis alba* the effect of light depends on the concentration of phytochrome in an all-or-none manner (MOHR and OELZE-KAROW 1976, OELZE-KAROW and MOHR 1973, OELZE-KAROW et al. 1970).

The competence of an enzyme to respond to light acting through phytochrome is a function of age and there are characteristic differences between individual enzymes with respect to the onset and shape of the kinetics of competence. In some cases the time course of competence is a continuous function (HUAULT 1974, MOHR 1972), whereas in the case of lipoxygenase in cotyledons of *Sinapis alba* threshold kinetics are observed. In this case there is no control by phytochrome up to 33.3 h after sowing (25 °C). At this point full repression of lipoxygenase is possible and 14.7 h later the system abruptly and completely emerges from phytochrome control (MOHR 1972, MOHR and OELZE-KAROW 1976, OELZE-KAROW and MOHR 1970). Starting and finishing points are not affected by light treatment.

The phytochrome system becomes active before the germinating embryo is competent to respond. However, the light signals perceived before the onset of competence can be stored either as active phytochrome or as an intermediate in the sequence of events leading to the final response, until the starting point of competence is achieved (STEINITZ et al. 1976).

An interesting example of the storage of a light signal is given by phytochrome control of peroxidase (SCHOPFER 1977). The level of this enzyme is controlled by phytochrome in leaves of *Zea mays* according to the kinetic pattern of photomodulation, that is full photoreversibility of the response is maintained up to the starting point of the response (SHARMA et al. 1976). In contrast in cotyledons of *Sinapis alba* the response involves two steps which do not overlap. The response can be induced by phytochrome only during the first 4 days after sowing, whereas the increase in peroxidase activity occurs only after this time (SCHOPFER and PLACHY 1973). Thus during the period in which phytochrome can act the response remains latent and this is followed by a realisation period during which phytochrome is no longer effective but enzyme activity increases, provided phytochrome has acted previously. Inhibitor and density labelling evidence indicates that in the first phase phytochrome induces the synthesis of the enzyme in an inactive form which is activated by a process not involving phytochrome in the second phase (SCHOPFER 1977).

It is well established that phytochrome effects organ-specific responses but relatively little is known of the spatial pattern of competence at the tissue level. Analysis of microtome sections of frozen cotyledons of *Sinapis alba* showed that the induction of phenylalanine ammonia-lyase was regulated quite differently by phytochrome in the upper and lower epidermis (WELLMANN 1974). The differential expression of the gene for ribulose bisphosphate carboxylase in mesophyll and bundle-sheath cells of *Zea mays* is a further example of this spatial competence and has been attributed at least in part, to failure of chloroplast mesophyll cells to accumulate mRNA for the large subunit of the enzyme (LINK et al. 1978). Recently it has been reported that in developing leaves of *Hordeum vulgare* the chlorophyll a/b binding protein complex appears first in the base of the leaf whereas ribulose bisphosphate carboxylase first appears in the apical sections (VIRO and KLOPPSTECH 1980). The distribution of mRNA activities for the two proteins correlate with this distribution.

In summary, it has become clear that in photomorphogenesis light acting through photoreceptors such as phytochrome can be regarded as a trigger of certain responses which are strictly pre-programmed in time and space by an endogenous regulatory system which is responsible for the specificity and competence of the cell towards external stimuli. However, there is as yet little information on the molecular processes involved in the interaction between the genetic programmes which regulate the spatial and temporal pattern of development and the environmental factors which epigenetically trigger realisation of these programmes.

One possible general mechanism might involve light control of the concentration or compartmentalisation of pathway-specific metabolites which in turn regulate the expression of genes for the enzymes of that pathway. For example phytochrome stimulation of anthocyanin accumulation in *Brassica oleracea* appears to involve changes associated with compartmentalisation of intermediates and membrane permeability in addition to increases in levels of biosynthetic enzymes (PECKETT and BASSIM 1974). Furthermore, exogenous (hydroxy)cinnamic acids regulate the level of specific enzymes of the pathway by both feedback (LAMB and RUBERY 1976c) and feedforward mechanisms (LAMB 1977b), and in some cases specific control over rates of de novo enzyme synthesis have been observed (JOHNSON et al. 1975, LAMB unpublished). Experiments with competitive inhibitors of phenylalanine ammonia-lyase in vivo show that feedback regulation of phenylalanine ammonia-lyase and cinnamic acid 4-hydroxylase occurs in the later stages of enzyme induction following accumulation of (hydroxy)cinnamic acids (AMRHEIN and GERHARDT 1979, BILLETT and SMITH 1980, LAMB unpublished). However, the role of changes in (hydroxy)cinnamic acid levels as a causal link in the chain of events during the initial light-induced increase in enzyme activity has not been established. Nonetheless many factors that induce phenylalanine ammonia-lyase cause, at least initially, decreases in the levels of (hydroxy)cinnamic acids (ENGELSMA 1974) and the hypothesis remains an attractive line of enquiry.

6 Summary and Future Prospects

In the last few years considerable advance has occurred in our understanding of the molecular mechanisms underlying photocontrol of gene expression. The picture emerging is of many diverse molecular mechanisms operating and in some cases at least, more than one mechanism may be involved in the photocontrol of the expression of a particular gene. Such diversity and multiplicity presumably reflect the biological requirements of the plant in relation to the perception of light. Thus theoretical considerations indicate that in slowly growing plant cells, concomitant but inverse changes in the rates of enzyme production and removal might enhance the flexibility and rapidity of changes in the levels of specific enzymes in response to an environmental stimulus (PASKIN and MAYER 1977, 1978). Light may also exert control over the processes governing location of, and movement of, specific gene products within the cell. Similarly the diversity and multiplicity of mechanisms may reflect the range of environmental information the plant assimilates, on a variety of time scales, from perception of the intensity and wavelength of incident light. It is clear that application of immunological and recombinant DNA techniques will rapidly lead to detailed elucidation of the molecular mechanisms underlying photocontrol of gene expression. Hopefully this will allow study at the molecular level of the interaction between light and internal metabolic and developmental control programmes which govern the biological response of the plant to the environment.

Acknowledgements. We thank the SRC for a Research Grant to CJL and a Research Studentship to MAL. CJL is Browne Research Fellow, The Queen's College, Oxford.

References

Acton GJ (1972) Phytochrome control of the level of extractable ribonuclease activity in etiolated hypocotyls. Nature 236:255–256

Acton GJ (1974) Phytochrome-controlled acid RNase: an "attached" protein of ribosomes. Phytochemistry 13:1303–1310

Acton GJ, Gupta S (1979) A relationship between protein-degradation rates in vivo, isoelectric points and molecular weights obtained using density labelling. Biochem J 184:367–377

Acton GJ, Schopfer P (1974) Phytochrome-induced synthesis of ribonuclease de novo in lupin hypocotyl sections. Biochem J 142:449–455

Acton GJ, Schopfer P (1975) Control over activation or synthesis of phenylalanine ammonia-lyase by phytochrome in mustard (*Sinapis alba*)? A contribution to eliminate some misconceptions. Biochim Biophys Acta 404:231–242

Acton GJ, Grumm H, Mohr H (1974) Control of synthesis de novo of ascorbate oxidase in the mustard seedling (*Sinapis alba* L.) by phytochrome. Planta 121:39–50

Amrhein N, Gerhardt J (1979) Superinduction of phenylalanine ammonia-lyase (PAL) in gherkin hypocotyls caused by the PAL-inhibitor L-α-aminooxy-β-phenylpropionic acid. Biochim Biophys Acta 583:434–442

Amrhein N, Godeke K-H (1976) The estimation of phenylalanine ammonia-lyase (PAL) activity in intact cells of higher plant tissue. II. Correlations and discrepancies between activities measured in intact cells and cell-free extracts. Planta 131:41–45

Amrhein N, Zenk MH (1970) Concomitant induction of phenylalanine ammonia-lyase and cinnamic acid 4-hydroxylase during illumination of excised buckwheat hypocotyls. Naturwissenschaften 57:312

Amrhein N, Gerhardt J, Godeke K-H (1976) The estimation of phenylalanine ammonia-lyase (PAL) activity in intact cells of higher plant tissue. I. Parameters of the assay. Planta 131:33–40

Apel K (1979) Phytochrome-induced appearance of mRNA activity for the apoprotein of the light-harvesting chlorophyll a/b protein of barley (*Hordeum vulgare*). Eur J Biochem 97:183–188

Apel K (1981) The protochlorophyllide holochrome of barley (*Hordeum vulgare* L.). Phytochrome-induced decrease of translatable mRNA coding for the NADPH: protochlorophyllide oxidoreductase. Eur J Biochem 128:89–93

Apel K, Bogorad L (1976) Light-induced increase in the activity of maize plastid DNA-dependent RNA polymerase. Eur J Biochem 67:615–620

Apel K, Kloppstech K (1978) The plastid membranes of barley (*Hordeum vulgare*). Light-induced appearance of mRNA coding for the apoprotein of the light-harvesting chlorophyll a/b protein. Eur J Biochem 85:581–588

Apel K, Kloppstech K (1980) The effect of light on the biosynthesis of the light-harvesting chlorophyll a/b binding protein. Evidence for the requirement of chlorophyll a for the stabilisation of the apoprotein. Planta 150:426–430

Arias IM, Doyle D, Schimke RT (1969) Studies on the synthesis and degradation of proteins of the endoplasmic reticulum of rat liver. J Biol Chem 244:3303–3315

Armond PA, Staehelin LA, Arntzen CJ (1977) Spatial relationship of photosystem I, photosystem II, and the light-harvesting complex in chloroplast membranes. J Cell Biol 73:400–418

Aslam M, Huffaker RC, Travis RL (1973) The interaction of respiration and photosynthesis in induction of nitrate reductase activity. Plant Physiol 52:137–141

Attridge TH (1974) Phytochrome-mediated synthesis of ascorbic acid oxidase in mustard cotyledons. Biochim Biophys Acta 362:258–265

Attridge TH, Smith H (1967) A phytochrome-mediated increase in the level of phenylalanine ammonia-lyase activity in terminal buds of *Pisum sativum*. Biochim Biophys Acta 148:805–807

Attridge TH, Smith H (1973) Evidence for a pool of inactive phenylalanine ammonia-lyase in *Cucumis sativus* seedlings. Phytochemistry 12:1569–1574

Attridge TH, Smith H (1974) Density-labelling evidence for the blue light-mediated activation of phenylalanine ammonia-lyase in *Cucumis sativus* seedlings. Biochim Biophys Acta 343:452–464

Attridge TH, Johnson CB, Smith H (1974) Density-labelling evidence for the phytochrome-mediated activation of phenylalanine ammonia-lyase in mustard cotyledons. Biochim Biophys Acta 343:440–451

Bajracharya D, Falk H, Schopfer P (1976) Phytochrome-mediated development of mitochondria in the cotyledons of mustard (*Sinapis alba* L.) seedlings. Planta 131:253–261

Bar-Nun S, Schantz R, Ohad I (1977) Appearance and composition of chlorophyll-protein complexes I and II during chloroplast membrane biogenesis in *Chlamydomonas rheinhardi* y-1. Biochim Biophys Acta 459:451–467

Barraclough R, Ellis RJ (1979) The biosynthesis of ribulose bisphosphate carboxylase. Uncoupling of the synthesis of large and small subunits in isolated soybean leaf cells. Eur J Biochem 94:165–177

Beale SI (1971) Studies on the biosynthesis and metabolism of δ-aminolevulinic acid in *Chlorella*. Plant Physiol 48:316–319

Bedbrook JR, Link G, Coen DM, Bogorad L, Rich A (1978) Maize plastid gene expressed during photoregulated development. Proc Natl Acad Sci USA 75:3060–3064

Bedbrook JR, Coen DM, Beaton AR, Bogorad L, Rich A (1979) Location of the single gene for the large subunit of ribulosebisphosphate carboxylase on the maize chloroplast chromosome. J Biol Chem 254:905–910

Bedbrook JR, Smith SM, Ellis RJ (1980) Molecular cloning and sequencing of cDNA encoding the precursor to the small subunit of chloroplast ribulose-1,5-bisphosphate carboxylase. Nature 287:692–697

Beevers L, Schrader LE, Flesher D, Hageman RH (1965) The role of light and nitrate in the induction of nitrate reductase in radish cotyledons and maize seedlings. Plant Physiol 40:691–698

Bellemare G, Bartlett SG, Chua N-H (1981) The requirement of chlorophyll b for the stabilisation of the 25 kd polypeptide of the LHCP within the thylakoid membrane. Plant Physiol Suppl 67:169

Beneveniste I, Saläun J-P, Durst F (1978) Phytochrome-mediated regulation of a monooxygenase hydroxylating cinnamic acid in etiolated pea seedlings. Phytochemistry 17:359–364

Betsche T, Gerhardt B (1978) Apparent catalase synthesis in sunflower cotyledons during the change in microbody function. A mathematical approach for the quantitative evaluation of density-labelling data. Plant Physiol 62:590–597

Betz B, Schäfer E, Hahlbrock K (1978) Light-induced phenylalanine ammonia-lyase in cell suspension cultures of *Petroselinum hortense*. Quantitative comparison of rates of synthesis and degradation. Arch Biochem Biophys 190:126–135

Billett EE, Smith H (1980) Control of phenylalanine ammonia-lyase and cinnamic acid 4-hydroxylase in gherkin tissues. Phytochemistry 19:1035–1041

Billett EE, Wallace W, Smith H (1978) A specific and reversible macromolecular inhibitor of phenylalanine ammonia-lyase and cinnamic acid 4-hydroxylase in gherkins. Biochim Biophys Acta 524:219–230

Blair GE, Ellis RJ (1973) Protein synthesis in chloroplasts. I. Light-driven synthesis of the large subunit of fraction 1 protein by isolated pea chloroplasts. Biochim Biophys Acta 319:223–234

Blobel G, Dobberstein B (1975) Transfer of proteins accross membranes.I. Presence of proteolytically processed and unprocessed nascent immunoglobulin light chains on membrane-bound ribosomes of murine myeloma. J Cell Biol 67:835–851

Boardman NK, Anderson JM, Goodchild DJ (1978) Chlorophyll-protein complexes and structure of mature and developing chloroplasts. Curr Top Bioenerg 8:35–109

Bogorad L, Jolly SO, Kidd G, Link G, McIntosh L (1980) Organisation and transcription of maize chloroplast genes. In: Leaver CJ (ed) Genome organisation and expression in plants. Plenum, New York, pp 291–304

Bolton GW, Quail PH (1981) Cell-free synthesis of *Avena* phytochrome. Plant Physiol Suppl 67:728

Bottomley W (1970) Deoxyribonucleic acid-dependent ribonucleic acid polymerase activity of nuclei and plastids from etiolated peas and their response to red and far-red light in vivo. Plant Physiol 45:608–611

Boudet A, Humphrey TJ, Davies DD (1975) The measurement of protein turnover by density labelling. Biochem J 152:409–416

Bradbeer JW (1971) development in primary leaves of *Phaseolus vulgaris*. The effects of short blue, red, far-red and white light treatments on dark-grown plants. J Exp Bot 22:382–390

Brooker JD, Russell DW (1974) Some properties of 3-hydroxy-3-methylglutaryl coenzyme A reductase from *Pisum sativum*. R Soc NZ Bull 12:365–370
Brooker JD, Russell DW (1979) Regulation of microsomal 3-hydroxy-3-methylglutaryl coenzyme A reductase from pea seedlings: rapid posttranslational phytochrome-mediated decrease in activity and in vivo regulation by iso-prenoid products. Arch Biochem Biophys 198:323–334
Brown RD, Haselkorn R (1971) Chloroplast RNA populations in dark grown, light grown, and greening *Euglena gracilis*. Proc Natl Acad Sci USA 68:2536–2539
Brulfert J, Guerrier D, Quieroz O (1973) Photoperiodism and enzyme activity: Balance between inhibition and induction of the crassulacean acid metabolism. Plant Physiol 51:220–222
Bruning K, Drumm H, Mohr H (1975) On the role of phytochrome in controlling enzyme levels in plastids. Biochem Physiol Pflanz 168:141–156
Burgess RJ, Walker JH, Mayer RJ (1978) Choice of precursors for the measurement of protein turnover by the double-isotope method. Application to the study of mitochondrial proteins. Biochem J 176:919–926
Butler LG, Bennett V (1969) Phytochrome control of maize leaf inorganic pyrophosphatase and adenylate kinase. Plant Physiol 44:1285–1290
Camm EL, Towers GHN (1973) Phenylalanine ammonia-lyase. Phytochemistry 12:961–963
Cashmore AR (1979) Reiteration frequency of the gene coding for the small subunit of ribulose-1,5-bisphosphate carboxylase. Cell 17:383–388
Cerff R (1973) Glyceraldehyde 3-phosphate dehydrogenases and glyoxylate reductase. I. Their regulation under continuous red and far-red light in the cotyledons of *Sinapis alba* L. Plant Physiol 51:76–81
Cerff R (1974) Inhibitor-dependent, reciprocal changes in the activities of glyceraldehyde 3-phosphate dehydrogenases in *Sinapis alba* coyledons. Z. Pflanzenphysiol 73:109–118
Chan P-H, Wildman SG (1972) Chloroplast DNA codes for the primary structure of the large subunit of fraction 1 protein. Biochim Biophys Acta 277:677–680
Chelm BK, Hobben PJ, Hallick RB (1977) Expression of the chloroplast ribosomal RNA genes of *Euglena gracilis* during chloroplast development. Biochemistry 16:776–778
Chua N-H, Schmidt DW (1978) Post-translational transport into intact chloroplasts of a precursor to the small subunit of ribulose-1,5-bisphosphate carboxylase. Proc Natl Acad Sci USA 75:6110–6114
Cobb AH, Wellburn AR (1973) Developmental changes in the levels of SDS-extractable polypeptides during plastid morphogenesis. Planta 114:131–142
Cobb AH, Wellburn AR (1974) Changes in plastid envelope polypeptides during chloroplast development. Planta 121:273–282
Coen DM, Bedbrook JR, Bogorad L, Rich A (1977) Maize chloroplast DNA fragment encoding the large subunit of ribulosebisphosphate carboxylase. Proc Natl Acad Sci USA 74:5487–5491
Cohen D, Schiff JA (1976) Events surrounding the early development of *Euglena* chloroplasts. Photoregulation of the transcription of chloroplastic and cytoplasmic ribosomal RNAs. Arch Biochem Biophys 177:201–216
Creasy LL (1976) Phenylalanine ammonia-lyase-inactivating system in sunflower leaves. Phytochemistry 15:673–675
Davidson EH (1976) Gene activity in early development, 2nd edn. Academic Press, London New York
Daussant J, Lauriere C, Carfantan N, Skakonn A (1977) Immunochemical approaches to questions concerning enzyme regulation in plants. Phytochem Soc Symp 14:197–223
Dharvan AK, Malik CP (1919) Phytochrome control of some oxidoreductases in germinating *Pinus roxburghii* Sarg pollen. Plant Cell Physiol 20:615–618
Dobberstein B, Blobel G, Chua N-H (1977) In vitro synthesis and processing of a putative precursor for the small subunit of ribulose-1,5-bisphosphate carboxylase of *Chlamydomonas reinhardtii*. Proc Natl Acad Sci USA 74:1082–1085
Drumm H, Mohr H (1973) Control by phytochrome of glutathione reductase levels in the mustard seedling. Z Naturforsch Teil C 28:559–563

Drumm H, Schopfer P (1974) Effect of phytochrome on development of catalase activity and iso-enzyme pattern in mustard (*Sinapis alba* L.) seedlings. A reinvestigation. Planta 120:13–20

Drumm H, Elchinger I, Moller J, Peter K, Mohr H (1971) Induction of amylase in mustard seedlings by phytochrome. Planta 99:265–274

Drumm H, Bruning K, Mohr H (1972) Phytochrome-mediated induction of ascorbate oxidase in different organs of a dicotyledonous seedling (*Sinapis alba* L.). Planta 106:259–267

Dugaiczyk A, Woo SLC, Lai EC, Mace ML, McReynolds L, O'Malley BW (1978) The natural ovalbumin gene contains several intervening sequences. Nature 274:328–333

Durst F, Duranton H (1970) Phytochrome et phenylalanine ammonia-lyase dans les tissus du topinambour (*Helianthus tuberosus* L. variété blanc commun) cultives in vitro. CR Acad Sci Ser D 270:2940–2942

Ebel J, Hahlbrock K (1977) Enzymes of flavone and flavonol-glycoside biosynthesis. Co-ordinated and selective induction in cell-suspension cultures of *Petroselinum hortense*. Eur J Biochem 75:201–209

Efstratiadis A, Kafatos FC (1976) The chorion of insects: techniques and perspectives. In: Last J (ed) Eukaryotes at the subcellular level. Dekker, New York, pp 1–124

Ellis RJ (1975) Inhibition of chloroplast protein synthesis by lincomycin and 2-(4-methyl-2,6-dinitroanilino)-N-methylpropionamide. Phytochemistry 14:89–93

Ellis RJ (1977) Protein synthesis by isolated chloroplasts. Biochim Biophys Acta 463:185–215

Ellis RJ (1979) The most abundant protein in the world. Trends Biochem Sci 4:241–244

Ellis RJ, McDonald IR (1970) Specificity of cycloheximide in higher plant systems. Plant Physiol 46:227–232

Engelsma G (1967) Photoinduction of phenylalanine deaminase in gherkin seedlings. II. Effect of red and far-red light. Planta 77:49–57

Engelsma G (1968) The influence of light of different spectral regions on the synthesis of phenolic compounds in gherkin seedlings, in relation to photomorphogenesis. IV. Mechanism of far-red action. Acta Bot Neerl 17:85–89

Engelsma G (1970) A comparative investigation of the control of phenylalanine ammonia-lyase in gherkin and red cabbage hypocotyls. Acta Bot Neerl 19:403–414

Engelsma G (1974) On the mechanism of the changes in phenylalanine ammonia-lyase activity induced by ultraviolet and blue light in gherkin hypocotyls. Plant Physiol 54:702–705

Engelsma G, Meijer G (1965) The influence of light of different spectral regions on the synthesis of phenolic compounds in gherkin seedlings in relation to photomorphogenesis. I. Biosynthesis of phenolic compounds. Acta Bot Neerl 14:54–72

Feierabend J (1969) Der Einfluß von Cytokininen auf die Bildung von Photosyntheseenzymen in Roggenkeimlingen. Planta 84:11–29

Feierabend J (1975) Developmental studies on microbodies in wheat leaves. III. On the photocontrol of microbody development. Planta 123:63–77

Feierabend J, Pirson A (1966) Die Wirkung des Lichts auf die Bildung von Photosyntheseenzymen in Roggenkeimlingen. Z Pflanzenphysiol 55:235–245

Filner B, Klein AO (1968) Changes in enzymatic activities in etiolated bean seedling leaves after a brief illumination. Plant Physiol 43:1587–1596

Fourcroy P, Lambert C, Rollin P (1979) Far-red-mediated polyribosome formation in radish cotyledons. Effect of endogenous ribonucleases on polyribosome recovery. Planta 147:1–5

French CJ, Smith H (1975) An inactivator of phenylalanine ammonia-lyase from gherkin hypocotyls. Phytochemistry 14:963–966

Fritzemeier K-H, Kindl H (1981) Coordinate induction by UV light of stilbene synthase, phenylalanine ammonia-lyase and cinnamate 4-hydroxylase in leaves of Vitaceae. Planta 151:48–52

Frosch S, Wagner E (1973a) Endogeneous rhythmicity and energy transduction. II. Phytochrome action and the conditioning of rhythmicity of adenylate kinase, NAD- and NADP-linked glyceraldehyde-3-phosphate dehydrogenase in *Chenopodium rubrum* by temperature and light intensity cycles during germination. Can J Bot 51:1521–1528

Frosch S, Wagner E (1973b) Endogenous rhythmicity and energy transduction. III. Time course of phytochrome action on adenylate kinase, NAD- and NADP-linked glyceraldehyde-3-phosphate dehydrogenase in *Chenopodium rubrum*. Can J Bot 51:1529–1535

Frosch S, Bergfeld R, Mohr H (1976) Light control of plastogenesis and ribulose bisphosphate carboxylase levels in mustard seedling cotyledons. Planta 133:53–56

Frosch S, Drumm H, Mohr H (1977) Regulation of enzyme levels by phytochrome in mustard cotyledons: multiple mechanisms? Planta 136:181–186

Funkhouser EA, Ramadoss CS (1980) Synthesis of nitrate reductase in *Chlorella*. II. Evidence for synthesis in ammonia-grown cells. Plant Physiol 65:944–949

Funkhouser EA, She T-C, Ackermann R (1980) Synthesis of nitrate reductase in *Chlorella*. I. Evidence for an inactive precursor. Plant Physiol 65:939–943

Furuya M, Galston AW, Stowe BB (1962) Isolation from peas of co-factors and inhibitors of indolyl-3-acetic acid oxidase. Nature 193:456–457

Gardiner SE, Schröder J, Matern U, Hamer D, Hahlbrock K (1980) mRNA-dependent regulation of UDP-apiose synthase activity in irradiated plant cells. J Biol Chem 255:10752–10757

Gerhardt B (1974) Studies on the formation of glycolate oxidase in developing cotyledons of *Helianthus annuus* L. and *Sinapis alba* L. Z Pflanzenphysiol 74:14–21

Giles AB, Grierson D, Smith H (1977) In-vitro translation of messenger-RNA from developing bean leaves. Evidence for the existence of stored messenger-RNA and its light-induced mobilisation into polyribosomes. Planta 136:31–36

Goatly MB, Smith H (1974) Differential properties of phosphoenolpyruvate carboxylase from etiolated and green sugar cane. Planta 117:67–73

Goatly MB, Coombs J, Smith H (1975) Development of C_4 photosynthesis in sugar cane: changes in properties of phosphoenolpyruvate carboxylase during greening. Planta 125:15–24

Gottmann K, Schäfer E (1982) In vitro synthesis of phytochrome apoprotein directed by mRNA from light and dark *Avena* seedlings. Photochem Photobiol 35:521–525

Graham D, Grieve AM, Smillie RM (1968) Phytochrome as the primary photoregulator of the synthesis of Calvin cycle enzymes in etiolated pea seedlings. Nature 218:89–90

Graham D, Hatch MD, Slack CP, Smillie RM (1970) Light-induced formation of enzymes of the C_4-dicarboxylic acid pathway of photosynthesis in detached leaves. Phytochemistry 9:521–532

Graham D, Grieve AM, Smillie RM (1971) Phytochrome-mediated plastid development in etiolated pea stem apices. Phytochemistry 10:2905–2914

Gray JC, Kekwick RGO (1974) The synthesis of the small subunit of ribulose 1,5-bisphosphate carboxylase in the French bean *Phaseolus vulgaris*. Eur J Biochem 44:491–500

Grebanier A, Coen DM, Rich A, Bogorad L (1978) Membrane proteins synthesised but not processed by isolated maize chloroplasts. J Cell Biol 78:734–746

Grebanier A, Steinback KE, Bogorad L (1979) Comparison of the molecular weights of proteins synthesised by isolated chloroplasts with those which appear during greening in *Zea mays*. Plant Physiol 63:436–439

Grierson DG, Covey S (1975) Changes in the amount of ribosomal RNA and poly(A)-containing RNA during leaf development. Planta 127:77–86

Guiz C, Hirel B, Shedlofsky G, Gadal P (1979) Occurrence and influence of light on the relative proportions of two glutamine synthetases in rice leaves. Plant Sci Lett 15:271–277

Gupta S, Acton GJ (1979) Purification to homogeneity and some properties of L-phenylalanine ammonia-lyase of irradiated mustard (*Sinapis alba* L.) cotyledons. Biochim Biophys Acta 570:187–197

Hachtel W (1972) Der Einfluß des Plasmotypus auf die Regulation der Aktivität der L-Phenylalanin-Ammonium-Lyase (Untersuchungen an *Raimannia-Oenotheren*). Planta 102:247–260

Haff LA, Bogorad L (1976) Hybridization of maize chloroplast DNA with transfer ribonucleic acids. Biochemistry 15:4105–4109

Hague DR, Sims TL (1980) Evidence for light-stimulated synthesis of phosphoenolpyruvate carboxylase in leaves of maize. Plant Physiol 66: 505–509

Hague DR, Sims TL (1981) Translatable mRNA for PEP carboxylase. Plant Physiol Suppl 67: 91 Entry 514

Hahlbrock K, Grisebach H (1979) Enzymic controls in the biosynthesis of lignin and flavonoids. Annu Rev Plant Physiol 30: 105–130

Hahlbrock K, Ragg H (1975) Light-induced changes of enzyme activities in parsley cell suspension cultures. Effects of inhibitors of RNA and protein synthesis. Arch Biochem Biophys 166: 41–46

Hahlbrock K, Sutter A, Wellmann E, Ortmann R, Grisebach H (1971a) Relationship between organ development and activity of enzymes involved in flavone glycoside biosynthesis in young parsley plants. Phytochemistry 10: 109–116

Hahlbrock K, Ebel J, Ortmann R, Sutter A, Wellmann E, Grisebach H (1971b) Regulation of enzyme activities related to the biosynthesis of flavone glycosides in cell suspension cultures of parsley (*Petroselinum hortense*). Biochim Biophys Acta 244: 7–15

Hahlbrock K, Knobloch K-H, Kreuzaler F, Potts JRM, Wellmann E (1976) Co-ordinated induction and subsequent activity changes of two groups of metabolically inter-related enzymes: Light-induced synthesis of flavonoid glycosides in cell suspension cultures of *Petroselinum hortense*. Eur J Biochem 61: 199–206

Hahlbrock K, Lamb CJ, Purwin C, Ebel J, Fautz E, Schäfer E (1981) Rapid response of suspension-cultured parsley cells to the elicitor from *Phytophthora megasperma* var. *sojae*. Induction of the enzymes of general phenylpropanoid metabolism. Plant Physiol 67: 768–773

Hampp R, Ziegler H (1975) Lichtabhängige Neusynthese von δ-Aminolaevulinsäure-Dehydratase in isolierten *Avena*-Etioplasten. Planta 124: 255–260

Harel E, Bogorad L (1973) Effect of light on ribonucleic acid metabolism in greening maize leaves. Plant Physiol 51: 10–16

Hartley MR, Wheeler A, Ellis RJ (1975) Protein synthesis in chloroplasts. V. Translation of messenger RNA for the large subunit of fraction 1 protein in a heterologous cell-free system. J Mol Biol 91: 67–77

Haslett BG, Cammack R (1974) The development of plastocyanin in greening bean leaves. Biochem J 144: 567–572

Haslett BG, Cammack R (1976) Changes in the activity of ferredoxin-NADP-reductase during the greening of bean leaves. New Phytol 76: 219–226

Haslett BG, Cammack R, Whatley FR (1973) Quantitative studies on ferredoxin in greening bean leaves. Biochem J 136: 697–703

Hayakawa S, Matsunaga K, Sagiyama T (1981) Light induction of phosphoenolpyruvate carboxylase in etiolated maize leaf tissue. Plant Physiol 67: 133–138

Heim B, Jabben M, Schäfer E (1981) Phytochrome destruction in dark- and light-grown *Amaranthus caudatus* seedlings. Photochem Photobiol 34: 89–94

Heinze H, Herzfeld F, Kiper M (1980) Light-induced appearance of polysomal poly(A)-rich messenger RNA during greening of barley plants. Eur J Biochem 111: 137–144

Heller JS, Fong WF, Canellakis ES (1976) Induction of a protein inhibitor to ornithine decarboxylase by the end products of its reaction. Proc Natl Acad Sci USA 73: 1858–1862

Henshall JD, Goodwin TW (1964) Amino acid activating enzymes in greening pea seedlings. Phytochemistry 3: 677–691

Hewitt EJ (1975) Assimilatory nitrate-nitrite reduction. Annu Rev Plant Physiol 26: 73–100

Highfield PE, Ellis RJ (1978) Synthesis and transport of the small subunit of chloroplast ribulose bisphosphate carboxylase. Nature 271: 420–424

Hoober JK, Stegeman WJ (1976) Kinetics and regulation of synthesis of the major polypeptides of thylakoid membranes in *Chlamydomonas reinhardtii* y-1 at elevated temperatures. J Cell Biol 70: 326–337

Hopkins DW, Briggs WR (1973) Phytochrome and NAD kinase: A re-examination. Plant Physiol Suppl 51: 52

Huault C (1974) Phytochrome regulation of phenylalanine ammonia-lyase (PAL) activity in radish cotyledons: A threshold mechanism. Plant Sci Lett 3:149–155

Huff AK, Ross CW (1975) Promotion of radish cotyledon enlargement and reducing sugar content by zeatin and red light. Plant Physiol 56:429–433

Huffaker RC, Peterson LW (1974) Protein turnover in plants and possible means of its regulation. Annu Rev Plant Physiol 25:363–392

Hunt RE, Pratt LH (1980) Radioimmunoassay of phytochrome content in green, light-grown oats. Plant Cell Environ 3:91–95

Hüttermann A, Wendlberger G (1976) Density labelling of proteins. Methods Cell Biol 13:153–170

Iwanij V, Chua NH, Siekevitz P (1975) Synthesis and turnover of ribulose bisphosphate carboxylase and of its subunits during the cell cycle of *Chlamydomonas reinhardtii*. J Cell Biol 64:572–585

Lynedjian PB (1979) Techniques for messenger RNA assay in enzyme induction studies. Tech Metab Res B209:1–27

Jabben M, Deitzer GF (1978) A method for measuring phytochrome in plants grown in white light. Photochem Photobiol 27:799–802

Jeffreys AJ, Flavell RA (1977) The rabbit β-globin gene contains a large insert in the coding sequence. Cell 12:1097–1108

Johnson CB (1976) Rapid activation by phytochrome of nitrate reductase in the cotyledons of *Sinapis alba*. Planta 128:127–131

Johnson CB (1977) The use of density labelling techniques in investigations into the control of enzyme levels. Phytochem Soc Symp 14:225–243

Johnson CB (1979) Activation, synthesis and turnover of nitrate reductase controlled by nitrate and ammonium in *Chlorella vulgaris*. Planta 147:63–68

Johnson CB, Smith H (1978) Phytochrome control of amino acid synthesis in cotyledons of *Sinapis alba*. Phytochemistry 17:667–670

Johnson CB, Attridge T, Smith H (1975) Regulation of phenylalanine ammonia-lyase synthesis by cinnamic acid. Its implications for the light-mediated regulation of the enzyme. Biochim Biophys Acta 385:11–19

Johnson CB, Whittington WJ, Blackwood GC (1976) Nitrate reductase as a possible predictive test of crop yield. Nature 262:133–134

Jolly SO, Tolbert NE (1978) NADH-nitrate reductase inhibitor from soybean leaves. Plant Physiol 62:133–134

Jolly SO, McIntosh L, Link G, Bogorad L (1981) Differential transcription in vivo and in vitro of two adjacent maize chloroplast genes. Proc Natl Acad Sci USA 78:6821–6825

Jones RW, Sheard RW (1975) Phytochrome, nitrate movement, and induction of nitrate reductase in etiolated pea terminal buds. Plant Physiol 55:954–959

Kadam SS, Gandhi AP, Sawnhey SK, Naik MS (1974) Inhibitor of nitrate reductase in the roots of rice seedlings and its effect on the enzyme activity in the presence of NADH. Biochim Biophys Acta 350:162–170

Kasemir H, Masoner M (1975) Control of chlorophyll synthesis by phytochrome. II. The effect of phytochrome on aminolevulinate dehydratase in mustard seedlings. Planta 126:119–126

Katunuma N, Katsunuma T, Kominami E, Suzuki K, Hamaguchi Y, Chichibu K, Kobayashi K (1972) Regulation of intracellular enzyme levels by group-specific proteases in various organisms. Adv Enzyme Regul 11:37–51

Kaveh D, Harel E (1973) Light-induced changes in the pattern of protein synthesis during the early stages of greening of etiolated maize leaves. Plant Physiol 51:671–676

Kawashima N, Wildman SG (1970) Fraction 1 Protein. Annu Rev Plant Physiol 21:325–358

Kawashima N, Wildman SG (1972) Studies on fraction 1 protein. IV. Mode of inheritance of primary structure in relation to whether chloroplast or nuclear DNA contains the code for a chloroplast protein. Biochim Biophys Acta 262:42–49

Keister DL, Jagendorf AT, San Pietro A (1962) Development of bean leaf transhydrogenase in etiolated leaves. Biochim Biophys Acta 62:332–337

Kelly GJ, Gibbs M (1973) Nonreversible D-glyceraldehyde 3-phosphate dehydrogenase of plant tissues. Plant Physiol 52:111–118

Klein AO (1969) Persistent photoreversibility of leaf development. Plant Physiol 44:897–902

Klein AO, Pine K (1977) Light-induced polysome formation in etiolated leaves. Kinetics of inhibition by antibiotics. Plant Physiol 59:767–770

Klein S, Katz E, Neeman E (1977) Induction of δ-aminolevulinic acid formation in etiolated maize leaves controlled by the two light systems. Plant Physiol 60:335–338

Kleinhopf GE, Huffaker RC, Matheson A (1970) Light-induced de novo synthesis of ribulose 1,5-diphosphate carboxylase in greening leaves of barley. Plant Physiol 46:416–418

Kominami E, Katunuma N (1976) Studies on new intracellular proteases in various types of rats. Participation of proteases in degradation of ornithine aminotransferase in vitro and in vivo. Eur J Biochem 62:425–430

Kung S (1976) Tobacco fraction 1 protein: A unique genetic marker. Science 191:429–434

Lamb CJ (1977a) Phenylalanine ammonia-lyase and cinnamic acid 4-hydroxylase: Characterisation of the concomitant changes in enzyme activities in illuminated potato tuber discs. Planta 135:169–175

Lamb CJ (1977b) trans-Cinnamic acid as a mediator of the light-stimulated increase in hydroxycinnamoyl-CoA: quinate hydroxycinnamoyl transferase. FEBS Lett 75:37–41

Lamb CJ (1979) Regulation of enzyme levels in phenylpropanoid biosynthesis: Characterisation of the modulation by light and pathway intermediates. Arch Biochem Biophys 192:311–317

Lamb CJ, Merritt TK (1979) Density labelling studies of the photocontrol of L-phenylalanine ammonia-lyase in discs of potato (*Solanum tuberosum*) tuber parenchyme. Biochim Biophys Acta 588:1–11

Lamb CJ, Rubery PH (1976a) Interpretation of the rate of density labelling of enzymes with ^2H_2O. Possible implications for the mode of action of phytochrome. Biochim Biophys Acta 421:308–318

Lamb CJ, Rubery PH (1976b) Photocontrol of chlorogenic acid biosynthesis in potato tuber discs. Phytochemistry 15:665–668

Lamb CJ, Rubery PH (1976c) Phenylalanine ammonia-lyase and cinnamic acid 4-hydroxylase: Product repression of the level of enzyme activity in potato tuber discs. Planta 130:283–290

Lamb CJ, Merritt TK, Butt VS (1979) Synthesis and removal of phenylalanine ammonia-lyase activity in illuminated discs of potato tuber parenchyme. Biochim Biophys Acta 582:196–212

Lawton MA, Dixon RA, Lamb CJ (1980) Elicitor modulation of the turnover of L-phenylalanine ammonia-lyase in French bean cell suspension cultures. Biochim Biophys Acta 633:162–175

Leto KJ, Miles D (1980) Characterisation of three photosystem II mutants in *Zea mays* L. lacking a 32,000 dalton lamellar polypeptide. Plant Physiol 66:18–24

Lett MC, Fleck J, Frisch C, Durr A, Hirth L (1980) Suitable conditions for characterisation, identification, and isolation of the mRNA of the small subunit of ribulose-1,5-bisphosphate carboxylase from *Nicotiana sylvestris*. Planta 148:211–216

Leung DWM, Bewley JD (1981) Immediate phytochrome action in inducing α-galactosidase in lettuce seeds. Nature 289:587–588

Link G (1981) Enhanced expression of a distinct plastid DNA region in mustard seedlings by continuous far-red light. Planta 152:379–380

Link G (1982) Phytochrome control of plastid mRNA in mustard (*Sinapis alba* L.). Planta 154:81–86

Link G, Bogorad L (1980) Sizes, locations, and directions of transcription of two genes on a closed maize chloroplast DNA sequence. Proc Natl Acad Sci USA 77:1832–1836

Link G, Coen DM, Bogorad L (1978) Differential expression of the gene for the large subunit of ribulose bisphosphate carboxylase in maize leaf cell types. Cell 15:725–731

Lodish HF (1976) Translational control of protein synthesis. Annu Rev Biochem 45:39–72

Loschke DC, Hadwiger LA, Schröder J, Hahlbrock K (1981) Effects of light and *Fusarium solani* on synthesis and activity of phenylalanine ammonia-lyase in peas. Plant Physiol 68:680–685

McClure JW (1974) Phytochrome control of oscillating levels of phenylalanine ammonia-lyase in *Hordeum vulgare* shoots. Phytochemistry 13:1065–1069

McClure JW, Gross GG (1975) Diverse photoinduction characteristics of hydroxycinnamate:coenzyme A ligase and phenylalanine ammonia-lyase in dicotyledonous seedlings. Z Pflanzenphysiol 76:51–55

Machold O, Aurich O (1972) Sites of synthesis of chloroplast lamellar proteins in *Vicia faba*. Biochem Biophys Acta 281:103–112

McIntosh L, Poulsen C, Bogorad L (1980) Chloroplast gene sequence for the large subunit of ribulose bisphosphate carboxylase of maize. Nature 288:556–560

McIntosh L, Steinback KE, Arntzen CJ, Bogorad L (1981) Characterisation of chloroplast gene mutants responsible for triazine herbicide resistance. Plant Physiol Suppl 67:111

Malcolm AA, Russell DW (1974) Early changes in RNA metabolism in etiolated bean leaves irradiated with red light. R Soc NZ Bull 12:333–338

Malnoe P, Rochaix J-D, Chua NH, Spahr P-F (1979) Characterisation of the gene and messenger RNA of the large subunit of ribulose 1,5-bisphosphate carboxylase in *Chlamydomonas reinhardtii*. J Mol Biol 133:417–434

Mappleston RE, Griffiths WT (1980) Light modulation of the activity of protochlorophyllide reductase. Biochem J 189:125–133

Masoner MH, Kasemir H (1975) Control of chlorophyll synthesis by phytochrome. I. The effect of phytochrome on the formation of δ-aminolevulinate in mustard seedlings. Planta 126:111–117

Minty AJ, Birnie GD (1981) Messenger RNA populations in eukaryotic cells-evidence from recent nucleic acid hybridisation experiments bearing on the extent and control of differential gene expression. In: Buckingham ME (ed) Biochemistry of cellular regulation Vol III. CRC Press, Boca Raton, pp 43–82

Mohr H (1957) Der Einfluß monochromatischer Strahlung auf das Längenwachstum des Hypocotyls und auf die Anthocyanbildung bei Keimlingen von *Sinapis alba*. Planta 49:389–405

Mohr H (1972) Lectures on photomorphogenesis. Springer, Berlin Heidelberg New York

Mohr H (1974) The role of phytochrome in controlling enzyme levels in plants. In: Paul J (ed) Biochemistry of Cell Differentiation. Butterworth, London

Mohr H, Oelze-Karow H (1976) Phytochrome action as a threshold phenomenon. In: Smith H (ed) Light and plant development. Butterworth, London, pp 257–284

Morgenthaler J-J, Mendiola-Morgenthaler L (1976) Synthesis of soluble, thylakoid and envelope membrane proteins by spinach chloroplasts purified from gradients. Arch Biochem Biophys 172:51–58

Muller M, Viro M, Balke C, Kloppstech K (1980a) Kinetics of the appearance of mRNA for light-harvesting chlorophyll a/b protein in polysomes of barley. Planta 148:448–452

Muller M, Viro M, Balke C, Kloppstech K (1980b) Polyadenylated mRNA for the light-harvesting chlorophyll a/b protein. Its presence in green and absence in chloroplast-free plant cells. Planta 148:444–447

Mullett JE, Burke JJ, Arntzen CJ (1980) A developmental study of photosystem I peripheral chloroplast proteins. Plant Physiol 65:823–827

Nadler K, Granick S (1970) Control of chlorophyll synthesis in barley. Plant Physiol 46:240–246

Newbury HJ, Smith H (1981) Immunochemical evidence for phytochrome regulation of the specific activity of ascorbate oxidase in mustard seedlings. Eur J Biochem 117:575–580

Nicholas JC, Harper JE, Hageman RH (1976) Nitrate reductase activity in soybeans (*Glycine max* L. Merr.). I. Effects of light and temperature. Plant Physiol 58:731–735

Nover L (1976) Density labelling of chloroplast-specific leucyl-tRNA synthetase in greening cells of *Euglena gracilis*. Plant Sci Lett 7:403–407

Oelze-Karow H, Mohr H (1970) Experiments regarding the problem of differentiation in multicellular systems. Z Naturforsch Teil B 25:1282–1286

Oelze-Karow H, Mohr H (1973) Quantitative correlation between spectrophotometric phytochrome assay and physiological response. Photochem Photobiol 18:319–330

Oelze-Karow H, Schopfer P, Mohr H (1970) Phytochrome-mediated repression of enzyme synthesis (lipoxygenase): A threshold phenomenon. Proc Natl Acad Sci USA 65:51–57

Paskin N, Mayer RJ (1977) The role of enzyme degradation in enzyme turnover during tissue differentiation. Biochim Biophys Acta 474:1–10

Paskin N, Mayer RJ (1978) A method for the analysis of protein turnover characteristics. Indirect estimation of rates of protein degradation. Biochem J 174:153–161

Patterson BM, Roberts BE, Kuff EL (1977) Structural gene identification and mapping by DNA:mRNA hybrid-arrest cell-free translation. Proc Natl Acad Sci USA 74:4370–4374

Payne PI (1976) The long-lived messenger ribonucleic acid of flowering plant seeds. Biol Rev 51:329–363

Peckett RC, Bassim TAH (1974) Mechanism of phytochrome action in the control of biosynthesis of anthocyanin in *Brassica oleracea*. Phytochemistry 13:815–821

Pelham HRB, Jackson RJ (1976) An efficient mRNA-dependent translation system from reticulocyte lysates. Eur J Biochem 67:247–256

Penel C, Greppin H (1974) Variation de la photostimulation de l'activité des peroxidases basiques chez l'épinard. Plant Sci Lett 3:75–80

Penel C, Greppin H, Boisard J (1976) In vitro photomodulation of a peroxidase activity through membrane bound phytochrome. Plant Sci Lett 6:117–121

Peterson LW, Huffaker RC (1975) Loss of ribulose 1,5-bisphosphate carboxylase and increase in proteolytic activity during senescence of detached primary barley leaves. Plant Physiol 55:1009–1015

Pine K, Klein AO (1972) Regulation of polysome formation in etiolated bean leaves. Dev Biol 28:280–289

Poulson R, Beevers L (1970) Nucleic acid metabolism during greening and unrolling of barley leaf segments. Plant Physiol 46:315–319

Quail PH, Schäfer E, Marmé D (1973a) De novo synthesis of phytochrome in pumpkin hooks. Plant Physiol 52:124–127

Quail PH, Schäfer E, Marmé D (1973b) Turnover of phytochrome in pumpkin cotyledons. Plant Physiol 52:128–131

Queiroz O (1969) Photoperiodisme et activité enzymatique (PEP carboxylase et enzyme malique) dans les feuilles de *Kalanchoë blossfeldiana*. Phytochemistry 8:1655–1663

Ragg H, Hahlbrock K (1980) Messenger RNA coding for phenylalanine ammonia-lyase. Characterisation and partial purification from cell suspension cultures of *Petroselinum hortense*. Eur J Biochem 103:323–330

Ragg H, Kuhn D, Hahlbrock K (1981) Co-ordinated regulation of 4-coumarate: CoA ligase and phenylalanine ammonia-lyase mRNAs in cultured plant cells. J Biol Chem 256:10061–10065

Rao LVM, Datta N, Sopory SK, Guha-Mukherjee S (1980) Phytochrome-mediated induction of malate reductase activity in etiolated maize leaves. Physiol Plant 50:208–212

Rich PR, Lamb CJ (1977) Biophysical and enzymological studies upon the interaction of trans-cinnamic acid with higher plant microsomal cytochromes P450. Eur J Biochem 72:353–360

Roop DR, Nordstrom JL, Tsai SY, Tsai M-J, O'Malley BW (1978) Transcription of structural and intervening sequences in the ovalbumin gene and identification of potential ovalbumin mRNA precursors. Cell 15:671–685

Roy H, Costa KA, Adari H (1978) Free subunit of ribulose 1,5-bisphosphate carboxylase in pea leaves. Plant Sci Lett 11:159–168

Russell DW (1971) The metabolism of aromatic compounds in higher plants. X. Properties of the cinnamic acid 4-hydroxylase of pea seedlings and some aspects of its metabolic and developmental control. J Biol Chem 246:3870–3878

Santel H-J, Apel K (1981) The protochlorophyllide holochrome of barley (*Hordeum vulgare* L.). The effect of light on the NADPH: protochlorophyllide oxidoreductase. Eur J Biochem 120:95–103

Sasakawa H, Yamamoto Y (1979) Effects of red, far-red and blue light on enhancement of nitrate reductase activity and on nitrate uptake in etiolated rice seedlings. Plant Physiol 63:1098–1101

Sasiki Y, Tomi H, Kamikubo T (1979) Effect of light on the solubulised RNA polymerase level in pea buds. Plant Sci Lett 14:355–364

Sawnhey SK, Naik MS (1972) Role of light in the synthesis of nitrate reductase and nitrite reductase in rice seedlings. Biochem J 130:475–485

Scherf H, Zenk MH (1967) Induction of anthocyanin and phenylalanine ammonia-lyase formation by a high energy light reaction and its control through the phytochrome system. Z Pflanzenphysiol 56:203–206

Schimke RT (1975) Methods for analysis of enzyme synthesis and degradation in animal tissues. Methods Enzymol 60:241–266

Schimke RT, Rhoads RE, McKnight GS (1974) Assay of ovalbumin mRNA in a reticulocyte lysate. Methods Enzymol 30:694–701

Schopfer P (1977) Phytochrome control of enzymes. Annu Rev Plant Physiol 28:223–252

Schopfer P, Mohr H (1972) Phytochrome-mediated induction of phenylalanine ammonia-lyase in mustard seedlings. A contribution to eliminate some misconceptions. Plant Physiol 49:8–10

Schopfer P, Plachy C (1973) Die organspezifische Photodetermination der Entwicklung von Perooxydaseaktivität im Senfkeimling (*Sinapis alba* L.) durch Phytochrom. I. Kinetische Analyse. Z Naturforsch Teil C 28:296–301

Schröder J, Schäfer E (1980) Radioiodinated antibodies, a tool in studies on the presence and role of inactive enzyme forms: Regulation of chalcone synthase in parsley cell suspension cultures. Arch Biochem Biophys 203:800–808

Schröder J, Betz B, Hahlbrock K (1977) Messenger RNA-controlled increase of phenylalanine ammonia-lyase activity in parsley. Light-independent induction by dilution of cell suspension cultures into water. Plant Physiol 60:440–445

Schröder J, Heller W, Hahlbrock K (1979a) Flavanone synthase: simple and rapid assay for the key enzyme of flavonoid biosynthesis. Plant Sci Lett 14:281–286

Schröder J, Kreuzaler F, Schäfer E, Hahlbrock K (1979b) Concomitant induction of phenylalanine ammonia-lyase and flavanone synthase mRNAs in irradiated plant cells. J Biol Chem 254:57–65

Schrott EL, Rau W (1977) Evidence for a photoinduced synthesis of poly(A) containing mRNA in *Fusarium aquaeductum*. Planta 136:45–48

Schutz G, Kieval S, Groner B, Sippel AE, Kurtz DT, Feigelson P (1977) Isolation of specific messenger RNA by adsorption of polysomes to matrix-bound antibody. Nucl Acid Res 4:71–84

Shapiro DJ, Taylor JM, McKnight GS, Palauos R, Gonzalez C, Kiely ML, Schimke RT (1974) Isolation of hen oviduct ovalbumin and rat liver albumin polysomes by indirect immunoprecipitation. J Biol Chem 249:3665–3671

Sharma R, Malik CP (1978) Effect of light on pollen germination, tuber elongation and IAA-oxidase in *Campsis grandiflora* Biochem Physiol Pflanz 173:451–455

Sharma R, Sopory SK, Guha-Mukherjee S (1976) Phytochrome regulation of peroxidase activity in maize. Plant Sci Lett 6:69–75

Sharma R, Sopory SK, Guha-Mukherjee S (1977) Phytochrome regulation of peroxidase activity in maize. II. Interaction with hormones, acetylcholine and cAMP. Z Pflanzenphysiol 82:417–427

Siegelman HW, Hendricks SB (1958) Photocontrol of anthocyanin formation in turnip and red cabbage. Plant Physiol 32:393–398

Sisler EC, Klein WH (1963) The effect of age and various chemicals on the lag phase of chlorophyll synthesis in dark grown bean seedlings. Physiol Plant 16:315–322

Sluiters-Scholten CMT (1973) Effect of chloramphenicol and cycloheximide on the induction of nitrate reductase and nitrite reductase in bean leaves. Planta 113:229–240

Smith H (1976) Phytochrome-mediated assembly of polyribosomes in etiolated bean leaves. Evidence for post-transcriptional regulation of development. Eur J Biochem 65:161–170

Smith H, Attridge TH (1970) Increased phenylalanine ammonia-lyase activity due to light treatment and its significance for the mode of action of phytochrome. Phytochemistry 9:487–495

Smith H, Billett EE, Giles AB (1977) The photocontrol of gene expression in higher plants. Phytochem Soc Symp 14:93–127

Smith MA, Criddle RS, Peterson L, Huffaker RC (1974) Synthesis and assembly of ribulosebisphosphate carboxylase enzyme during greening of barley plants. Arch Biochem Biophys 165:494–504

Smith SM, Ellis RJ (1979) Processing of small subunit precursor of ribulose bisphosphate carboxylase and its assembly into whole enzyme are stromal events. Nature 278:662–664

Sobel ME, Yamamoto T, Adams SL, Di Lauro R, Avvedimento VE, de Crombrugghe B, Pastan I (1978) Construction of a recombinant bacterial plasmid containing a chick pro- 2 collagen gene sequence. Proc Natl Acad Sci USA 75:5846–5850

Steer BT, Gibbs M (1969a) Changes in succinyl CoA synthetase activity in etiolated bean leaves caused by illumination. Plant Physiol 44:775–780

Steer BT, Gibbs M (1969b) Changes in succinyl CoA synthetase activity in etiolated bean leaves caused by illumination. Plant Physiol 44:775–780

Steinitz B, Drumm H, Mohr H (1976) The appearance of competence for phytochrome-mediated anthocyanin synthesis in the cotyledons of *Sinapis alba* L. Planta 130:23–31

Stout ER, Parenti R, Mans RJ (1967) An increase in RNA polymerase activity after illumination of dark-grown maize seedlings. Biochem Biophys Res Commun 29:322–326

Sun SM, Slightom JL, Hall TC (1981) Intervening sequences in a plant gene. Comparison of the partial sequence of cDNA and genomic DNA of French bean phaseolin. Nature 278:37–40

Surrey K (1967) Action and interaction of red and far-red radiation on lipoxidase metabolism of squash seedlings. Plant Physiol 42:421–424

Tanaka Y, Uritani I (1977) Synthesis and turnover of phenylalanine ammonia-lyase in root tissue of sweet potato injured by cutting. Eur J Biochem 73:255–260

Tanaka Y, Matsushita K, Uritani I (1977) Some investigations on inactivation of phenylalanine ammonia-lyase in cut-injured sweet potato tissue. Plant Cell Physiol 18:1209–1216

Taylor JM (1979) The isolation of eukaryotic messenger RNA. Annu Rev Biochem 48:681–717

Tezuka T, Yamamoto Y (1975) Photoactivation of NAD kinase through phytochrome. Phosphate donors and cofactors. Plant Physiol 56:728–730

Thien W, Schopfer P (1975a) Control by phytochrome of cytoplasmic and plastid rRNA accumulation in cotyledons of mustard seedlings in the absence of photosynthesis. Plant Physiol 56:660–664

Thien W, Schopfer P (1975b) Stimulation of precursor rRNA synthesis in the cotyledons of mustard seedlings by phytochrome. Planta 124:215–217

Thornber JP (1975) Chlorophyll-proteins: light-harvesting and reaction center components of plants. Annu Rev Plant Physiol 26:127–158

Ting IP, Osmond CB (1973) Multiple forms of plant phosphoenolpyruvate carboxylase associated with different metabolic pathways. Plant Physiol 51:448–453

Tischner R, Hüttermann A (1978) Light-mediated activation of nitrate reductase in synchronous *Chlorella*. Plant Physiol 62:284–286

Tischner R, Hüttermann A (1980) Regulation of glutamine synthetase by light and during nitrogen deficiency in synchronous *Chlorella sorokiniana*. Plant Physiol 66:805–806

Tobin EM (1978) Light regulation of specific mRNA species in *Lemna gibba* L. G-3. Proc Natl Acad Sci USA 75:4749–4753

Tobin EM (1981) White light effects on the mRNA for the light-harvesting chlorophyll a/b binding protein in *Lemna gibba* L. G-3. Plant Physiol 67:1078–1083

Tobin EM, Suttie JL (1980) Light effects on the synthesis of ribulose-1,5-bisphosphate carboxylase in *Lemna gibba*. L. G-3. Plant Physiol 65:641–647

Tome F, Campedelli L, Bellini E (1975) Distribution of phenylalanine transaminase and phenylalanine ammonia-lyase in etiolated and light irradiated radish seedlings (*Raphanus sativus* L.) Experientia 31:1119–1121

Tong W-F, Schopfer P (1976) Phytochrome-mediated de novo synthesis of phenylalanine ammonia-lyase: An approach using pre-induced mustard seedlings. Proc Natl Acad Sci USA 43:4017–4021

Travis RL, Jordan WR, Huffaker RC (1969) Evidence for an inactivating system for nitrate reductase in *Hordeum vulgare* L. during darkness that requires protein synthesis. Plant Physiol 44:1150–1156

Travis RL, Key JL, Ross CW (1974) Activation of 80S maize ribosomes by red light treatment of dark-grown seedlings. Plant Physiol 53:28–31

Tsai M-J, Trug AC, Nordstrom JL, Zimmer W, O'Malley BW (1980) Processing of high molecular weight ovalbumin and ovamucoid precursor RNAs to messenger RNA. Cell 22:219–230

Unser G, Masoner M (1972) Kinetics of monogalactosyltransferase in mustard seedlings. Naturwissenschaften 59:39

Van Poucke M, Barthe F (1970) Induction of glycollate oxidase activity in mustard seedlings under the influence of continuous irradiation with red and far-red light. Planta 94:308–318

Van Poucke M, Cerff R, Barthe F, Mohr H (1970) Simultaneous induction of glycollate oxidase and glyoxylate reductase in white mustard seedlings by phytochrome. Naturwissenschaften 56:132–133

Vijayaraghavan SJ, Sopory SK, Guha-Mukherjee S (1979) Role of light in the regulation of the nitrate reductase level in wheat (*Triticum aestivum*). Plant Cell Physiol 20:1251–1261

Viro M, Kloppstech K (1980) Differential expression of the genes for ribulose-1,5-bisphosphate carboxylase and light-harvesting chlorophyll a/b protein in the developing barley leaf. Planta 150:41–45

Wallace W (1974) Purification and properties of a nitrate reductase-inactivating enzyme. Biochim Biophys Acta 341:265–276

Wallace W (1975) Effects of a nitrate reductase-inactivating enzyme and NAD(P)H on the nitrate reductase from higher plants and *Neurospora*. Biochim Biophys Acta 377:239–250

Weinbaum SA, Gressel J, Reisfeld A, Edelman M (1979) Characterisation of the 32,000 dalton chloroplast membrane protein. III. Probing its biological function in *Spirodela*. Plant Physiol 64:828–832

Wellman E (1974) Gewebespezifische Kontrolle von Enzymen des Flavonoidstoffwechsels durch Phytochrom in Kotyledonen des Senfkeimlings (*Sinapis alba* L.). Ber Dtsch Bot Ges 87:275–279

Wellmann E, Baron D (1974) Durch Phytochrom kontrollierte Enzyme der Flavonoidsynthese in Zellsuspensionkulturen von Petersilie (*Petreselinum hortense* Hoffm.). Planta 119:161–164

Wellmann E, Schopfer P (1975) Phytochrome-mediated de novo synthesis of phenylalanine ammonia-lyase in cell suspension cultures of parsley. Plant Physiol 55:822–827

Wheeler AM, Hartley MR (1975) Major mRNA species from spinach chloroplasts do not contain poly(A). Nature 257:66–67

Williams LE, Kennedy RA (1978) Photosynthetic carbon metabolism during leaf ontogeny in *Zea mays* L.: enzyme studies. Planta 142:269–274

Williamson PJ (1976) Analysis of data from DNA-DNA or DNA-RNA hybridization experiments. In: Last J (ed) Eukaryotes at the subcellular level. Dekker, New York, pp 125–159

Wittenbach VA (1978) Breakdown of ribulose bisphosphate carboxylase and change in proteolytic activity during dark-induced senescence of wheat seedlings. Plant Physiol 62:604–608

Wu R (ed) (1979) Recombinant DNA. Methods Enzymol Vol 68. Academic Press, London New York

Yamamoto N, Slasaki S, Asakawa S, Hasegawa M (1974) Polysome formation induced by light in *Pinus thunbergii* seed embryos. Plant Cell Physiol 15:1143–1146

Yamamoto N, Hasegawa M, Sasaki S, Asakawa S (1975) Red-far-red reversible effect on polysome formation in the embryos of *Pinus thunbergii* seeds. Plant Physiol 56:734–737

Yamaya T, Ohira K (1978) Reversible inactivation of nitrate reductase by its inactivating factor from rice cells in suspension culture. Plant Cell Physiol 19:1085–1089

Yamaya T, Solomonson LP, Oaks A (1980) Action of corn and rice inactivating proteins on a purified nitrate reductase from *Chlorella vulgaris*. Plant Physiol 65:146–150

Zielke HR, Filner P (1971) Synthesis and turnover of nitrate reductase induced by nitrate in cultured tobacco cells. J Biol Chem 246:1772–1779

Zouaghi M (1976) Phytochrome-induced changes of β-fructosidase activity in radish cotyledons. Planta 131:27–31

Zouaghi M, Rollin P (1976) Phytochrome control of β-fructosidase activity in radish. Phytochemistry 15:897–901

Zouaghi M, Klein-Eude D, Rollin P (1979) Phytochrome-regulated transfer of fructosidase from cytoplasm to cell wall in *Raphanus sativis* L. Hypocotyls. Planta 147:7–13

Zucker M (1963) The influence of light on synthesis of protein and of chlorogenic acid in potato tuber tissue. Plant Physiol 38:575–580

Zucker M (1972) Light and enzymes. Annu Rev Plant Physiol 23:133–156

11 Intracellular Photomorphogenesis

P. SCHOPFER and K. APEL

1 Introduction

Growth, differentiation and the integration of these processes (=morphogenesis) in the multicellular plant is profoundly influenced by light acting through photomorphogenetic sensor and effector molecules, from which phytochrome is the one most widely known. It is obvious that a similar situation exists at the next lower level in the hierarchy of complexity, i.e., at the level of the cell which functions as an integrated system of numerous subcellular compartments. A major reason for studying the functional and structural changes within the cell during photomorphogenesis is the hope that the regulatory processes governing the coordinate development of subcellular compartments may, at the present state of experimental arts, be more open for a successful attack than those directing photomorphogenesis of the whole plant. (In its strict sense this problem has been hardly addressed up till now). Indeed the last few years have seen the advent of powerful new techniques for the in vitro investigation of basic cellular processes such as DNA transcription and RNA translation by the various genetic systems, or the specific transport of functional proteins from one cell compartment to another. These techniques are now being explored in the research on intracellular photomorphogenesis. It appears appropriate, therefore, to devote this chapter mainly to a critical review of the recent molecular approaches in this field. Until now, the research on intracellular photomorphogenesis has been almost exclusively concerned with the *plastid*, the *mitochondrion*, and the *peroxisome* which are dealt with here.

2 Photomorphogenesis of Plastids

Angiosperms are unable to form chloroplasts in the absence of light. When seedlings are grown in the dark the proplastids of the developing leaves differentiate into another plastid type, the etioplast. This plastid is photosynthetically inactive and possesses no extensive thylakoid membranes. Instead of chlorophyll, the etioplast accumulates protochlorophyll(ide), which seems to be stored within the primary thylakoids and the prolamellar body (BOARDMAN et al. 1978, BOGORAD 1976, HAREL 1978, LÜTZ 1978). Upon illumination protochlorophyll(ide) is photoreduced to chlorophyll(ide), the prolamellar body desintegrates, and an activation of the porphyrin pathway leads to the accumulation of chlorophyll. Parallel to chlorophyll accumulation, a rapid formation of the thylakoid membranes occurs, leading to the assembly of the typical membrane system of the mature chloroplast (e.g., BOARDMAN and ANDERSON 1978). In addition to these processes, which are directly related to the light-induced chlorophyll accumulation, other light-dependent changes occur as well within the plastid. Etioplasts already possess ample levels of ribosomes and all other components of a protein synthesis system. Transference of etiolated seedlings to the light generally stimulates both the formation and the activity of this system. The levels of total protein, lipids, and nucleic acids within the plastids are increased on illumination (e.g., BOGORAD 1967) and many examples for specific light effects on the activity of the synthesis of defined plastid constituents have been reported (KIRK and TILLNEY-BASSETT 1978, MOHR 1977, SMILLIE and SCOTT 1969).

The dramatic changes during the light-induced transformation of etioplasts to chloroplasts (see Chaps. 12 and 25, this Vol.), which occur within a relatively short period of time, have made this system an attractive model for studies of the molecular mechanisms which govern light-dependent processes during the development of higher plants. One of the most intriguing problems is the identification of the photoreceptors which regulate the light-dependent changes within the etioplast. The reaction which has been most frequently implicated as the primary target for the action of light is the photoreduction of protochlorophyllide to chlorophyllide, protochlorophyllide being the photoreceptor of its own reduction (KOSKI et al. 1951). Until very recently protochlorophyllide has been regarded, at least in some instances, as the only photoreceptor which triggers light-dependent chloroplast formation (e.g., OGAWA et al. 1973). Since the work of WITHROW et al. (1956), however, it has become evident that besides protochlorophyllide the phytochrome system plays a crucial role during the etioplast transformation in higher plants. The stimulation of the formation of various chloroplast constituents by phytochrome is not only part of a general stimulatory effect that phytochrome exerts on leaf development, but is highly specific for the plastid (e.g., KOBAYASHI et al. 1980, MOHR 1977, SCHOPFER et al. 1975, SMILLIE and SCOTT 1969). A summary of reported phytochrome-controlled processes related to light-induced chloroplast formation in higher plants is given in Table 1.

The occurrence of at least two different photoreceptors, both of which operate simultaneously during light-induced chloroplast formation, leads to the question of how and to what extent the various developmental processes within the plastid are controlled by each of them. Such an analysis is further complicated by the fact that at least two different genetic systems are involved in the control of plastid development (BOGORAD 1975, ELLIS 1977, KIRK and TILLNEY-BASSET 1978). Plastids possess their own DNA and protein synthesis system. However, less than 10% of the plastid protein is synthesized within the plastid compartment. The majority of plastid proteins is coded by nuclear DNA, synthesized outside the plastid on cytoplasmic ribosomes and post-translationally transported into the plastid (CHUA and SCHMIDT 1979). Thus, an analysis of the function of the two photoreceptors during plastid development cannot be restricted to light-dependent changes within the plastid, but must also consider processes which occur in the rest of the cell.

Table 1. Some phytochrome-dependent responses related to the etioplast/chloroplast transformation of angiosperms. (Modified after SCHOPFER et al. 1975)

Response	Reference
Lag-phase of chlorophyll accumulation	WITHROW et al. (1956)
Rate of chlorophyll accumulation	MOHR et al. (1974), KASEMIR et al. (1973)
Formation of chlorophyll b	OELZE-KAROW and MOHR (1978a)
Shibata shift of chlorophyll a	JABBEN and MOHR (1975)
Accumulation of the light-harvesting chlorophyll a/b protein	APEL (1979)
Accumulation of protochlorophyll(ide)	AUGUSTINUSSEN and MADSEN (1965)
Decrease of protochlorophyll(ide)	JABBEN et al. (1974)
Disappearance of the NADPH-protochlorophyllide oxidoreductase	SANTEL and APEL (1981)
Synthesis of 5-aminolevulinate	MOHR et al. (1974)
Activity of aminolevulinate dehydratase	MOHR et al. (1974)
Accumulation of plastid protein	MEGO and JAGENDORF (1961)
Increase of plastid volume	MEGO and JAGENDORF (1961)
Volume and structure of the prolamellar body	BRADBEER et al. (1974a, b)
Accumulation of carotenoids	COHEN and GOODWIN (1962), SCHNARRENBERGER and MOHR (1970)
Accumulation of galactolipids	UNSER and MOHR (1970)
Activity of the monogalactosyltransferase	UNSER and MASONER (1972)
Degradation of starch	KLEIN et al. (1963)
Activity of plastocyanin	HASLEFF and CAMMACK (1974)
Activity of enzymes of the Calvin cycle	MARCUS (1960), FEIERABEND and PIRSON (1966), GRAHAM et al. (1968, 1971), BRÜNING et al. (1975)
Activity of enzymes of the C_4 cycle	QUEIROZ (1969), GRAHAM et al. (1970), KOBAYASHI et al. (1980)
Accumulation of fraction I protein	GRAHAM et al. (1968)
Photophosphorylation	OELZE-KAROW and MOHR (1978b)
Accumulation of plastid rRNA	THIEN and SCHOPFER (1975)
Accumulation of plastid mRNA	LINK (1981, 1982)
Induction of translatable mRNA activity for the light-harvesting chlorophyll a/b protein	APEL (1979), TOBIN (1981)
Induction of translatable mRNA activity for the small subunit of the ribulose-bisphosphate carboxylase	TOBIN (1981)
Decrease of translatable mRNA activity for the NADPH-protochlorophyllide oxidoreductase	APEL (1981)

It is not the purpose of this article to give a complete summary of known light effects on plastid development. Instead, a few examples have been selected in which the molecular mechanisms underlying the light-induced responses have been studied in some detail. Summarizing articles on the same subject emphasizing other aspects of plastid development have appeared (BOARDMAN et al. 1978, BOGORAD 1976, ELLIS 1977, 1981, MOHR 1981).

2.1 Formation of Ribulosebisphosphate Carboxylase

Ribulose bisphosphate carboxylase is the most abundant plastid enzyme. In higher plants it occurs as an oligomer of eight large subunits of 55,000 and eight small subunits of 12,000–14,000 molecular weight (ELLIS 1979). The large subunit is encoded in the plastid DNA (COEN et al. 1977, MALNOE et al. 1979) and synthesized inside the plastid on 70S ribosomes (ELLIS 1977). The small subunit is encoded in nuclear DNA and synthesized on 80S ribosomes in the cytosol as a precursor 4,000–6,000 molecular weight larger than the mature polypeptide. When this precursor enters the chloroplast it is shortened to its final size (CHUA and SCHMIDT 1978, 1979, HIGHFIELD and ELLIS 1978).

The enzyme protein is present in etioplasts in varying amounts. Upon illumination of leaves a 3- to 15-fold increase in the enzyme activity has been observed (KIRK and TILLNEY-BASSETT 1978). This increase in activity can at least in part be attributed to the synthesis of the enzyme protein (KANNANGARA 1969, KLEINKOPF et al. 1970, SMITH et al. 1974). Synthesis of the enzyme is not correlated with the synthesis of chlorophyll. The photoreceptor of the light-induced enzyme synthesis has been identified as phytochrome (GRAHAM et al. 1971, BRÜNING et al. 1975, KOBAYASHI et al. 1980). Similarly, many other photosynthetic enzymes (e.g., those of the Calvin cycle) are inducible by phytochrome (Table 1).

Recently, one point of control has been characterized at which phytochrome regulates the synthesis of the small subunit of the ribulosebisphosphate carboxylase (TOBIN 1978). In dark-grown *Lemna gibba* fronds the level of mRNA activity for the small subunit of the ribulosebisphosphate carboxylase is extremely low and synthesis of the enzyme could not be detected. However, upon illumination with white light a dramatic increase in the level of mRNA activity was induced and the synthesis of the enzyme began rapidly. Later work demonstrated that phytochrome was responsible for these effects (TOBIN 1981). However, it was not possible with the methods used to distinguish between a phytochrome-dependent activation of preformed mRNA sequences (e.g., by initiation of processing), and a phytochrome-induced de novo synthesis of this mRNA. Recent work by BEDBROOK et al. (1980) and SMITH and ELLIS (1981) with pea plants has offered a possible answer to this problem. Using cloned cDNA encoding the mRNA for the small subunit as a hybridization probe, the amount of specific mRNAs was determined under various physiological conditions. Upon illumination of etiolated plants the level of mRNA sequences, coding for the small subunit, increased manyfold. These results, together with those presented by TOBIN (1981) suggest, but do not yet prove, that a phytochrome-dependent transcription of specific genes in the nucleus exists.

The synthesis of the large subunit is also increased by light (SMITH et al. 1974, SMITH and ELLIS 1981). It has been proposed that, after its transport into the plastid, the small subunit may control the synthesis of the large subunit by influencing the transcription of plastid DNA (ELLIS 1977). In this case the phytochrome-controlled induction of mRNA coding for the small subunit could be the only light-dependent step required for the induced synthesis of the complete enzyme. However, this model has not been verified so far and other possible modes of interaction between the biosynthetic pathways of the small and large subunits have been discussed (BARRACLOUGH and ELLIS 1979, HAGEMANN and BÖRNER 1978). It is conceivable that, in addition to its effect on the synthesis of the small subunit, phytochrome might also directly control the synthesis of the large subunit. This leaves the question open as to how the coordination between the syntheses of the small and the large subunit is accomplished.

Besides the phytochrome-mediated induction of mRNA, other light-dependent processes could modify the amount of the enzyme as well. For instance, it has been shown by GROSSMAN et al. (1980) that light stimulates the uptake of polypeptides synthesized in vitro into intact isolated chloroplasts. This light effect is attributed to the production of ATP by photophosphorylation. However, in vivo, light is not required for the uptake of the small subunit. The complete enzyme is formed also in etiolated plants (e.g., KANNANGARA 1969). Furthermore, the light-dependent increase of the enzyme level can be accomplished by phytochrome in the absence of photosynthesis. The enzyme level is increased to the same extent by far-red light treatment as by continuous white light (FROSCH et al. 1976). It therefore appears that the transport of polypeptides through the plastid envelope is not a rate-limiting process in vivo.

2.2 Formation of Photosynthetically Active Chlorophyll

When an etiolated leaf is transferred to light, the protochlorophyllide initially present is immediately photoreduced to chlorophyllide which subsequently becomes esterified to chlorophyll a. After this very rapid photoreaction, there is commonly a lag-phase of 1 to 2 h until steady-state accumulation of chlorophylls commences, which may continue for several days (BOARDMAN et al. 1978, BOGORAD 1976, HAREL 1978). In white light which saturates the protochlorophyllide photoreduction the accumulation rate appears to be controlled exclusively by phytochrome. Because of the absolute requirement of light absorption by protochlorophyllide during chlorophyll accumulation, phytochrome control is difficult to demonstrate under these conditions. However, if an etiolated leaf is preirradiated with a red light pulse (or a longer period of far-red light) several hours before the onset of white light, phytochrome control can be demonstrated by an earlier establishment of the steady state (i.e., an elimination or shortening of the lag-phase) and an increase in the rate of accumulation. Both responses are reversible by far-red light pulses of sufficiently long wavelength (WITHROW et al. 1956, PRICE and KLEIN 1961, SISLER and KLEIN 1963, VIRGIN 1972, KASEMIR et al. 1973). The length of the lag-phase during light-induced chlorophyll accumulation is influenced by the photodestruction of newly formed chlorophyllide which can be minimized by using medium or low fluence rates of light (VIRGIN 1972).

The decisive question is at which point(s) phytochrome controls chlorophyll accumulation and how the two photoreceptors cooperate in the coordinate production of chlorophylls and other constituents of the photosynthetically active thylakoid membrane. Basically, there are two models which consider different sites at which the accumulation of chlorophyll might be controlled. One of these hypotheses postulates a limiting enzyme activity (5-aminolevulinate synthetase) at the beginning of the biosynthetic chain of chlorophyll, the other postulates limiting amounts of chlorophyll-binding proteins which are necessary for a stable incorporation of chlorophyll in the thylakoid membrane (BOGORAD 1976, HAREL 1978, KIRK 1974, SCHNEIDER 1975).

2.2.1 The 5-Aminolevulinate-Synthesizing Enzyme(s)

It is widely held that the pathway of chlorophyll synthesis is regulated at the step leading to 5-aminolevulinate formation. There is extensive evidence that synthesis of 5-aminolevulinate is limiting for the synthesis of Mg porphyrins in etiolated leaves (GRANICK 1959). This limitation can be overcome by light (e.g., SISLER and KLEIN 1963). It has been suggested that light may affect the synthesis of 5-aminolevulinate in two ways: (1) by inducing the synthesis of enzymes required for its formation, and (2) by removal of an inhibitor affecting the activity of the enzyme system.

1. Based on studies using inhibitors of protein synthesis a 5-aminolevulinate-synthesizing enzyme with a deduced half-life of 90 min has been implicated in the regulation of 5-aminolevulinate formation (FLUHR et al. 1975, KLEIN et al. 1977, NADLER and GRANICK 1970). It has been proposed that the synthesis of the enzyme requires light that is absorbed by phytochrome. For instance, the reduction of the lag-phase of chlorophyll accumulation in plants that have been treated with a red light pulse prior to continuous illumination might be explained by a phytochrome-induced synthesis of a 5-aminolevulinate-synthesizing enzyme that was absent in dark-grown plants. Recent work on 5-aminolevulinate-synthesizing enzyme fractions from barley could not verify some of the proposals made previously (KANNANGARA and GOUTH 1979): During the first 12 h of illumination the activity is increased threefold. However, also in dark-grown plants significant amounts of activity were present and no evidence for a rapid turnover could be found.

2. If greening plants are returned to darkness, synthesis of 5-aminolevulinate ceases almost immediately (SCHNEIDER 1973, FLUHR et al. 1975, KLEIN et al. 1977). This inhibition is rapidly reversed upon reillumination. Such rapid switching on and off of 5-aminolevulinate synthesis during dark-to-light and light-to-dark transitions is more plausible in terms of direct metabolic control than in terms of enzyme synthesis and breakdown. Metabolic feedback control of chlorophyll biosynthesis could be exerted by an intermediate in the pathway that accumulates in the dark and disappears in the light. Protochlorophyllide has been regarded as a candidate for such a regulatory substance (BOGORAD 1976, MASONER and KASEMIR 1975, KLEIN et al. 1977). But also earlier intermediates of chlorophyll biosynthesis have been implicated as possible regulators of the 5-aminolevulinate-synthesizing enzyme (CASTELFRANCO and JONES 1975, DUGGAN and GASSMAN 1974, HENDRY and STOBBART 1978). In etiolated mustard seedlings a linear correlation between the accumulation of 5-aminolevulinate (in seedlings with blocked chlorophyll synthesis) and the extent of protochlorophyllide photoconversion has been described (FORD and KASEMIR 1980). This result is in agreement with a modulation of the 5-aminolevulinate-synthesizing

enzyme activity by the level of phototransformable protochlorophyllide. The molecular mechanism for operating this control is not yet known. Other reactions within the biosynthetic pathway of chlorophyll are also affected by phytochrome: the amount of protochlorophyllide reduced by a light flash, the regeneration of phototransformable protochlorophyllide during the subsequent dark period, the spectral in vivo shifts of the freshly synthesized chlorophyll(ide), the esterification of chlorophyllide, and the formation of chlorophyll b (JABBEN and MOHR 1975, JABBEN et al. 1974, KASEMIR and PREHM 1976, OELZE-KAROW and MOHR 1978a, SUNDQVIST 1978).

2.2.2 Protochlorophyllide Holochrome

According to present models the photoreduction of protochlorophyllide, as well as the transfer of the freshly formed chlorophyllide from its site of synthesis to its ultimate site of function within the thylakoid membrane, includes the participation of a photoactive protein complex, the protochlorophyllide holochrome (BOARDMAN et al. 1978, BOGORAD 1976, BOGORAD et al. 1967, HAREL 1978). Until recently none of the molecular events underlying these processes have been known. One reason for this ignorance was the lack of information on the molecular properties of the protochlorophyllide holochrome. Even though this protein has been known and studied intensively for a number of years (BOARDMAN 1966), it has not been purified to such an extent that its exact structure and subunit composition could be determined. The finding that NADPH can be used as a hydrogen donor for the photoreduction of protochlorophyllide in vitro has enabled GRIFFITHS (1974, 1978) to set up an assay for the determination of the protochlorophyllide-reducing enzyme activity in etioplasts. Based on this work it became possible to solubilize the active enzyme from plastid preparations and to purify it to apparent homogeneity (APEL et al. 1980, BEER and GRIFFITHS 1981). The isolated enzyme of barley consists of a single polypeptide of 36,000 molecular weight. It is present in heat-bleached barley plants which lack plastid ribosomes (BATSCHAUER et al. 1982). The enzyme is encoded by poly(A)-containing mRNA and is synthesized outside the plastids on cytoplasmic ribosomes as a higher molecular weight precursor before it is transported into the etioplast (APEL 1981).

It has been suggested that the protochlorophyllide holochrome is responsible for the overall chlorophyll synthesis during light-induced greening (BOARDMAN 1966). Thus, one would expect to find protochlorophyllide reductase present throughout the greening period. However, MAPLESTON and GRIFFITHS (1980) found a rapid decrease of the reductase activity in the light. They interpreted their findings as a reversible inactivation of the enzyme by a light-induced alteration of the redox state of the plastid NADP pool.

SANTEL and APEL (1981) have confirmed this observation. Moreover, it was found that the light-induced decrease of the enzyme activity was accompanied by a rapid decline of the enzyme protein and of the translatable mRNA coding for it. In dark-grown barley plants this mRNA species is one of the most abundant translatable mRNAs. The effect of light on the translatable mRNA coding for the protochlorophyllide reductase is controlled by phytochrome (APEL 1981, SANTEL and APEL 1981).

Thus it appears that phytochrome, although accelerating chlorophyll accumulation, at the same time causes a faster decrease of an enzyme that is so far the only candidate for catalyzing the reduction of protochlorophyllide to chlorophyllide.

2.2.3 The Light-Harvesting Chlorophyll a/b Protein

Within the thylakoid membrane all of the chlorophyll seems to be attached to specific proteins (MARKWELL et al. 1979). As discussed by KIRK (1974), it seems mandatory that the insertion of chlorophyll into the membrane depends on the availability of pigment-binding proteins. It is thus conceivable that the accumulation of chlorophyll can be limited not only by the biosynthesis of chlorophyll but also by the provision of binding sites for the pigment. Among the chlorophyll-binding proteins the most abundant one, the light-harvesting chlorophyll a/b protein, has been selected as a model system to study the interdependence between chlorophyll and protein syntheses during greening. Encoded in the nucleus and synthesized by cytoplasmic ribosomes in precursor form, the apoprotein is taken up into the plastid, shortened to a molecular weight of 25,000, and inserted into the thylakoid membrane where it binds chlorophyll b and most chlorophyll a (KUNG et al. 1972, MACHOLD and AURICH 1972, SCHMIDT et al. 1980, THORNBER 1975).

The apoprotein is not detectable in the etioplast; it accumulates within the thylakoid membrane only during continuous illumination (APEL and KLOPPSTECH 1978, 1980, ARMOND et al. 1977, HØYER-HANSEN and SIMPSON 1977). The kinetics for the appearance of this protein closely resemble those of chlorophyll accumulation, exhibiting, for example, the same lag-phase. Because of these similarities, one might assume that protochlorophyllide is also the primary photoregulator for the synthesis of the light-harvesting chlorophyll a/b protein. Recent work has shown, however, that the synthesis of the apoprotein is controlled by phytochrome (APEL 1979, TOBIN 1981). In dark-grown barley plants not only the complete chlorophyll a/b protein is not detectable, but also the translatable mRNA coding for it. The appearance of this mRNA is inducible by white light or by a single red light pulse which produces a maximum response after 4 to 5 h. The induction is reversible by far-red light (APEL 1979).

The phytochrome-induced mRNA is taken up into the polysomal fraction and most probably also translated into the polypeptide. However, an accumulation of this protein cannot be detected in the dark period following a red light pulse. Thus it appears that in the absence of chlorophyll the freshly synthesized apoprotein is subject to immediate degradation (APEL and KLOPPSTECH 1980, BENNETT 1981). It is known that the massive accumulation of the chlorophyll a/b protein occurs only in the presence of chlorophyll b (ARMOND et al. 1977, THORNBER and HIGHKIN 1974). However, as shown by work with a chlorophyll b-less mutant of barley, chlorophyll a alone is sufficient to stabilize the apoprotein to a certain extent (APEL and KLOPPSTECH 1980). This result indicates a control mechanism for chlorophyll accumulation in which the formation of chlorophylls is the rate-limiting factor. However, there is evidence that under different experimental conditions the accumulation of chlorophyll may also be regulated by the availability of the pigment-binding protein. A pretreatment of etiolated barley plants with a red light pulse triggers the appearance of translatable mRNA for the apoprotein of the chlorophyll a/b protein during the next 4–5 h of darkness. This treatment almost completely abolishes the normal delay in chlorophyll accumulation in white light (VIRGIN, 1972). Consequently the chlorophyll a/b protein, as well as the bulk of chlorophyll, begins to accumulate within the thylakoid membrane at a much earlier time than in dark-grown control plants (APEL 1979). The effect of the preirradiation on chlorophyll accumulation shows a maximum when the dark period after the preirradiation has a length roughly corresponding to the time of maximal mRNA accumulation (APEL 1979, VIRGIN 1972). This temporal correlation is in agreement with the assumed rate-limiting role of the chlorophyll-binding protein for chlorophyll accumulation. These results, however, do not solve the problem of how the synthesis of pigment and protein might be coordinated so that neither of the two constituents is synthesized in excess and thus wasted.

2.2.4 The Interaction Between Phytochrome and Protochlorophyllide During Chloroplast Development

So far, several rate-limiting steps for chlorophyll accumulation have been described which are controlled by phytochrome: the synthesis of 5-aminolevulinate, the decline of the protochlorophyllide reductase, and the synthesis of the light-harvesting chlorophyll a/b protein. All these processes are interlocked with the light-dependent reduction of protochlorophyllide. Without this photoreaction the expression of these phytochrome-induced processes at the level of membrane assembly is blocked. The function of protochlorophyllide is different from that of phytochrome in that it does not act as a typical sensor pigment. Instead, its light-dependent reduction is the prerequisite for the formation of chlorophyll which is a crucial constituent for the assembly of photosynthetic units within the thylakoid membrane. Together with the close interaction with protochlorophyllide it might be the overlap of several, separate phytochrome-induced reactions, each of which contributes to chlorophyll accumulation and might become rate-limiting under certain environmental conditions, which coordinates the various pathways and narrows down the range in which the rate of chlorophyll accumulation may vary. A finer control of chlorophyll and protein insertion might be accomplished either by the breakdown of protein in the absence of chlorophyll or by the destruction of chlorophyll (e.g., photooxidation) when the availability of pigment-binding proteins becomes limiting.

2.3 Outlook for Coordination Mechanisms

The light-induced changes within the developing plastids are based on an intimate interaction of the two genetic systems within the nucleus and the plastid. Light exerts its control not only directly on processes which occur within the

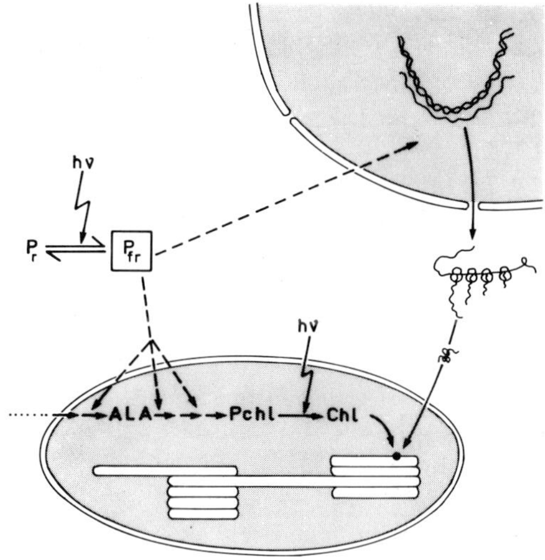

Fig. 1. Scheme illustrating the controlling points at which phytochrome influences the formation of photosynthetically active chlorophyll (light-harvesting chlorophyll a/b protein) in the thylakoid membrane

plastid but simultaneously triggers reactions in the cytoplasm which contribute to chloroplast formation. It is not surprising, therefore, that not only protochlorophyllide within the plastid but also at least one additional photoreceptor, phytochrome, is involved in the light-dependent control of the complex interaction between plastids and the nucleus. The light-dependent assembly of the light-harvesting chlorophyll a/b protein demonstrates the close interdependence of the two photoreceptors (Fig. 1). The two constituents of the complex, chlorophyll and the apoprotein, are synthesized within different compartments of the cell and are both under the control of phytochrome. The assembly of the complete chlorophyll protein depends on a close coordination of both biosynthetic pathways by light. Despite the detailed analysis of the effect of light on the biosynthesis of chloroplast constituents, we are still far from understanding the complete pattern of control mechanisms which coordinate the interaction between plastids and the surrounding cytoplasm.

One mode of control by which phytochrome regulates the light-dependent development of the chloroplast has been identified: three plastid proteins, the light-harvesting chlorophyll a/b protein, the protochlorophyllide reductase and the small subunit of the ribulosebisphosphate carboxylase are synthesized outside the plastid and are affected by light at the level of translatable mRNA. It is possible that also the other plastid proteins which are encoded in nuclear DNA and affected by phytochrome are regulated in the same way.

The regulation of specific mRNA levels outside the plastid is not the only process by which phytochrome controls plastid development in higher plants. In several instances multiple sites and modes of phytochrome action have been described (e.g. Mohr, 1977).

As shown by the work of Smith and Ellis (1981) white light stimulates the accumulation of transcripts of the gene for the large subunit of the ribulosebisphosphate carboxylase. First direct evidence for a phytochrome control of plastid mRNA in mustard seedlings has been presented recently by the work of LINK (1981, 1982). Phytochrome regulates the steady state levels of the mRNA for a 35,000 molecular weight plastid polypeptide which is encoded by the plastid DNA. Furthermore, evidence has been presented also for a stimulation of plastid rRNA accumulation by phytochrome (THIEN and SCHOPFER 1975).

So far, we have presented examples for a light-controlled development of plastid components which are also determined by nuclear genes and depend on the control of cytoplasmic protein synthesis. It is also conceivable, however, that the plastid is involved in the photocontrol of developmental processes in the extraplastidic part of the cell. The existence of such a control is suggested, for instance, by the work of HAGEMANN and BÖRNER (1978) on the influence of plastid protein synthesis on the accumulation of the small subunit of ribulosebisphosphate carboxylase. Even though the nature of the postulated transmitter has not been identified, there is at least one example of how such a control by the plastid may operate. A close association of phytochrome with the etioplast envelope has been described (COOKE and SAUNDERS 1975, COOKE and KENDRICK 1976, EVANS and SMITH 1976) and it has been suggested by these authors that phytochrome may alter the permeability of the envelope membrane, leading for instance to the release of hormones into the surrounding cytoplasm. It is not yet known whether this mode of control indeed operates in the intact plant.

So far, only two photoreceptors which control the chloroplast development in higher plants have been clearly identified. In *Euglena* (SCHIFF 1978) and in various green algae (VESK and JEFFREY 1977) a third, blue-light-absorbing photoreceptor has been described which functions in a way very similar to phytochrome in higher plants. Indirect evidence for the occurrence of a blue-light-absorbing pigment system in higher plants has been given and in some instances specific blue light effects on plastid development have also been observed in angiosperms (GRESSEL 1978, RICHTER and DIRKS 1978). Thus, the pattern of complex interactions between phytochrome and protochlorophyllide during plastid development could be further complicated by the involvement of this third photoreceptor.

3 Photomorphogenesis of Mitochondria

In contrast to plastids, mitochondria are usually considered as relatively conservative cell organelles which demonstrate at best slight quantitative structural and functional modifications during cell differentiation in higher plants (e.g., an increase in volume or folding of the inner membrane). Because of this estimation and the absence of any obvious relationship to light-dependent metabolism the mitochondrion has clearly been outrivalled by the plastid as a model system for developmental studies with genetically semi-autonomic organelles, at least in autotrophic plants. Correspondingly, the information on the morphogenesis and molecular development of mitochondria and the participating regulatory factors is extremely scanty (ÖPIK 1974, SCHOPFER et al. 1975). Although mitochondrial fractions have been repeatedly chosen for studies of binding and action of phytochrome in vitro (MANABE and FURUYA 1973, 1974, 1975, SCHMIDT and HAMPP 1977, CEDEL and ROUX 1980a, b), no physiologically meaningful picture has emerged so far from these investigations. Crude mitochondrial fractions, isolated from oat leaves, increase rapidly in electron transport capacity if the plants are irradiated with a brief light pulse (or a longer period of continuous light) prior to homogenization. This effect is mediated by phytochrome (HAMPP and WELLBURN 1979). However, since this response returns to the dark level within a few hours even in continuous white or far-red light, it seems to reflect a transient shift in the adenylate metabolism of light-treated tissue rather than a photomorphogenetic event.

In the autotrophic plant, mitochondrial differentiation has been investigated to some extent in maturing and germinating embryos. In fact, the periods of seed formation and germination are probably the only stages during the ontogeny of a plant which are characterized by profound morphogenetic changes in mitochondria. Towards the end of the ripening period of the seed, which is correlated to a dramatic dehydration of the embryo tissues, the mitochondria undergo a controlled disorganization indicated, for example, by the disappearance of respiratory enzymes and membrane structures (BAIN and MERCER 1966a, KLEIN and POLLOK 1968, KOLLÖFFEL 1970). This process can be described as a developmental regression of metabolically active mitochondria to poorly organized promitochondria in analogy to the very similar morphogenetic changes of the chloroplasts occurring simultaneously in these cells. (SCHOPFER et al. 1975). This regressive morphogenetic process is strikingly reminiscent of the adaptive changes of the mitochondria of yeast cells after transfer from aerobic to anaerobic conditions (SCHATZ 1970). Since the promitochondrion serves no obvious metabolic task, its major function seems to be the conservation of the mitochondrial genetic system through a phase of adverse environmental conditions. As in re-aerated yeast cells, this regression is fully reversible after germination of the seed (BAIN and MERCER 1966b, MALHOTRA and SPENCER 1973), and it is this developmental stage where light appears to exert the most striking morphogenetic effect on mitochondriogenesis.

In the cotyledons of young mustard seedlings O_2 consumption as well as CO_2 production demonstrate a typical "grand period of respiration" concomitant with a rise of respiratory enzyme activities that completely disappeared during the seed stage. Both processes are modulated by phytochrome (HOCK and MOHR 1964, WEISCHET 1971). More recently it has been shown that the activities of cytochrome oxidase, succinate dehydrogenase, and fumarase are significantly promoted by phytochrome during this period (BAJRACHARYA et al. 1976). Interestingly, although the light-mediated time course of these enzymes appears to be similar, there are characteristic differences in the extent of the light effect which is strongest in cytochrome oxidase and weakest in succinate dehydrogenase. Thus, these phytochrome effects are to some extent enzyme-specific and therefore do not merely represent the general growth or proliferation of the mitochondria. Inhibitors of cytoplasmic protein synthesis suppress the increase of these enzyme activities. The quantitative changes in respiratory en-

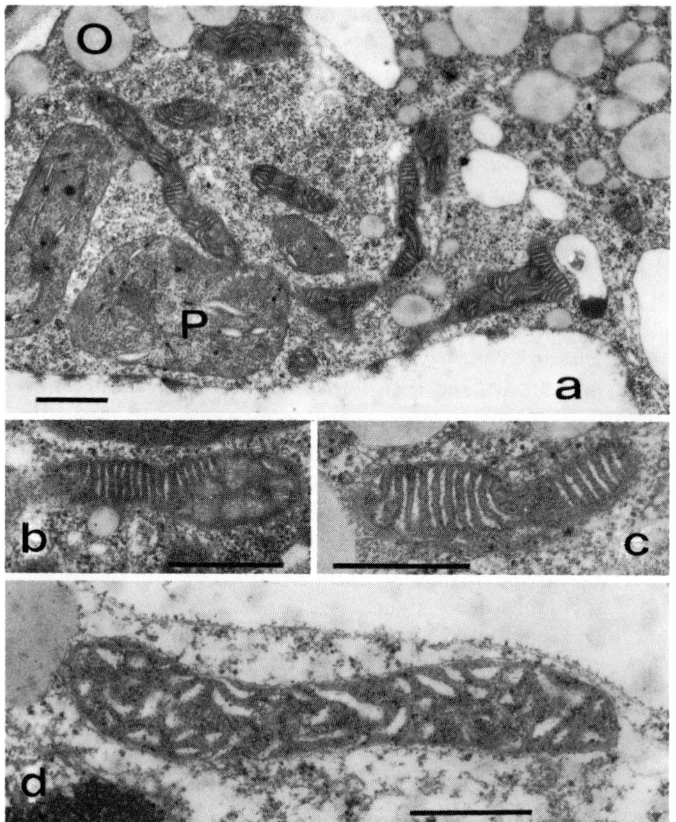

Fig. 2 a–d. Typical mitochondria from palisade parenchyma cells of mustard cotyledons demonstrating the effect of phytochrome on the configuration of the inner membrane. The seedlings were either grown for 72 h in darkness (**a–c**) or irradiated with far-red light (3.5 Wm^{-2}) from 36 until 72 h after sowing (**d**). Note the parallel orientation of the plate-like invaginations of the inner mitochondrial membrane in etiolated cotyledons and the irregular orientation of tubular invaginations in the mitochondria of far-red light-grown cotyledons. This configuration is also expressed in white light. *P* plastid; *O* oleosome. *Bars* represent 0.5 μm. (Electron micrographs courtesy of Dr. H. FALK)

zyme activities are accompanied by a significant increase in the buoyant density of isolated mitochondria on a sucrose gradient and a rather drastic structural differentiation of the inner mitochondrial membrane (BAJRACHARYA et al. 1976). The typical configuration of plant mitochondria characterized by irregularly arranged, curved tubules ("sacculi") develops only in the presence of active phytochrome. In the dark-grown mustard cotyledons the promitochondria instead develop into larger, more or less cylindrical particles with an electron-dense matrix and parallel oriented sheets of "cristae". Favorable sections show large stacks of regularly spaced, plate-like invaginations of the inner membrane with a narrow neck that can rarely be seen on electron micrographs (Fig. 2). The structure of these organelles is strikingly similar to that of the mitochondria of some animal cells (e.g., in the mammalian kidney).

So far this unusual type of mitochondrion has not been described for other plant cells, with one notable exception. MORISSET (1973, 1974) has reported that anaerobiosis produces mitochondria of very similar fine-structural appearance in certain regions of isolated tomato roots. Upon reaeration, the inner mitochondrial membrane again assumed the typical "sacculi" configuration. These observations have been confirmed in anaerobically grown rice seedlings and pumpkin roots (VARTAPETIAN et al. 1977). Despite the striking similarity it is not clear, however, whether the structural "normalization" induced by phytochrome in the mustard cotyledons and these oxygen-induced changes do have a common basis.

In summary, these observations indicate that intracellular photomorphogenesis also involves the mitochondria. The physiological significance of these regulatory modifications is not obvious at present, mainly because of our general ignorance of the relationship between structure and function in these organelles. However, the available evidence is in agreement with the view that mitochondrial photomorphogenesis is controlled essentially through differential, phytochrome-mediated synthesis of mitochondrial proteins derived from the nucleo-cytoplasmic gene expression system which are post-translationally transferred to the organelle (SCHATZ 1979).

4 Photomorphogenesis of Microbodies/Peroxisomes

4.1 Functional Types of Peroxisomes

Microbodies are particles about 1 µm in diameter belonging to the family of ER-derived cell compartments. They are cytologically defined by a single limiting membrane and a fine-granular proteinaceous matrix which gives a positive reaction with 3,3′-diaminobenzidine under conditions favoring the peroxidatic activity of catalase rather than true peroxidase activity. Biochemically, it has become clear in recent years that these organelles are the major, if not the only sites of cytoplasmic H_2O_2 production and destruction in the eukaryotic cell (GERHARDT 1978, BÖCK et al. 1980). The microbodies of both higher plants and animals have been shown to house catalase together with oxidative sections of three independent catabolic sequences:

1. The β-oxidation pathway of fatty acid breakdown (including H_2O_2-producing acyl-CoA oxidase). In the peroxisomes of germinating seeds and at least some animal embryos (nematodes, KHAN and MCFADDEN 1980; guinea pigs, JONES 1980) the resulting acetyl-CoA is further metabolized via the glyoxylate cycle.

2. A part of the glycolate pathway leading to amino acids and carbohydrates (including H_2O_2-producing glycolate oxidase).

3. A part of the catabolic purine pathway (including H_2O_2-producing urate oxidase).

These three metabolic functions are coupled to O_2 consumption and, at least indirectly, to CO_2 production and therefore contribute to the respiratory

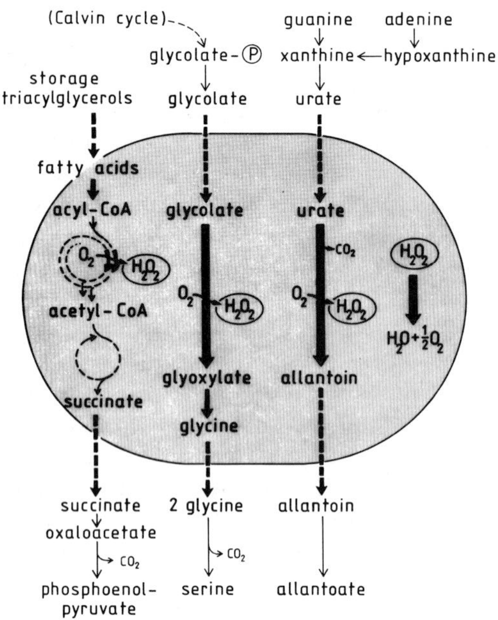

Fig. 3. Generalized scheme indicating the three degradative pathways partly localized in the peroxisomes of higher plants
1. (*left*) Breakdown of fatty acids via *β*-oxidation and glyoxylate cycle ("glyoxysomal" function; the majority of enzymes is bound to the peroxisomal membrane). 2. (*center*) Transformation of glycolate into glycine ("leaf peroxisomal" function; in the green leaf this pathway is essential for the removal of the photosynthetic by-product glycolate). 3. (*right*) Oxidative decarboxylation or urate ("uricosomal" function). The scheme emphasizes the common sequestration of H_2O_2-producing oxidases (acyl-CoA oxidase, glycolate oxidase, urate oxidase) together with catalase (*peroxisomal concept,* De Duve 1965), but disregards a variety of accessory enzymes also localized in the peroxisome (see Gerhardt 1978 for a full account). Mettler and Beevers (1980) have recently suggested a modified pathway for the transfer of carbon out of the peroxisome operating in the glyoxysomal mode which is not included in this simplified scheme

gas exchange of the cell (Fig. 3). Because of the striking functional similarities between plant and animal microbodies it appears justified today to extend the term peroxisome—originally coined for animal microbodies by De Duve (1965) and later adopted by Tolbert et al. (1968) specifically for the glycolate-metabolizing microbodies of green leaves—to all types of microbodies containing catalase and at least one H_2O_2-producing oxidase, irrespective of their particular enzyme composition (Gerhardt 1978, Böck et al. 1980).

Plant peroxisomes have been extensively reviewed (Beevers 1979, Gerhardt 1978, Frederick et al. 1975). In our context the most interesting feature of plant peroxisomes is their amazingly broad functional differentiation which is closely related to the metabolic specialization of the particular tissue. Briefly, plant peroxisomes can be grossly classified into three categories according to the predominant functional complex of enzymes they contain (see Fig. 3):

1. The peroxisomes in fat-storing tissues (e.g., endosperm, scutellum, aleurone) of germinating seeds are predominantly involved in the mobilization of

storage fat. This type of peroxisome has been discovered in the endosperm of germinating *Ricinius* seeds and christened "glyoxysome" by BREIDENBACH and BEEVERS (1967). The specific function of peroxisomes from fatty tissues is clearly indicated by their high content of lipase, β-oxidation and glyoxylate cycle enzymes and by their intimate association with oleosomes (storage lipid bodies) during the period of fat degradation in these highly specialized storage tissues. Isocitrate lyase and malate synthase are specific markers of this functional complex of peroxisomal enzymes. It should not be disregarded, however, that these organelles also possess, beside the "glyoxysomal" group of enzymes, considerable amounts of the enzymes from the two other peroxisomal functions, including, for instance, glycolate oxidase and urate oxidase.

2. The peroxisomes of green leaves predominantly exhibit the enzymes of glycolate metabolism which are essential for the elimination of the photosynthetic by-product, glycolate, and the recovery of most of its carbon as carbohydrate (photorespiratory glycolate pathway). Convenient peroxisomal marker enzymes of this pathway are glycolate oxidase and hydroxypyruvate reductase. On electron micrographs leaf peroxisomes are often found closely associated with chloroplasts. In general, these organelles are considered to lack glyoxysomal enzymes. However, leaf peroxisomes have been reported to contain, at least in some plants, traces of enzymes from the two other peroxisomal functions (e.g., isocitrate lyase and malate synthase in lentil leaves, urate oxidase in spinach leaves).

3. Heterotrophic plant tissues (from roots, flowers, fruits, tubers) and tissue cultures generally possess peroxisomes with a more limited enzyme complement compared to fatty tissue and leaf peroxisomes. These organelles, however, almost invariably contain glycolate oxidase and urate oxidase in addition to catalase. It can be assumed, therefore, that these peroxisomes are involved in the breakdown of glycolate and urate. In view of its poorly established metabolic function this type of peroxisome has been unjustifyably named "non-specialized" microbodies (HUANG and BEEVERS 1971).

In summary, the admittedly fragmentary information on the overlapping enzyme compositions of peroxisome types nevertheless creates the impression that the functional modifications of this organelle result from a quantitative rather than a qualitative variation of the basic peroxisome concept depicted in Fig. 3. In other words, peroxisomes may represent a group of closely related multifunctional organelles which are differentiated by preferential expression of a certain part of their total enzymic potential, depending on the metabolic specialization of the cell. It is to be expected therefore that the metabolic affiliation of these organelles is neither absolute nor irreversible, but can be extensively modulated in accordance with the changing metabolic needs of the developing cell.

4.2 Functional Transformations of Peroxisomes

The functional plasticity of peroxisomes is impressively documented in the epigeal, fat-storing cotyledons of many dicotyledonous plants (e.g., from

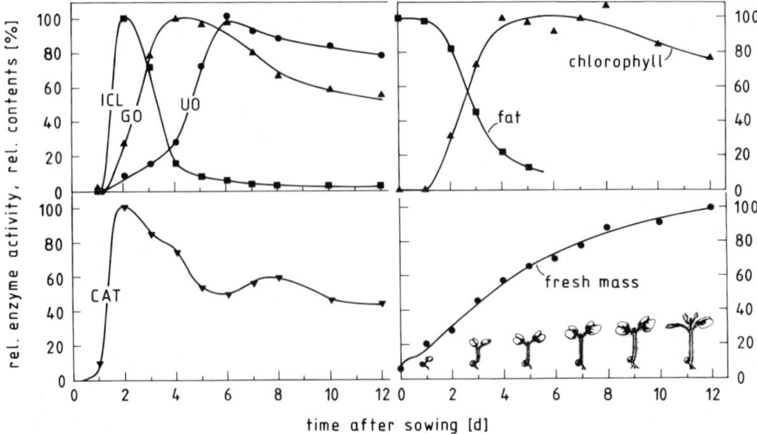

Fig. 4. Time courses of peroxisomal enzymes (isocitrate lyase, glycolate oxidase, catalase), fat content, chlorophyll (a+b) content, and fresh mass in the cotyledons during the development of mustard seedlings in continuous white light (14,000 lux). The seedlings were grown on vermiculite soaked with distilled water (25 °C). Normalization: 100% represents (per pair of cotyledons): 0.26 nkat (glyoxylate) for isocitrate lyase, 0.88 nkat (glyoxylate) for glycolate oxidase, 25 pkat (− urate) for urate oxidase, 3.6 μkat (− H_2O_2) for catalase, 1.7 μmol (triacylglycerol) for fat, 30 nmol for chlorophyll, and 30 mg for fresh mass. In the case of urate oxidase the value for dry seeds (0.025 nkat) is subtracted. Qualitatively similar results have been obtained with seedlings grown on Hoagland's nutrient solution instead of distilled water (HONG and SCHOPFER 1981). (Unpublished data from HONG and SCHOPFER)

Cucumis, Helianthus, or *Sinapis*). Storage fat is the only significant source of energy for the first few days of seedling development initiated after germination. During this short period of self-sustained growth, a transient burst of peroxisomes with glyoxysomal functions takes place which are actively engaged in the breakdown of the fat reserves. Normally, the cotyledons of the developing seedling reach the light before complete depletion of the fat and are now competent to develop chloroplasts which can take over the supply of energy for the growing plant. During this changeover from a storage organ to an assimilatory organ, a dramatic change in the metabolic properties of the mesophyll peroxisomes takes place (Fig. 4). The glyoxysomal markers like isocitrate lyase rapidly disappear and the peroxisomes acquire the enzymic equipment of leaf peroxisomes, represented, e.g., by a strong increase of glycolate oxidase activity. This transition in enzyme content is accompanied by a characteristic change of intracellular localization of the peroxisomes (Fig. 5). Originally intimately attached to oleosomes, the microbodies now gain close membrane contact with plastids (GRUBER et al. 1970, TRELEASE et al. 1971, SCHOPFER et al. 1976). A few days later, urate oxidase also increases strongly, indicating that the uricosomal function (AFZELIUS 1965) is now also present in the peroxisomes (Fig. 4; HONG and SCHOPFER 1981). At this time the fat reserves are largely depleted and isocitrate lyase has fallen to a very low level. Hence, the peroxisomes exhibit almost exclusively leaf peroxisomal and uricosomal properties for the rest of the life of the cotyledon.

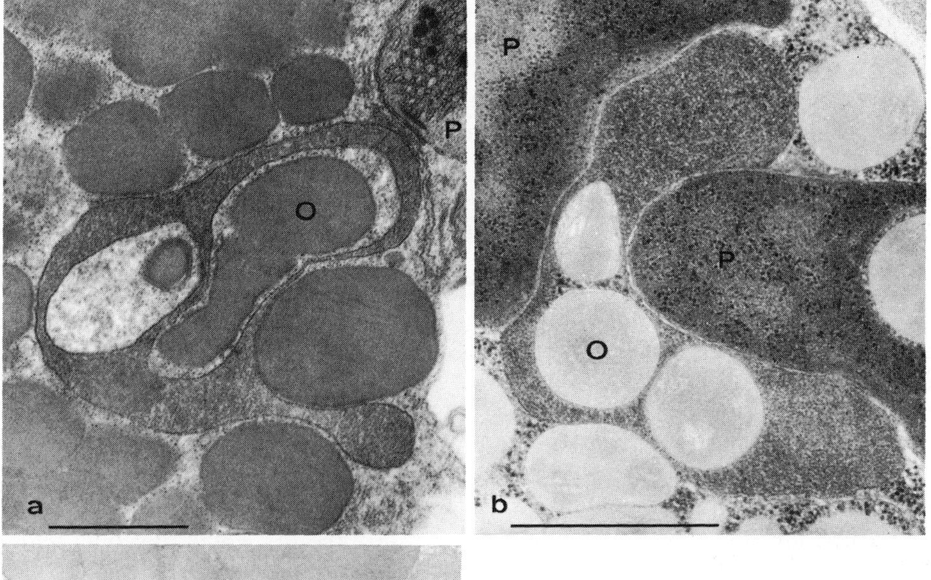

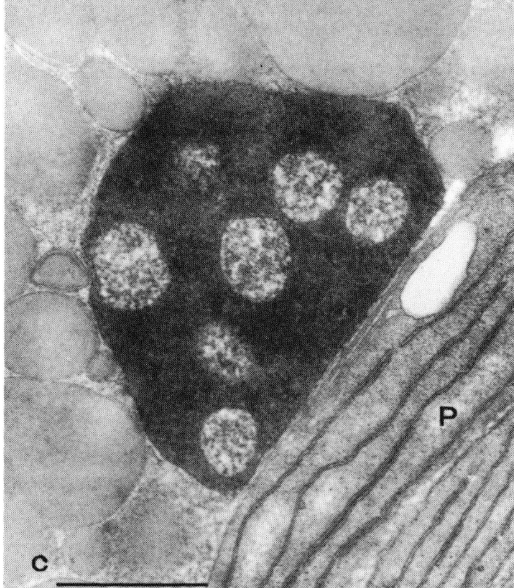

Fig. 5. Typical microbodies (peroxisomes) from etiolated (**a**) and far-red light-treated (**b**, **c**) mustard cotyledons. **a** Seedlings grown in darkness for 84 h. The microbody shows an intimate association with oleosomes (*O*) which appear to be degraded within microbody pockets. **b** Seedlings grown for 36 h in darkness followed by 24 h of standard far-red light (3.5 Wm^{-2}). The microbody shows intimate associations to oleosomes as well as to two plastids (*P*) indicating its intermediate nature. **c** Seedlings grown for 36 h in darkness followed by 48 h of far-red light. The microbody shows the intimate association to a plastid, a situation typical for leaf peroxisomes. *Bars* represent 1 μm. [Electron micrographs courtesy of Dr. R. BERGFELD (**a**, **c**) and Dr. H. FALK (**b**)]

4.3 The Role of Light in Peroxisome Transformation

4.3.1 Peroxisomes of Leaves

The development of peroxisomes with glyoxysomal function in fatty endosperm or related tissues occurs independently of light and darkness. In contrast, the development of peroxisomes with glycolate-metabolizing function in leaves is a component of the photomorphogenesis of these organs. Etiolated foliar leaves usually contain considerable amounts of catalase, but only very low levels of leaf peroxisomal marker enzymes. These enzymes are bound to a labile particle which forms a broad equilibrium sedimentation band below 1.20 kg l^{-1} on a sucrose gradient (FEIERABEND and BEEVERS 1972b, FEIERABEND 1975). Using wheat shoots, FEIERABEND and BEEVERS (1972a) showed that the leaf peroxisomal enzymes can be increased severalfold by continuous white light. Moreover, light was required to produce a particulate cell fraction containing these enzymes which could be recovered at the density characteristic for peroxisomes (about 1.25 kg l^{-1}) after equilibrium centrifugation on a sucrose density gradient (FEIERABEND and BEEVERS 1972b). To produce these effects, continuous light could largely be replaced by brief light pulses which did not allow significant photosynthesis to occur, and inhibition of chloroplast development by aminotriazole did not impair the light responses. These results strongly suggested that the light effect on peroxisome development in greening leaves is not causally dependent on the advent of photosynthetic activity (e.g., on photosynthesis of the peroxisomal substrate glycolate) but is a photomorphogenetic process that is not directly related to chloroplast development. Later investigations using red light induction/far-red light reversion techniques (FEIERABEND 1975) have confirmed the suspicion that phytochrome (and possibly a separate blue light-absorbing pigment) is mediating the effect of light in replacing the less distinct forerunner particles of the etiolated leaf by typical leaf peroxisomes. Fragmentary evidence from other plants indicates that the development of peroxisomes in normal foliar leaves is generally controlled by light (GRUBER et al. 1972, GRUBER et al. 1973, KINDL and MAJUNKE 1973, MURRAY et al. 1973) absorbed by phytochrome (KLEIN 1969, TOMOMATSU and ASAHI 1978).

4.3.2 Peroxisomes of Fatty Cotyledons

The photocontrol of peroxisome development is intimately related to the mechanism of peroxisome transformation in these potentially photosynthetic organs. While the development of the glyoxysomal function takes place in light as well as in darkness, the subsequent transformation into typical leaf peroxisomes requires light (GERHARDT 1973, KAGAWA and BEEVERS 1975, BECKER et al. 1978, KÖLLER and KINDL 1978, TRELEASE et al. 1971, SCHNARRENBERGER et al. 1971, KAGAWA et al. 1973). In the cotyledons of mustard seedlings it has been shown that the rise and fall of the glyoxysomal marker enzymes is independent of light acting through phytochrome (e.g., continuous red or far-red light; KAROW and MOHR 1967, BAJRACHARYA and SCHOPFER 1979) while the light-mediated increase of leaf-peroxisomal and uricosomal marker enzymes is a typical phytochrome-mediated event (VAN POUCKE and BARTHE 1970, VAN POUCKE et al. 1970, HONG and SCHOPFER 1981). Notably, catalase, an enzyme characteristic for both the phytochrome-dependent and -independent peroxisomal functions (Fig. 3), demonstrates, during the changeover, an intermediate kinetic behavior (Fig. 4; HONG and SCHOPFER 1981). Furthermore, the existing catalase isoenzyme pattern is preserved but supplemented by an additional set of isoenzymes (DRUMM and SCHOPFER 1974).

White light has an inhibitory effect in the final stage of accumulation of glyoxysomal enzymes (e.g., HOCK 1969, SCHNARRENBERGER et al. 1971, KAGAWA and Beevers 1975, Hong and Schopfer 1981). The reason for this phytochrome-independent light effect, that can be traced back to a light-mediated decrease of respective mRNAs (Weir et al. 1981), is possibly a consequence of the photosynthetic production of sugars which are known to depress the formation of this enzyme (e.g., Lado et al. 1968). At any rate, the absence of such a precocious decrease of isocitrate lyase during the induction of leaf-peroxisomal enzymes by red or far-red light (Bajracharya and Schopfer 1979) shows that this light effect is not an essential component of peroxisome transformation.

As usual, phytochrome action is characterized by a distinct temporal pattern of competence which in this case is broad enough to allow the light-mediated induction of leaf-peroxisomal properties either before (Schopfer et al. 1976, Köller and Kindl 1978) or after (Kagawa et al. 1973, Kagawa and Beevers 1975) surpassing the peak of glyoxysomal activity.[1] Thus, the rise and fall of glyoxysomal activity and the induction of the other peroxisomal functions by phytochrome are causally unrelated processes, overlapping strongly only in cotyledons which are transferred to light at an early stage of development (Fig. 4). In retrospect, it is interesting to realize that the selection of the time for the onset of light in these time-course studies has strongly influenced the modeling of hypotheses on the mechanism of peroxisome differentiation.

Because of its peculiar photomorphogenetic flexibility, the epigeal, fat-storing cotyledons provide a unique experimental system to study the developmental relationships between the different types of peroxisomes within a constant population of cells. The problem to be solved can be posed in the form of three competitive hypothetical models which have been extensively discussed recently by Beevers (1979). Briefly, the crucial question is whether the changes in the enzyme complement during functional transition is brought about by (1) selective introduction and removal of certain enzymes in existing peroxisomes (*repackaging model*), or (2) de novo formation of distinct new populations of peroxisomes accompanied by the selective destrcution of the previously existing ones (*replacement model*), or (3) a gradual changeover from the synthesis of one set of peroxisomal enzymes to the synthesis of another set of peroxisomal enzymes in the presence of an unspecific (random) turnover of the whole organelles (*enzyme synthesis changeover model*). Fortunately, these three models are unequivocal enough to allow clear-cut predictions that can be tested experimentally. The critical test, of course, would be to check whether the peroxisomes during the functional transition behave as an absolutely homogeneous, more or less static population of structural entities with selective exchange of certain proteins (1), or as a heterogeneous, highly dynamic mixture of particles with either distinctly different (2), or gradually overlapping (3) properties. The pertinent experimental evidence available so fat is mainly of the negative type, i.e., intended to disprove certain aspects of the competing models rather than to prove the favorite model of the particular investigator.

The *repackaging model* of Trelease and collaborators has been largely based on findings such as the apparent lack of finestructural indications for microbody

[1] The significance of cellular competence in phytochrome action and the evidence for a temporal displacement of phytochrome action and increase of enzyme activity ("transmitter concept", Hong and Schopfer 1981) is treated in Chapter 14, this Vol.

formation and degradation (TRELEASE et al. 1971), or the lack of detectable heterogeneity in microbodies subjected to qualitative cytochemical assays for glycolate oxidase and malate synthase (BURKE and TRELEASE 1975), during the transition from glyoxysomal to leaf-peroxisomal function. In fact, there is so far no physical or chemical feature of the isolated organelles or its membranes demonstrating significant differences that could be used for a discrimination between different peroxisome types by biochemical or physical means (BROWN et al. 1974, HONG and SCHOPFER 1980, THEIMER and THEIMER 1975). Positive evidence for the repackaging model, especially for the postulated *specific* introduction of enzymes into and the simultaneous *specific* disappearance of others from the peroxisomal matrix and membrane without loss of the organelle's structural integrity is notably lacking. However, while the latter is rather difficult to imagine in biochemical terms at present, some recent findings suggest that a specific import of proteins from the cytosolic fraction of the cytoplasm into the peroxisomes is, in principle; conceivable. In rat liver, catalase and urate oxidase polypeptides are synthesized exclusively by cytosolic polysomes and post-translationally transferred in their original size to the nascent (?) peroxisomes via cytosolic pools (GOLDMAN and BLOBEL 1978). Similar results have been obtained with isocitrate lyase of acetate-induced *Neurospora* cells ZIMMERMANN and NEUPERT 1980). Also with respect to plant peroxisomes there is evidence for cytosolic precursor pools of peroxisomal enzymes (KINDL et al. 1980a, b, KÖLLER and KINDL 1980, FREVERT et al. 1980, WALK and HOCK 1978, RIEZMAN et al. 1980). Furthermore, RIEZMAN and BECKER (1980) recently reported on successful attempts to introduce polypeptides, synthesized in vitro on cotyledon RNA, into intact cucumber peroxisomes and KRUSE et al. (1981) presented similar evidence for malate synthase. It appears likely, therefore, that at least some matrix proteins of the peroxisome are post-translationally transported through the peroxisome membrane. On the other hand, it is hard to circumvent the assumption that at least the glycosylated polypeptides of the peroxisomal membrane are supplied via the ER route. At any rate, the recent findings probably demand a modification of the classical view of peroxisome biogenesis (BEEVERS 1979), according to which these organelles originate from vesiculating buds of proliferating ER, loaded exclusively with proteins that have been first discharged by membrane-bound polysomes into the membrane or the intracisternal space of the ER (LORD and BOWDEN 1978).

Verification of the *replacement model* of BEEVERS and collaborators has been obstructed so far by the amazing morphological and biochemical similarity of intact peroxisomes in vitro. Especially the absence of any utilizable shift in the buoyant density on a sucrose gradient during the functional transition (HONG and SCHOPFER 1980, GERHARDT 1973, KÖLLER and KINDL 1978) has prevented the straightforward testing of this model, which instead had to be based on indirect evidence such as the demonstration of turnover of peroxisome constituents (KAGAWA et al. 1975). Particular emphasis was directed to the finding that aged etiolated watermelon cotyledons, upon irradiation with white light, are capably or producing peroxisomes with increased leaf-peroxisomal enzyme levels although the glyoxysomal enzyme activities and most of the peroxisomal protein had largerly disappeared (KAGAWA et al. 1973, KAWAGA and BEEVERS 1975). These results are obviously in conflict with the repackaging model. On the other hand, however, the proponents of the replacement model are forced to ignore the apparent enzymic homogeneity of peroxisomes in the histochemical experiments of BURKE and TRELEASE (1975). Furthermore, one should be aware that this model in its strict form negates the basic fact that all known types of peroxisomes contain more than just one of the three independent functional sets of enzymes shown in Fig. 3. This is especially true for the most typical "glyoxysome" of castor bean endosperm which simultaneously contains marker enzymes of all three peroxisomal functions in about constant proportions throughout its existence (SCHNARRENBERGER et al. 1971, THEIMER and THEIMER 1975). Of course, this argument applies equally to the rat liver peroxisome which appears to differ from the endosperm "glyoxysome" primarily with respect to the historic succession of disclosure of the various enzymic functions.

The *enzyme synthesis changeover model* of SCHOPFER and collaborators shifts the emphasis to the control of the synthesis of specific groups of enzymes. The basic idea is that a changeover from the synthesis of one group of enzymes to the synthesis of another group of enzymes—both channeled to the nascent peroxisomes—necessarily leads to a gradual shift in the enzymic properties of the newly made organelles. However, because of a basic organelle turnover, each particle contributes to the peroxisomal population of the cell for only a limited period of time. Depending on the peroxsiome half-life, there will eventually be a corresponding shift of the population from one functional state to another. Biochemical testimony for this model, postulating a fluent instead of a distinct heterogeneity of cotyledon peroxisomes during the transition, obviously also depends on the availability of criteria which would allow the physical separation of peroxisomes with differing protein compositions. In the apparent absence of such criteria a cytological approach has been introduced, exploiting the dramatic effect of phytochrome (continuous far-red light), in the absence of greening, on leaf-peroxisomal enzymes and the intracellular localization of peroxisomes in mustard cotyledons (Fig. 5). It has previously been shown (GRUBER et al. 1970, TRELEASE et al. 1971) that the transition of the peroxisome population of greening cotyledons from the (preferentially) glyoxysomal to the (preferentially) leaf-peroxisomal state is accompanied by a decrease in the number of peroxisome/oleosome associations and a concomitant increase in the number of peroxisome/plastid associations. Accepting these spatial associations as criteria for functional peroxisome specification it was then possible, by means of electron microscopy, to discriminate between peroxisomes engaged in glyoxysomal function and peroxisomes (potentially) engaged in leaf-peroxisomal function, and to study the dynamics of their appearance and disappearance. Pursuing the time courses of the frequencies of these associations in darkness and far-red light revealed that phytochrome induces, concomitant to pertinent enzymic changes, a decrease in the fraction of peroxisomes associated exclusively with oleosomes. Simultaneously, there was a strong phytochrome-mediated increase in the fraction of peroxisomes associated with both oleosomes and plastids (see Fig. 5b), reaching a level of about 40% after 24 h. However, peroxisomes which appeared associated exclusively with plastids did not increase beyond the 5% level before 36 h after the onset of phytochrome action (SCHOPFER et al. 1976). These results led to the conclusion that phytochrome induces initially the formation of peroxisomes of intermediary character (so-called "glyoxyperoxisomes"), followed, in due course, by the formation of peroxisomes preferentially designed for leaf-peroxisomal function. It is the transitory appearance, with proper timing, of this intermediary stage together with the undisputable existence of peroxisomal turnover in plant and animal cells that has led to the proposal of this model which, incidentally, encompasses major aspects of the replacement model as a limiting case realized specifically in aged etiolated cotyledons (KAGAWA and BEEVERS 1975). In support of this concept, KÖLLER and KINDL (1978) have demonstrated that the accumulation, as well as the synthesis, of glyoxysomal malate synthase continues undistrubed when the transition to leaf-peroxisomal function is induced by light before the peak of glyoxysomal activity is reached in cucumber cotyledons. Although compatible with the possibility that there is a temporally overlapping biogenesis of two different peroxisomal populations (a case not provided for in the original replacement model), this result strongly argues against any selective degradation of glyoxysomal enzymes or particles as the leaf-peroxisomal enzymes are being produced. Pertinent experiments concerning the synthesis of catalase (BETSCHE and GERHARDT 1978) and isocitrate lyase (FRANZISKET and GERHARDT 1980) during the transition in sunflower cotyledons have been reported. It should be emphasized that, according to the enzyme synthesis changeover model, the turnover of peroxisomes is not a prerequisite for the formation of a different sort of organelles. In fact, this turnover may be completely absent during the initial period of increasing glyoxysomal activity. Turnover is required, however, to account for the disappearance of glyoxysomal features after this point.

Although probably not yet fully pleasing to the biochemist, it should be kept in mind that the enzyme synthesis changeover model appears to be the only one that is compatible with all present data, accounting unforcibly in particular for those pieces of evidence which are in conflict with either one of the two other models. Furthermore, by displacing the decisive regulatory switch operated by phytochrome to the sites of differential enzyme synthesis—which are well established targets of phytochrome action (SCHOPFER 1977; see Chap. 10, this Vol.)—, the enzyme synthesis changeover model requires the least complicated mechanistic assumptions to overcome the problems inherent in the specificity of peroxisomal enzyme replacement. Finally, it should be pointed out that this model, although illustrated originally using current ideas of peroxisome assembly and desintegration (SCHOPFER et al. 1976), is basically independent of the biochemical details of these processes.

In conclusion, the experimental evidence, as well as Occam's razor, favor the idea that the profound functional changes of this organelle can be accomplished by phytochrome through an amplification in the formation of specific enzymes which are fed into an unaltered line of organelle assembly and desintegration.

References

Afzelius BA (1965) The occurrence and structure of microbodies. A comparative study. J Cell Biol 26:835–843

Apel K (1979) Phytochrome-induced appearance of mRNA activity for the apoprotein of the light-harvesting chlorophyll a/b protein of barley (*Hordeum vulgare*). Eur J Biochem 97:183–188

Apel K (1981) The protochlorophyllide holochrome of barley (*Hordeum vulgare* L.). Phytochrome induced decrease of translatable mRNA coding for the NADPH-protochlorophyllide oxidoreductase. Eur J Biochem 120:89–93

Apel K, Kloppstech K (1978) The plastid membranes of barley (*Hordeum vulgare*). Light-induced appearance of mRNA coding for the apoprotein of the light-harvesting chlorophyll a/b protein. Eur J Biochem 85:581–588

Apel K, Kloppstech K (1980) The effect of light on the biosynthesis of the light-harvesting chlorophyll a/b protein. Evidence for the requirement of chlorophyll a for the stabilization of the apoprotein. Planta 150:426–430

Apel K, Santel HJ, Redlinger TE, Falk H (1980) The protochlorophyllide holochrome of barley (*Hordeum vulgare* L.). Isolation and characterization of the NADPH:protochlorophyllide oxidoreductase. Eur J Biochem 111:251–258

Armond PA, Staehelin LA, Arntzen CJ (1977) Spatial relationship of photosystem I, photosystem II, and the light-harvesting complex in chloroplast membranes. J Cell Biol 73:400–418

Augustinussen E, Madsen A (1965) Regeneration of protochlorophyll in etiolated barley seedlings following different light treatments. Physiol Plant 18:828–837

Bain JB, Mercer FV (1966a) Subcellular organization of the developing cotyledons of *Pisum sativum* L. Aust J Biol Sci 19:49–67

Bain JM, Mercer FV (1966b) Subcellular organization of the cotyledons in germinating seeds and seedlings of *Pisum sativum* L. Aust J Biol Sci 19:69–84

Bajracharya D, Schopfer P (1979) Effect of light on the development of glyoxysomal functions in the cotyledons of mustard (*Sinapis alba* L.) seedlings. Planta 145:181–186

Bajracharya D, Falk H, Schopfer P (1976) Phytochrome-mediated development of mitochondria in the cotyledons of mustard (*Sinapis alba* L.) seedlings. Planta 131:253–261

Barraclough R, Ellis RJ (1979) The biosynthesis of ribulosebisphosphate carboxylase. Uncoupling of the synthesis of the large and small subunits in isolated soybean leaf cells. Eur J Biochem 94:165–177

Batschauer A, Santel HJ, Apel K (1982) The presence and synthesis of the NADPH: protochlorophyllide oxidoreductase in barley leaves with a high temperature-induced deficiency of plastid ribosomes. Planta 154:459–464

Becker WM, Leaver CJ, Weir EM, Riezman H (1978) Regulation of glyoxysomal enzymes during germination of cucumber. I. Developmental changes in cotyledonary protein, RNA, and enzyme activities during germination. Plant Physiol 62:542–549

Bedbrook JR, Smith SM, Ellis RJ (1980) Molecular cloning and sequencing of cDNA encoding the precursor to the small subunit of chloroplast ribulose-1,5-bisphosphate carboxylase. Nature 287:692–697

Beer NS, Griffiths WT (1981) Purification of the enzyme NADPH-protochlorophyllide oxidoreductase. Biochem J 195:83–92

Beevers H (1979) Microbodies in higher plants. Annu Rev Plant Physiol 30:159–193

Bennett J (1981) Biosynthesis of the light-harvesting chlorophyll a/b protein. Polypeptide turnover in darkness. Eur J Biochem 118:61–70

Betsche T, Gerhardt B (1978) Apparent catalase synthesis in sunflower cotyledons during the change in microbody function. A mathematical approach for the quantitative evaluation of density-labeling data. Plant Physiol 62:590–597

Boardman NK (1966) Protochlorophyll. In: Vernon LP, Seely GR (eds) The chlorophylls. Academic Press, London New York, pp 437–476

Boardman NK, Anderson JM (1978) Composition, structure and photochemical activity of developing and mature chloroplasts. In: Akoyunoglou G, Argyroudi-Akoyunoglou (eds) Chloroplast development. Elsevier North-Holland, New York, pp 1–14

Boardman NK, Anderson JM, Goodchild DJ (1978) Chlorophyll-protein complexes and structure of mature and developing chloroplasts. Curr Top Bioenerg 8:35–109

Böck P, Kramar R, Pavelka M (1980) Peroxisomes and related particles in animal tissues. Cell Biology Monographs Vol 7. Springer, Wien New York

Bogorad L (1967) Biosynthesis and morphogenesis in plastids. In: Goodwin TW (ed) Biochemistry of chloroplasts. Academic Press, London New York, pp 615–631

Bogorad L (1975) Evolution of organelles and eukaryotic genomes. Science 188:891–898

Bogorad L (1976) Chlorophyll biosynthesis. In: Goodwin TW (ed) Chemistry and biochemistry of plant pigments, 2nd edn, Vol 1. Academic Press, London New York, pp 64–148

Bogorad L, Laber L, Gassman M (1967) Aspects of chloroplast development: Transitory pigment-protein complexes and protochlorophyllide regeneration. In: Shibata K, Takamiya A, Jagendorf AT, Fuller RC (eds) Comparative biochemistry and biophysics of photosynthesis. Univ Park Press, Baltimore, pp 299–312

Bradbeer JW, Ireland HMM, Smith JW, Rest J, Edge HJW (1974a) Plastid development in primary leaves of *Phaseolus vulgaris*. VII. Development during growth in continuous darkness. New Phytol 73:263–270

Bradbeer JW, Gyldenholm AO, Smith JW, Rest J, Edge HJW (1974b) Plastid development in primary leaves of *Phaseolus vulgaris*. IX. The effects of short light treatments on plastid development. New Phytol 73:281–290

Breidenbach RW, Beevers H (1967) Association of the glyoxylate cycle enzymes in a novel subcellular particle from castor bean endosperm. Biochem Biophys Res Commun 27:462–469

Brown RH, Lord JM, Merrett MJ (1974) Fractionation of the proteins of plant microbodies. Biochem J 144:559–566

Brüning K, Drumm H, Mohr H (1975) On the role of phytochrome in controlling enzyme levels in plastids. Biochem Physiol Pflanz 168:141–156

Burke JJ, Trelease RN (1975) Cytochemical demonstration of malate synthase and glycolate oxidase in microbodies of cucumber cotyledons. Plant Physiol 56:710–717

Castelfranco PA, Jones OTG (1975) Protoheme turnover and chlorophyll synthesis in greening barley tissue. Plant Physiol 55:485–490

Cedel TE, Roux SJ (1980a) Further characterization of the in vitro binding of phytochrome to a membrane fraction enriched for mitochondria. Plant Physiol 66:696–703

Cedel TE, Roux SJ (1980b) Modulation of a mitochondrial function by oat phytochrome in vitro. Plant Physiol 66:704–769

Chua NH, Schmidt GW (1978) Post-translational transport into intact chloroplasts of a precursor to the small subunit of ribulose-1,5-bisphosphate carboxylase. Proc Natl Acad Sci USA 75:6110–6114

Chua NH, Schmidt GW (1979) Transport of proteins into mitochondria and chloroplasts. J Cell Biol 81:461–483

Coen D, Bedbrock JR, Bogorad L, Rich A (1977) Maize chloroplast DNA fragment encoding the large subunit of ribulosebisphosphate carboxylase. Proc Natl Acad Sci USA 74:5487–5491

Cohen RZ, Goodwin TW (1962) The effect of red and far-red light on carotenoid synthesis by etiolated maize seedlings. Phytochemistry 1:67–72

Cooke RJ, Saunders PF (1975) Phytochrome-mediated changes in extractable gibberellin activity in a cell-free system from etiolated wheat leaves. Planta 123:299–302

Cooke RJ, Kendrick RE (1976) Phytochrome controlled gibberellin metabolism in etioplast envelopes. Planta 131:303–307

De Duve C (1965) Function of microbodies (peroxisomes). J Cell Biol 27:25 A–26 A

Drumm H, Schopfer P (1974) Effect of phytochrome on development of catalase activity and isoenzyme pattern in mustard (*Sinapis alba* L.) seedlings. A reinvestigation. Planta 120:13–30

Duggan J, Gassman ML (1974) Induction of porphyrin synthesis in etiolated bean leaves by chelators of iron. Plant Physiol 53:206–215

Ellis RJ (1977) Protein synthesis by isolated chloroplasts. Biochim Biophys Acta 463:185–215

Ellis RJ (1979) The most abundant protein in the world. Trends Biochem Sci 4:241–244

Ellis RJ (1981) Chloroplast proteins: synthesis, transport and assembly. Annu Rev Plant Physiol 32:111–137

Evans A, Smith H (1976) Localization of phytochrome in etioplasts and its regulation in vitro of gibberellin levels. Proc Natl Acad Sci USA 73:138–142

Feierabend J (1975) Developmental studies on microbodies in wheat leaves. III. On the photocontrol of microbody development. Planta 123:63–77

Feierabend J, Beevers H (1972a) Developmental studies on microbodies in wheat leaves. I. Conditions influencing enzyme development. Plant Physiol 49:28–32

Feierabend J, Beevers H (1972b) Developmental studies on microbodies in wheat leaves. II. Ontogeny of particulate enzyme associations. Plant Physiol 49:33–39

Feierabend J, Pirson A (1966) Die Wirkung des Lichts auf die Bildung von Photosyntheseenzymen in Roggenkeimlingen. Z Pflanzenphysiol 55:235–245

Fluhr R, Harel E, Klein S, Meller E (1975) Control of aminolevulinic acid and chlorophyll accumulation in greening maize leaves upon light–dark transitions. Plant Physiol 56:497–501

Ford MJ, Kasemir H (1980) Correlation between 5-aminolaevulinate accumulation and protochlorophyll photoconversion. Planta 50:206–210

Franzisket U, Gerhardt B (1980) Synthesis of isocitrate lyase in sunflower cotyledons during the transition in cotyledonary microbody function. Plant Physiol 65:1081–1084

Frederick SE, Gruber PJ, Newcomb EH (1975) Plant Microbodies. Protoplasma 84:1–29

Frevert J, Köller W, Kindl H (1980) Occurrence and biosynthesis of glyoxysomal enzymes in ripening cucumber seeds. Hoppe-Seyler's Z Physiol Chem 361:1557–1565

Frosch S, Bergfeld R, Mohr H (1976) Light control of plastogenesis and ribulosebisphosphate carboxylase levels in mustard seedlings cotyledons. Planta 133:53–56

Gerhardt B (1973) Untersuchungen zur Funktionsänderung der Microbodies in den Kotyledonen von *Helianthus annuus* L. Planta 110:15–28

Gerhardt B (1978) Microbodies/Peroxisomen pflanzlicher Zellen. Cell Biology Monographs Vol 5. Springer, Wien New York

Goldman BM, Blobel G (1978) Biogenesis of peroxisomes: Intracellular site of synthesis of catalase and uricase. Proc Natl Acad Sci USA 75:5066–5070

Graham D, Grieve AM, Smillie RM (1968) Phytochrome as the primary photoregulator of the synthesis of Calvin cycle enzymes in etiolated pea seedlings. Nature 218:89–90

Graham D, Hatch MD, Slack CR, Smillie RM (1970) Light-induced formation of enzymes of the C_4-dicarboxylic acid pathway of photosynthesis in detached leaves. Phytochemistry 9:521–532

Graham D, Grieve AM, Smillie RM (1971) Phytochrome-mediated plastid development in etiolated pea stem apices. Phytochemistry 10:2905–2914

Granick S (1959) Magnesium prophyrin is formed in barley seedlings treated with delta-aminolevulinic acid. Plant Physiol Suppl 34:18

Gressel J (1978) Light requirements for the enhanced synthesis of a plastid mRNA during *Spirodela* greening. Photochem Photobiol 27:167–169

Griffiths WT (1974) Source of reducing equivalents for the in vitro synthesis of chlorophyll from protochlorophyll. FEBS Lett 46:301–304

Griffiths TW (1978) Reconstitution of chlorophyllide formation by isolated etioplast membranes. Biochem J 174:681–692

Grossman A, Bartlett S, Chua NH (1980) Energy-dependent uptake of cytoplasmically synthesized polypeptides by chloroplasts. Nature 285:625–628

Gruber PJ, Trelease RN, Becker WM, Newcomb EH (1970) A correlative ultrastructural and enzymatic study of cotyledonary microbodies following germination of fat-storing seeds. Planta 93:269–288

Gruber PJ, Becker WM, Newcomb EH (1972) The occurrence of microbodies and peroxisomal enzymes in achlorophyllous leaves. Planta 105:114–138

Gruber PJ, Becker WM, Newcomb EH (1973) The development of microbodies and peroxisomal enzymes in greening bean leaves. J Cell Biol 56:500–518

Hagemann R, Börner T (1978) Plastid-ribosome-deficient mutants of higher plants as a tool in studying chloroplast biogenesis. In: Akoyunoglou G, Argyroudi-Akoyunoglou JH (eds) Chloroplast development. Elsevier North-Holland, Amsterdam, pp 709–720

Hampp R, Wellburn AR (1979) Control of mitochondrial activities by phytochrome during greening. Planta 147:229–235

Harel E (1978) Chlorophyll biosynthesis and its control. In: Reinhold L, Harbone JB, Swain T (eds) Progress in phytochemistry. Pergamon, New York, pp 127–180

Haslett BG, Cammack R (1974) The development of plastocyanin in greening bean leaves. Biochem J 144:567–572

Hendry GAF, Stobbart AK (1978) The effect of haem on chlorophyll synthesis in barley leaves. Phytochemistry 17:73–77

Highfield PE, Ellis RJ (1978) Synthesis and transport of the small subunit of chloroplast ribulose bisphosphate carboxylase. Nature 271:420–424

Hock B (1969) Die Hemmung der Isocitrat-Lyase bei Wassermelonenkeimlingen durch Weißlicht. Planta 85:340–350

Hock B, Mohr H (1964) Die Regulation der O_2-Aufnahme von Senfkeimlingen (*Sinapis alba* L.) durch Licht. Planta 61:209–228

Hong YN, Schopfer P (1980) Density of microbodies on sucrose gradients during phytochrome-mediated glyoxysome peroxisome transformation in cotyledons of mustard seedlings. Plant Physiol 66:194–196

Hong YN, Schopfer P (1981) Control by phytochrome of urate oxidase and allantoinase activities during peroxisome development in the cotyledons of mustard (*Sinapis alba* L.) seedlings. Planta 152:325–335

Høyer-Hansen G, Simpson DJ (1977) Changes in the polypeptide composition of internal membranes of barley plastids during greening. Carlsberg Res Commun 42:379–389

Huang AHC, Beevers H (1971) Isolation of microbodies from plant tissues. Plant Physiol 48:637–641

Jabben M, Mohr H (1975) Stimulation of the Shibata shift by phytochrome in the cotyledons of the mustard seedling *Sinapis alba* L. Photochem Photobiol 22:59–61

Jabben M, Masoner M, Kasemir H, Mohr H (1974) Phytochrome-stimulated regeneration of protochlorophyll in cotyledons of the mustard seedlings. Photochem Photobiol 20:233–239

Jones CT (1980) Is there a glyoxylate cycle in the liver of the fetal guinea pig? Biochem Biophys Res Commun 95:849–856

Kagawa T, Beevers H (1975) The development of microbodies (glyoxysomes and leaf peroxisomes) in cotyledons of germinating watermelon seedlings. Plant Physiol 55:258–264

Kagawa T, McGregor DI, Beevers H (1973) Development of enzymes in the cotyledons of watermelon seedlings. Plant Physiol 51:66–71

Kawaga T, Lord JM, Beevers H (1975) Lecithin synthesis during microbody biogenesis in watermelon cotyledons. Arch Biochem Biophys 167:45–53

Kannangara CG (1969) The formation of ribulose diphosphate carboxylase protein during chloroplast development in barley. Plant Physiol 44:1533–1537

Kannangara CG, Gough SP (1979) Biosynthesis of δ-aminolevulinate in greening barley leaves II: Induction of enzyme synthesis by light. Carlsberg Res Commun 44:11–20

Karow H, Mohr H (1967) Aktivitätsänderungen der Isocitritase (EC 4.1.3.1.) während der Photomorphogenese beim Senfkeimling (*Sinapis alba* L.). Planta 72:170–186

Kasemir H, Prehm G (1976) Control of chlorophyll synthesis by phytochrome. III. Does phytochrome regulate the chlorophyllide esterification in mustard seedlings? Planta 132:291–295

Kasemir H, Oberdorfer U, Mohr H (1973) A twofold action of phytochrome in controlling chlorophyll a accumulation. Photochem Photobiol 18:481–486

Khan FR, McFadden BA (1980) Embryogenesis and the glyoxylate cycle. FEBS Lett 115:312–314

Kindl H, Majunke G (1973) Glyoxysomale Enzyme in Laubblättern von *Lens culinaris*. Hoppe-Seyler's Z Physiol Chem 354:999–1005

Kindl H, Köller W, Frevert J (1980a) Cytosolic precursor pools during glyoxysome biosynthesis. Hoppe-Seyler's Z Physiol Chem 361:465–467

Kindl H, Schiefer S, Löffler HG (1980b) Occurrence and biosynthesis of catalase at different stages of seed maturation. Planta 148:199–207

Kirk JTO (1974) The relation of chlorophyll synthesis to protein synthesis in the growing thylakoid membrane. Port Acta Biol Ser A 14:127–152

Kirk JTO, Tilney-Bassett RAE (1978) The plastids. Their chemistry, structure, growth and inheritance, 2nd edn. Elsevier North-Holland, Amsterdam

Klein AO (1969) Persistent photoreversibility of leaf development. Plant Physiol 44:897–902

Klein S, Pollok BM (1968) Cell fine structure of developing lima bean seeds related to seed desiccation. Am J Bot 55:658–672

Klein WH, Price L, Mitrakos K (1963) Light-stimulated starch degradation in plastids and leaf morphogenesis. Photochem Photobiol 2:233–240

Klein S, Katz E, Neeman E (1977) Induction of δ-aminolevulinic acid formation in etiolated maize leaves controlled by two light systems. Plant Physiol 60:335–338

Kleinkopf GE, Huffaker RC, Matheson A (1970) Light-induced de novo synthesis of ribulose 1,5-diphosphate carboxylase in greening leaves of barley. Plant Physiol 46:416–418

Kobayashi H, Asami S, Akazawa T (1980) Development of enzymes involved in photosynthetic carbon assimilation in greening seedlings of maize (*Zea mays*). Plant Physiol 65:198–203

Köller W, Kindl H (1978) Studies supporting the concept of glyoxyperoxisomes as intermediary organelles in transformation of glyoxysomes into peroxisomes. Z Naturforsch 33c:962–968

Köller W, Kindl H (1980) 19S cytosolic malate synthase. A small pool characterized by rapid turnover. Hoppe Seyler's Z Physiol Chem 361:1437–1444

Kollöffel C (1970) Oxidative and phosphorylative activity of mitochondria from pea cotyledons during maturation of the seed. Planta 91:321–328

Koski VM, French SL, Smith FHC (1951) The action spectrum for the transformation of protochlorophyll to chlorophyll a in normal and albino corn seedlings. Arch Biochem Biophys 31:1–17

Kruse C, Frevert J, Kindl H (1981) Selective uptake by glyoxysomes of in vitro translated malate synthase. FEBS-Lett 129:36–38

Kung SD, Thornber JP, Wildman SG (1972) Nuclear DNA codes for the photosystem II chlorophyll-protein of chloroplast membranes. FEBS Lett 24:185–188

Lado P, Schwendimann M, Marrè E (1968) Repression of isocitrate lyase synthesis in seeds germinated in the presence of glucose. Biochim Biophys Acta 157:140–148

Lichtenthaler HK, Kleudgen HK (1975) Phytochromsystem und Lipochinonsynthese in den Plastiden etiolierter *Hordeum*-Keimlinge. Z Naturforsch 30c:64–66

Link G (1981) Enhanced expression of a distinct plastid DNA region in mustard seedlings by continuous far-red light. Planta 152:379–380

Link G (1982) Phytochrome control of plastid mRNA in mustard (*Sinapis alba* L.). Planta 154:81–86

Lord JM, Bowden L (1978) Evidence that glyoxysomal malate synthase is segregated by the endoplasmic reticulum. Plant Physiol 61:266–270

Lütz C (1978) Separation and comparison of prolamellar bodies and prothylakoids of etioplasts from *Avena sativa* L. In: Akoyunoglou G (ed) Chloroplast development. Elsevier North-Holland, Amsterdam, pp 481–488

Machold O, Aurich O (1972) Sites of synthesis of chloroplast lamellar proteins in *Vicia faba*. Biochim Biophys Acta 281:103–112

Malhotra SS, Spencer M (1973) Structural development during germination of different populations of mitochondria from pea cotyledons. Plant Physiol 52:575–579

Malnoe P, Rochaix JD, Chua NH, Spahr PF (1979) Characterization of the gene and messenger RNA of the large subunit of ribulose 1,5-diphosphate carboxylase in *Chlamydomonas reinhardii*. J Mol Biol 133:417–434

Manabe K, Furuya M (1973) A rapid phytochrome-dependent reduction of nicotinamide

adenine dinucleotide phosphate in particle fraction from etiolated bean hypocotyl. Plant Physiol 51:982–983
Manabe K, Furuya M (1974) Phytochrome-dependent reduction of nicotinamide nucleotides in the mitochondrial fraction isolated from etiolated pea epicotyls. Plant Physiol 53:343–347
Manabe K, Furuya M (1975) Experimentally induced binding of phytochrome to mitochondrial and microsomal fractions in etiolated pea shoots. Planta 123:207–215
Mapleston RE, Griffiths WT (1980) Light modulation of the activity of protochlorophyllide reductase. Biochem J 189:125–133
Marcus A (1960) Photocontrol of formation of red kidney bean leaf triphosphopyridine nucleotide linked triosephosphate dehydrogenase. Plant Physiol 35:126–128
Markwell JP, Thornber JP, Boogs RT (1979) Higher plant chloroplasts: Evidence that all the chlorophyll exists as chlorophyll–protein complexes. Proc Natl Acad Sci USA 76:1233–1235
Masoner M, Kasemir H (1975) Control of chlorophyll synthesis by phytochrome. I. The effect of phytochrome on the formation of 5-aminolevulinate in mustard seedlings. Planta 126:111–117
Mego JL, Jagendorf AT (1961) Effect of light on growth of black valentine bean plastids. Biochim Biophys Acta 53:237–254
Mettler IJ, Beevers H (1980) Oxidation of NADH in glyoxysomes by a malate-aspartate shuttle. Plant Physiol 66:555–560
Mohr H (1977) Phytochrome and chloroplast development. Endeavour, New Ser 1:107–114
Mohr H (1983) Phytochrome and chloroplast development. In: Barber J, Baker NR (eds) Topics in photosynthesis Vol. 5. Elsevier, Amsterdam (in press)
Mohr H, Drumm H, Kasemir H (1974) Licht und Farbstoffe. Ber Dtsch Bot Ges 87:49–69
Morisset C (1973) Comportement des racines isolées de *Lycopersicum esculentum* (Solanacées) cultivées in vitro et soumises à l'anoxie. CR Acad Sci (Paris) Sér D, 276:311–314
Morisset C (1974) Variations ultrastructurales comparées des mitochondries dans des racines de *Lycopersicum esculentum* Mill. (Solancées), isolées, cultivées in vitro et soumises à divers traitements: anoxie prolongée, agent découplant, plasmolyse. CR Acad Sci (Paris) Ser D, 279:1257–1260
Murray DR, Wara-Aswapati, O, Ireland MM, Bradbeer JW (1973) The development of activities of some enzymes concerned with glycolate metabolism in greening bean leaves. J Exp Bot 24:175–184
Nadler K, Granick S (1970) Controls on chlorophyll synthesis in barley. Plant Physiol 46:240–246
Oelze-Karow H, Mohr H (1978a) Control of chlorophyll b biosynthesis by phytochrome. Photochem Photobiol 27:189–193
Oelze-Karow H, Mohr H (1978b) Phytochrome control of the development of photophosphorylation. Photochem Photobiol 27:255–258
Ogawa T, Inoue V, Kitajina M, Shibata K (1973) Action spectra for biosynthesis of chlorophylls a and b and β-carotene. Photochem Photobiol 18:229–235
Öpik H (1974) Mitochondria. In: Robards AW (ed) Dynamic aspects of plant ultrastructure. McGraw-Hill, New York, pp 52–83
Price L, Klein WH (1961) Red, far-red response and chlorophyll synthesis. Plant Physiol 36:733–735
Queiroz O (1969) Photopériodisme et activité enzymatique (PEP carboxylase et enzyme malique) dans les feuilles de *Kalanchoë blossfeldiana*. Phytochemistry 8:1655–1663
Richter G, Dirks W (1978) Blue-light-induced development of chloroplasts in isolated seedling roots. Preferential synthesis of chloroplast ribosomal RNA species. Photochem Photobiol 27:155–160
Riezman H, Becker WM (1980) Post-translational processing of glyoxysomal enzymes. Plant Physiol Suppl 65:32
Riezman H, Weir EM, Leaver CJ, Titus DE, Becker WM (1980) Regulation of glyoxysomal enzymes during germination of cucumber. 3. In vitro translation and characterization of four glyoxysomal enzymes. Plant Physiol 65:40–46

Santel HJ, Apel K (1981) The protochlorophyllide holochrome of barley (*Hordeum vulgare* L.). The effect of light on the NADPH: protochlorophyllide oxidoreductase. Eur J Biochem 120:95–103

Schatz G (1970) Biogenesis of mitochondria. In: Racker E (ed) Membranes of mitochondria and chloroplasts. Van Nostrand, New York, pp 251–314

Schatz G (1979) How mitochondria import proteins from the cytoplasm. FEBS Lett 103:203–211

Schiff JA (1978) Photocontrol of chloroplast development in *Euglena*. In: Akoyunoglou G, Argyroudi-Akoyunoglou (eds) Chloroplast development. Elsevier North-Holland, Amsterdam, pp 747–768

Schmidt GW, Bartlett S, Grossman AR, Cashmore AR, Chua NH (1980) In vitro synthesis, transport, and assembly of the constituent polypeptides of the light-harvesting chlorophyll a/b protein complex. In: Leaver CJ (ed) Genome organization and expression in plants. NATO ASI Ser A: Life Sci Vol 29. Plenum, New York London, pp 337–351

Schmidt HW, Hampp R (1977) Regulation of membrane properties of mitochondria and plastids during chloroplast development. II. The action of phytochrome in a cell-free system. Z Pflanzenphys 82:428–434

Schnarrenberger C, Mohr H (1970) Carotenoid synthesis in mustard seedlings as controlled by phytochrome and inhibitors. Planta 94:296–307

Schnarrenberger C, Oeser A, Tolbert NE (1971) Development of microbodies in sunflower cotyledons and castor bean endosperm during germination. Plant Physiol 48:566–574

Schneider HAW (1973) Regulation der Chlorophyllbiosynthese. Licht- und entwicklungsbedingte Aktivitätsveränderungen von vier aufeinanderfolgenden Enzymen der Porphyrin- und Chlorophyllbiosynthesekette. Z Naturforsch 28c:45–58

Schneider HAW (1975) Chlorophylle: Aspekte der Biosynthese und ihrer Regulation. Ber Dtsch Bot Ges 88:83–123

Schopfer P (1977) Phytochrome control of enzymes. Annu Rev Plant Physiol 28:223–252

Schopfer P, Bajracharya D, Falk H, Thien W (1975) Phytochrom-gesteuerte Entwicklung von Zellorganellen (Plastiden, Microbodies, Mitochondrien). Ber Dtsch Bot Ges 88:245–268

Schopfer P, Bajracharya D, Bergfeld R, Falk H (1976) Phytochrome-mediated transformation of glyoxysomes into peroxisomes in the cotyledons of mustard (*Sinapis alba* L.) seedlings. Planta 133:73–80

Sisler EC, Klein WH (1963) The effect of age and various chemicals on the lag phase of chlorophyll synthesis in dark grown bean seedlings. Physiol Plant 16:315–322

Smillie RM, Scott NS (1969) Organelle biosynthesis: The chloroplast. Progress in molecular and subcellular biology Vol 1. Springer, Berlin Heidelberg New York, pp 137–202

Smith MA, Criddle RS, Peterson L, Huffaker RC (1974) Synthesis and assembly of ribulosebisphosphate carboxylase enzyme during greening of barley plants. Arch Biochem Biophys 165:494–504

Smith SM, Ellis RJ (1981) Light-stimulated accumulation of transcripts of nuclear and chloroplast genes for ribulosebisphosphate carboxylase. J Mol Appl Genet 1:127–137

Sundqvist C (1978) Red light stimulated accumulation of protochlorophyllide in dark-grown leaves treated with δ-aminolevulinic acid. Plant Sci Lett 12:69–76

Theimer RR, Theimer E (1975) Studies on the development and localization of catalase and H_2O_2-generating oxidases in the endosperm of germinating castor beans. Plant Physiol 56:100–104

Thien W, Schopfer P (1975) Control by phytochrome of cytoplasmic and plastid rRNA accumulation in the cotyledons of mustard seedlings in the absence of photosynthesis. Plant Physiol 56:660–664

Thornber JP (1975) Chlorophyll-proteins: Light-harvesting and reaction center components of plants. Annu Rev Plant Physiol 26:127–158

Thronber JP, Highkin HR (1974) Composition of the photosynthetic apparatus of normal barley leaves and a mutant lacking chlorophyll b. Eur J Biochem 41:109–116

Tobin EM (1978) Light regulation of specific mRNA species in *Lemna gibba* L. G-3. Proc Natl Acad Sci USA 75:4749–4753

Tobin EM (1981) Phytochrome-mediated regulation of messenger RNAs for the small subunit of ribulose 1,5-bisphosphate carboxylase and the light-harvesting chlorophyll a/b protein in *Lemna gibba*. Plant Molecular Biology 1:35–51

Tolbert NE, Oeser A, Kisaki T, Hageman RH, Yamazaki RK (1968) Peroxisomes from spinach leaves containing enzymes related to glycolate metabolism. J Biol Chem 243:5179–5184

Tomomatsu A, Asahi T (1978) Non-synchronous increases in activities of peroxisomal enzymes in etiolated mung bean seedling leaves after illumination. Plant Cell Physiol 19:183–188

Trelease RN, Becker WM, Gruber PJ, Newcomb EH (1971) Microbodies (glyoxysomes and peroxysomes) in cucumber cotyledons. Plant Physiol 48:461–475

Unser G, Masoner M (1972) Kinetics of monogalactosyltransferase in mustard seedlings. Naturwissenschaften 59:39

Unser G, Mohr H (1970) Phytochrome-mediated increase of galactolipids in mustard seedlings. Naturwissenschaften 57:358–359

Vesk M, Jeffrey SW (1977) Effect of blue-green light on photosynthetic pigments and chloroplast structure in unicellular marine algae from six classes. J Phycol 13:280–288

Virgin HJ (1972) Chlorophyll biosynthesis and phytochrome action. In: Mitrakos K, Shropshire W (eds) Phytochrome. Academic Press, London New York, pp 371–404

Van Poucke M, Barthe F (1970) Induction of glycollate oxidase activity in mustard seedlings under the influence of continuous irradiation with red and far-red light. Planta 94:308–318

Van Poucke M, Cerff R, Barthe F, Mohr H (1970) Simultaneous induction of glycolate oxidase and glyoxylate reductase in white mustard seedlings by phytochrome. Naturwissenschaften 56:132–133

Vartapetian BB, Andreeva IN, Kozlova GI, Agapova LP (1977) Mitochondrial ultrastructure in roots of mesophyte and hydrophyte at anoxia and after glucose feeding. Protoplasma 91:243–256

Walk, RA, Hock B (1978) Cell-free synthesis of glyoxysomal malate dehydrogenase. Biochem Biophys Res Commun 81:636–643

Weir EM, Riezman H, Grienenberger JM, Becker WM, Leaver CJ (1981) Regulation of glyoxysomal enzymes during germination of cucumber. Temporal changes in translatable mRNAs for isocitrate lyase and malate synthase. Eur J Biochem 112:469–477

Weischet W (1971) cited in Friederich KE, Mohr H (1975) Adenosine 5′-triphosphate content and energy charge during photomorphogenesis of the mustard seedling *Sinapis alba* L. Photochem Photobiol 22:49–53

Withrow RB, Wolff JB, Price L (1956) Elimination of the lag phase of chlorophyll synthesis in dark-grown bean leaves by a pretreatment with low irradiances of monochromatic energy. Plant Physiol Suppl 31:13–14

Zimmermann R, Neupert W (1980) Biogenesis of glyoxysomes. Synthesis and intracellular transfer of isocitrate lyase. Eur J Biochem 112:225–233

12 Control of Plastid Development in Higher Plants

H.I. VIRGIN and H.S. EGNÉUS

1 Introduction

The term plastid (-plast) is used in the biological literature to define different kinds of organelles present in plant cells. Naming of the organelles has mainly been based on morphological characteristics, chemical composition, and on where in the cells they are found.

The present article will deal with plastid development in spermatophytes where one stage is characterized by having a photosynthetic competence.

Plastid development which does not include a chloroplast stage will not be treated. The development of non-green plastids has recently been treated in KIRK and TILNEY-BASSETT (1978), THOMSON and WHATLEY (1980).

Most schemes on plastid development, whether seen as linear (FREY-WYSSLING et al. 1956) or cyclic (CLOWES and JUNIPER 1968, WHEATLEY 1978) are based on data from ultrastructural work. Relatively little work has been done where ultrastructural changes have been correlated to the functional and compositional changes taking place during plastid development. The exception being the work done on the plastid development in etiolated dark-grown plants (see e.g. BISHOP 1974, BOARDMAN 1977a, KIRK and TILNEY-BASSETT 1978, WELLBURN and HAMPP 1979).

Plastid development has often been studied in short-term experiments, e.g., during the greening of dark-grown seedlings, during the development of a new leaf, but changes taking place during longer periods of time, months or even years should be considered when discussing control of plastid development (SENSER et al. 1975). The necessity of having a valid time perspective when plastid development is studied has been pointed out by among others LEECH et al. (1972), ROBERTSON and LAETSCH (1974).

From a physiological/ecological point of view it is often difficult to interpret data on control of plastid development because of the variable conditions under which the experiments were performed. To put it crudely, how relevant are data on chloroplast development which are obtained from very old leaves of dark-grown etiolated seedlings which have been nutrient-depleted? What does the excision of a leaf mean for chloroplast development?

The regulation of plastid development is determined by internal (e.g., hormones) and external factors (e.g., light), and it is therefore often difficult to observe the effect of a single factor. The interpretation of data on plastid development is furthermore complicated by the fact that the site of action of the factors is found on different levels (e.g., metabolic, structural). The relationship between data observed on different levels is in many cases not even pointed out.

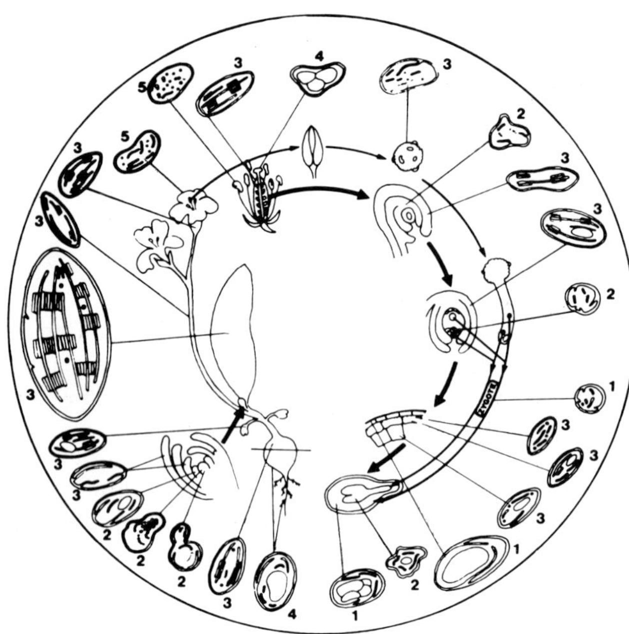

Fig. 1. A diagram of the type and distribution of plastids found in organs of a higher plant during successive stages in the plant life cycle from seed development to fertilization. *1* amyloplast; *2* proplastid; *3* chloroplast; *4* leucoplast; *5* chromoplast. (Leech 1976)

Plastid development in different plants and parts of plants is so diverse that one may question the existence of one common pathway. This is graphically described in Fig. 1, taken from Leech (1976), showing the site and type of plastid development which takes place in different parts of a plant during a growth period. The development of plastids can be driven in different directions depending on changing environmental conditions (Kirk and Tilney-Bassett 1978, Thomson and Whatley 1980, Schnepf 1980).

2 The Main Plastid Developmental Sequences

Development of a plastid can mean two things (Parthier et al. 1975, Leech 1976).

a) Plastid morphogenesis, where functions are developed, and chemical constituents are synthesized, inserted or removed from structures which are built up or broken down as the plastid changes in size. For detailed treatments of various aspects of chloroplast development see Kirk and Tilney-Bassett (1978), Reinert (1980).

b) A production of new plastids by division of plastids in various stages of development (Leech 1976, Kirk and Tilney-Bassett 1978, Possingham 1980).

We will only treat plastid development according to (a) although from an ecological point of view, development seen as division might be more relevant.

2.1 The Normal Sequence (Sequence I)

This sequence has been described for many different species, in all aboveground parts (stems, leaves, flowers, cotyledons) of green higher plants and can be summarized:

proplastid → intermediate stages → chloroplast → senescing chloroplast
(amyloplastidic)

In leaves of angiosperms at least the last steps of plastid development require light to form a chloroplast in the true sense, i.e., a chlorophyll-containing organelle. In gymnosperms it can take place in the dark (see below). The development should not be viewed as unidirectional (WHATLEY 1978). As an example, it was found that fully differentiated cells which contain chloroplasts can form proplastids from these chloroplasts when they regain meristematic activity (SCHNEPF 1980). In senescing cells, different types of plastids can be formed, partly depending upon the organ in which the development takes place and partly depending upon the age of the plant. Chromoplasts are often the final stage in the developmental sequence (BUTLER and SIMON 1971, SCHNEPF 1980). In an aging plant, the senescing chloroplast can revert to an active chloroplast if senescence is stopped, or has not gone too far (Fig. 2).

Ultrastructurally the development of plastids of varying age in a leaf is characterized by great changes in plastid size and internal structure (Fig. 3). A membrane system consisting of stroma and grana lamellae (thylakoids) is built up during the development from proplastids to chloroplasts. Initially the thylakoids are formed from the inner of the two proplastid envelopes and seen as invaginations from this membrane (KIRK and TILNEY-BASSETT 1978). In the later stages the thylakoid growth takes place from existing stroma and grana lamellae (ANDERSON 1981). The number of thylakoids and the ratio between the amount of stroma and grana lamellae are determined by external and internal factors.

Plants grown under different light regimes show large differences in thylakoid structure (BORDMAN 1977b, MEIER and LICHTENTHALER 1981). In extreme cases only one type of thylakoid can be observed. Almost only grana lamellae were found in plants growing at low light intensities in a tropical rain forest (ANDERSON et al. 1973). During senescence the lamellar system breaks down and plastoglobuli of varying size and number appear (Fig. 2). Finally the two plastid envelopes disintegrate. At different stages of development, ribosomes, DNA-containing fibrils, and several types of crystalline bodies can be observed.

The senescing chloroplast can in some instances be considered either as a storage organelle, where compounds of lipid nature are stored in plastoglobuli, or in some cases a preliminary stage for new chloroplasts (WHATLEY 1978).

Functionally the normal sequence is characterized by the development of photosynthetic competence although the data on photosynthesis of proplastids in green leaves are limited. With photosynthetic competence is not only meant electron transport and carbon dioxide assimilation, but also lipid, pigment, amino acid, and protein synthesis (GIVAN and HARWOOD 1976, ELLIS 1977).

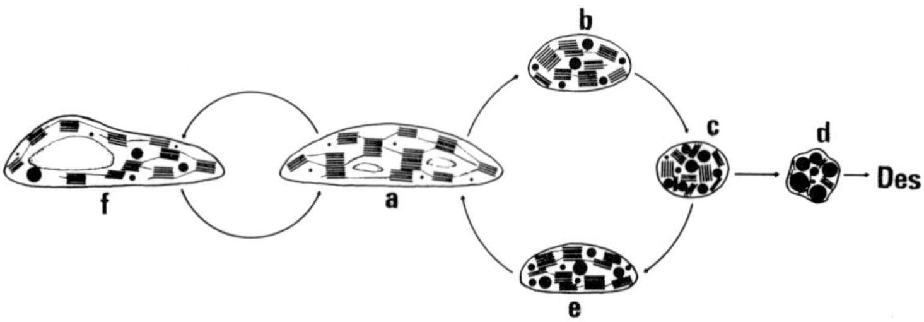

Fig. 2. A diagram of ultrastructural changes taking place during senescence and re-greening (*Nicotiana rustica*). **a** Normal chloroplast; **b, c** different stages of senescence (yellow stages); **d** irreversible senescence stage. **Des** desintegration; **e** beginning of re-greening; **f** chloroplast with starch formation. (LJUBEŠIĆ 1968)

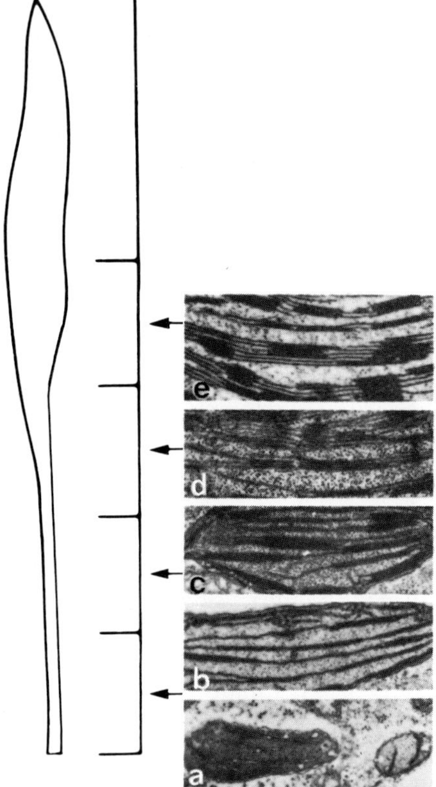

Fig. 3a–e. Plastids in progressively older cells of a green maize leaf. The plastids are in cells at measured distances from the leaf base. **a** 0 cm; **b** 0.5 cm; **c** 1.5 cm; **d** 3 cm; **e** 5 cm, (× 30000). (BAKER and LEECH 1977)

2.2 The Etioplast Sequence (Sequence II)

This type of plastid development can be found in stems and leaves of dark-grown plants, and in parts of plants which are hidden from light during the first stages and then irradiated. It can be summarized as:

proplastid → etioplast → chloroplast → senescing chloroplast

In darkness the development stops at the etioplast stage. Ultrastructurally and functionally the difference between this type of development and the one observed in green plants exposed to light is that an intermediate stage, the etioplast, is present. In the etioplast one or several prolamellar bodies (PLBs) and so-called prothylakoids (LÜTZ 1978) can be found. Upon irradiation stroma and grana lamellae are formed, prolamellar bodies disappear, and a photosynthetic competence is built up after a certain time in light, see, e.g. BOARDMAN (1977a), WELLBURN and HAMPP (1979). The etioplast has two functions; it is a preliminary stage to chloroplasts, and a carrier of chemical constituents necessary for chloroplast appearance. The fact that an etioplast very quickly turns into a photosynthetically competent organelle after irradiation indicates that all components necessary for the appearance of a chloroplast are present in the etioplast and only some regulatory factor(s) has to be operated upon by the light. All Calvin cycle enzymes (BRADBEER 1973), components such as cytochrome f, cytochrome b-563, plastocyanin (PLESNIČAR and BENDALL 1973), chloroplast ATPase (LOCKSHIN et al. 1971), coupling factor CF_1 (HERRMANN et al. 1980) and RuBPCase (BOTTOMLEY 1980) are present in etioplasts. Almost all polypeptides found in normal chloroplasts are found in etioplasts (GREBANIER et al. 1979). The etioplast lacks chlorophyll pigments but contains the pigment precursor (protochlorophyllide).

2.3 The Amyloplast Sequence (Sequence III)

This type of plastid development has been described for roots and underground organs (BJÖRN 1967, BADENHUIZEN 1971, THOMSON and WHATLEY 1980)

proplastids → amyloplasts → chloroplasts

In darkness the development stops at the proplastid stage. Ultrastructurally the proplastids do not differ from proplastids in the other sequences. Under greening conditions (long irradiation times), thylakoid formation takes place by invaginations from the inner of the two plastid envelopes.

2.4 The Sequence in Gymnosperms

Contrary to most higher plants, gymnosperms can form chlorophyll in darkness. This formation is restricted to the young seedlings and is low as compared to the formation in light. As is the case for angiosperms phytochrome in its far-red absorbing form is a prerequisite for maintaining a high rate of chlorophyll a accumulation in continuous light (KASEMIR and MOHR 1981). Morphologically the plastid development is similar to that described in sequence I.

3 Prolamellar Bodies

Prolamellar bodies are structures particularly characteristic for etioplasts. Their structure has been described by among others GUNNING and JAGOE (1967). PLBs are taxonomically widespread (KIRK and TILNEY-BASSETT 1978). Prolamellar bodies can be found in

a) dark-grown plants (GUNNING 1965)
b) in light-grown plants growing under a light/dark regime (SIGNOL 1961)
c) in plants which can form chlorophyll in darkness, i.e., gymnosperms (NIKOLIĆ and BOGDANOVIĆ 1972, WALLES and HUDÁK 1975)
d) in light-grown plants where the light intensity is low (WEIER and BROWN 1970, LEECH et al. 1972).

The appearance of PLBs is dependent on the age of the plant material. In young leaves a cycle of appearance and disappearance can be observed under dark/light transitions (SIGNOL 1961). In old leaves PLBs have not been observed. The correlation between PLB disappearance and thylakoid formation upon irradiation was for many years seen as two events directly coupled to each other. The preparative techniques which now allow a separation of PLBs from prothylakoids, which are seen as protruding from the PLBs (LÜTZ 1978), and investigations on the synthetic capacities (functions) of the inner of the two plastid envelopes (DOUCE and JOYARD 1979) have made this picture untenable. There seems to be very little protein associated with the crystalline part of PLBs. They seem to consist mainly of lipids and saponins (KESSELMEIER and BUDZIKIEWICZ 1979). The protochlorophyllide is in largest measure found in the prothylakoids. The function of the PLB is now considered to be a carrier for lipid components necessary for the development of the stroma and grana membranes.

4 Factors Affecting the Development of Plastids

4.1 Introduction

The development of plastids is affected by:

a) external factors, principally light, temperature and chemical agents, having an effect either directly or indirectly, through one or several internal agents
b) internal factors of which the genetic systems are the most important.

A distinction between the two groups is in many cases not meaningful, as cooperation between external and internal factors is more often the case than non-cooperation. Many papers have been published on the effects of internal and external factors on plastid development. The topic is comprehensively covered in the monumental works of KIRK and TILNEY-BASSETT (1967, 1978).

4.2 Light

4.2.1 Introduction

The control of chloroplast development in respect to light is very complex as it involves several photoreceptors, among which phytochrome, protochlorophyllide, cryptochrome, and chlorophylls play important roles. Light affects several steps in chloroplast development. The effects depend both on light intensity and light quality. The mode of action also varies depending on when, during plastid development, a response to light is observed. Thus, the regulation of chloroplast growth as well as its function during photosynthesis is dependent on light for the activation of a number of enzymes which have either low activities or are totally inactive in the dark (BUCHANAN 1980).

4.2.2 Spectral Dependence

Chloroplast formation in roots and underground organs (sequence III above) requires light between 350 and 550 nm with a peak around 450 nm to start the development. Red light (600–700 nm) fails here to initiate chloroplast formation (BJÖRN 1976). In complete darkness not even etioplasts are formed. Red light given before blue light enhances the blue light effect. Lower intensities of blue light are therefore needed when red light is given before blue light. The action spectrum for the red light effect is similar both to the spectrum of protochlorophyllide absorption, and the red peak of phytochrome absorption. It is clear that the main effect of red light is a transformation of existing protochlorophyllide to chlorophyllide which, within about half an hour at room temperature, will become esterified to chlorophyll. However, it cannot be excluded that also phytochrome-governed reactions are involved (see below). The closeness between the main peak for protochlorophyllide absorption (650 nm) and that of phytochrome in its red-absorbing form (660 nm) makes it difficult to distinguish between the two responses.

The requirement for blue light seems to be connected with RNA metabolism, as the formation of many components necessary for the synthesis of RNA is triggered by blue light (RICHTER and DIRKS 1978). In the dark RNA synthesis (DIRKS and RICHTER 1975) and synthesis of ribosomes and structural protein (NEWCOMB 1967, DYER et al. 1971) in proplastids is low. The chemical nature of the blue light receptor is not known.

In dark-grown leaves (sequence II), the blue light requirement for plastid development to the etioplast stage is replaced by non-light-requiring enzymatic reactions. This pattern can be compared to the difference in chlorophyll formation between leaves from angiosperms and gymnosperm seedlings. In leaves, the protochlorophyllide → chlorophyllide transformation requires light, while the reaction in gymnosperm seedlings is in part enzymatic. In both cases, however, the development in darkness progresses further than the stages reached in non-irradiated tissues following sequence III. The development of etioplasts to normal chloroplasts (the second stage in sequence II) only requires red light

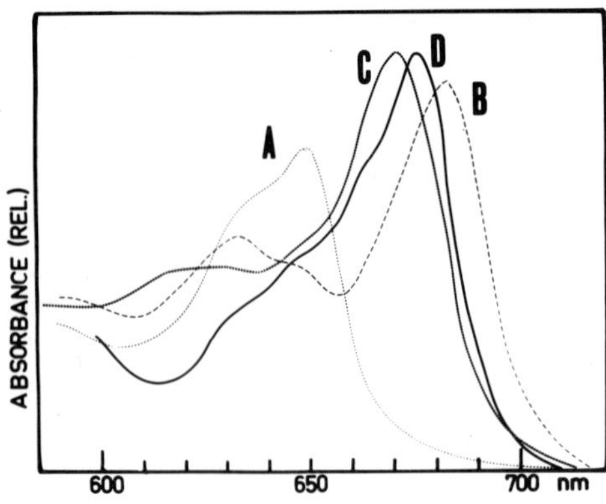

Fig. 4. Absorption spectra of barley leaves. *A* dark-grown, non-irradiated; *B* ca. 1 s after an electronic flash; *C* 6 min after *B; D* a light-grown green leaf. *D* changed to match the peak height in *C*. The initial changes taking place during the first fraction of a second are not shown. (Original)

(SUNDQVIST et al. 1980). The red light works here on two systems. On one hand light is used for the phototransformation of protochlorophyllide to chlorophyllide. In this process protochlorophyllide itself is the light absorber. On the other hand light is needed to build the structural framework of chloroplasts (MOHR and KASEMIR 1975) (see below).

4.2.3 Chlorophyll Formation

Light-dependent chlorophyll formation can be followed by in vivo spectrophotometry as the absorption maximum of chlorophyll differs from that of its precursor, protochlorophyllide.

Upon irradiation of a dark-grown, etiolated leaf the absorption spectrum of newly formed chlorophyll passes through a series of fast and slow changes until the normal in vivo spectrum of a leaf is observed (DUJARDIN and SIRONVAL 1977, INOUE et al. 1981). Some of the spectral differences (Fig. 4) are probably due to changes in aggregation states of the pigment molecules, as well as of the apoprotein attached to the pigments (VIRGIN 1981). Simultaneously with these changes, a phytylation of the chlorophyllides takes place (LILJENBERG 1977).

4.2.4 Phytochrome as Mediator of Light Effects

Phytochrome in its far-red absorbing form accelerates pigment formation in the plastid. This effect can be brought about by a short pulse of red light (transforming P_r to P_{fr}) (VIRGIN 1961, MOHR and KASEMIR 1975). After a period in darkness (3–4 h) and upon irradiation chlorophyll formation starts immediately with a high rate. Without such a pre-irradiation a lag phase in the chlorophyll formation is usually seen. The effect of phytochrome on chlorophyll formation is caused by an acceleration of protochlorophyllide synthesis which in its turn is caused by increased rate of ALA synthesis (FORD and KASEMIR 1980). In etiolated seedlings there is a linear correlation between the rate of formation of ALA and the degree of photoconversion of protochlorophyllide. If pretreated with a red pulse the seedlings show a non-linear correlation in this respect.

Experimental results suggest that there may be an interaction between P_{fr} and the protochlorophyllide–chlorophyllide conversion in controlling ALA synthesis (FORD and KASEMIR 1980). In seedlings of gymnosperms P_{fr} strongly increases the rate of synthesis of chlorophyll even though some chlorophyll synthesis takes place in the dark. Also carotenoid formation is accelerated by such a pretreatment (VIRGIN 1972, FROSCH and MOHR 1980). P_{fr} seems also to be necessary for the formation of chlorophyll b. In mustard seedlings (OELZE-KAROW and MOHR 1978) a short red light treatment followed by at least 2 h of darkness before the onset of continuous light has the effect that chlorophyll b appears immediately upon beginning continuous irradiation. In non-pretreated material the chlorophyll b level hardly shows any increase between 20 and 70 min after the onset of continuous light. P_r also increases the rate of phytylation of chlorophyllide (KASEMIR and PREHM 1976, LILJENBERG 1977).

The increased rate of pigment synthesis caused by phytochrome in its far-red absorbing form is reflected in a more rapid differentiation and growth of the etio/chloroplast. The mechanism for this control is however obscure (BRADBEER et al. 1974, KASEMIR et al. 1975). Particularly evident is a more rapid grana formation (MOHR 1977, RYBERG and VIRGIN 1978).

4.2.5 Other Light Effects

Photosynthesis per se is not necessary for the formation of the chloroplast. Photosynthesis is, however, by production of structural elements a controlling factor in the development of the plastids. Etioplasts which are exposed to intermittent light do not form fully developed grana and do not accumulate chlorophyll in the same amount as found in etioplasts given continuous light (AKOYUNOGLOU and ARGYROUDI-AKOYUNOGLOU 1969).

Phytochrome is also involved in mitochondrial activities, expecially during the process of greening. Chloroplast development is highly dependent on oxidative phosphorylation during this process being stimulated by P_{fr} (HAMPP and WELLBURN 1979). However, for plants which are dependent on light for their greening, only white light together with red light induces the complete developmental sequence of the etioplast to chloroplast.

4.2.6 Sun and Shade Plants

Flowering plants are either obligate sun- or shade plants, whereas some species are able to adapt phenotypically to changes in environmental conditions (BOARDMAN 1977b). Single leaves of trees may also develop into sun- or shade types, depending on the light intensities at the time of formation of the buds. The plastids in shade plants and leaves are generally larger than those in sun plants and leaves. Chloroplasts prepared from sun plants differ in photosynthetic activity from those of shade plants (BJÖRKMAN et al. 1972, BOARDMAN 1977b). Sun plant chloroplasts have at least three- to four fold higher ribulose-1,5-bis-phosphate carboxylase content than corresponding shade plant chloroplasts. Similar differences also exist in the content of cytochrome f, cytochrome b-559 and b-6 (BJÖRKMAN et al. 1972).

4.3 Temperature

4.3.1 Introduction

The development of photosynthetic membranes is highly dependent on temperature. The rate of chlorophyll formation increases rapidly from 0° to 40 °C above which synthesis sharply decreases (SMILLIE 1976). Studies on mutants suggest that changes of the temperature deeply influence not only the formation of the lamellar system but also the ultrastructure and composition of plastids (SMILLIE et al. 1975, BOARDMAN et al. 1974).

4.3.2 Effects of Low Temperature

Changing the temperature in lettuce and spinach chloroplasts from about +5° to −35 °C leads to lipid phase changes (FORK et al. 1977). Such temperature-induced changes in membrane properties have also been described for mitochondrial and chloroplast membranes in chilling-sensitive plants (LYONS 1973, RAISON 1973).

The lower limit for chlorophyll formation in barley is 2°–5 °C (SMILLIE et al. 1978). At this growth temperature the leaves are either pale yellow and contain plastids with vesicles but hardly recognizable thylakoids, or pale green plastids containing stromal thylakoids and small grana of usually two or three thylakoids. In darkness, below 12 °C, no formation of prolamellar bodies takes place, but at higher temperatures (26 °C) they are reformed within a few hours (IKEDA 1971). These structures can therefore appear every night in chloroplasts of developing leaves. Variations in light intensity at constant temperature seem to have smaller effects on mesophyll chloroplast structures than variations in temperature at constant light intensity (FORDE et al. 1975). However, the behavior can differ between species (SMILLIE et al. 1978). Seasonal changes in chloroplast structure and shape have been reported for spruce (SENSER et al. 1975).

Frost hardening affects chloroplast structure. The chloroplasts in cold-acclimatized leaves of potato show an increase in the number of osmiophilic globuli with a disappearance of starch grains (CHEN et al. 1977). The chlorophyll a/chlorophyll b ratio of different plants changes (TAYLOR and ROWLEY 1971). In *Sorghum* and *Paspalum* low temperature reduces the size of the starch grains followed by a closing of the individual stromal thylakoids, reducing the space between them, at the same time a swelling of the whole chloroplast takes place (TAYLOR and CRAIG 1971). At low temperatures, bleaching (photooxidation) of the leaf pigments in *Cucumis* has been reported (VAN HASSELT 1972). This is probably a general phenomenon often visible as a tendency to chlorosis during cold periods in evergreen leaves (ÖQUIST et al. 1978). The effect of low temperatures on the morphology of the chloroplasts differs from one species to another. Thus, plastids of rye respond differently than do plastids of subtropical and tropical species (KIMBALL and SALISBURY 1973). In respect to the effect of low temperatures and other factors, it is important to distinguish between effects due to either changes from one condition to another or to effects obtained after cultivating plants under various constant conditions. A rapid change in the intensity of a factor often results in more drastic morphological changes than will be obtained during a slower change of the conditions, where gradual adaptation to the new conditions occurs (BALLANTINE and FORDE 1970).

4.3.3 Effects of High Temperature

Chloroplasts prepared from plants grown at high temperatures (30°–40 °C) have reduced photoreductive activity, a tendency towards formation of large grana, and a disorientation of the lamellar systems (SMILLIE et al. 1978). High temperatures (32 °C) prevent the formation of 70S chloroplast ribosomes and their rRNA (SCHÄFERS and FEIERABEND 1976). Ribosome-deficient leaves are chlorotic. Ultrastructurally such leaves have abnormal grana stacks not regularly connected by stroma thylakoids. In completely chlorotic leaf parts no grana are found in the plastids (SCHÄFERS and FEIERABEND 1976).

The ultrastructures obtained after treatment with inhibitors such as chloramphenicol and streptomycin have many features in common with those obtained at high temperatures (DÖBEL 1963). Such structural changes are amoeboid, irregular morphology with invaginations or protrusions, the occurrence of single large grana, single or aggregated vesicles and tubules, aberrant prolamellar bodies, and the accumulation of plastoglobuli which may contain precursor lipids or breakdown products of chloroplast lamellae (RÖBBELEN 1966, SPREY 1972, VALANNE and VALANNE 1972, VON WETTSTEIN 1974).

5 Hormonal Regulation

The development of plastids is, like other parts of the plant, affected by phytohormones. They influence not only the biosynthesis of thylakoids, but also the rate of pigment turnover (LICHTENTHALER and GRUMBACH 1975).

Kinetin, the most studied hormone in this respect, has effects similar to those of phytochrome in its far-red form. There are indications that the factors work independently of each other and on different sites in the chain leading to the differentiated chloroplast (FORD et al. 1981). Thus, effects by the two factors are additive.

The effect of applied cytokinins on chloroplast development indicates that this hormone has a dominant role in regulating the senescence and thus the life span of chloroplasts (STOBART et al. 1972, SUNDQVIST et al. 1980). Applied GA decreases the amounts of chlorophyll, carotenoids and vitamin K_1 in greening seedlings (LICHTENTHALER and BECKER 1970) and inhibits the activity of glutamate dehydrogenase (HUBER and SANKHLA 1974, SUNDQVIST et al. 1980). The concentration of P-700 (photosystem I reaction center chlorophyll) and of Hill activity is increased by IAA-treatment (BUSCHMANN and LICHTENTHALER 1975). As the effects of external factors (light, temperature) and added phytohormones are often similar to each other, it is thought that external factors in many cases regulate plastid development by controlling phytohormone levels in plants. However, as already mentioned above, effects of phytohormones are additative to those of external factors, indicating effects on different sites in the chain leading to an effect. Comparatively few studies have been made on changes in internal levels of phytohormones as a result of external factors. Water stress increases the amount of free GA in the plant (LOVERS 1977), showing a direct correlation between external and internal factors.

Common for all these factors is that they affect plastid development under optimal supply of nutrients to the plants. They are all in one way or another necessary during the various stages of the ontogeny of plastids.

6 Genetic Regulation and Control of Plastid Development

6.1 Introduction

The developmental potential of plastids is tremendous. This is seen from the fact that there are so many possible ways by which different types of plastids, with varying functions, can be formed.

Basically, it is the type of plant and part of the plant where the proplastid might be present which determines which plastid developmental sequence (cf. Fig. 1) will take place. The facts that different types of plastids can be formed in adjacent cell layers, and that the etioplast only needs light for turning into a chloroplast, are evidence for the validity of this statement. But the concept of a developmental potential also indicates that an extreme stress, for example roots that are irradiated, can change one development sequence into another. Environmental pressures are essential for explaining certain types of development. The type of tissue to a large extent determines which type of plastid development will take place and is an effect of the evolutionary process. This fact also implies that the final regulation of a plastid sequence must be sought on the gene level, and be associated with DNA/RNA/enzyme metabolism. An idea of how this mechanism may work can be found in the works by APEL (1979), APEL and KLOPPSTECH (1980). Phytochrome induces synthesis of the light harvesting chlorophyll a/b apoprotein. However, a massive insertion of the light harvesting chlorophyll protein complex, takes place only after synthesis of chlorophyll in the developing thylakoid membrane. Red light induces activity of mRNA for synthesis of the apoprotein of this chlorophyll protein. Such a synthesis, however, only takes place during a subsequent period of continuous irradiation preceded by a dark period. It is thus clear that the red light induces the synthesis of mRNA, which is taken up into the polysomes during the ensuing dark period.

6.2 Genetic Control

It is well established that the phenotype of the chloroplast is under control of nuclear genes and factors showing non-Mendelian inheritance (YOSHIDA 1959, von Wettstein and Eriksson 1964). The chromosomal control of chloroplast structures and their function is recognized from a great number of recessive, mostly lethal mutants. In barley there are around 200 recessive lethal chloroplast mutants whose genetic characters have been assigned to 86 diallellic nuclear gene loci (Wettstein 1974). A long series of albino mutants have been studied in which the gene blockage has been shown to be present at various loci on the chromosomes. Thus, in the so-called *xantha 10* mutant in barley there is a block between the synthesis of protoporphyrin IX and protochlorophyll (Wettstein 1961). Another such blockage has been found in a mutant of *Vicia faba* unable to reduce pyridine nucleotides (Heber and Gottschalk 1963). Mutants in which the synthesis of single amino acids is blocked have also been described, e.g., a mutant lacking the capacity to synthesize leucine, forms giant grana (Walles 1963). The development of chloroplasts has also been found to be controlled by carotenoid synthesis in plants. Mutants with disturbed carotenoid synthesis, for example, without β-carotene, do not form normal chloroplast in the light (Walles 1967, Bachmann et al. 1969). Carotenoids protect the chlorophylls against photooxidation and are normally present in small amounts already before chlorophylls are formed. There is evidence (Frosch and Mohr 1980) that larger amounts of carotenoids can only accumulate in connection with the formation of grana and stroma which in turn requires the formation of chlorophylls. There should be two mechanisms involved in light-dependent carotenoid accumulation: a push regulation exerted by P_{fr} on carotenogenesis and a pull regulation exerted by the formation of chlorophyll carotenoid protein complexes.

The sequence and appearance of photosynthetic components, structures, and functions during the development of plastids support the concept of a stepwise process of membrane assembly, the genetic control being on nuclear, chloroplast and mitochondrial genomes. The regulation of the membrane assembly is, among other factors, affected by the earlier mentioned three light receptors (protochlorophyllide, phytochrome and cryptochrome). During the developmental sequence of the plastid, stroma lamellae are first formed and later grana stacks. The control of grana formation is not associated with the appearance of photosystem II activity in greening leaves (Hiller et al. 1973). The stacking of grana is probably due to the formation of chlorophyll–protein complex II which is inserted in the membranes (Anderson 1975). The control of the developmental sequence is thus associated with protein synthesis, and consequently is under nuclear and plastid genetic control and the factors which are expressed in the genetic material.

There are two genetic systems (mitochondrial genomes not being considered), working in a cooperative mode in most plant cells. One system is for transcription and translation in the cytoplasm and one is in the plastid. Genetic material, e.g., DNA fibrils have been found not only in chloroplast but also in proplastids, etioplasts and chromoplasts (Thompson 1980). Polypeptides as found in proteins in membranes and enzymes are coded for both by plastid and nuclear DNA. As the synthesis of structural elements such as polypeptides is regulatory for plastid development some of the gene products in chloroplasts are presented in Table 1.

Table 1. Some identified chloroplast gene products. (Modified after PARTHIER 1979)

Class	Species
Ribosomal RNAs	23S, 16S, 5S, 4.5S
Transfer RNAs	20–30 species
Messenger RNA	Large subunit ribulose-1,5-bisphosphate carboxylase 32,000 m.w. protein
Ribosomal proteins	2–5 species
Soluble proteins	Large subunit ribulose-1,5-bisphosphate carboxylase elongation factors EFG, EFT_u
Membrane proteins	Cytochrome f 32,000 m.w. protein 3 or 4 of the subunits of the coupling factor CF_1 cyt b-559 polypeptides (subunits) of photosystem II

The informational content of plastid DNA is now being mapped (BEDBROOK and BOGORAD 1976, HOBOM et al. 1977). It is essential for an understanding of the replication and synthetic capacities of plastids, to know that the DNA is found in many copies in the chloroplasts. The organelle is a conservative genetic system (HERRMANN and POSSINGHAM 1980).

Although protochlorophyllide and phytochrome can be assigned regulatory roles and the final effect can be seen as an accelerated de novo synthesis of proteins, almost nothing is known about how the signals from the light receptors are transmitted to the genetic systems. But the fact that there are plants where entire plastid development can take place in darkness shows that either some fundamental differences exist between development in normal chloroplasts and the greening of the etioplast, or that repressor molecules are present in dark-grown plants. Part of the light effect in these plants would then depend on repressor removal through the phytochrome system. Another explanation being the activation of genes by phytochrome (MOHR 1977). As has been pointed out earlier, almost all data on sequential events are obtained from dark-grown leaves, and there might be qualitative differences between the sequence in light- and dark-grown leaves. During the last step in plastid development ribosomal proteins, ribosomes, polysomes, transcriptional and translational enzymes are synthesized in large amounts, with synthesis taking place both in the cytoplasm and the plastid (SCHOPFER et al. 1975). Before that a change, in among other things the synthesis of plastid mRNA, has taken place (COEN et al. 1977). Everything must have started with some reaction which triggers the transcription of, for example, mRNA on nuclear (or plastid) DNA, or both.

A summary of the effects of various factors on DNA synthesis, RNA (all species) synthesis, transcription and translation taking place in the nucleus, plastids, and cytoplasm is found in review articles by PARTHIER (1979), HERRMANN and POSSINGHAM (1980), WOLLGIEHN and PARTHIER (1980) (Fig. 5).

Although it has been shown that there are effects of light and phytohormones on almost all organization levels, there is no solid evidence for interaction.

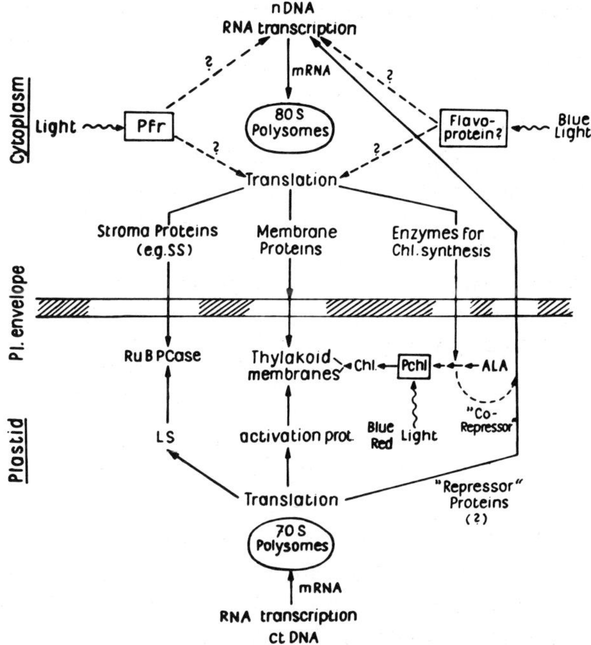

Fig. 5. A diagram illustrating a tentative scheme for the intracellular cooperation of plastid and nucleo-cytoplasmic gene expression in the biogenesis of the chloroplast. P_{fr} phytochrome; *LS, SS* large and small subunits of ribulose-1,5-bisphosphatecarboxylase; *ct* chloroplastic; *n* nuclear; *pl* plastid; *chl* chlorophyll. (PARTHIER 1979)

In the case of the influence of kinetin on chlorophyll accumulation only additative effects have been found, indicating that phytochrome and kinetin act on different causal sequences leading to chlorophyll accumulation.

6.3 Control on Membrane Level

The demand for structural and metabolic components during plastid development makes regulation by transport across organelle envelopes essential. The two envelopes of plastids can have several functions.

a) The inner envelope of plastids functions as a selective barrier to low-molecular compounds. The transport of metabolites across the envelopes is regulated by different translocators. For example, ATP and NADPH cannot be directly transported in or out of the chloroplast to any large extent. Such transport occurs indirectly (HEBER 1974, HEBER and HELDT 1981). Such regulation is not only typical for chloroplasts but recently it was shown that both phosphate and adenylate translocators are present in chromoplast membranes (LIEDVOGEL and KLEINIG 1980).

During the development of an etioplast to a chloroplast, it is possible that this regulatory function is set aside by changed envelope characteristics. HAMPP and SCHMIDT (1976), HAMPP and WELLBURN (1980) have shown that the etioplast and mitochondrial envelopes have a changed permeability for ATP and various other metabolites, e.g., sucrose, during the initial stages of plastid development. Metabolites and ATP could be supplied to the plastid during this period when photosynthesis cannot provide these substances. Mitochondria are at this stage of plastid development often seen as bordering plastids. The onset of electron transport in the plastids is accompanied by a change in permeability of the plastid envelope.

The transport of large molecular weight compounds, such as polypeptides and proteins probably also is regulated at the envelope. Synthesis of RuBPCase which consists of large and small subunits, takes place in the plastids, but the small subunit is synthesized on cytoplasmic ribosomes and transported into the plastid (ELLIS 1977). The passage through the envelope is therefore of regulatory character. There are several ways by which proteins can be transported through a membrane (DOUCE and JOYARD 1979, CHUA and SCHMIDT 1979, WOLLGIEHN and PARTHIER 1980). Most evidences indicate that a post-translational transport (i.e., the polypeptide or protein is synthesized outside the plastid and then transported inside) (CHUA and SCHMIDT 1978, HIGHFIELD and ELLIS 1978). A translational mechanism cannot be excluded. The association of the endoplasmic reticulum and ribosomes with intermediate stages of developing plastids (WHATLEY 1977, 1978) could be part of a transport system for the plastid.

b) The inner envelope also regulates plastid development by being the site or coordinating system for the synthesis of many components (structural and functional) necessary for membrane formation and photosynthetic activity (LEECH and MURPHY 1976, DOUCE and JOYARD 1979). Phytochrome has been found to be associated with the envelopes of plastids in some plants (EVANS and SMITH 1976, MARMÉ 1977).

6.4 Control by "Energy" Metabolism

ATP formed in photosynthetic phosphorylation is used for many biosynthetic purposes in the plastids, such as phosphorylation of light-harvesting chlorophyll pigments (ALLEN et al. 1981) and polypeptide formation within the chloroplast (BLAIR and ELLIS 1973, BOTTOMLEY et al. 1974).

Protochlorophyllide is one of the regulators of chloroplast development. In greening leaves (ROBERTSON and LAETSCH 1974, MACKENDER 1978), the development of PLBs and the prothylakoids associated with them, is correlated to protochlorophyllide accumulation. The formation of protochlorophyllide is under phytochrome control (see above). Its reduction to chlorophyll requires ATP (HORTON and LEECH 1972) and NADPH as hydrogen donor (GRIFFITHS 1980). The reduction is mediated through the enzyme system NADPH: protochlorophyllide oxidoreductase (APEL et al. 1980). During the initial stages of plastid development most data indicate that energy is supplied by respiratory processes (KLEIN and NEUMANN 1966, DODGE et al. 1971, DE GREEF and VERBELEN 1977). It is probable that this is the case also for later stages, as photosynthesis per se does not seem to be necessary to make a chloroplast. The results of HAMPP and WELLBURN (1980) on the changing properties of plastid envelopes is consistent with this conclusion. Mitochondrial respiration increases during the initial stages of greening and the increase in the level of the cytoplasmic

ATP pool (HAMPP and WELLBURN 1979, HAMPP 1979). Mitochondrial activities at this stage are also influenced by phytochrome either acting directly on the mitochondria or indirectly via the phytochrome mediated etio-chloroplast development (HAMPP and WELLBURN 1979).

References

Akoyunoglou G, Argyroudi-Akoyunoglou JH (1969) Effects of intermittent and continuous light on the chlorophyll formation in etiolated plants at various ages. Physiol Plant 22:288–295

Allen JF, Bennett J, Steinback KE, Arntzen CJ (1981) Chloroplast protein phosphorylation couples plastoquinone redox state to distribution of excitation energy between photosystems. Nature 291:25–29

Anderson JM (1975) The molecular organization of chloroplast thylakoids. Biochim Biophys Acta 416:191–235

Anderson JM (1981) Consequences of spatial separation of photosystem 1 and 2 in thylakoid membranes of higher plant chloroplasts. FEBS Lett 124:1–10

Anderson JM, Goodchild DJ, Boardman NK (1973) Composition of the photosystems and chloroplast structure in extreme shade plants. Biochim Biophys Acta 325:573–585

Apel K (1979) Phytochrome-induced appearance of mRNA activity for the apoprotein of the light-harvesting chlorophyll a/b protein of barley (*Hordeum vulgare*). Eur J Biochem 97:183–188

Apel K, Kloppstech K (1980) The effect of light on the biosynthesis of the light-harvesting chlorophyll a/b protein. Evidence for the requirement of chlorophyll a for the stabilization of the apoprotein. Planta 150:426–430

Apel K, Santel H-J, Redlinger TE, Falk H (1980) The protochlorophyllide holochrome of barley (*Hordeum vulgare* L.). Isolation and characterization of the NADPH: protochlorophyllide oxidoreductase. Eur J Biochem 111:251–258

Bachmann MD, Robertson DS, Bowen CC (1969) Thylakoid anomalies in relation to grana structure in pigment-deficient mutants of *Zea mays*. J Ulstrastruc Res 28:435–451

Badenhuizen NP (1971) The biogenesis of starch granules in higher plants. Appleton-Century-Crofts, New York

Baker NR, Leech RM (1977) Development of photosystem I and photosystem II activities in leaves of light-grown maize (*Zea mays*). Plant Physiol 60:640–644

Ballantine JEM, Forde BJ (1970) The effect of light intensity and temperature on plant growth and chloroplast ultrastructure in soybean. Am J Bot 57:1150–1159

Bedbrook JR, Bogorad L (1976) Endonuclease recognition sites mapped on *Zea mays* chloroplast DNA. Proc Natl Acad Sci USA 73:4309–4313

Bishop DG (1974) Lamellar structure and composition of chloroplasts in relation to photosynthetic electron transfer. Photochem Photobiol 20:281–299

Björkman O, Boardman NK, Anderson JM, Thorne SW, Goodchild DJ, Pyliotis NA (1972) Effect of light intensity during growth of *Atriplex patula* on the capacity of photosynthetic reactions, chloroplast components and structure. Carnegie Inst Wash Year Book 71:115–135

Björn LO (1967) Chloroplast in roots. In: Sironval C (ed) Le chloroplaste. Masson, Paris, pp 313–322

Björn LO (1976) The state of protochlorophyll and chlorophyll in corn roots. Physiol Plant 37:183–184

Blair GE, Ellis RJ (1973) Protein synthesis in chloroplasts I. Light-driven synthesis of the large subunit of fraction I protein by isolated pea chloroplasts. Biochim Biophys Acta 319:223–234

Boardman NK (1977a) Development of chloroplast structure and function. In: Trebst A, Avron M (eds) Photosynthesis I. Encyclopedia of plant physiology new ser Vol 5. Springer, Berlin Heidelberg New York, pp 583–600

Boardman NK (1977b) Comparative photosynthesis of sun and shade plants. Annu Rev Plant Physiol 28:355–377

Boardman NK, Anderson JM, Björkman O, Goodchild DJ, Grimme LH, Thorne SW (1974) Chloroplast differentiation in sun and shade plants: Relationship between chlorophyll content, grana formation, photochemical activity and fraction of the photosystems. Port Acta Biol Ser A 14:213–236

Bottomley W (1980) Fraction I protein. In: Reinert J (ed) Chloroplasts. Results and problems in cell differentiation Vol 10. Springer, Berlin Heidelberg New York, pp 179–199

Bottomley W, Spencer D, Whitfeld PR (1974) Protein synthesis in isolated spinach chloroplasts: Comparison of light-driven and ATP-driven synthesis. Arch Biochem Biophys 164:106–117

Bradbeer JW (1973) The synthesis of chloroplast enzymes. In: Milborrow BV (ed) Biosynthesis and its control in plants. Academic Press, London New York, pp 279–302

Bradbeer JW, Gyldenholm AO, Ireland HMM, Smith JW, Rest J, Edge HJW (1974) Plastid development in primary leaves of *Phaseolus vulgaris* VIII. The effects of the transfer of dark-grown plants to continuous illumination. New Phytol 73:271–279

Buschmann C, Lichtenthaler HK (1975) Hill-reaction of chloroplasts from *Raphanus seedlings* grown with β-indoleacetic acid and kinetin. In: Avron M (ed) Proc 3rd Int Congr Photosynthesis 1974. Elsevier, Amsterdam New York, pp 753–756

Buchanan BB (1980) Role of light in the regulation of chloroplast enzymes. Annu Rev Plant Physiol 31:341–374

Butler RD, Simon EW (1971) Ultrastructural aspects of senescence in plants. In: Strehler BL (ed) Advances in gerontological research Vol 3. Academic Press, London New York, pp 73–129

Chen P, Li PH, Cunningham WP (1977) Ultrastructural differences in leaf cells of some *Solanum* species in relation to their frost resistance. Bot Gaz 138:276–285

Chua NH, Schmidt GW (1978) Post-translational transport into intact chloroplasts of a precursor to the small subunit of ribulose-1,5-bisphosphate carboxylase. Proc Natl Acad Sci USA 75:6110–6114

Chua NH, Schmidt GW (1979) Transport of proteins into mitochondria and chloroplasts. J Cell Biol 81:461–483

Clowes FAL, Juniper BE (1968) Plant cells. Blackwell, Oxford

Coen DM, Bedbrook JR, Bogorad L, Rich A (1977) Maize chloroplast DNA fragment encoding the large subunit of three ribulosebisphosphate carboxylase. Proc Natl Acad Sci USA 74:5487–5491

Dirks W, Richter G (1975) Effect of blue light on synthesis of RNA components during differentiation of chloroplasts in isolated roots (*Pisum sativum*). Biochem Physiol Pflanz 168:157–166

Döbel P (1963) Untersuchung der Wirkung von Streptomycin-Chloramphenicol- und 2-Thiouracil-Behandlung auf die Plastidenentwicklung von *Lycopersicum esculentum*. Biol Zentralbl 82:275–295

Dodge AD, Alexander DJ, Blackwood GC (1971) The contribution of photosynthesis to chlorophyll formation in etiolated mung bean leaves. Physiol Plant 25:71–74

Douce R, Joyard J (1979) Structure and function of the plastid envelope. In: Woolhouse HW (ed) Advances in botanical research. Academic Press, London New York, pp 1–116

Dujardin E, Sironval C (1977) Transitory-pigment-protein complexes similar to photosynthesis active centers during protochlorophyll(ide) photoreduction. Plant Sci Lett 10:347–355

Dyer TA, Miller RH, Greenwood AD (1971) Leaf nucleic acids I. Characteristics and role in the differentiation of plastids. J Exp Bot 22:125–136

Ellis RJ (1977) Protein synthesis by isolated chloroplasts. Biochim Biophys Acta 463:185–215

Evans A, Smith H (1976) Spectrophotometric evidence for the presence of phytochrome in the envelope membranes of barley etioplasts. Nature 259:323–325

Ford MJ, Kasemir H (1980) Correlation between 5-aminolaevulinate accumulation and protochlorophyll photoconversion. Planta 150:206–210

Ford MJ, Kasemir H, Mohr H (1981) The influence of phytochrome and kinetin on chlorophyll accumulation in mustard cotyledons: A two-factor analysis. Ber Dtsch Bot Ges 94:35–41

Forde BJ, Whitehead HCM, Rowley JA (1975) Effect of light intensity and temperature on photosynthetic rate, leaf starch content and ultrastructure of *Paspalum dilatatum*. Aust J Plant Physiol 2:185–195

Fork DC, Murata N, Avron M (1977) The effect of temperature on the physical phase of chloroplast membrane lipids and photosynthesis. Carnegie Inst Wash Year Book 76:220–226

Frey-Wyssling A, Ruch F, Berger X (1956) Monotrope Plastiden—Metamorphose. Protoplasma 45:97–114

Frosch S, Mohr H (1980) Analysis of light-controlled accumulation of carotenoids in mustard (*Sinapis alba* L.) seedlings. Planta 148:279–286

Givan CV, Harwood JL (1976) Biosynthesis of small molecules in chloroplasts of higher plants. Biol Rev 51:365–406

Grebanier AE, Steinback KE, Bogorad L (1979) Comparison of the molecular weights of proteins synthesized by isolated chloroplasts with those which appear during greening in *Zea mays*. Plant Physiol 63:536–539

Greef JA de, Verbelen JP (1977) Plastid development in etiolated bean leaves under uncoupling conditions of oxidative phosphorylation. Ann Bot 41:1371–1373

Griffiths WT (1980) Substrate-specificity studies on protochlorophyllide reductase in barley (*Hordeum vulgare*) etioplast membranes. Biochem J 186:267–278

Gunning BES (1965) The greening process in plastids 1. The structure of the prolamellar body. Protoplasma 60:111–130

Gunning BES, Jagoe MP (1967) The prolamellar body. In: Goodwin TW (ed) Biochemistry of chloroplasts, Vol II. London, New York: Academic Press, pp 655–676

Hampp R (1979) Kinetics of mitochondrial phosphate transport and rates of respiration and phosphorylation during greening of etiolated *Avena* leaves. Planta 144:325–332

Hampp R, Schmidt HW (1976) Changes in envelope permeability during chloroplast development. Planta 129:69–73

Hampp R, Wellburn AR (1979) Control of mitochondrial activities by phytochrome during greening. Planta 147:229–235

Hampp R, Wellburn AR (1980) Translocation and phosphorylation of adenine nucleotides by mitochondria and plastids during greening. Z Pflanzenphysiol 98:289–303

Hasselt PR van (1972) Photo-oxidation of leaf pigments in *Cucumis* leaf discs during chilling. Acta Bot Neerl 21:539–548

Heber U (1974) Metabolite exchange between chloroplasts and cytoplasm. Annu Rev Plant Physiol 25:393–421

Heber U, Gottschalk W (1963) Die Bestimmung des genetisch fixierten Stoffwechselblockes einer Photosynthese-Mutante von *Vicia faba*. Z Naturforsch 18b:36–44

Heber U, Heldt HW (1981) The chloroplast envelope: Structure, function, and role in leaf metabolism. Annu Rev Plant Physiol 32:139–168

Herrmann RG, Possingham JV (1980) Plastid DNA — The plastome. In: Reinert J (ed) Chloroplasts. Results and problems in cell differentiation Vol 10. Springer, Berlin Heidelberg New York, pp 45–96

Herrmann FH, Börner Th, Hagemann R (1980) Biosynthesis of thylakoids and the membrane-bound enzyme systems of photosynthesis. In: Reinert J (ed) Chloroplasts. Results and problems in cell differentiation Vol 10. Springer, Berlin Heidelberg New York, 147–177

Highfield PE, Ellis RJ (1978) Synthesis and transport of the small subunit of chloroplast ribulose bisphosphate carboxylase. Nature 271:420–424

Hiller RG, Pilger D, Genge S (1973) Photosystem II activity and pigment-protein complexes in flashed bean leaves. Plant Sci Lett 1:81–88

Hobom G, Bohnert HJ, Driesel A, Herrmann RG (1977) Restriction fragment map of the circular plastid DNA from *Spinacia oleracea*. In: Bogorad L, Weil JH (eds) Acides Nucléiques et Synthèse de Proteines chez les Végétaux. CNRS, Paris, pp 63–69

Horton, P, Leech RM (1972) The effect of ATP on photoconversion of protochlorophyllide into chlorophyllide in isolated etioplasts. FEBS Lett 26:277–280

Huber W, Sankhla N (1974) Abscisic acid–kinetin interaction in growth and activities of enzymes of amino-acid metabolism in *Pennisetum typhoides* seedlings. Z Pflanzenphysiol 73:160–166

Ikeda T (1971) Prolamellar body formation under different light and temperature conditions. Bot Mag Tokyo 84:363–375

Inoue Y, Koboyashi T, Ogawa T, Shibata K (1981) A short-lived intermediate in the photoconversion of protochlorophyllide to chlorophyllide a. Plant Cell Physiol 22:197–204

Kasemir H, Mohr H (1981) The involvement of phytochrome in controlling chlorophyll and 5-aminolevulinate formation in a gymnosperm seedling (*Pinus sylvestris*). Planta 152:369–373

Kasemir H, Prehm G (1976) Control of chlorophyll synthesis by phytochrome III. Does phytochrome regulate the chlorophyllide esterification in mustard seedlings? Planta 132:291–295

Kasemir H, Bergfeld R, Mohr H (1975) Phytochrome-mediated control of prolamellar body reorganization and plastid size in mustard cotyledons. Photochem Photobiol 21:111–120

Kesselmeier J, Budzikiewicz (1979) Identification of saponins as structural building units in isolated prolamellar bodies from etioplasts of *Avena sativa* L. Z Pflanzenphysiol 91:333–344

Kimball SL, Salisbury FB (1973) Ultrastructural changes of plants exposed to low temperatures. Am J Bot 60:1028–1033

Kirk JTO, Tilney-Bassett RAE (1967) The plastids. Their chemistry, structure, growth and inheritance. Freeman, London, San Francisco

Kirk JTO, Tilney-Bassett RAE (1978) The plastids. Their chemistry, structure, growth and inheritance, 2nd edn. Elsevier/North-Holland, Amsterdam

Klein S, Neuman J (1966) The greening of etiolated bean leaves and the development of chloroplast fine structure in absence of photosynthesis. Plant Cell Physiol 7:115–123

Leech RM (1976) The replication of plastids in higher plants. In: Yeoman MM (ed) Cell division in higher plants. Academic Press, London New York, pp 135–159

Leech RM, Murphy DJ (1976) The cooperative function of chloroplasts in the biosynthesis of small molecules. In: Barber J (ed) The intact chloroplast. Elsevier/North-Holland, Amsterdam New York, pp 365–401

Leech RM, Rumsby MG, Thomson WW, Crosby W, Wood P (1972) Lipid changes during plastid differentiation in developing maize leaves. In: Forti G, Avron M, Melandri A (eds) Proc 2nd Int Congr Photosynthesis 1971. Junk, The Hague, pp 2479–2488

Lichtenthaler HK, Becker K (1970) Inhibition of the light-induced vitamin K_1 and pigment synthesis by abscisic acid. Phytochemistry 9:2109–2113

Lichtenthaler HK, Grumbach KH (1975) Observations on the turnover of thylakoids and their prenyl lipids in *Hordeum vulgare* L. In: Avron M (ed) Proc 3rd Int Congr Photosynthesis 1974. Elsevier, Amsterdam New York, pp 2007–2015

Liedvogel B, Kleinig H (1980) Phosphate translocator and adenylate translocator in chromoplast membranes. Planta 150:170–173

Liljenberg C (1977) Chlorophyll formation: The phytylation step. In: Tevini M, Lichtenthaler HK (eds) Lipids and lipid polymers in higher plants. Springer, Berlin Heidelberg New York, pp 259–270

Ljubešić N (1968) Feinbau der Chloroplasten während der Vergilbung und Wiederergrünung der Blätter. Protoplasma 66:369–379

Lockshin A, Falk RH, Bogorad L, Woodcock CLF (1971) A coupling factor for photosynthetic phosphorylation from plastids of light- and dark-grown maize. Biochim Biophys Acta 226:366–382

Loveys BR (1977) The intracellular location of abscisic acid in stressed and non-stressed leaf tissue. Physiol Plant 40:6–10

Lütz C (1978) Separation and comparison of prolamellar bodies and prothylakoids of etioplasts from *Avena sativa* L. In: Akoyunoglou G, Argyroudi-Akoyunoglou JH (eds) Chloroplast development. Elsevier/North-Holland, Amsterdam, pp 481–488

Lyons JM (1973) Chilling injury in plants. Annu Rev Plant Physiol 24:445–466

Mackender RO (1978) Etioplast development in dark-grown leaves of *Zea mays* L. Plant Physiol 62:499–505

Marmé D (1977) Phytochrome: membranes as possible sites of primary action. Annu Rev Plant Physiol 28:173–198

Meier D, Lichtenthaler HK (1981) Ultrastructural development of chloroplasts in radish seedlings grown at high- and low-light conditions and in the presence of the herbicide Bentazon. Protoplasma 107:195–207

Mohr H (1977) Phytochrome and chloroplast development Endeavour NS 36:107–114

Mohr H, Kasemir H (1975) Control of plastid development and chlorophyll synthesis by phytochrome. Proc Indian Natl Sci Acad 41:503–525

Newcomb EH (1967) Fine structure of protein-storing plastids in bean root tips. J Cell Biol 33:143–163

Nikolić, D, Bogdanović M (1972) Plastid differentiation and chlorophyll synthesis in cotyledons of black pine seedlings grown in the dark. Protoplasma 75:205–213

Oelze-Karow H, Mohr H (1978) Control of chlorophyll b biosynthesis by phytochrome. Photochem Photobiol 27:189–193

Öquist G, Mårtensson O, Martin B, Malmberg G (1978) Seasonal effects on chlorophyll-protein complexes isolated from *Pinus silvestris*. Physiol Plant 44:187–192

Parthier B (1979) The role of phytohormones (cytokinins) in chloroplast development. Biochem Physiol Pflanz 174:173–214

Parthier B, Krauspe R, Munsche D, Wollgiehn R (1975) The biogenesis of chloroplasts. In: Harborne JB, van Sumere CF (eds) The chemistry and biochemistry of plant proteins. Ann Proc Phytochem Vol 11. Academic Press, London New York, pp 167–210

Plesničar M, Bendall DS (1973) The photochemical activities and electron carriers of developing barley leaves. Biochem J 136:803–812

Possingham JV (1980) Plastid replication and development in the life cycle of higher plants. Annu Rev Plant Physiol 31:113–129

Raison JK (1973) Temperature-induced phase changes in membrane lipids and their influence on metabolic regulation. Symp Soc Exp Biol 27:485–512

Reinert J (ed) (1980) Chloroplasts. Results and problems in cell differentiation Vol 10. Springer, Berlin Heidelberg New York

Richter G, Dirks W (1978) Blue-light induced development of chloroplasts in isolated seedling roots. Preferential synthesis of chloroplast ribosomal RNA species. Photochem Photobiol 27:155–160

Robertson D, Laetsch WM (1974) Structure and function of developing barley plastids. Plant Physiol 54:148–159

Röbbelen G (1966) Gestörte Thylakoidbildung in Chloroplasten einer Xantha-Mutante vor *Arabidopsis thaliana* L. Hyenh. Planta 69:1–26

Ryberg H, Virgin HI (1978) Red light (phytochrome)—induced acceleration of chloroplast differentiation. In: Akoyunoglou G, Argyroudi-Akoyunoglou JH (eds) Chloroplast development. Elsevier/North-Holland, Amsterdam New York, pp 793–800

Schäfers HA, Feierabend J (1976) Ultrastructural differentiation of plastids and other organelles in rye leaves with a high-temperature-induced deficiency of plastid ribosomes. Cytobiologie 14:75–90

Schnepf E (1980) Types of plastids: Their development and interconversions. In: Reinert J (ed) Chloroplasts. Results and problems in cell differentiation Vol 10. Springer, Berlin Heidelberg New York, pp 1–27

Schopfer P, Bajracharya D, Falk H, Thien W (1975) Phytochromgesteuerte Entwicklung von Zellorganellen (Plastiden, Microbodies, Mitochondrien). Ber Dtsch Bot Ges 88:245–268

Senser M, Schötz F, Beck E (1975) Seasonal changes in structure and function of spruce chloroplasts. Planta 126:1–10

Signol M (1961) Action de l'acide 3-(α-iminoéthyl)-5 méthyltétronique sur l'infrastructure des chloroplastes de Mais. CR Acad Sci Paris 252:1645–1646

Smillie RM (1976) Temperature control of chloroplast development. In: Bücher T (ed) Genetics and biogenesis of chloroplasts and mitochondria. Elsevier/North-Holland, Amsterdam New York, pp 103–110

Smillie RM, Nielsen NC, Henningsen KW, von Wettstein D (1975) Ontogeny and environmental regulation of photochemical activity in chloroplast membranes. In: Avron M (ed) Proc 3rd Int Congr Photosynthesis 1974. Elsevier, Amsterdam New York, pp 1841–1860

Smillie RM, Critchley C, Bain JM, Nott R (1978) Effect of growth temperature on chloroplast structure and activity in barley. Plant Physiol 62:191–196

Sprey B (1972) Ribosomale RNA und Thylakoidmembranen in Plastiden von Chlorophylldefektmutanten der Gerste. Z Pflanzenphysiol 67:223–243

Stobart AK, Shewry PR, Thomas DR (1972) The effect of kinetin on chlorophyll synthesis in ageing etiolated barley leaves exposed to light. Phytochem 11:571–577

Sundqvist C, Björn LO, Virgin HI (1980) Factors in chloroplast differentiation. In: Reinert J (ed) Chloroplasts. Results and problems in cell differentiation Vol 10. Springer, Berlin Heidelberg New York, pp 201–224

Taylor AO, Craig AS (1971) Plants under climatic stress II. Low temperature, high light effects on chloroplast ultrastructure. Plant Physiol 47:719–725

Taylor AO, Rowley JA (1971) Plants under climatic stress I. Low temperature, high light effects on photosynthesis. Plant Physiol 47:713–718

Thompson JA (1980) Isolation and characterization of DNA from different plastid types of *Tropaeolum majus*. Eur J Cell Biol 21:37–42

Thomson WW, Whatley JM (1980) Development of nongreen plastids. Annu Rev Plant Physiol 31:375–394

Valanne N, Valanne T (1972) Structure of plastids of a variegated *Betula pubescens* mutant. Can J Bot 50:1835–1839

Virgin HI (1961) Action spectrum for the elimination of the lag phase in chlorophyll formation in previously dark grown leaves of wheat. Physiol Plant 14:439–452

Virgin HI (1972) Chlorophyll biosynthesis and phytochrome action. In: Mitrakos K, Shropshire jr W (eds) Phytochrome. Academic Press, London New York, pp 371–404

Virgin HI (1981) The physical state of protochlorophyll(ide) in plants. Annu Rev Plant Physiol 32:451–463

Walles B (1963) Macromolecular physiology of plastids IV. On amino acid requirements of lethal chloroplast mutants in barley. Hereditas 50:317–344

Walles B (1967) Use of biochemical mutants in analyses of chloroplast morphogenesis. In: Goodwin TW (ed) Biochemistry of chloroplasts Vol 2. Academic Press, London New York, pp 633–653

Walles B, Hudák J (1975) A comparative study of chloroplast morphogenesis in seedlings of some conifers (*Larix decidua, Pinus sylvestris* and *Picea abies*). Stud For Suec 127:1–22

Weier TE, Brown DL (1970) Formation of the prolamellar body in 8-day, dark-grown seedlings. Am J Bot 57:267–275

Wellburn AR, Hampp R (1979) Appearance of photochemical function in prothylakoids during plastid development. Biochim Biophys Acta 547:380–397

Wettstein von D (1961) Nuclear and cytoplasmic factors in development of chloroplast structure and function. Can J Bot 39:1537–1545

Wettstein von D (1974) Protein synthesis in organelles and assembly of energy conservation systems. Structural and regulatory streams for the assembly of photosynthetic membranes. Biochem Soc Trans 544:176–179

Wettstein von D, Eriksson G (1964) The genetics of chloroplasts. In: Geerts SJ (ed) Genetics today Vol 3. Pergamon, Oxford, pp 591–612
Whatley JM (1977) Variations in the basic pathway of chloroplast development. New Phytol 78:407–420
Whatley JM (1978) A suggested cycle of plastid developmental interrelationships. New Phytol 80:489–502
Wollgiehn R, Parthier B(1980) RNA and protein synthesis in plastid differentiation. In: Reinert J (ed) Chloroplasts. Results and problems in cell differentiation Vol 10., Springer, Berlin Heidelberg New York, pp 97–145
Yoshida Y (1959) The correlation of nucleus with the chloroplast activity. Bot Mag Tokyo 72:397–403

13 Control of Plastogenesis in Euglena

S.D. SCHWARTZBACH and J.A. SCHIFF

1 Introduction

In dark-grown cells of *Euglena gracilis,* plastid development is arrested and only a proplastid is formed. The formation of a chloroplast from this proplastid is a substrate induction in that light, the substrate for photosynthesis by the mature chloroplast, induces the proplastid to develop into the chloroplast. There are at least two major components involved in control of chloroplast development by light. One is a generalized step-up in cellular metabolism and biosynthesis brought about largely by a blue light receptor system. The other is a series of specific inductions that are required to form the enzymes that are necessary for chloroplast development and, therefore, to form the components that are synthesized or controlled by these proteins; these specific inductions appear to be under the control of both the blue light receptor system and a red–blue absorbing system (SCHIFF 1980b).

Light, however, is only one of several substrates to which the cell can respond. Substrates like ethanol and acetate require the induction of various enzymes necessary for their utilization (COLLINS and MERRETT 1975, HORRUM and SCHWARTZBACH 1981, WOODCOCK and MERRETT 1980). Since these substrates repress light-induced chloroplast development in *Euglena,* (APP and JAGENDORF 1963, BUETOW 1967, GARLASCHI et al. 1974, HARRIS and KIRK 1969, HORRUM and SCHWARTZBACH 1980a, SCHWELITZ et al. 1978), light occupies an inferior position in the hierarchy of inducing substrates; the cells would rather expend less energy to make the systems to utilize acetate or ethanol than make the large investment in chloroplast development necessary to use light. Certain substrates such as the members of the tricarboxylic acid cycle, succinate and malate, do not require the induction of enzymes for their utilization. These substrates do not repress chloroplast development since not additional resources need be invested in order to utilize them.

This review will be concerned with the photocontrol of plastid development in *Euglena,* with a few comparative remarks to relate this system to others discussed in this volume (see Chap. 12, this Vol.). It will not be possible to provide an exhaustive survey in the space available but since the field has been reviewed at intervals, most of the material can be found in the following publications: NIGON and HEIZMANN (1978), SCHMIDT and LYMAN (1976), SCHIFF (1973, 1974, 1978, 1980a, b), LEEDALE (1967), BUETOW (1968, 1982), BUETOW and WOOD (1978), and PARTHIER (1981).

2 Arrested Development of the Plastid in Darkness and Chloroplast Development in the Light

Light is indispensable for plastid development in *Euglena gracilis* var. *bacillaris* and the Z strain. During growth in darkness plastid development is arrested at the stage of a small proplastid (KLEIN et al. 1972, OSAFUNE et al. 1980). On exposure to light, the proplastid in either dividing or non-dividing cells is induced to develop into a chloroplast (STERN et al. 1964a, b). Cells dividing in darkness maintain about ten proplastids and those dividing in the light about ten chloroplasts (KLEIN et al. 1972) through plastid division. Light-grown cells placed in darkness on a complete medium divide and the chloroplasts become reduced to proplastids (BEN SHAUL et al. 1965, PELLEGRINI 1980). If the cells are placed on a resting medium lacking carbon and nitrogen, they cease division and the chloroplasts remain much as they were; they are not reduced to proplastids in the absence of cell division (BEN-SHAUL et al. 1965). Although our information on loss of plastid structure and plastid pigments during the return of the chloroplast to the proplastid in darkness is rather meagre (see SCHOCH et al. 1981), there is considerable information on the development of the proplastid into the chloroplast in the light.

2.1 The Developmental System

Although light-induced chloroplast development from the proplastid occurs in dividing cells on a complete medium, many investigators have chosen to work with non-dividing cells maintained on a resting medium devoid of utilizable carbon and nitrogen (e.g., see STERN et al. 1964a). While these conditions provide a developmental system in which variables due to cell division are absent, the cells are starved for essential metabolites and for this reason metabolic turnover is high. The lack of external substrates has important consequences for the response of the cells to light and other inducing substrates.

The proplastid of *Euglena* in dark-grown resting cells is rather small and undifferentiated, like the proplastids of dark-grown *Ochromonas danica* (GIBBS 1962) and the younger etiolated leaves of higher plants, when compared with the etioplasts of older angiosperm leaves (KLEIN and SCHIFF 1972). The highly crystalline prolamellar bodies and extensive prothylakoids characteristic of etioplasts are not present, but what it lacks in quantity of internal membranes, the *Euglena* proplastid more than makes up in the three-dimensional complexity of its form. Based on reconstructions from serial sections (OSAFUNE et al. 1980) the *Euglena* proplastid resembles a polyp with many protuberances extending out into the surrounding cytoplasm; at the extremity of each of these protuberances is a non-crystalline prolamellar body connected with prothylakoids which extend through the plastid. Each proplastid contains several prolamellar bodies and this probably explains why several red-fluorescing centers were seen in proplastids by fluorescence microscopy (KLEIN et al. 1972). The proplastid may be somewhat plastic in extent and form since it has been reported that plastids at various stages in the life cycle of *Euglena* may come together or separate to form larger or smaller structures (LEFORT-TRAN 1975, PELLEGRINI 1980). The proplastid is closely associated with other cellular organelles such as mitochondria, the Golgi apparatus and microbody-like organelles (SCHIFF 1973, OSAFUNE et al. 1980).

On exposure to light (OSAFUNE et al. 1980, KLEIN et al. 1972), the proplastid expands somewhat and becomes more globular. Mitochondria, the Golgi apparatus, and microbody-like objects are seen in very close association with the developing plastid, and membrane whorls are much in evidence sometimes as connections to other organelles. As development proceeds, there is a progressive pairing of thylakoids with the deposition of a dense matrix between them. By 24 h of light exposure, paired thylakoids are the rule and the plastid has expanded considerably. This expansion continues concommitantly with the differentiation of the pyrenoid. The pyrenoid in *Euglena,* as in other algae, may contain ribulose bisphosphate (RuBP) carboxylase and perhaps other enzymes of the Calvin cycle in condensed form (SCHIFF 1973, 1980a).

The fully developed chloroplast at 72 to 96 h of development still has apressed mitochondria but has expanded considerably in its development from the proplastid. There has been an extensive formation of thylakoids which are clearly seen (KLEIN et al. 1972, BEN-SHAUL et al. 1964).

The morphological development of the chloroplast is correlated with a series of biochemical changes which occur concommitantly. Both soluble and membrane proteins are made de novo or increase during chloroplast development: the same is true of lipids including the plastid thylakoid pigments, nucleic acids and small molecules (see reviews cited in Sect. 1). These constituents are synthesized within the developing chloroplast or are made outside the organelle and are transported in. Although certain soluble plastid enzymes, such as the NADP triose phosphate dehydrogenase, which are synthesized outside the plastid, appear without a lag during light-induced development and reach their peak concentrations early (after about 24–36 h) (SCHIFF 1973, BOVARNICK et al. 1974b) many other constituents such as chlorophyll and cytochrome c-552 (FREYSSINET et al. 1979) show a lag in development and only reach their maximum levels after 72–96 h of development. Thus there are reasons to suppose that there are early and late events in plastid development, but of course these may overlap to form a sequence of inductions that favor the proper assembly of the organelle. The length of the lag period in non-dividing cells induced by light to form chloroplasts is dependent on the previous nutritional history of the cells. Growth on a medium containing a high carbon to nitrogen ratio favors a greater accumulation of paramylum and a shorter lag period than growth on a medium containing a lower carbon to nitrogen ratio (FREYSSINET 1976a, b). Net photosynthesis makes its appearance at about 3 h of development (STERN et al. 1964a). The formation of the various plastid constituents is highly coordinated through tight control mechanisms which undoubtably serve to insure that the various structures are assembled correctly. This is reflected in the close correlation of chlorophyll and carotenoid synthesis, and the incoming of photosynthetic function, as well as in the tight mutual control of formation of plastid lipids, plastid thylakoid polypeptides and other plastid thylakoid constituents (to be discussed later) which lead to the proper assembly of the photosynthetic membranes.

There is a concomitant activation of synthesis outside the developing chloroplast as well. For example, microbody enzymes such as glycolate dehydrogenase increase after 24 h of light-induced chloroplast development and are thought to be the consequence of induction by products of photosynthesis such as glycolate (LOR and COSSINS 1978, MERRETT and LORD 1973, HORRUM and SCHWARTZBACH 1981). In non-dividing cells undergoing light-induced chloroplastid development, mitochondrial enzymes such as fumarase and succinate dehydrogenase (HORRUM and SCHWARTZBACH 1980b, 1982) are observed to increase transiently; maximum activity is seen at the end of the lag period of development. Similar transients are observed in the activities of enzymes of early carbohydrate metabolism such as β-1, 3 glucan phosphorylase and hydrolase (DWYER and SMILLIE 1970). Enzymes of glycolysis such as phosphofructokinase, NAD glyceraldehyde-3-phosphate dehydrogenase, and aldolase, which are presumably free in the cytoplasm, decrease during light-induced chloroplast development (DOCKERTY and MERRET 1979, KARLAN and RUSSEL 1976, BOVARNICK et al. 1974b). These changes are selective in that other enzymes in the same compartments do not change in activity such as the hydroxypyruvate reductase of the microbodies and serine hydroxymethyltransferase of mitochondria and cytoplasm (HORRUM and SCHWARTZBACH 1980d). It is likely that those non-plastid enzymes which increase in activity during light-induced chloroplast development are concerned with the general step-up in cellular activity already mentioned, while those which decrease may undergo metabolic turnover to provide the amino acids for the synthesis of enzymes necessary for biosynthesis and function of the developing chloroplast. Those activities which do not change are probably necessary for general cellular housekeeping whether or not the chloroplast is being formed (i.e., they are constitutive). The extensive turnover observed in cyt rRNAs (D. COHEN and SCHIFF 1976) during light-induced chloroplast development, which is under control of the blue light receptor external to the developing chloroplast, is another example of the means available to a carbon limited cell for the reutilization of constituents for the synthesis of macromolecules of high priority.

2.2 Origin and Photocontrol of Energy and Metabolites for Chloroplast Development

There are several lines of evidence which indicate that the developing chloroplast need not obtain energy and essential metabolites from its own photosynthetic functions. The light intensity that is optimal for chloroplast development is lower than that required to saturate photosynthesis and photosynthetic activity is low during early development when many new constituents are being synthesized (STERN et al. 1964a, b). Concentrations of inhibitors of photosynthesis, such as DCMU, which block carbon dioxide fixation completely, have little effect on chloroplast development in otherwise nutritionally sufficient cells (SCHIFF et al. 1967, DWYER and SMILLIE 1971, HORRUM and SCHWARTZBACH

1981). On the other hand, inhibitors of mitochondrial respiration, such as azide (FONG and SCHIFF 1977) and anaerobiasis (KLEIN et al. 1972), and of mitochondrial phosphorylation, such as dinitrophenol (EVANS 1971), are extremely effective inhibitors of chloroplast development; an increase in O_2 uptake occurs on light induction of chloroplast development (SCHIFF 1963, FONG and SCHIFF 1977). Also, mutants of various photosynthetic organisms including *Euglena* blocked in some essential step in photosynthesis make otherwise normal chloroplasts (see SCHIFF 1980a, SCHIFF et al. 1971, SCHNEYOUR and AVRON 1975). Many organisms make chlorophyll and chloroplasts in the dark and, therefore, do not require photosynthesis for the formation of their photosynthetic machinery (see SCHIFF 1980a). All of this evidence indicates that photosynthetic metabolism is gratuitous for chloroplast development. It can be dispensed with as long as the rest of the cell can supply the energy, reducing power and metabolites required. The close association of the developing plastid with other cellular organelles such as mitochondria, microbodies, and the Golgi apparatus and with the surrounding cytoplasm (OSAFUNE et al. 1980) already mentioned is undoubtably part of the means by which this feeding of the developing plastid is accomplished. We have also mentioned that one important area of control of chloroplast development by light is the general step-up induction of the cell to provide the energy and constituents for chloroplast development. This activation of central metabolism external to the developing chloroplast by light is one means of providing the necessary metabolites to the developing organelle. It also insures that the organism is able to respond to substrate induction by light even under starvation conditions. The ability to make a chloroplast in order to utilize light energy when it becomes available, even though no carbon or nitrogen is available from the environment, would be strongly selected during evolution to insure that the cell always has enough internal reserves to make the machinery to use the light before it starves to death.

2.2.1 Endogenous Sources of Energy and Metabolites

Although turnover of proteins and nuclei acids in non-dividing cells of *Euglena* provides a source of constituents for light-induced chloroplast development (BOVARNICK et al. 1974b, HORRUM and SCHWARTZBACH 1980a, D. COHEN and SCHIFF 1976), carbon storage compounds such as paramylum (SCHWARTZBACH et al. 1975) and wax (ROSENBERG and PECKER 1964) are also mobilized on exposure of the dark-grown cells to light; wax can also be utilized for limited development in darkness (SHIHIRA-ISHIKAWA et al. 1977). Since light-induced breakdown of paramylum takes place at a lower rate in dark-grown bleached mutants such as W_3BUL and W_{10}BSmL (SCHWARTZBACH et al. 1975, HORRUM and SCHWARTZBACH 1980b) compared with dark-grown wild-type cells (SCHWARTZBACH et al. 1975, FREYSSINET 1976a, FREYSSINET et al. 1972, DWYER and SMILLIE 1970, 1971) this process may be controlled, at least in part, by a non-plastid photoreceptor system. The higher rate in wild-type cells may be due to control of paramylum degradation by both the non-plastid and plastid photoreceptors, or because the presence in wild-type cells of a proplastid capable of development provides a metabolic sink for the products formed from paramylum.

Cycloheximide, a specific inhibitor of cytoplasmic protein synthesis in *Euglena* (KIRK 1970) or levulinic acid, a competative inhibitor of 5-aminolevulinic acid (ALA) utilization for chlorophyll synthesis (RICHARD and NIGON 1973) can replace light as an inducer of paramylum degradation (SCHWARTZBACH et al. 1975, HORRUM and SCHWARTZBACH 1980b). A model based on negative transcriptional control has been proposed which suggests that levulinic acid may act as an analog of a normal corepressor (ALA) in controlling the formation of a protein on cytoplasmic ribosomes which inhibits paramylum degradation (SCHWARTZBACH et al. 1975, SCHIFF 1978). Another model, based on light-induced changes in membrane transport, is also applicable. Although cycloheximide does not inhibit respiration in *Euglena* (KIRK 1970) it does appear to inhibit membrane transport in a manner unrelated to its action on protein synthesis (EVANS 1971, PARTHIER 1974). Evans suggested that cycloheximide produces a redistribution of intracellular ions which causes a decrease in the pH gradient between the inside and the outside of the cell; this altered pH gradient is thought to be responsible for the inhibition of transport of various compounds into the cells. Levulinic acid might act in a similar manner since it is a weak acid which is taken up by the cells. Thus light might induce paramylum breakdown in *Euglena* by altering membrane transport in such a manner as to change the intracellular pH. A similar model has been offered to explain light-induced starch degradation in *Scenedesmus* (see BRINKMANN and SENGER in SENGER 1980).

2.2.2 Influence of Exogenous Sources of Energy and Metabolites

A number of utilizable carbon sources have been reported to repress light-induced chloroplast development (APP and JAGENDORF 1963, BUETOW 1967, GARLASCHI et al. 1974, HARRIS and KIRK 1969, HORRUM and SCHWARTZBACH 1980a, SCHWELITZ et al. 1978). In some cases, repression occurs even when cells grown in darkness on the repressing carbon source are exposed to light in the absence of the carbon source.

The presence of a utilizable nitrogen source can reverse the apparent repression of chloroplast development due to growth on a particular carbon source (HARRIS and KIRK 1969, SCHWELITZ et al. 1978). Since the amount of chlorophyll formed on reaching the stationary phase is proportional to the nitrogen content of the medium (HORRUM and SCHWARTZBACH 1980a), the repression of chloroplast development observed during growth of the cells on media containing a high carbon to nitrogen ratio probably results from an inability of the nitrogen-deficient cells to synthesize chlorophyll rather than from a specific repression by a particular carbon source. By using nitrogen-sufficient dark-grown resting cells supplemented with various carbon sources at the time of light exposure, it has been shown that ethanol or acetate specifically represses light-induced chloroplast development (HORRUM and SCHWARTZBACH 1980a, 1981).

Light, ethanol and/or malate induce fumarase, succinic dehydrogenase and phosphoglycerate phosphatase in dark-grown resting cells (HORRUM and

SCHWARTZBACH 1980b, JAMES and SCHWARTZBACH unpublished) but these do not induce fumarase in cells maintained in balanced growth (HORRUM and SCHWARTZBACH 1982). Enzymes induced by light or organic carbon are required for growth on all carbon sources, suggesting that their induction represents a general step-up response, a consequence of providing a starving cell with a utilizable carbon source. Enzymes induced by light and repressed by ethanol, however, are specifically required for phototrophic growth. The catabolite repression of light-induced chloroplast development by those organic substrates which require enzyme induction for their utilization indicates the inferior position of light in the hierarchy of inducing substrates.

2.3 Origin and Photocontrol of Genetic Information for Chloroplast Development

2.3.1 Mutants Blocked in Chloroplast Development

Mutants of *Euglena* can be obtained which are blocked in chloroplast development in various ways (SCHIFF et al. 1971, 1980, BINGHAM and SCHIFF 1979a). These are thought to be mutants in plDNA because the nucleus appears to be 2n or higher in ploidy and nuclear mutants would not be expressed unless dominant. Mitochondrial mutants affecting function also are unlikely since *Euglena* is an obligate aerobe and such mutants would be lethal.

The bleached mutants of *Euglena* (many of these are designated W for white) seem to lack much or all of their plDNA. These mutants are produced by treatment with bleaching agents such as ultraviolet light, temperatures above 34 °C (the chloroplast appears to be a heat-sensitive organelle in a more resistant cell), and antibiotics such as streptomycin (see references in SCHIFF 1980a, SCHIFF et al. 1971, 1980; and reviews cited in Sect. 1). As the cells divide during or after treatment with these agents, replication of plDNA appears to cease and the remaining plDNA is diluted among the progeny; in some cases it may be more or less destroyed. On plating these cells in the light, white colonies are recovered which after cloning never revert to green (SRINIVAS and LYMAN 1980, GOLDSTEIN et al. 1974, NICOLAS et al. 1977).

In one of the best-studied of these mutants, W_3BUL, plDNA is undetectable by analytic ultracentrifugation in cesium chloride (EDELMAN et al. 1965), separation on columns (E. STUTZ personal communication) and by fluorescence microscopy using a DNA-specific fluorochrome (COLEMAN 1979). It has been suggested that some bleached mutants may contain very small amounts of residual plDNA limited perhaps to some ribosomal cistrons or somewhat more (HEIZMANN et al. 1981). It has been reported that most of our laboratory strains of *Euglena* are infected with a *Mycoplasma* (RIGGS 1979, RIGGS personal communication). This should be considered when interpreting the presence of minor DNA components since the base compositions of the mycoplasma DNAs are very close to those of pl and mtDNAs of *Euglena* (A+T=60–77 mol %: BUCHANAN and GIBBONS 1974).

Although wild-type cells can make colored carotenoids in darkness and do not accumulate ζ carotene, certain bleached mutants accumulate this intermediate of carotenoid biosynthesis in the cis form. Illumination of the cells with blue light causes a cis to trans photoisomerization of the ζ carotene. There may be two means of isomerizing ζ carotene in wild-type cells, one a direct photochemical reaction, the other a dark enzymatic reaction (FONG and SCHIFF 1979, STEINITZ et al. 1980, SCHIFF et al. 1982).

Dark-grown cells of the bleached mutant W_3BUL contain a proplastid remnant which is reduced to a thin leaflet lacking plastid ribosomes and most of the stroma and prothylakoids (OSAFUNE and SCHIFF 1980a, PARTHIER and NEUMANN 1977, SCHIFF and EPSTEIN 1968, PALISANO and WALNE 1976). As in wild type, close associations with other cellular organelles such as mitochondria are seen. When the cells are exposed to light, a limited expansion of the plastid occurs with the formation of a large vacuole and a crystalline prolamellar body, but plastid development does not proceed beyond this point. Since no plastid ribosomes are seen and no plrRNA is detected (D. COHEN and SCHIFF 1976) in W_3BUL, if any plastid DNA remnants are present it is unlikely that they provide translatable information for the formation of plastid proteins. Therefore, it is likely that the prothylakoids and other structures seen in this proplastid remnant are nuclear-coded and that the vacuole represents an area which would ordinarily be filled with materials formed when plDNA was present and active, as in wild-type cells.

Various properties of bleached mutant W_3BUL relevant to chloroplast development are summarized in Table 1. In general, chloroplast components are low in concentration (are present at repressed levels comparable to those in dark-grown wild-type cells) or are undetectable. Since plDNA is undetectable in W_3 by conventional methods, this mutant has been useful as an indicator of where various plastid components may be coded. Those present in W_3 (even at repressed levels) are thought to be nuclear coded since there is no conclusive evidence that the limited mitochondrial genome contributes genetic information for plastid development. Those plastid components that are not detectable in W_3 may be coded in plDNA, but is should be recalled that a structural gene for a protein may be elsewhere than in plDNA while a regulatory gene in plDNA regulates its expression.

Various photoresponses associated with light-induced chloroplast development, particularly those thought to be localized outside the plastid, persist in W_3. The induction of transcription of cyt rRNA from nDNA by blue light (D. COHEN and SCHIFF 1976), light-induced paramylum breakdown (SCHWARTZBACH et al. 1975), a light-induced change in the rate of leucine incorporation into protein (SCHWARTZBACH and SCHIFF 1979), induction of the photoreactivating enzyme (DIAMOND et al. 1975), induction of the chloramphenicol-reducing system (VAISBERG et al. 1976, SCHIFF 1974), and a light-induced increase in cellular respiration (SCHIFF 1963, FONG and SCHIFF 1977) are examples. As a general rule, W_3 appears to retain (at least partially) those blue light photoresponses connected with the general step-up of the central metabolism of the cell necessary for feeding the developing chloroplast. This mutant, however, appears to lack the photoresponses which are concerned with the specific induction of various

Table 1. Selected properties of mutant W_3BUL compared with dark and light-grown wild-type *Euglena* cells (Schiff 1980b)

Parameter	W_3BUL	Wild type	
		Dark-grown	Light-grown
Plastid DNA	Not detectable	Present	Present
Chlorophyll	Not detectable	Not detectable	Present
Carotenoids	Low	Low	High
Photosynthesis	Not detectable	Not detectable	Present
Various plastid enzymes thought to be Nuclear-Coded[a]	Low	Low	High
Various plastid enzymes thought to be plastid-Coded, at least in part[b]	Not detectable	Not detectable or low	High
Non-plastid enzymes[c]	High	High	High
Plastid rRNA's, ribosomes and tRNA's	Not detectable	Low	High
Cytoplasmic rRNA's, ribosomes and tRNA's	High	High	High

[a] e.g., NADP triose phosphate dehydrogenase, alkaline DNAase, leucyl tRNA Synthetase, enzymes forming sulfolipid, proplastid thylakoid polypeptides, etc.
[b] e.g., RuBP Carboxylase, etc.
[c] e.g., NAD triose phosphate dehydrogenase, etc.

plastid components such as those enzymes which remain low or undetectable in illuminated cells of W_3 (Table 1).

2.3.2 Sources of Genetic Information for the Formation of Plastid Constituents

There are three known potential sources of genetic information in *Euglena* cells (Edelman et al. 1965). Nuclear DNA constitutes about 90% of cell DNA with mt and plDNA making up the total with about 5% each, although the actual amounts vary with the growth conditions (Chelm et al. 1977, Rawson and Boerma 1976). The nuclear genome is thought to be diploid based on renaturation kinetics (Rawson 1975, Rawson et al. 1979), but octaploidy has been suggested based on target theory (Hill et al. 1966, see Schiff et al. 1980). Mitochondrial DNA is thought to have a molecular weight of about 40×10^6 based on renaturation experiments and hybridization with mt rRNA (Crouse et al. 1974, Talen et al. 1974). Plastid DNA is circular with a molecular weight of 92×10^6 (Manning and Richards 1972, Manning et al. 1971, Slavik and Hershberger 1975). Hybridization experiments indicate that cyt rRNA is coded in nDNA (Gruol and Haselkorn 1976) and that mt rRNA is coded in mtDNA (Crouse et al. 1974); the pl rRNAs are coded in plDNA (Gray and Hallick 1978). The plDNA codes for 26–30 tRNAs (Gruol and Haselkorn 1976,

McCrea and Hershberger 1976, Mubumbila et al. 1980, Schwartzbach et al. 1976a); the cyt tRNAs are coded in nDNA (Gruol and Haselkorn 1976).

Consistent with earlier genetic experiments in various organisms (see Schiff 1980a), genetic information for the formation of plastid proteins originates in both the plastid and nuclear genomes. Plastid proteins which are present in bleached mutants in which plDNA is undetectable are thought to be nuclear-coded; those whose synthesis is also sensitive to cycloheximide but insensitive to streptomycin and chloramphenicol are thought to be translated on cytoplasmic ribosomes. These include: NADP glyceraldehyde-3-phosphate dehydrogenase (Bovarnick et al. 1974b); photoreactivating enzyme (Diamond et al. 1975); chloroplast-specific aminoacyl tRNA synthetases (Hecker et al. 1974); chloroplast ribosomal proteins (Freyssinet 1977, 1978); chloroplast elongation factors G and EF-TS (Breitenberger et al. 1979, Breitenberger and Spremulli 1980, Fox et al. 1980), alkaline DNAase (Egan and Carrell 1972), class I aldolases (Karlan and Russel 1976, Russel et al. 1978, Schmidt and Lyman 1974), phosphoglycerate kinase (Schmidt and Lyman 1974, Russel et al. 1978); phosphoglycolate phosphatase (Horrum and Schwartzbach 1980c) and a number of thylakoid polypeptides (Bingham and Schiff 1979a, b). The enzymes of carotenoid biosynthesis (Vaisberg and Schiff 1976), of sulfolipid formation (Bingham and Schiff 1979a) and of sulfate activation and reduction (Brunold and Schiff 1976) also appear to belong to this group.

Studies with specific inhibitors of chloroplast translation such as streptomycin and chloramphenicol (Avadhani and Buetow 1972, Schwartzbach and Schiff 1974) and labeling of various proteins in isolated chloroplasts supplied with radioactive amino acids have served to identify proteins that are translated (and might be coded) in the chloroplast. Absence of a protein from a mutant in which plDNA is undetectable may be an indication of chloroplast coding but further evidence is necessary since the structural gene could be in nDNA (or mDNA) and be regulated by a gene product of plDNA. One fairly unambiguous case is the large subunit of ribulose bisphosphate carboxylase. This polypeptide is undetectable in mutants in which plDNA is undetectable (Sagher et al. 1976), the increase in enzyme activity during chloroplast development is inhibited by streptomycin (Bovarnick et al. 1974b), the mRNA for the polypeptide is found in the non-poly (A) fraction of plRNA (Sagher et al. 1976), the polypeptide is synthesized in isolated chloroplasts (Vasconcelos 1976) and the gene that codes for it has been tentatively localized to a site in the plDNA genome (Hallick et al. 1979). The following plastid proteins may be chloroplast-coded and translated but more evidence is necessary to be certain: various thylakoid membrane proteins (Bingham and Schiff 1979a, b, Vasconcelos 1976), a number of plastid ribosomal proteins (Freyssinet 1977) and various unidentified soluble plastid proteins (Vasconcelos 1976).

Although the first enzyme(s) of chlorophyll synthesis is made in the cytoplasm (Beney and Nigon 1975, Richard and Nigon 1973), it is not known with certainty where the other enzymes of chlorophyll biosynthesis are coded and translated. A long pathway of this sort would be expected to have enzymes from both the nuclear-cytoplasmic group and the chloroplast group, and this would explain why chlorophyll synthesis is inhibited both by inhibitors of cytoplasmic and chloroplastic translation (see FREYSSINET et al. 1979). It should be remembered, however, that this could also be a consequence of the tight co-regulation of the formation of one thylakoid constituent by another (BINGHAM and SCHIFF 1979a, b).

2.4 Photocontrol of Formation of Thylakoid Membrane Constituents

The work carried out so far points to an extremely tight control of the formation of each thylakoid membrane constituent by the others. Unusual light regimes have been used to defeat this control in some instances (GUREVITZ et al. 1977) but under normal continuous illumination the formation of thylakoid polypeptides, lipids, and pigments appear to be closely coupled, insuring that constituents will be present at the right time and in the correct amounts to achieve an orderly assembly of the membranes (BINGHAM and SCHIFF 1979a, b).

2.4.1 Chlorophyll Synthesis and the Consequences of Preillumination

As already indicated, chlorophyll synthesis begins slowly on illumination of dark-grown cells and after a lag of some 12 h enters the period of rapid synthesis (STERN et al. 1964a). Chlorophyll synthesis during the lag period does not require chloroplast protein synthesis, but cytoplasmic protein synthesis is necessary (BOVARNICK et al. 1974a, SCHWARTZBACH et al. 1976b). Protein synthesis in both compartments is necessary during the subsequent period of rapid synthesis of chlorophyll. There is also a lag in the synthesis of thylakoids and their polypeptides, a period during which newly synthesized chlorophyll appears to fill already existing membrane sites (BINGHAM and SCHIFF 1979b, KLEIN et al. 1972, BOVARNICK et al. 1974a).

The enzyme(s) producing the first committed precursor of chlorophyll, 5-aminolevulinic acid (ALA) is synthesized on cytoplasmic ribosomes upon light induction (BENEY and NIGON 1975, RICHARD and NIGON 1973). Light is required for maintenance of the enzyme level, which falls if cytoplasmic protein synthesis is inhibited or the cells are darkened. Synthesis of ALA is required for chlorophyll synthesis during both the lag and the period of rapid synthesis of chlorophyll (SCHWARTZBACH et al. 1976b), but conditions which eliminate the lag period do not produce an accumulation of ALA or increase the capacity of the cell for ALA synthesis.

The length of the lag period in chlorophyll synthesis depends on the previous nutritional history of the cells. Cells exposed to light in the presence of ethanol or acetate have an indefinite lag period and never enter the period of rapid chlorophyll synthesis (HORRUM and SCHWARTZBACH 1980a). The length of the lag period in cells grown on other substrates (FREYSSINET 1976a) is inversely related to their paramylum content and consumption. Cells differing in paramylum content undoubtedly differ in their content of other macromolecules and various reserves of carbon, nitrogen and phosphorous (FREYSSINET et al. 1972); paramylum content may be only one indicator of the general nutritional state of the cells. Nutrient-rich cells would, in general, be expected to synthesize the machinery required for rapid chlorophyll formation more speedily than nutrient-poor cells.

The lag period in chlorophyll synthesis can be eliminated by exposing the dark-grown cells to a 2-h preillumination period followed by a 12-h dark period prior to placing them in continuous light for chloroplast development (HOLOWINSKY and SCHIFF 1970). Synthesis of proteins on chloroplastic and cytoplasmic ribosomes during the preillumination period and during the subsequent dark period are required for lag elimination (SCHWARTZBACH et al. 1976b).

The optimal length of the dark period required for lag elimination is approximately equal to the length of the lag period (FREYSSINET 1976b, HOLOWINSKY and SCHIFF 1970); consequently, cells with higher paramylum contents require shorter dark periods. Although the degradation of paramylum is strictly light-dependent (FREYSSINET et al. 1972, SCHWARTZBACH et al. 1975), the optimal length of the preillumination period for lag elimination is not related to the paramylum content of the cells. Cells which have received a preillumination period followed by a dark period and cells not so treated show similar rates of paramylum degradation when subsequently exposed to continuous light to allow chloroplast development to take place. Thus it appears that the preillumination period provides the carbon and energy for the synthesis of proteins in the subsequent dark period that are required for rapid chlorophyll synthesis; the same events must occur during the normal lag period in non-preilluminated cells, since the end result in both cases is rapid chlorophyll synthesis without a further lag.

Other early strictly light-dependent events of chloroplast development that are not influenced by preillumination include the kinetics of NADP-glyceraldehyde-3-phosphate dehydrogenase synthesis (BOVARNICK 1969), alkaline DNAase synthesis (EGAN and SCHIFF unpublished experiments), fumarase synthesis (HORRUM and SCHWARTZBACH 1980b) and cyt rRNA synthesis (D. COHEN and SCHIFF 1976). As might be expected from the observation that carotenoid levels do not change significantly during the lag period of normal development, (STERN et al. 1964a, BOVARNICK et al. 1974a), preillumination does not eliminate the lag in carotenoid synthesis (BOVARNICK 1969); the lag period in appearance of carbon dioxide fixation is also not eliminated by preillumination (SCHWARTZBACH et al. 1976b).

Preillumination does increase the capacity of the developing chloroplast to synthesize proteins. Thus preillumination induces the synthesis of pl rRNA (D. COHEN and SCHIFF 1976) and chloroplast valyl tRNA synthetase (MCCARTHY et al. 1981) and this continues through the dark period. Preillumination increases the rate of cytochrome c-552 synthesis on subsequent illumination (FREYSSINET et al. 1979) and shortens the lag period for thylakoid membrane formation (KLEIN et al. 1972). Preillumination also increases the rate of formation of cytoplasmic polysomes (FREYSSINET 1977).

Both photoreceptor systems known to participate in the regulation of chloroplast development are involved in the regulation of chlorophyll biosynthesis. The action spectrum for chlorophyll synthesis in preilluminated or non-preuilluminated cells exposed to continuous light indicates that a protochlorophyll(ide) holochrome is the light receptor molecule (EGAN et al. 1975, EGAN and SCHIFF

1974); blue and red light are both effective. The preillumination action spectrum indicates the participation of a blue light receptor system augmented to some extent by the red–blue protochlorophyll(ide) system (EGAN et al. 1975, HOLOWINSKY and SCHIFF 1970).

The experiments just described lend themselves to the following interpretation. The blue light system controls the central metabolism of the cell and other machinery largely external to the developing plastid; this system is thought to be predominantly involved in the step-up response that provides the materials and energy for feeding the developing chloroplast. For this reason this system should be the first to be activated. The lag period in plastid development appears to be the interval in which these externally supplied constituents are entering the developing chloroplast from the rest of the cell. The end of the lag period would signal the accumulation of sufficient imports to allow the machinery within the plastid to operate at full capacity and development enters the rapid period of synthesis and expansion where induction of formation of specific plastid constituents by the red–blue receptor becomes the predominant feature.

2.4.2 Protochlorophyll(ide) and Related Pigments

The proplastid thylakoids of *Euglena* contain protochlorophyll and protochlorophyllide which are transformable to chlorophyll(ide) in the light (C. COHEN and SCHIFF 1976, KINDMAN et al. 1978). Large amounts of protopheophytin and protopheophorbide are found as well (C. COHEN and SCHIFF 1976), but it is not known for certain whether these are normal constituents or are formed during extraction and purification of the pigments. Control experiments and the constancy of protopheophorbide/protopheophytin levels, however, indicate that they may not be wholely artifacts. Studies using fluorescence spectroscopy indicate the possibility of magnesium insertion into protopheophytin/propheophrobide to form protochlorophyll(ide) (FREY et al. 1979). The growing evidence that pheophytins of chlorophyll and bacterochlorophyll are normal functional constituents of photosynthetic reaction centers (STRALEY et al. 1973, KLIMOV et al. 1980, RUTHERFORD et al. 1981) should keep us alert to the possibility that not all of the protopheophytin, protopheophorbide, pheophytin and bacteriopheophytin found on extraction of cells and photosynthetic membranes is artifactual and that biosynthetic pathways may exist for the formation of these magnesium-free pigments.

The absorption spectrum of dark-grown cells of *Euglena* indicates that they contain only a protochlorophyll(ide) species absorbing at 635 nm but lack the form of protochlorophyll(ide) absorbing at 650 nm (KINDMAN et al. 1978). On illumination, the 635 nm absorption decreases and a new absorption due to chlorophyll(ide) appears at 676 nm. This absorption very rapidly shifts a few nm to shorter wavelengths to lie at about 673 nm. In darkness, after illumination, protochlorophyll(ide) absorbing at 635 nm is resynthesized to the same levels as before illumination, indicating a tight control of its synthesis. When dividing cells of *Euglena* are placed under non-dividing conditions on a medium lacking available carbon and nitrogen, protochlorophyll(ide) is progressively lost from the cells, undoubtedly being subject to turnover into other cellular constituents needed for survival under starvation conditions (ALHADEFF et al. 1979, 1980). However, on illumination of the cells, or addition of a carbon source such as glutamate or malate, the protochlorophyll(ide) is rapidly regenerated. Since light is known to induce paramylum breakdown, (SCHWARTZBACH et al. 1975), light or an exogenous carbon source appears to provide needed carbon for protochlorophyll(ide) biosynthesis and for a general step-up in metabolism, signaling that nutrients are available and that turnover and loss of certain constituents can cease.

On the whole, the proplastids and protopigments of *Euglena* are more like those of younger etiolated seedlings than they are like the better-studied older etiolated seedlings (LANCER et al. 1976, KLEIN and SCHIFF 1972). Thus *Euglena* cells have proplastids rather than etioplasts, lack crystalline prolamellar bodies, contain protochlorophyll(ide) absorbing at 635 rather than 650 nm, etc. The smaller less highly developed proplastid system of *Euglena* [like that in *Ochromonas* (GIBBS 1962)] may be the more normal route of chloroplast development; the more highly developed etioplasts of older higher plants appear to be an adaptation to prolonged etiolation where faster development on illumination is an asset to allow rapid formation of photosynthesis before seed reserves are exhausted (SCHIFF 1980a).

2.4.3 Plastid Thylakoid Polypeptides, Sulfolipid, and Carotenoids

Like similar preparations from other organisms, plastid thylakoid membranes from chloroplasts or from wild-type light-grown cells of *Euglena gracilis* var. *bacillaris* or strain Z display 30–40 polypeptides on SDS polyacrylamide gel electrophoresis (BINGHAM and SCHIFF 1979a). Some of the more prominent bands among them (those thought to be associated with the pigment–protein complexes) are not detectable in similar preparations from dark-grown wild-type cells. The patterns from thylakoid membranes of wild-type dark-grown cells are identical with those obtained from either light or dark-grown bleached mutants such as W_3BUL, indicating that most, if not all, of the thylakoid polypeptides of the proplastid may be coded elsewhere than in plDNA, probably in nDNA. Certain plastid proteins are present at low levels in dark-grown cells of wild type, but are undetectable in bleached mutants such as W_3BUL. Such molecules, at least in part, are candidates for coding in plDNA; some of these polypeptides can be found among those labeled in isolated chloroplasts supplied with radioactive amino acids (BINGHAM and SCHIFF 1979a, VASCONCELOS 1976).

When dark-grown wild-type cells are exposed to light, some new plastid thylakoid polypeptides appear, but all plastid thylakoid polypeptides increase concommitantly with development of the plastid membranes seen by electron microscopy (BINGHAM and SCHIFF 1979a, b, KLEIN et al. 1972, BEN-SHAUL et al. 1964).

The thylakoid lipids, among them the sulfolipid and the polyunsaturated fatty acids, increase during chloroplast development (ROSENBERG and PECKER 1964, ERWIN 1968). The carotenoids are at low levels in the dark-grown cells and also increase during light-induced chloroplast development (STERN et al. 1964a). The selective inhibitor of carotenoid biosynthesis Norflurazon (SAN

9789), which blocks the pathway just after phytoene formation, has been useful in studying the developmental consequences of carotenoid depletion. In Norflurazon-treated cells, the levels of plastid thylakoid polypeptides, chlorophyll, and sulfolipid formed during light-induced chloroplast development are closely related to the amount of carotenoid formed, another example of the coregulation of synthesis of plastid thylakoid components that insures correct assembly of the thylakoid structure (VAISBERG and SCHIFF 1976). Lack of synthesis of chlorophyll and other constituents when carotenoid synthesis is blocked would have been highly selected during evolution since carotenoids are known to protect the cells from deleterious photoxidations sensitized, in some cases, by chlorophyll. In higher plants, for some reason, this control is not as stringent, higher plants make normal proplastids and protochlorophyll(ide) in the absence of carotenoids (PARDO and SCHIFF 1980) and, in some cases can even undergo a limited amount of light-induced chloroplast development in dim light. Carotenoids and light are necessary for the differentiation of the stigma in *Euglena* (OSAFUNE and SCHIFF 1980b).

The addition of inhibitors of translation on either 87S cytoplasmic ribosomes of *Euglena* (such as cycloheximide) or on 70S chloroplast ribosomes (such as streptomycin or chloramphenicol) blocks the incoming of all plastid thylakoid polypeptides during light-induced chloroplast development (BINGHAM and SCHIFF 1979b); the same pattern of inhibition is observed for cytochrome C-552 and chlorophyll (FREYSSINET et al. 1979). Cytochrome c-552 formation is also correlated with the presence of protochlorophyll(ide) (RUSSEL et al. 1978). The uniformity of these responses points once again to a high degree of coregulation of the biosynthesis of various membrane constituents.

3 Photoreceptors and Levels of Control

It is apparent from the foregoing discussion that at least two photoreceptors control chloroplast development in *Euglena*. Although action spectra are not easy to measure or to interpret, they constitute the best evidence, at present, for the nature of the molecules involved. Mutants blocked in each of the photoreceptors or in the expression of their function (RUSSEL et al. 1978, SCHMIDT and LYMAN 1974) should also help in their identification and in sorting out the developmental systems they control.

3.1 The Red-Blue Photoreceptor System

The action spectrum for chlorophyll biosynthesis during continuous greening is very similar to the absorption spectrum of the protochlorophyll(ide) holochrome from beans, within the precision of the measurements (EGAN et al. 1975, EGAN and SCHIFF 1974). Thus the action spectrum shows a ratio of blue to red peaks of about 5:1. The action spectrum is in accord with the information we have on the photoconversion of protochlorophyll(ide) to chlorophyll(ide) in intact dark-grown cells (KINDMAN et al. 1978). Detailed action spectra during greening for the formation of various plastid enzymes thought to be nuclear-coded and cytoplasmically synthesized such as alkaline DNAase and NADP triose phosphate dehydrogenase also resemble the absorption of protochlorophyll(ide) holochrome (EGAN et al. 1975). Action spectra exist for the formation of plastid rRNA and for chlorophyll synthesis in the postillumination period of potentiation which are consistent with the same type of photoreceptor (EGAN et al. 1975, HOLOWINSKY and SCHIFF 1970, D. COHEN and SCHIFF 1976). It should be noted, however, that the data do not rule out the participation of other photoreceptors as well. For example, the action spectra are frequently rather broad, the dose response curves are complex enough to suggest that more than one photoreceptor system is operating, and red and far-red light are sometimes additive. It is not conclusively shown that all red-blue phenomena are the same in *Euglena*. For example, although protochlorophyll(ide) photoreduction to chlorophyll(ide) is an obligatory step in chlorophyll biosynthesis, it is not known whether protochlorophyll(ide) itself or other pigments which mimic its absorption are active in the other red-blue phenomena such as photocontrol of protein and nucleic acid synthesis (D. COHEN and SCHIFF 1976).

3.2 The Blue Receptor System

In *Euglena,* this photoreceptor makes itself known as an unusually high blue peak in action spectra for the preillumination which brings about the elimination of the lag in chlorophyll synthesis (EGAN et al. 1975, HOLOWINSKY and SCHIFF 1970); the blue to red ratio is about 40:1. It appears to be unmixed with the red–blue receptor in broad band action spectra for the formation of cytoplasmic rRNAs during chloroplast development (D. COHEN and SCHIFF 1976) and in mutants lacking the red–blue response (SCHMIDT and LYMAN 1974, RUSSEL et al. 1978). Since the same action spectrum can be measured in mutant W_3BUL in which plastid DNA, plastid ribosomes and protochlorophyll(ide) are undetectable, and since it appears to control the transcription of cyt rRNA genes in nDNA, this photoreceptor probably occurs externally to the plastid. Whether this receptor has peaks in regions of the spectrum other than the blue is not known with certainty; the absorption of this system in the near ultraviolet appears to be quite low (EGAN et al. 1975). This blue light system, therefore, joins a long list of other similar blue receptors of unknown composition referred to collectively as cryptochrome (SCHIFF 1980b). Blue light receptors are known to be of importance in controlling chloroplast development in other organisms such as light-requiring mutants of *Scenedesmus* and *Chlorella* (see SENGER 1980), organisms that ordinarily make chlorophyll and chloroplasts in darkness.

3.3 Co-Regulation by the Two-Photoreceptor Systems

Although phenomena controlled by each of the two photoreceptor systems can be identified, there is also evidence for coordinate control by both systems.

For example, a number of plastid enzymes thought to be nuclear-coded and cytoplasmically translated are repressed in dark-grown wild-type cells and are induced to increase by light. The action spectra indicate that the red–blue photoreceptor controls these inductions (EGAN et al. 1975) and consistent with this (Table 1), mutant W_3BUL which lacks protochlorophyll(ide) and the red–blue induction of pl rRNA synthesis shows no derepression of these enzymes on illumination (BOVARNICK et al. 1974b). However, another mutant (Y_9ZNalL), lacking protochlorophyll(ide) but thought to retain plDNA, shows a derepression of these enzymes by light absorbed by the blue receptor (RUSSEL et al. 1978, SCHMIDT and LYMAN 1974). This evidence indicates a coordinate control of the induction of these enzymes in wild-type cells by the blue and red–blue photoreceptors, as in lag elimination in chlorophyll synthesis (HOLOWINSKY and SCHIFF 1970). Consistent with this interpretation, red light alone can bring about normal chloroplast development in *Euglena* (BOVARNICK et al. 1969, BINGHAM and SCHIFF 1979b). If we assume that the blue system has no red absorption whatever, the absorption of light by the red–blue system must somehow generate a signal that induces the blue-controlled system in the absence of blue light. This co-regulation may be related to the co-regulation of one plastid thylakoid constituent by another already discussed.

3.4 Levels of Control

Since rather low levels of light are effective in inducing chloroplast development in *Euglena* (STERN et al. 1964b, HOLOWINSKY and SCHIFF 1970), some sort of cellular amplification must occur to convert the photoresponse into an effective metabolic signal. By analogy to other prokaryotic and eukaryotic systems, one might expect control at any level of cellular function including information processing (transcription, translation and post-transational modification) and modulation of membrane transport. So far, in *Euglena,* evidence exists only for transcriptional control by light [e.g., the control of rRNA transcription from nDNA and plDNA (D. COHEN and SCHIFF 1976)]. A detailed model has been offered based on control of prokaryotic transcription in which pigmented repressors are modified by light in such a manner as to derepress the operons they control (SCHIFF 1981a, b). Since light is an inducing substrate for the formation of the machinery to use light for photosynthesis resulting in chloroplast development, this is analogous to substrate induction by organic molecules in prokaryotes.

We should not overlook the possibility of control at the level of membrane transport by either of the two photoreceptors. The control of movement of ions or substrates by photoreceptors localized in membranes could achieve the amplification needed for control (see BRINKMANN and SENGER in SENGER 1980). Although the photoreduction of protochlorophyll(ide) to chlorophyll(ide) serves as a control point in chlorophyll biosynthesis, it is not known whether the other manifestations of control by the red–blue receptor are due to protochlorophyll(ide) itself or to another molecule with similar absorption properties. Since protochlorophyll(ide) is membrane-localized and requires two hydrogens for reduction, it could play a strategic role in the photocontrol of processes such as transcription of plDNA by modulating the levels of hydrogen and other ions within the chloroplast. This would appear to be the way for bulky, and often insoluble, molecules to control processes some distance away from their resident sites.

4 Conclusion

The picture of chloroplast development in *Euglena* we have tried to assemble is as follows. We consider light-induced chloroplast development to be a substrate induction by light where an entire organelle is constructed to utilize the inducing substrate. This substrate induction has at least two components: a blue light system and a red–blue system. Illumination of the blue receptor, thought to be localized outside the chloroplast, produces a step-up in the central metabolism of the cell to provide the energy and metabolites necessary to feed the developing chloroplast, particularly during the lag period early in development. It also controls transcription of nuclear DNA. In higher plants, phytochrome usually assumes these functions; phytochrome has not been detected in *Euglena* (HOLOWINSKY and SCHIFF 1970, BOUTIN and KLEIN 1972). The red–blue system is probably localized in the plastid and is thought to control the formation of chlorophyll and the transcription in *Euglena* of plDNA. The blue light and red–blue systems appear to act coordinately in the induction of plastid proteins thought to be coded in nDNA. In this way, the specific substrate induction of individual chloroplast proteins is achieved.

These various control systems absorb in the blue and red regions of the spectrum, whatever their chemistry or level of control, because the substrate to be utilized by the fully induced chloroplast for photosynthesis is blue and red light. There would have been a strong selection during evolution for control pigments which absorbed in the same regions of the spectrum as the sensitizers of photosynthesis to insure that cellular resources would be expended to form a chloroplast only when light of the appropriate quality for photosynthesis was availbable.

Acknowledgement. This work was supported by grants from the National Institutes of Health: GM 14595 (JAS) and GM 26994 (SDS).

References

Alhadeff M, Coronado R, Figueroa N, Schiff JA (1979) Loss and re-formation of protochlorophyll(ide) in dark grown non-dividing *Euglena gracilis* var. *bacillaris*. Plant Physiol 63:S-98

Alhadeff M, Coronado R, Figueroa N, Schiff JA (1980) The protochlorophyll(ide) system in chloroplast development In: deGreef J (ed) Photoreceptors and plant development. Antwerpen Univ Press, Antwerpen pp 175–177

App AA, Jagendorf AT (1963) Repression of chloroplast development in *Euglena gracilis* by substrates. J Protozool 10:340–343

Avadhani NG, Buetow DE (1972) Isolation of active polyribosomes from cytoplasm, mitochondria and chloroplast of *Euglena gracilis*. Biochem J 128:353–365

Beney G, Nigon V (1975) Effect of cycloheximide (CHI) on the production of δ-aminolevulinic acid (ALA) during greening of dark-grown *Euglena gracilis*. In: Avron M(ed) Proc 3rd Intern Cong on Photosynthesis. Elsevier, Amsterdam New York, pp 1801–1808

Ben-Shaul Y, Schiff JA, Epstein HT (1964) Studies of chloroplast development in *Euglena* VII. Fine structure of the developing plastid. Plant Physiol 39:231–240

Ben-Shaul Y, Epstein HT, Schiff JA (1965) Studies of chloroplast development in Euglena 10. The return of the chloroplast to the proplastid condition during dark adaptation. Can J Bot 43:129–136

Bingham S, Schiff JA (1979a) Events surrounding the early development of *Euglena* chloroplasts 15. Origin of plastid thylakoid polypeptides in wild-type and mutant cells. Biochim Biophys Acta 547:512–530

Bingham S, Schiff JA (1979b) Events surrounding the early development of *Euglena* chloroplasts 16. Plastid thylakoid polypeptides during greening. Biochim Biophys Acta 547:531–543

Boutin ME, Klein RM (1972) Absence of phytochrome participation in chlorophyll synthesis in *Euglena*. Plant Physiol 49:656–657

Bovarnick JG (1969) Early events of chloroplast development in *Euglena gracilis* var. *bacillaris* as studied with streptomycin and actinomycin D. PhD Dissertation, Brandeis Univ

Bovarnick JG, Zeldin MH, Schiff JA (1969) Differential effects of actinomycin D on cell division and light-induced chloroplast development in *Euglena*. Dev Biol 19:321–340

Bovarnick JG, Chang SW, Schiff JA, Schwartzbach SD (1974a) Events surrounding the early development of *Euglena* chloroplasts: Experiments with streptomycin in non-dividing cells. J Gen Microbiol 83:51–62

Bovarnick JG, Schiff JA, Freedman Z, Egan JM Jr (1974b) Events surrounding the early development of *Euglena* chloroplasts: Cellular origins of chloroplast enzymes in *Euglena*. J Gen Microbiol 83:63–71

Breitenberger CA, Spremulli LL (1980) Purification of *Euglena gracilis* chloroplast elongation factor G and comparison with other prokaryotic and eukaryotic translocases. J Biol Chem 255:9814–9820

Breitenberger CA, Graves MC, Spremulli LL (1979) Evidence for the nuclear location of the gene for chloroplast elongation factor G. Arch Biochem Biophys 194:265–270

Brunold C, Schiff JA (1976) Studies of sulfate utilization by algae 15. Enzymes of assimilatory sulfate reduction in *Euglena* and their cellular localization. Plant Physiol 57:430–436

Buchanan RE, Gibbons NE (eds) (1974) Bergey's manual of determinative bacteriology, 8th edn. Williams and Wilkins, Baltimore

Buetow DE (1967) Acetate repression of chlorophyll synthesis in *Euglena gracilis*. Nature 213:1127–1128

Buetow DE (ed) (1968) The biology of *Euglena* Vols I, II. Academic Press, London New York

Buetow DE (ed) (1982) The biology of *Euglena* Vols III, IV. Academic Press, London New York

Buetow DE, Wood WM (1978) The mitochondrial translation system. In: Roodyn DB (ed) Subcellular biochemistry. Plenum, New York

Chelm BK, Hoben PJ, Hallick RB (1977) Cellular content of chloroplast DNA and chloroplast ribosomal RNA in *Euglena gracilis* during chloroplast development. Biochemistry 16:782–785

Cohen C, Schiff JA (1976) Events surrounding the early development of *Euglena* chloroplasts-XI. Protochlorophyll(ide) and its photoconversion. Photochem Photobiol 24:555–566

Cohen D, Schiff JA (1976) Events surrounding the early development of *Euglena* chloroplasts; photoregulation of the transcription of chloroplastic and cytoplasmic ribosomal RNAs. Arch Biochem Biophys 177:201–216

Coleman AW (1979) Use of the fluorochrome 4′6-diamidino-2-phenylindole in genetic and developmental studies of chloroplast DNA. J Cell Biol 82:299–305

Collins N, Merrett MJ (1975) Microbody marker enzymes during transition from phototrophic growth in *Euglena*. Plant Physiol 55:1018–1022

Crouse EJ, Vandrey JP, Stutz E (1974) Hybridization studies with RNA and DNA isolated from *Euglena gracilis* chloroplasts and mitochondria. FEBS Lett 42:262–266

Diamond J, Schiff JA, Kelner A (1975) Photoreactivating enzyme from *Euglena* and the control of its intracellular level. Arch Biochem Biophys 167:603–614

Dockerty A, Merrett MJ (1979) Isolation and enzymatic characterization of *Euglena* proplastids. Plant Physiol 63:468–473

Dwyer M, Smillie RM (1970) A light-induced B 1,3-glucan breakdown associated with the differentiation of chloroplasts in *Euglena gracilis*. Biochim Biophys Acta 216:392–401

Dwyer MR, Smillie RM (1971) Bl,3-glucan: A source of carbon and energy for chloroplast development in *Euglena gracilis*. Aust J Biol Sci 24:15–22

Edelman M, Schiff JA, Epstein H (1965) Studies of chloroplast development in *Euglena* 12. Two types of satellite DNA. J Mol Biol 11:769–774

Egan JM Jr, Carrell EF (1972) Studies on chloroplast development and replication in *Euglena* III. A study of the site of synthesis of alkaline deoxyribonuclease induced during chloroplast development in *Euglena gracilis*. Plant Physiol 50:391–395

Egan J, Schiff JA (1974) A reexamination of the action spectrum for chlorophyll synthesis in *Euglena gracilis*. Plant Sci Lett 3:101–105

Egan JM Jr, Dorsky D, Schiff JA (1975) Events surrounding the early development of *Euglena* chloroplasts VI. Action spectra for the formation of chlorophyll, lag elimination in chlorophyll synthesis, and appearance of TPN-dependent triose phosphate dehydrogenase and alkaline DNase activities. Plant Physiol 56:318–323

Erwin JA (1968) Lipid metabolism. In: Buetow DE (ed) The biology of *Euglena*. Academic Press London New York, pp 133–148

Evans WR (1971) The effect of cycloheximide on membrane transport in *Euglena*. A comparative study with nigericin. J Biol Chem 246:6144–6151

Fong F, Schiff JA (1977) Mitochondrial respiration and chloroplast development in *Euglena gracilis* var *bacillaris*. Plant Physiol 59:S-92

Fong F, Schiff JA (1979) Blue-light-induced absorbance changes associated with carotenoids in *Euglena*. Planta 146:119–127

Fox L, Erion J, Tarnowski J, Spremulli L, Brot N, Weissbach H (1980) *Euglena gracilis* chloroplast EF-Ts evidence that it is a nuclear coded gene product. J Biol Chem 255:6018–6019

Frey MA, Alberte RS, Schiff JA (1979) Studies by fluorescence of protochlorophyll(ide) and its phototransformation in dark-grown *Euglena gracilis* var. *bacillaris*. Biol Bull 157:368–369

Freyssinet G (1976a) Relation between paramylum content and the length of the lag period of chlorophyll synthesis during greening of dark-grown *Euglena gracilis*. Plant Physiol 57:824–830

Freyssinet G (1976b) Influence of culture conditions on the length of the lag period of chlorophyll synthesis in preilluminated dark-grown *Euglena*. Plant Physiol 57:831–835

Freyssinet G (1977) The protein synthesizing system of *Euglena:* synthesis of ribosomal proteins in vivo and their characterization. Physiol Veg 15:519–550

Freyssinet G (1978) Determination of the site of synthesis of some *Euglena* cytoplasmic and chloroplast ribosomal proteins. Exp Cell Res 115:207–219

Freyssinet G, Heizmann P, Verdier G, Trabuchet G, Nigon V (1972) Influence des conditions nutritionnelles sur la réponse a l'éclairement chez les Euglènes étiolées. Physiol Veg 10:421–442

Freyssinet G, Harris GC, Nasitir M, Schiff JA (1979) Events surrounding the early development of *Euglena* chloroplasts 14. Biosynthesis of cytochrome c-552 in wild type and mutant cells. Plant Physiol 63:908–915

Garlaschi FM, Garlaschi AM, Lombardi A, Forti G (1974) Effect of ethanol on the metabolism of *Euglena gracilis*. Plant Sci Lett 2:29–39

Gibbs SP (1962) Nuclear envelope-chloroplast relationships in algae. J Cell Biol 14:433–444

Goldstein NH, Schwartzbach SD, Schiff JA (1974) Conservation of plastid-forming units (PFU) during bleaching in *Euglena gracilis* var. *bacillaris*. J Protozool 21:443

Gray PW, Hallick RB (1978) Physical mapping of the *Euglena gracilis* chloroplast DNA and ribosomal RNA gene region. Biochemistry 17:284–289

Gruol DJ, Haselkorn R (1976) Counting the genes for stable RNA in the nucleus and chloroplasts of *Euglena*. Biochim Biophys Acta 447:82–95

Gurevitz M, Kratz H, Ohad I (1977) Polypeptides of chloroplastic and cytoplasmic origin required for development of photosystem II activity and chlorophyll-protein complexes in *Euglena gracilis* Z chloroplast membranes Biochim Biophys Acta 461:475–488

Hallick RB, Rushlow KE, Orozco EM Jr, Stiegler GL, Gray PW (1979) Chloroplast DNA of *Euglena gracilis* gene mapping and selective in vitro transcription of the ribosomal RNA region. In: Cummings DJ, Borst P, Dawid IB, Weissman SM, Fox CF (eds) ICN-UCLA Symposia on Molecular and Cellular Biology Vol XV, Extrachromosomal DNA. Academic Press, London New York

Harris RC, Kirk JTO (1969) Control of chloroplast formation in *Euglena gracilis:* Antagonism between carbon and nitrogen sources. Biochem J 113:195–205

Hecker LI, Egan J, Reynolds RJ, Nix CE, Schiff JA, Barnett WE (1974) The sites of transcription and translation for *Euglena* chloroplastic aminoacyl tRNA synthetases. Proc Natl Acad Sci 71:1910–1944

Heizmann P, Doly J, Hussein Y, Nicolas P, Nigon V, Bernardi G (1981) The chloroplast genome of bleached mutants of *Euglena gracilis*. Biochim Biophys Acta 653:412–415

Hill HZ, Schiff JA, Epstein HT (1966) Studies of chloroplast development in *Euglena* XIII. Variation of ultraviolet sensitivity with extent of chloroplast development. Biophys J 6:125–133

Holowinsky AW, Schiff JA (1970) Events surrounding the early development of Euglena chloroplasts 1. Induction by preillumination. Plant Physiol 45:339–347

Horum MA, Schwartzbach SD (1980a) Nutritional regulation of organelle biogenesis in *Euglena*. Repression of chlorophyll and NADP-glyceraldehyde-3-phosphate dehydrogenase synthesis. Plant Physiol 55:382–386

Horrum MA, Schwartzbach SD (1980b) Nutritional regulation of organelle biogenesis in *Euglena:* Photo and metabolite induction of mitochondria. Planta 149:376–383

Horrum MA, Schwartzbach SD (1980c) Photo and nutritional regulation of four photorespiratory enzymes in greening *Euglena*. Plant Physiol 65:S-19

Horrum MA, Schwartzbach SD (1980d) Absence of photo and nutritional regulation of two glycolate pathway enzymes in *Euglena*. Plant Sci Lett 20:133–139

Horrum MA, Schwartzbach SD (1981) Nutritional regulation of organelle biogenesis in *Euglena*. Induction of microbodies. Plant Physiol 68:430–434

Horrum MA, Schwartzbach SD (1982) Induction of fumarase in resting *Euglena*. Biochim Biophys Acta 714:407–414

Karlan AW, Russell GK (1976) Aldolase levels in wild-type and mutant *Euglena gracilis*. J Protozool 23:176–179

Kindman LA, Cohen CE, Zeldin MH, Ben-Shaul Y, Schiff JA (1978) Events surrounding the early development of *Euglena* chloroplasts 12. Spectroscopic examination of the protochlorophyll(ide) phototransformation in intact cells. Photochem Photobiol 27:787–794

Kirk JTO (1970) Failure to detect effects of cycloheximide on energy metabolism in *Euglena gracilis*. Nature 226:182

Klein S, Schiff JA, Holowinsky AW (1972) Events surrounding the early development of *Euglena* chloroplasts II. Normal development of fine structure and the consequences of preillumination. Dev Biol 28:253–273

Klein SH, Schiff JA (1972) The correlated appearance of prolamellar bodies, protochlorophyll(ide) species and the Shibata shift during development of bean etioplasts in the dark. Plant Physiol 49:619–626

Klimov VV, Dolan E, Ke B (1980) EPR properties of an intermediary electron acceptor (pheophytin) in photosystem-II reaction centers at cryogenic temperatures. FEBS Lett 112:97–100

Lancer H, Cohen CE, Schiff JA (1976) Changing ratios of phototransformable protochlorophyll and protochlorophyllide of bean seedlings developing in the dark. Plant Physiol 57:369–374

Leedale GF (1967) Euglenoid flagellates. Prentice-Hall, Englewood Cliffs

Lefort-Tran M (1975) Mitochondries et chloroplastes chez *Euglena* en culture synchrone. In: Les cycles cellulaires et leur blockage chez plusieurs protistes. Cent Nat Rech Sci, Paris, pp 297–308

Lor KL, Cossins EA (1978) Relationship between glycolate and folate metabolism in *Euglena gracilis*. Photochem 17:659–665

Manning JE, Richards OC (1972) Isolation and molecular weight of circular chloroplast DNA from *Euglena gracilis*. Biochim Biophys Acta 259:285–296

Manning JE, Wolstenholme DR, Ryan RS, Hunter JA, Richards OC (1971) Circular chloroplast DNA from *Euglena gracilis*. Proc Natl Acad Sci 68:1169–1173

McCarthy S, James L, Schwartzbach SD (1981) Induction of chloroplast valyl tRNA synthetase in *Euglena*. Plant Physiol 67:S-33

McCrea JM, Hershberger CL (1976) Chloroplast DNA codes for transfer RNA. Nucl Acids Res 3:2005–2018

Merrett MJ, Lord JM (1973) Glycollate formation and metabolism by algae. New Phytol 72:751–767

Mubumbila M, Burkard G, Keller M, Steinmetz A, Crouse E, Weil JH (1980) Hybridization of bean, spinach, maize and *Euglena* chloroplast transfer RNA, with homologous and heterologous chloroplast DNAs. An approach to the study of homology between chloroplast tRNAs from various species. Biochim Biophys Acta 609:31–39

Nicolas P, Innovent JP, Nigon V (1977) Somatic segregation and rate of greening after ultraviolet irradiation of *Euglena gracilis*. Mol Gen Genet 155:123–129

Nigon V, Heizmann P (1978) Morphology, biochemistry and genetics of plastid development in *Euglena gracilis*. Int Rev Cytol 53:212–290

Osafune T, Schiff JA (1980a) Events surrounding the early development of *Euglena* chloroplasts 17. Light-induced changes in a proplastid remnant in mutant W_3BUL. J Ultrastruct Res 73:64–76

Osafune T, Schiff JA (1980b) Stigma and flagellar swelling in relation to light and carotenoids in *Euglena gracilis* var. *bacillaris*. J Ultrastruct Res 73:336–349

Osafune T, Klein SH, Schiff JA (1980) Events surrounding the early development of *Euglena* chloroplasts 18. Structure of the developing proplastid in the first hours of illumination from serial sections of wild-type cells. J Ultrastruct Res 73:77–90

Palisano JR, Walne PL (1976) Light and electron microscopy of two permanently bleached cell lines of *Euglena gracilis* (Euglenophyceae). Nova Hedwigia 27:455–481

Pardo A, Schiff JA (1980) Plastid and seedling development in SAN-9789 (4-chloro-5-(methylamino)-2-(α, α, α-trifluoro-m-tolyl)-3-(2H)-pyridasinone) treated etiolated bean seedlings. Can J Bot 58:25–35

Parthier B (1974) Stimulation of amino acid uptake by *Euglena gracilis* and the effect of cycloheximide. Biochem Physiol Pflanz 166:555–560

Parthier B, Neumann D (1977) Structural and functional analysis of some plastid mutants of *Euglena gracilis*. Biochem Physiol Pflanz 171:547–562

Parthier B (1981) Chloroplast development in *Euglena:* Regulatory aspects. In: Levandowsky M, Hutner SH, Provasoli L (eds) Biochemistry and physiology of protozoa, 2nd edn. Academic Press, London New York, pp 261–300

Pellegrini M (1980) Three-dimensional reconstruction of organelles in *Euglena gracilis* Z. II. Qualitative and quantitative changes of chloroplasts and mitochondrial reticulum in synchronous cultures during bleaching. J Cell Sci 46:313–340

Rawson JRY (1975) The characterization of *Euglena gracilis* DNA by its reassociation kinetics. Biochim Biophys Acta 402:171–178

Rawson JRY, Boerma CL (1976) Influence of growth conditions upon the number of chloroplast DNA molecules in *Euglena gracilis*. Proc Natl Acad Sci 73:2401–2404

Rawson JRY, Eckenrode VK, Boerma CL, Curtis S (1979) DNA sequence organization in *Euglena gracilis*. Biochim Biophys Acta 563:1–16

Richard F, Nigon V (1973) La synthèse de l'acide-aminolevulinique et de la chlorophylle lors de l'éclairement d'*Euglena gracilis* étiolées. Biochem Biochys Acta 313:130–149

Riggs WAC (1979) Intracellular Mycoplasma infection in *Euglena gracilis* Z. J Protozool 26:7 A

Rosenberg A, Pecker M (1964) Lipid alterations in *Euglena gracilis* cells during light-induced greening. Biochemistry 3:254–258

Russell GK, Draffan AG, Schmidt GW, Lyman H (1978) Light-induced enzyme formation in a chlorophyll-less mutant of *Euglena gracilis*. Plant Physiol 62:678–682

Rutherford AW, Mullet JE, Paterson DR, Robinson HH, Arntzen CJ, Crofts AR (1981) Some biophysical properties of purified PSI and PSII particles. In: Akoyounouglou G (ed) Intern Cong on Photosynthesis

Sagher D, Grosfeld H, Edelman M (1976) Large subunit ribulosebisphosphate carboxylase messenger RNA from *Euglena* chloroplasts. Proc Natl Acad Sci 73:722–726

Schiff JA (1963) Oxygen exchange by *Euglena* cells undergoing chloroplast development. Carnegie Inst Wash Yearbook 62:375–378

Schiff JA (1973) The development, inheritance and origin of the plastid in *Euglena*. Adv Morphog 10:265–312

Schiff JA (1974) The control of chloroplast differentiation in *Euglena*. In: Avron M(ed) Proc 3rd Int Cong Photosynthesis. Elsevier, Amsterdam New York

Schiff JA (1978) Photocontrol of chloroplast development in *Euglena*. In: Akoyounouglou G et al. (eds) Chloroplast development. Elsevier, Amsterdam New York

Schiff JA (1980a) Development, inheritance, and evolution of plastids and mitochondria. In: Stumpf P, Conn E (eds) The biochemistry of plants vol 1. Academic Press, London New York, pp 209–272

Schiff JA (1980b) Blue light and the photocontrol of chloroplast development in *Euglena*. In: Senger H (ed) The Blue light syndrome. Springer, Berlin Heidelberg New York

Schiff JA (1981a) Origin and evolution of the plastid and its function. Ann NY Acad Sci 361:166–192

Schiff JA (1981b) Evolution of the control of pigment and plastid development. In: Baltscheffsky H (ed) Molecular origins and evolution of photosynthesis. Biosystems 14:123–147

Schiff JA, Epstein HT (1968) The continuity of the chloroplast in *Euglena*. In: Buetow DE (ed) The biology of *Euglena*. Academic Press, London New York

Schiff JA, Zeldin MH, Rubman J (1967) Chlorophyll formation and photosynthetic competence in *Euglena* during light-induced chloroplast development in the presence of 3,(3,4-dichlorophenyl)1,1-dimethyl urea (DCMU). Plant Physiol 42:1716–1725

Schiff JA, Lyman H, Russel GK (1971) Isolation of mutants from *Euglena gracilis*. Methods Enzymol 23:143–162

Schiff JA, Lyman H, Russel GK (1980) Isolation of mutants from *Euglena gracilis*. An addendum. Methods Enzymol 69:23–29

Schiff JA, Cunningham FX, Green MS (1982) Carotenoids in relation to chloroplasts and other organelles. In: Britton G, Goodwin TE (eds) 6th Int Symposium on carotenoids, Liverpool. Pergamon, New York, pp 329–338

Schmidt G, Lyman H (1974) Photocontrol of chloroplast enzyme synthesis in mutant and wild-type *Euglena gracilis* In: Avron M (ed) 3rd Int Cong Photosynthesis, Rehovot, Israel. Elsevier, Amsterdam New York, pp 1755–1764

Schmidt G, Lyman H (1976) Inheritance and synthesis of chloroplasts and mitochondria of *Euglena gracilis*. In: Lewin RA (ed) The genetics of algae. Univ Calif Press, Berkeley, pp 257–299

Schneyour A, Avron M (1975) Properties of photosynthetic mutants isolated from *Euglena gracilis*. Plant Physiol 55:137–141

Schoch S, Scheer H, Schiff JA, Rüdiger W, Siegelman HW (1981) Pyropheophytin a accompanies pheophytin a in darkened light-grown cells of *Euglena*. Z Naturforsch 36c:827–833

Schwartzbach SD, Schiff JA (1974) The chloroplast and cytoplasmic ribosomes of *Euglena:* Selective binding of dihydrostreptomycin to chloroplast ribosomes. J Bacteriol 120:334–341

Schwartzbach SD, Schiff JA (1979) Events surrounding the early development of *Euglena* chloroplasts: 13. Photocontrol of protein synthesis. Plant Cell Physiol 20:827–838

Schwartzbach SD, Schiff JA, Goldstein NH (1975) Events surrounding the early development of *Euglena* chloroplasts V. Photocontrol of paramylum degradation. Plant Physiol 56:313–317

Schwartzbach SD, Hecker LI, Barnett WE (1976a) Transcriptional origin of *Euglena* chloroplast tRNAs. Proc Natl Acad Sci 73:1984–1988

Schwartzbach SD, Schiff JA, Klein S (1976b) Biosynthetic events required for lag elimination in chlorophyll synthesis in *Euglena*. Planta 131:1–9

Schwelitz FD, Cisneros PL, Jagielo JA, Comer JL, Butterfield KA (1978) The relationship of fixed carbon and nitrogen sources to the greening process in *Euglena gracilis* strain Z. J Protozool 25:257–261

Senger H (ed) (1980) The blue light syndrome. Springer, Berlin Heidelberg New York

Shihira-Ishikawa I, Osafune T, Ehara T, Ohkuro I, Hase E (1977) An early light independent phase of chloroplast development in dark-grown cells of *Euglena gracilis* Z. I. Dependence of the plastid development on previous culture conditions. Plant Cell Physiol (Special Issue: Photosynthetic Organelles) pp 445–454

Slavik NS, Hershberger CL (1975) The kinetic complexity of *Euglena gracilis* chloroplast DNA. FEBS Lett 52:171–174

Srinivas U, Lyman H (1980) Photomorphogenic regulation of chloroplast replication in *Euglena*. Enhanced loss of chloroplast DNA in red light. Plant Physiol 66:295–301

Steinitz YL, Schiff JA, Osafune T, Green MS (1980) Cis to trans photoisomerization of γ-carotene in *Euglena gracilis* var. *bacillaris* W_3BUL: further purification and characterization of the photoactivity. In: Senger H (ed) The blue light syndrome. Springer, Berlin Heidelberg New York, pp 269–280

Stern AI, Schiff JA, Epstein HT (1964a) Studies of chloroplast development in *Euglena* V. Pigment biosynthesis, photosynthetic oxygen evolution and carbon dioxide fixation during chloroplast development. Plant Physiol 39:220–226

Stern AI, Epstein HT, Schiff JA (1964b) Studies of chloroplast development in *Euglena* VI. Light intensity as a controlling factor in development. Plant Physiol 39:226–231

Straley SC, Parson WW, Mauzerall DC, Clayton RK (1973) Pigment content and molar extinction coefficients of photochemical reaction centers from *Rhodopseudomonas spheroides*. Biochim Biophys Acta 305:597–609

Talen JL, Sanders JPM, Flavell RA (1974) Genetic complexity of mitochondrial DNA from *Euglena gracilis*. Biochim Biophys Acta 374:129–135

Vaisberg AJ, Schiff JA (1976) Events surrounding the early development of *Euglena* chloroplasts 7. Inhibition of carotenoid biosynthesis by the herbicide SAN 9789 (4-choro-5-(methylamino)-2-(α, α, α-trifluoro-m-tolyl)-3(2H) pyridazinone) and its developmental consequences. Plant Physiol 57:260–269

Vaisberg AJ, Schiff JA, Li L, Freedman Z (1976) Events surrounding the early development of Euglena chloroplasts 8. Photocontrol of the source of reducing power for chloramphenicol reduction by the ferredoxin-NADP reductase system. Plant Physiol 57:594–601

Vasconcelos AC (1976) Synthesis of proteins by isolated *Euglena gracilis* chloroplasts. Plant Physiol 58:719–721

Woodcock E, Merrett MJ (1980) Malate synthase messenger RNA in relation to enzyme derepression in *Euglena gracilis*. Arch Microbiol 124:33–38

14 Pattern Specification and Realization in Photomorphogenesis

H. MOHR

1 Introduction

1.1 The Significance of Pattern Formation in Development

Development in multicellular organisms is usually discussed from the point of view of growth, differentiation, and morphogenesis. "Growth" refers to quantitative changes; "differentiation" refers to qualitative differences which appear between different cells and tissues within a given organism; "morphogenesis" refers to the obvious changes in form and structure during the process of development.

*Photo*morphogenesis refers to that strategy of development which a plant follows if light is available. Photomorphogenesis (development in the presence of light) is conspicuously different from skotomorphogenesis (or etiolation, development in the absence of light) (see Chap. 3, this Vol.). In this chapter a fourth aspect is emphasized: the appearance of "patterns" during development (MOHR 1978). In development of multicellular organisms we observe not only the emergence of a diversity of cell types, but these differentiated cells are precisely localized within the developing embryo or organ. The resulting spatial organization can be referred to as the appearance of patterns: Cells become arranged in regular, reproducible, non-random fashion to form specific tissues and organs. The mechanisms involved in the generation of patterns have intrigued investigators ever since developmental biology emerged. However, how these spatial patterns are formed is still one of the great enigmas of developmental biology. The study of photomorphogenesis in higher plants has at least led us to realize (1) that the process of pattern formation does, in fact, occur stepwise, (2) that pattern specification and pattern realization must be kept apart, (3) that not only spatial but also temporal patterns must be considered, and (4) that the theoretical framework which originated from the study of photomorphogenesis may also be applicable to the action of hormones during development (see MOHR and SITTE 1971).

Thus, the study of photomorphogenesis offers the advantage of a fresh approach to problems of pattern formation in development of multicellular organisms; this approach must first be a search for formal rules (or "principles"). Only when we understand the formal rules underlying the phenomena of development can we ask appropriate questions on the molecular level (MOHR 1982). In animal development the present situation is precisely the same, as pointed out recently by BRYANT et al. (1981): "The way in which the activities of the cells of an embryo are coordinated in space and time is one of the great remaining enigmas of biology, and it is a problem which has not, so far, benefited significantly from recent advances in the methodology of molecular biology. Instead, our efforts at understanding the formation of spatial patterns in developing animals are still comparable to the pre-Mendelian stage of genetics; we are still searching for the formal 'rules' by which we can predict the behaviour of embryos under various experimental treatments. Only when we understand these principles at a formal level, do we expect to be able to frame appropriate questions for a molecular analysis".

But even then two basic problems remain: First, the present understanding of genome function in eukaryotic cells is not sufficient to provide appropriate answers even if we could frame appropriate questions. Second, there is notoriously a bewildering diversity in biology which makes it difficult to rise above it and to develop the appropriate level of abstraction. Can we expect at all basic and comprehensive spatial and dynamic laws (principles) of morphogenesis which could be elucidated? This question has been dealt with elsewhere (MOHR 1982); in this chapter we follow a different strategy, namely, to describe a particular case history—skoto=morphogenesis vs. photo=morphogenesis of the mustard (*Sinapis alba*) seedling—as precisely as the limited space permits, and to draw conclusions which can be checked for their general validity. The mustard seedling was chosen because it has been intensely investigated by a number of researchers on the formal as well as on the molecular level in a coherent and coordinated research program (see MOHR 1972, 1978). Moreover, it can be considered a representative of the wide-spread epigeal type of dicotyledonous seedlings.

1.2 Historical Perspectives of the Problem

Description of the appearance of form and pattern in the course of development of cells, organisms and societies has had a long tradition in biology. The reader is reminded only of D'ARCY THOMPSON'S (1917, 1961) treatise *On Growth and Form,* of EDMUND SINNOTT'S (1960) chapter on *Organisation* in his monograph on plant morphogenesis and of ERWIN BÜNNING'S (1948) essay on the formation of stomatal patterns. J.H. WOODGER'S (1930, 1931) discussions on the *Concept of Organism* are particularly useful, even 50 years after they were written. However, in spite of thorough analysis, including mathematical descriptions of developmental events and d'Arcy Thompson's impressive formal and physical analogies of organic forms, we still lack an answer—to be given in the scientific language of our days—to the question of how *control* of pattern appearance is achieved.

However, a mass of data accumulated recently from studies of regeneration tells us that all cells in the plant body, at least at their beginnings, are totipotent and that the total genome (and thus the full potential of the organism) is present in the living matter of every vegetative cell. It is *differential* gene expression (and possibly some physical constraints) which underlie differentiation in general and pattern formation in particular.

"Morphogenetic fields" of various sorts have often been postulated in animal development in a descriptive sense, as the developmental influence of a given region over differentiation and pattern formation in it. However, as SINNOTT (1960) pointed out: "The formation of a morphogenetic field recognizes the formative influences acting within a given region or throughout the embryo, but offers little explanation of this action". In plant morphogenesis the field concept has also been employed without much explanatory success, e.g., by the SNOWS (1934) and by WARDLAW (1952).

More recently WOLPERT (1969) has suggested that spatial patterns in animal embryology result from cells acquiring information about their physical position in a developing cell population by detecting their position in a "developmental field". This positional information was considered (see BRYANT et al. 1981, for review) to be specified along polar coordinates in two dimensions (that is, in a cell layer) so that each cell would have a positional value with an angular and a radial component in the "epimorphic field". In fact, various experiments on regenerating newt limbs, cockroach legs and *Drosophila* imaginal discs have produced results supporting the view (FRENCH et al. 1976) that cells in these developing systems form structures in accordance with their positional information which is specified in two dimensions along polar coordinates. However, a recent revision of the "polar coordinate model" (BRYANT et al. 1981) has led to the conclusion that pattern formation in diverse regenerating systems can be understood in terms of strictly *local* cell interactions. BRYANT et al. (1981) suggest that pattern regulation may depend upon interactions between adjacent cells rather than on long-range diffusion systems which have been postulated previously by TURING (1952) and his many followers to be the material equivalent of morphogenetic fields. Even though the conclusion that positional information may be a property of the cell surface can be considered significant progress in our knowledge, BRYANT et al. (1981) leave no doubt that "there are still very few clues indeed about the molecular basis for positional information".

A particularly interesting prokaryotic model for pattern specification in eukaryotes are certain filamentous cyanobacteria, in which heterocysts inhibit nearby vegetative cells from differentiating into heterocysts and induce nearby vegetative cells to differentiate into spores (WOLK 1979). According to WOLK, glutamine produced by heterocysts from N_2 which they fix and from glutamate synthesized in vegetative cells, may be the inhibitory-field morphogen produced by mature heterocysts. In any case, the processes underlying pattern formation in cyanobacteria seem to be analogous to embryonic fields: the heterocyst pattern corresponds to a field of inhibition, and the juxtaposition of spores and heterocysts corresponds to an embryonic induction.

On the other hand, it does not seem that elegant models developed to account for polarity *fixation* in *Fucus* zygotes (a cytoplasmic potential gradient or a mechanical, contractible mechanism involving actin and microfilaments; QUATRANO et al. 1979) will help us to identify the factors underlying pattern *specification*.

Pattern formation *as a two-step process* was recognized (in principle, at least) by SINNOTT (1960): "In an exploration of the problem of organic form the obvious hypothesis ... is that there are formative substances, ... various 'organ-forming substances': chemical substances that make roots or shoots or flowers ... All such ideas, if carried far enough, face the serious problem of how it is possible for a *substance* to become translated into a *form*. This was the difficulty on which Spemann's 'organizer' came to grief. Today it seems much more likely that these various substances, the effects of which undoubtedly result in the production of form changes, act rather as evocators, releasers, or triggers which call forth *specific* responses by the organized living system".

1.3 Timing in Development

Developmental processes in multicellular organisms are strictly coordinated, both spatially and temporally. The precision of timing of events is most impressive, in particular when the decisive environmental variables are controlled. Certainly the synchrony observed between germinating seedlings of one species (or one genotype) growing in a defined environment seems much more than mere coincidence. Rather, the suggestion is unavoidable that a very precise timing mechanism ("developmental clock") operates to control the emergence of developmental events.

While the models to explain timing in development are still largely speculative, e.g., the phase-shift model for spatial and temporal organization proposed by GOODWIN and COHEN (1969), it remains undisputed (GOODWIN 1970) that development in multicellular organisms takes place in *four* dimensions and requires the existence of time- and space-measuring processes within an embryo or a seedling.

It remains undisputed as well that genes affect the timing of developmental processes. Moreover, it is argued (see LEWIN 1981, for a review) that small genetic adjustments of the developmental clock could account for conspicuous differences between taxa, e.g., for many of the important physical differences between humans and chimpanzees.

2 The Multiple Action of Phytochrome

2.1 A Convenient System

Let us consider two mustard (*Sinapis alba*) seedlings of the same chronological age (Fig. 1), a dark-grown etiolated seedling and the corresponding light-grown normal seedling. All photoresponses of the light-grown seedling (except phototropism and protochlorophyllide photoreduction) are consequences of the operation of phytochrome in the cells of the seedling (MOHR 1972, MOHR and DRUMM-HERREL 1981). Conspicuous photoresponses include inhibition of hypocotyl lengthening, enlargement of cotyledons, synthesis of anthocyanin in hypocotyl and cotyledons, and hair formation along the hypocotyl.

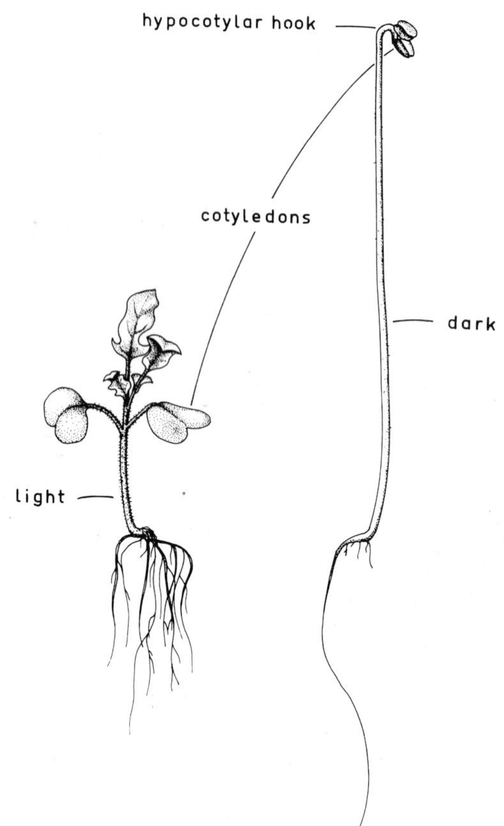

Fig. 1. These two mustard seedlings (*Sinapis alba*) have the same chronological age and are virtually identical genetically. The differences in morphogenesis are due to light. The drawings emphasize the point that light causes the cotyledons to develop from storage organs (*right*) to photosynthetically active leaves (*left*). Unlike, for instance, bean cotyledons, the mustard cotyledon is not devoid of function after depletion of lipid and protein reserves, since in the light it expands, becomes green and persists as a photosynthetic organ. (MOHR 1972)

We have every reason to assume that phytochrome, as a molecule, is the same in all the cells of the seedling in which it occurs. But as we observe, the different organs and tissues of the seedling respond differently to phytochrome. We have been calling this phenomenon the "multiple action" of phytochrome. This is to say that phytochrome releases a great number of photoresponses in a plant whereby the specificity of the response does not depend on phytochrome but on the particular state ("competence") of a cell or tissue at the moment when active phytochrome (P_{fr}) is formed within this cell. This crucial aspect of our argument is illustrated in Fig. 2. This figure shows segments of cross-sections through the hypocotyl of mustard seedlings on the left from a dark-grown seedling, on the right from a seedling of the same age which has been kept in continuous far-red light for some time, i.e., under conditions where the phytochrome-dependent HIR operates. One sees that under the influence of phytochrome certain cells of the epidermis (trichoblasts) have formed long hairs and that all the cells of the subepidermal layer—but no other cells—have formed anthocyanin. It is obvious from this simple drawing that phytochrome functions only as a "releasing signal"; the specificity of the response—hair formation vs. anthocyanin synthesis—depends on the specific state of the

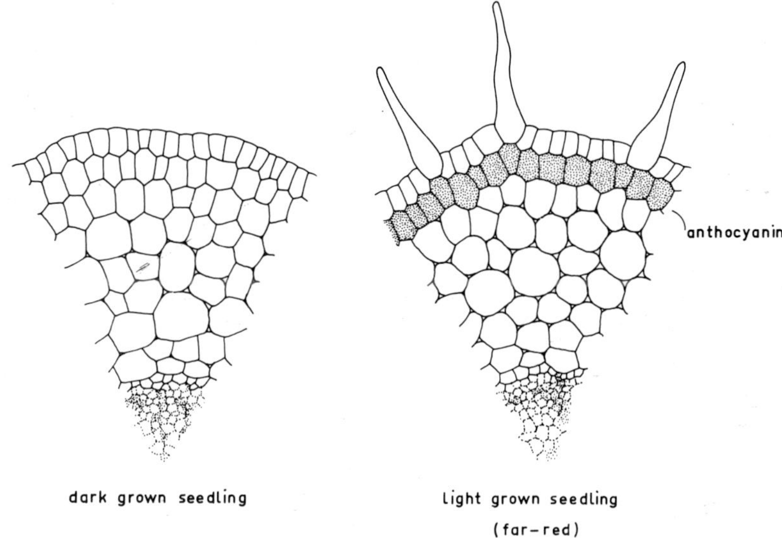

Fig. 2. Segments of cross-sections through the hypocotyl of mustard seedlings (*Sinapis alba*) grown in the dark or in standard far-red light. (WAGNER and MOHR 1966)

epidermal and subepidermal cells at the moment when physiologically active phytochrome is formed in the seedling. An obviously weak point in the present argument is that we do not know the spatial pattern, if any, of phytochrome distribution within a hypocotyl or a cotyledon. For lack of knowledge we assume that there is no spatial pattern in the sense that a particular cell contains phytochrome and its neighbor contains none. In any event, if a particular cell responds to phytochrome with the formation of a hair and its neighbor with the formation of anthocyanin, this *difference* cannot be due to phytochrome.

2.2 A Convenient Terminology

The following terminology has proved useful in describing the situation: all cells which respond to the formation of physiologically active phytochrome (P_{fr})[1] are "competent" with regard to this releasing signal.

The striking fact that the cells differ with respect to the specificity of their response shows that differently responding cells were dissimilar from each other before the formation of P_{fr}. In other words: the cells were strictly "determined"

[1] Regarding the action of the phytochrome system, i.e., the coupling of the phytochrome system to cell functions, it can be stated that there is no phytochrome action without P_{fr}. In some extensively investigated photoresponses such as anthocyanin formation, it was found in light-pulse experiments that within the range of the validity of the reciprocity law, the extent of the response is determined by the amount of P_{fr} made by the light pulse (DRUMM and MOHR 1974). Even though in continuous light (e.g., in continuous far-red light) cycling of the phytochrome system ($P_r \leftrightarrow P_{fr}$) is believed by some investigators (e.g., JOHNSON 1980) to contribute to the extent of phytochrome action, this action depends on the presence of P_{fr} as well

to respond to P_{fr} in a specific manner before the formation of P_{fr}. In view of the fact that the different states of determination of the cells are invisible before the formation of P_{fr} and become visible only after the formation of the releasing signal, we cannot but conceive anthocyanin (or hair) formation as a two-step process: "pattern specification" (involving all processes which determine a particular cell to respond to the formation of P_{fr} with anthocyanin synthesis or hair formation) and "pattern realization" (including those processes which lead to actual anthocyanin synthesis or hair formation after the formation of the releasing signal, P_{fr}). The term "pattern" is clearly justified: the cells that are able to form anthocyanin in the light are never distributed randomly within the organism, but always form a rigid spatial pattern. The epidermal trichoblasts show a non-random distribution as well which can be described as the spatial pattern of trichoblasts within a cellular matrix of atrichoblasts (for descriptive and argumentative details see WAGNER and MOHR 1966, MOHR 1972).

Two questions arise immediately: (1) Is pattern specification a temporal process which includes several steps? (2) Is the specification of the spatial pattern light (P_{fr})-dependent? These questions were studied in the mustard seedling cotyledons (WAGNER and MOHR 1966, STEINITZ et al. 1976, STEINITZ and BERGFELD 1977).

3 Appearance of Spatial and Temporal Patterns in Phytochrome-Mediated Anthocyanin and Chlorophyll Synthesis in the Mustard Seedling Cotyledons (a Case Study)

3.1 Starting Point and Appearance of Competence

Anthocyanin synthesis in the mustard cotyledons takes place only in the epidermal cells (WAGNER and MOHR 1966). There is no anthocyanin synthesis without P_{fr} (see BÜHLER 1977). The amount of anthocyanin synthesized is strictly correlated to the amount of P_{fr} produced by the inductive light treatment (DRUMM and MOHR 1974, STEINITZ et al. 1979, SCHMIDT and MOHR 1982). However, even if P_{fr} is present in the cotyledons from the time of sowing onwards, anthocyanin synthesis only starts at 27 h after sowing (at 25 °C) (STEINITZ et al. 1976). This "starting point" (i.e., the point of time when the epidermal cells become capable of forming anthocyanin as a response to P_{fr} formation) cannot be shifted by any light treatment or by the application of nutrients. It is, at a given temperature parameter, a system's constant.

The second crucial point on the time axis is the point at which competence appears. STEINITZ et al. (1976) have shown that the epidermal cells of the mustard cotyledons become competent for P_{fr} with respect to anthocyanin synthesis only shortly before the starting point. Before this time the phytochrome system (which functions immediately after sowing of the seed) is not yet coupled or connected to that cell function which leads to anthocyanin synthesis. Operationally, a 24-h treatment with red light (starting with sowing of the seeds) was found to be totally ineffective with regard to anthocyanin synthesis, provided that P_{fr} is reverted to P_r by a saturating pulse with 756 nm light at 24 h which lowers the level of P_{fr} to less than 1%, possibly to less than $^1/_{10}$ of a per cent ($\varphi_{756\ nm} < 0.01$, see SCHÄFER et al. 1975).

3.2 Specification of the Spatial Pattern

Pattern specification is in itself a multistep process. The latter conclusion is based on the following argument: In mustard seedling cotyledons, anthocyanin synthesis is restricted to the epidermal cells. Determination of the future epidermal cells takes place during early embryogenesis (see RUTISHAUSER 1969). It is an irreversible determination since plant cells neither migrate nor move relative to each other within a given tissue or organ. On the other hand, competence—with regard to P_{fr} and anthocyanin synthesis—does not appear in the epidermal cells prior to approximately 26 or 27 h after sowing.

The process of pattern specification was described in detail by STEINITZ and BERGFELD (1977). When mustard seedlings are irradiated from the time of sowing onwards for 27 h, no pigmentation can be observed (Fig. 3a), although P_{fr} was always present in the cotyledons from the time of sowing. Neither is any pigmentation detectable 25 h later if a 5-min pulse with 756 nm light is given after 27 h of red light irradiation (Fig. 3b). These drawings illustrate the previous conclusion (STEINITZ et al. 1976) that both the competence for P_{fr} with regard to the initial action of P_{fr} (with regard to anthocyanin formation), as well as the ability for anthocyanin synthesis proper, appear in the epidermal cells only after 27 h after sowing (at 25 °C). If, however, the seedlings are irradiated for 29 h, slight and scattered pigmentation appears, which is restricted to some marginal regions of the cotyledonar lamina (Fig. 3c). When you continue to irradiate the seedling, or alternatively place it in the dark, and inspect the pigmentation pattern at 52 h after sowing, you will find a homogeneous pigmentation spread over the epidermis of the abaxial side of the cotyledons (Fig. 3d).

The situation is entirely different if a 5-min 756-nm light pulse follows the 29-h red light irradiation before the seedlings are placed in the dark. As already stated the 756-nm light pulse removes virtually all the P_{fr} from the system ($\varphi_{756\ nm} < 1\%$). Under these circumstances, an intermediate pattern of pigmentation develops during the successive dark incubation period (Fig. 3e). New cells within the marginal regions and in some central regions of the cotyledonar lamina accumulate anthocyanin. Characteristically one finds pigmented cells also localized in the epidermis beneath the vascular bundles. Since the 756 nm light pulse had removed virtually all the P_{fr} from the system, the participation of new cells in anthocyanin synthesis during the dark incubation period (29 to 52 h) had taken place in the absence of P_{fr}. This indicates that the initial action of P_{fr}, with regard to anthocyanin synthesis, had occurred in these cells between 27 and 29 h after sowing. The result of the initial action was stored in these cells until the ability for anthocyanin synthesis had developed in these

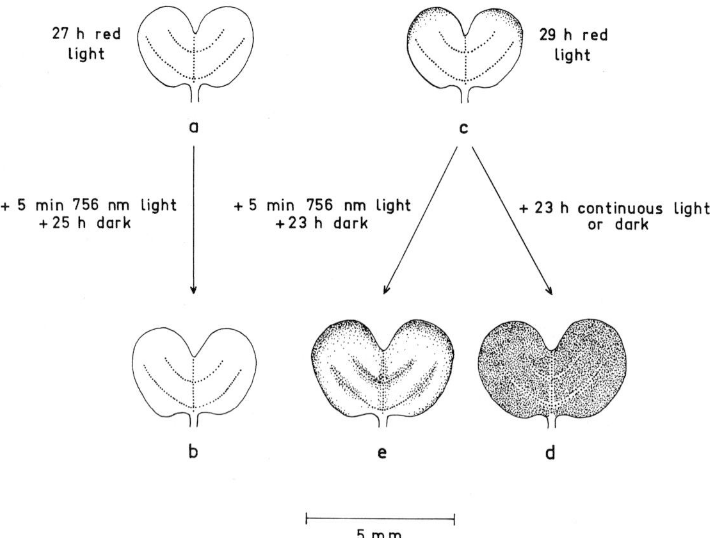

Fig. 3a–e. Abaxial view of the mustard cotyledon lamina showing the distribution pattern of anthocyanin. *Dotted regions* represent anthocyanin pigmentation in the epidermis tissue. Seedlings were irradiated for 27 h (**a, b**) or 29 h (**c, e, d**) with continuous standard red light from the time of sowing onwards. Pigmentation pattern was recorded either at the end of the continuous irradiation (**a, c**) or 52 h after sowing (**b, e, d**). (STEINITZ and BERGFELD 1977)

cells somewhere during the dark incubation period between 29 and 52 h after sowing. In comparing Fig. 3e and 3d we notice that many cells potentially capable of forming anthocyanin between 29 and 52 h remain unpigmented if virtually all the P_{fr} is returned to P_r at 29 h (Fig. 3e). These cells were not able to perform the initial action of P_{fr} during the 2 h period between 27 and 29 h after sowing. By this kind of experiment one can follow the time course of appearance of competence for P_{fr} (with regard to the initial action of P_{fr} in anthocyanin synthesis) and separate this phenomenon from the time course of actual anthocyanin synthesis of the pertinent cells.

Quite obviously a spatial pattern exists with regard to the starting point (in the above sense): the starting point with regard to P_{fr}-mediated anthocyanin synthesis is reached earlier in the marginal epidermal cells than in the intercostal epidermal cells in the center of the lamina. On the basis of this information an attempt has been made to identify relevant ultrastructural changes in those epidermal cells which actually approach the starting point (STEINITZ and BERGFELD 1977). It was found that a correlation exists between the differentiation of central cell vacuoles, originating from the aleurone vacuoles, and the appearance of the ability to accumulate anthocyanin. The authors suggested that the formation of a central cell vacuole is a prerequisite for anthocyanin accumulation in the epidermal cells of the mustard seedling cotyledons.

The temporal and spatial patterns of pigmentation as illustrated in Fig 3c, d, for red light-treated seedlings were found to be precisely the same for seedlings irradiated with continuous far-red light and in seedlings grown with Hoagland's nutrient solution in either red or far-red light. This fact shows that pattern specification is not influenced specifically by environmental factors, including light, even in those cases where pattern realization depends entirely on the light factor. We concluded that it is "developmental homeostasis"[2] which governs pattern specification.

Let us repeat the major point we wanted to make in this paragraph: Fig. 3e, d clearly shows that the ability of the intercostal cells to perform the initial action of P_{fr} can develop earlier than the ability to form anthocyanin. This means that the appearance of competence for P_{fr}, with regard to the initial action of P_{fr} in anthocyanin synthesis, precedes the "starting point" (i.e., the point of time at which a particular epidermal cell becomes capable of forming anthocyanin as a response to P_{fr} formation) considerably. Even after the "starting point" has been reached, this time gap between the onset of the initial action and the onset of the response remains (approximately 1.5 h; R. SCHMIDT personal communication). We will return to this important aspect once again when we discuss the transmitter concept (see Sect. 5).

3.3 Time Courses of Responsiveness in Phytochrome-Mediated Anthocyanin Synthesis

MOHR et al. (1979), investigating control of phenylalanine ammonia lyase and anthocyanin synthesis in mustard cotyledons, used red light pre-treatments, prior to competence to P_{fr}, to study the effect of a strong reduction in P_{tot} on the effectiveness of subsequent light treatments. In the case of a red light pulse, given at 24 h, it was found that certain red light pretreatments *increased* the effectiveness of the pulse (Fig. 4) even though the red light pretreatment caused a strong reduction (approximately 80%) of P_{tot}. However, if continuous far-red light (operating through the phytochrome-dependent High Irradiance Reaction = HIR) was given from 24 h onwards the effectiveness of the far-red light was reduced by a red light pretreatment. Thus, a red light pre-treatment has opposing effects on the effectiveness of a subsequent pulse and a subsequent HIR. Clearly the effect of the pre-treatment is not merely to cause a decrease in P_{tot}. MOHR et al. (1979) proposed that the light-pretreatments which operate through phytochrome cause changes in the availability of receptors for P_{fr}.

Further studies of the pre-treatment effect have shown (SCHMIDT and MOHR 1981b) that the time course of light sensitivity, operationally, the responsiveness towards a saturating red light pulse or towards continuous far-red light, is strongly and specifically influenced by a light pre-treatment prior to competence (approximately 25 h after sowing), while the starting point of the response (ap-

[2] The term "developmental homeostasis" as originally conceived by BERG (1959) implies that control over development is exerted by a balance of functions *within* the organism. Factors from outside the organism do not *specifically* control development even though they may non-specifically retard, inhibit or accelerate development, such as temperature (see MOHR 1972)

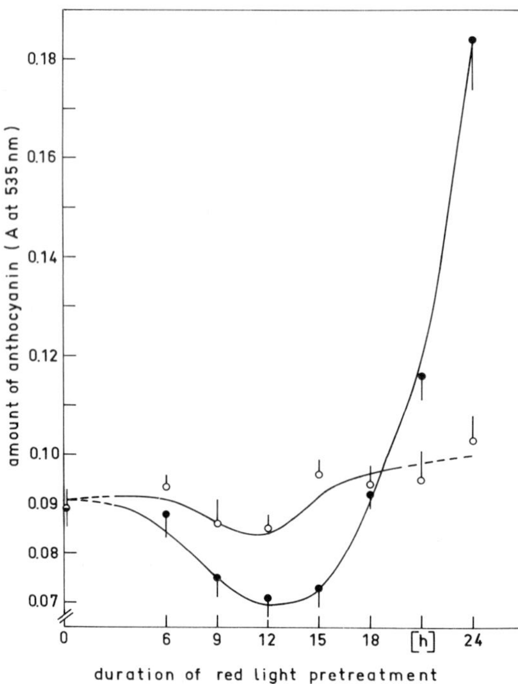

Fig. 4. The effect of a saturating red light pulse given at 24 h after sowing, on anthocyanin synthesis in the cotyledons of the mustard seedling as influenced by the duration of a red light pretreatment given from time zero (=time of sowing) onwards. The red light pretreatment was terminated with a 5-min 756-nm light pulse to prevent action of P_{fr} during the subsequent dark incubation period separating the pretreatment from the 5-min red light pulse given 24 h after sowing. o control experiments to test the effect of a 756 nm light pulse (5 min), given at the times indicated on the abscissa, on the effectiveness of a saturating red light pulse, given at 24 h after sowing. Anthocyanin was assayed 52 h after sowing. (MOHR et al. 1979)

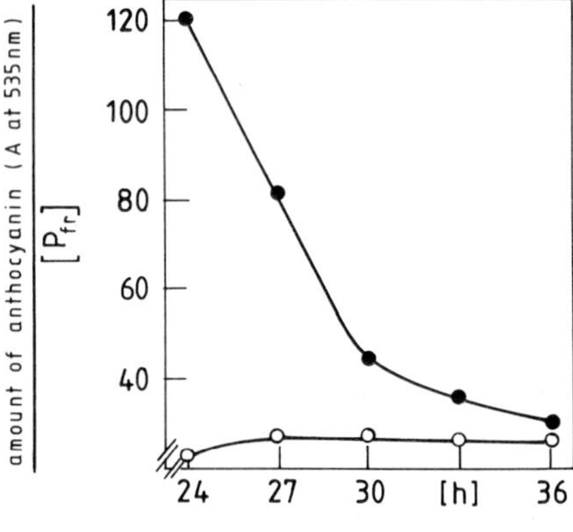

Fig. 5. Time courses of responsiveness to a saturating red light pulse. Light-induced anthocyanin synthesis in the mustard cotyledons was measured. The light pulses were applied at the points indicated. Anthocyanin of the cotyledons was assayed after completion of synthesis. Pretreatments: ● 24 h red light, from sowing onwards, terminated by a saturating 756 nm light pulse; o no light pretreatment. The response (amount of anthocyanin) is referred to the amount of P_{fr} established by the red light pulse (SCHMIDT and MOHR 1982)

proximately 27 h after sowing) and the lag-phase of the response (2 h) are insensitive to light pre-treatments prior to competence (SCHMIDT and MOHR 1981 a).

The stimulation by a light-pretreatment of responsiveness towards P_{fr} (produced by a saturating red light pulse) is only transient (Fig. 5). The half-life of the sensitizing effect is less than 5 h. Nevertheless, we must take into account in every experiment with light that responsiveness towards P_{fr} is strongly affected by preceding light treatments which also operate through phytochrome. Thus,

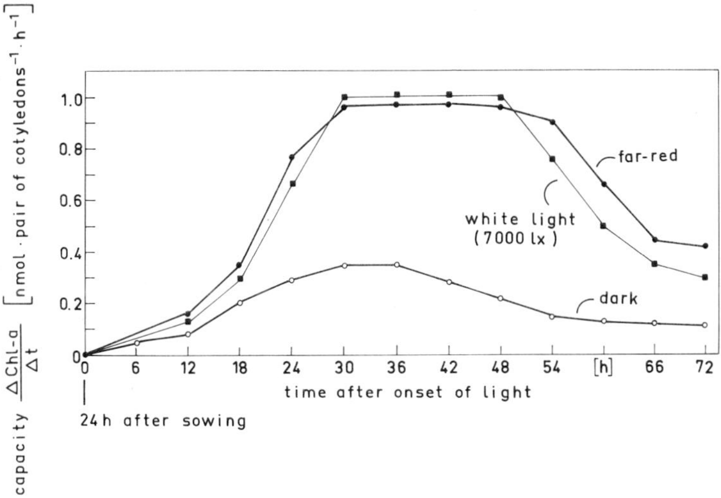

Fig. 6. Time course of the "capacity" of the chlorophyll a synthesizing pathway in the mustard seedling cotyledons in continuous darkness, continuous far-red or continuous white light. (GEHRING et al. 1977)

the action of light not only depends on the state of phytochrome at a particular time but also (and most significantly) on the responsiveness of the cell function on which P_{fr} acts, whereby this responsiveness is determined by light pretreatment.

3.4 Time Course of the "Capacity" for Chlorophyll Formation

The process of chlorophyll (Chl) biogenesis is light-dependent in two respects. (1) Photoreduction of protochlorophyll(ide) = (PChl) to Chlide-a is an integral step of the biosynthetic sequence leading to Chl-a and -b. (2) Phytochrome controls the potential rate ("capacity") of the metabolic sequence leading to PChl. By "capacity" we designate the *maximum* flow in the biosynthetic channel leading to Chl, i.e., the flow under light conditions which saturate the PChl → Chlide-a photoconversion. Figure 6 shows that the "capacity" depends on the developmental state (age) of the seedlings and that it is strongly increased by phytochrome, operationally, continuous far-red light operating through the phytochrome dependent HIR. The time course of the "capacity" under continuous *white* light (which permits Chl synthesis and photosynthesis) deviates only slightly from the time course of the "capacity" under continuous standard far-red light (which does not permit significant Chl accumulation). It was concluded that the capacity for Chl synthesis is determined by phytochrome irrespective of whether or not Chl is actually being synthesized. Experiments to localize the site of regulation of the "capacity" have led to the conclusion (GEHRING et al. 1977) that phytochrome operates at the level of 5-aminolevulinate synthesis.

With regard to the theme of the present chapter it is clear from Fig. 6 that the time course of the "capacity" for Chl formation is in principle the same in light and dark. While light via phytochrome exerts a strong modulatory effect it seems that the basic temporal change of the "capacity" is determined by endogenous factors rather than by light.

4 Temporal Patterns in Phytochrome-Mediated Enzyme Induction

Phytochrome (P_{fr}) controls the appearance of many enzymes in the mustard cotyledons (see MOHR 1974, SCHOPFER 1977, HONG and SCHOPFER 1981). Among those which are inducible by phytochrome we originally selected PAL, Carboxylase and GR (for abbreviations see Table 1) for an analysis of the temporal pattern of inducibility. PAL is a major enzyme of phenylpropanoid biogenesis, located in the cytoplasm (JABBEN and DEITZER 1979); Carboxylase is a major Calvin-cycle enzyme, only occurring within the plastid compartment and inducible by light via phytochrome (BRÜNING et al. 1975). Moreover, the enzyme is "activated" by photosynthetically effective light (HELDT et al. 1978). GR

Table 1. Anthocyanin and six enzymes from mustard seedling cotyledons are compared with regard to starting points and points of loss of full reversibility. (FROSCH et al. 1977, MOHR 1978, HONG and SCHOPFER 1981)

Enzyme[a]	Starting point[b] [h]	Point of loss of full reversibility[c] [h]
Anthocyanin	27	26
PAL	27	26
Carboxylase	42	15
β-amylase[e]	45	36
GR	48	18
Allantoinase	48	24
Urate oxidase	72	48
PChl[f]	30	18

[a] Phenylalanine ammonia-lyase (PAL), very low dark level; ribulosebisphosphate carboxylase (carboxylase), low dark level; glutathione reductase (GR), relatively high dark level

[b] The point on the time scale[d] where a phytochrome-mediated increase of the enzyme level is detectable

[c] The point on the time scale[d] up to which the inductive effect of continuous red light (given from the time of sowing) remains *fully* reversible by a saturating 756-nm light pulse

[d] The time scale starts with the sowing of the seeds (time after sowing)

[e] R. SHARMA and P. SCHOPFER (1982)

[f] The values for protochlorophyll(ide) are added here to help the reader to understand Sect. 5.1

is also located as a stromal enzyme in the plastid compartment (FOYER and HALLIWELL 1976). While its level is increased by light via phytochrome (DRUMM and MOHR 1973) the enzyme is not activated by light (HALLIWELL and FOYER 1978).

The major findings obtained with these enzymes and relevant for our theme are summarized in Table 1.

The following conclusions were drawn from this series of investigations: (1) The starting points are enzyme-specific. They are strictly determined by developmental homeostasis and, thus, independent of light. (2) There is no feedback from pattern realization to pattern specification. As an example, it was found that the onset and the rate of carboxylase synthesis or the time course of GR synthesis in the mustard cotyledons is always the same irrespective of the light treatment the seedling received prior to 36 h after sowing (DRUMM and MOHR 1973, BRÜNING et al. 1975, FROSCH et al. 1976). While these different light treatments between sowing and 36 h after sowing lead to different time courses of PAL or anthocyanin levels, the starting points and the rates of carboxylase or GR synthesis are not at all affected. In brief, the specification of the temporal pattern with regard to carboxylase or GR synthesis is not influenced by the extent of pattern realization with regard to PAL or anthocyanin synthesis.

5 The Transmitter Concept

5.1 Control of Protochlorophyll(ide) Accumulation

Significant accumulation of photoconvertible protochlorophyll(ide) (=PChl) in the cotyledons of the mustard seedling takes place from 24 h after sowing onwards (25 °C). The rate of accumulation in darkness is greatly increased by a pretreatment with red light (Fig. 7). The strong effect of continuous red light, given from the time of sowing, remains fully reversible by a 756-nm light pulse up to slightly less than 18 h after sowing (Fig. 8). This means that up to this point phytochrome (P_{fr}) has not yet coupled to that cell function which produces PChl. However, at approximately 18 h after sowing the coupling point (=point of loss of full reversibility, see Table 1) is reached. This means that even the elimination of P_{fr} (by a saturating 756-nm light pulse in our case) no longer prevents the expression of the response completely. How can the light signal be stored in the competent cells from the point of loss of full reversibility up to the actual starting point (a time interval of approximately 12 h in the case of PChl and 30 h in the case of GR, see Table 1)?

5.2 The Transmitter Concept in Phytochrome-Mediated Enzyme Induction

The term "transmitter" was introduced by Schopfer in connection with his studies on peroxidase appearance to designate this storage intermediate in

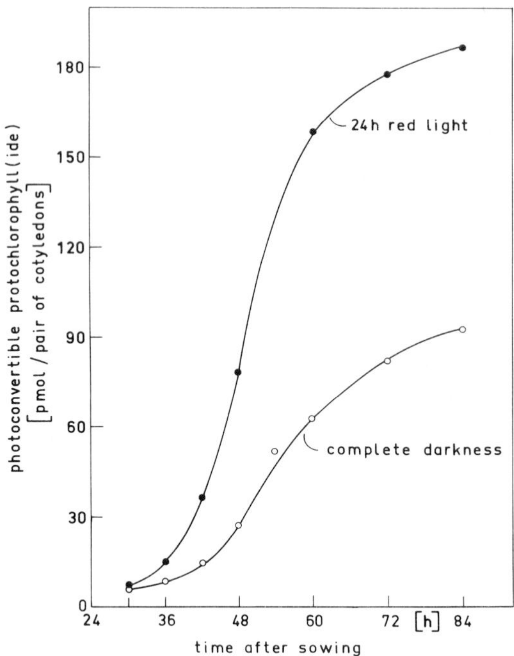

Fig. 7. Accumulation of photoconvertible protochlorophyll(ide) in the mustard seedling cotyledons in complete darkness (○) and in darkness following a red light treatment during the first 24 h after sowing (●). (Modified from KASEMIR et al., 1976)

the light signal-response chain (SCHOPFER 1972, SCHOPFER and PLACHY 1973). In mustard cotyledons, control of peroxidase levels by phytochrome involves two subsequent steps which do not overlap. The response can be induced by light via phytochrome only during the first 4 days after sowing while the increase of peroxidase activity occurs only after this time. Thus, an induction period is clearly separated from a realization period. SCHOPFER (1977) reports that it is very probably the synthesis of inactive enzyme protein which is induced by phytochrome in the first phase. During the second phase the inactive proenzyme seems to become activated by a process not involving phytochrome. Thus, the stable transmitter seems to be the enzyme protein in this particular case.

Irrespective of the molecular nature of the transmitter, we must keep two steps in the process of temporal pattern specification apart as soon as a transmitter comes into play: (1) The appearance of competence in the determined cells for P_{fr} with respect to transmitter formation (in the case of GR synthesis at approximately 18 h) and (2) the appearance of competence in the determined cells for the transmitter with regard to the performance of the particular cell function (in the present case, actual GR synthesis). We have seen previously that even in the case of anthocyanin synthesis the two steps could easily be identified during the small time interval between 27 and 29 h after sowing (see Fig. 3). In the case of urate oxidase (a particularly well investigated peroxisomal enzyme of the mustard cotyledons, see HONG and SCHOPFER 1981) the point of loss of full reversibility (=coupling point=onset of transmitter formation) was at 48 h after sowing while the starting point (=appearance of competence for the transmitter) was at 72 h.

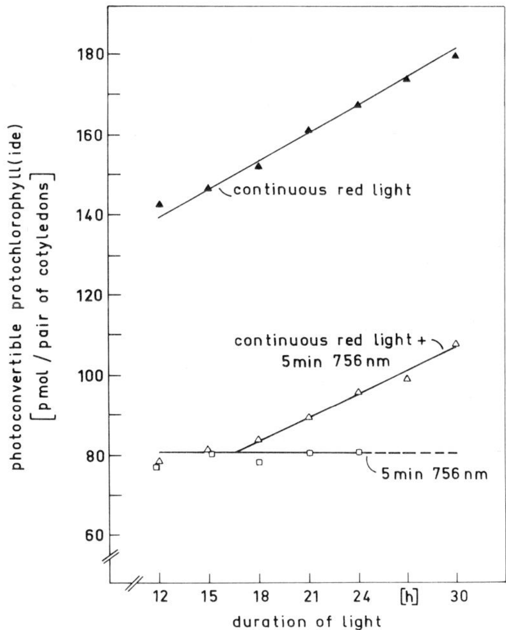

Fig. 8. Action of continuous red light of different durations and of 756-nm light pulses on the amount of photoconvertible protochlorophyll(ide) 72 h after sowing. Onset of red light at time of sowing. ▲ continuous red light only; △ continuous red light followed by 5 min 756-nm light; □ 5 min 756-nm light only. These pulses have no detectable effects as compared to the dark control [82 pmol photoconvertible protochlorophyll(ide)/pair of cotyledons]. (Modified from KASEMIR et al. 1976)

6 Temporal Pattern in Phytochrome-Mediated Enzyme Suppression

Increase of the level (=extractable activity) of the enzyme lipoxygenase (LOG) in the mustard cotyledons is controlled by phytochrome (P_{fr}). LOG is localized in the mustard seedling (see Fig. 1) exclusively in the laminae (95%) and petioles (5%) of the cotyledons. No LOG can be extracted from the hook region of the seedling or from the lower parts of the hypocotyl and the tap root.

A threshold (all-or-none mechanism) had to be postulated for the action of P_{fr} in this case (Fig. 9). If the actual level of P_{fr} exceeds the threshold level, the increase in LOG level is immediately and completely arrested. If the actual level of P_{fr} decreases below the threshold level, the increase in LOG activity is immediately resumed at full speed. In the course of these investigations, it has been found (OELZE-KAROW and MOHR 1973) that the suppression of LOG synthesis in the cotyledons requires the presence of the hypocotylar hook. Moreover, it was argued on good grounds that suppression of LOG synthesis in the cotyledons is controlled by phytochrome located in the hypocotylar hook (see Fig. 1), in other words that an interorgan signal transfer is involved. This startling but apparently inevitable concept has been substantiated by both spectrophotometric measurements of phytochrome (SCHÄFER et al. 1973) and physiological (OELZE-KAROW and MOHR 1974) evidence measuring LOG.

As far as our present theme is concerned, the problem has been to ascertain whether LOG synthesis responds to P_{fr} throughout the whole period of development of the mustard seedling. The result is given in Fig. 10: if one irradiates

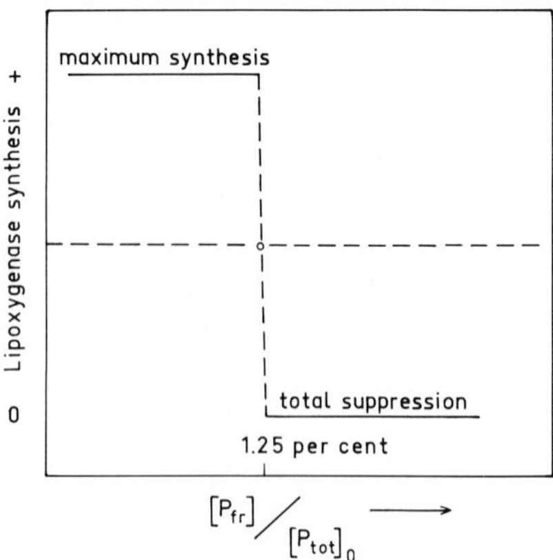

Fig. 9. Scheme to describe the concept of a threshold regulation of lipoxygenase synthesis in the mustard cotyledons by P_{fr}. $[P_{tot}]_o$ is the total phytochrome at time zero (36 h after sowing); this value is a constant. The amount of P_{fr}, $[P_{fr}]$, is always expressed as a fraction or percentage of $[P_{tot}]_o$. Expressed in this way, the threshold level, $[P_{fr}]_{th}$, is approximately 1.25% (0.0125). (MOHR and OELZE-KAROW 1976)

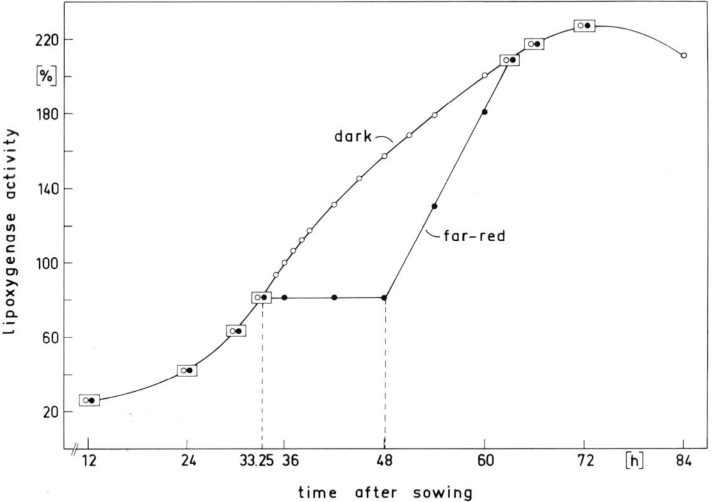

Fig. 10. Time-courses of lipoxygenase levels in mustard seedling cotyledons in continuous darkness (o) and under continuous standard far-red light (equivalent to 718-nm light) (●). The onset of far-red light was at the time of sowing of the seeds. (OELZE-KAROW and MOHR 1970)

with standard far-red or red light (both of which maintain a P_{fr} level above the threshold level) from the time of sowing, there is no control of LOG synthesis up to approximately 33 h. At this point, the full suppression of LOG synthesis by far-red or red light, i.e., by a P_{fr} level above the threshold level, suddenly comes into play, while at 48 h after sowing, the LOG synthesis of the seedling suddenly and completely escapes from P_{fr} control. LOG synthesis is resumed even under continuous red or far-red light. The time course of the LOG level after resumption of increase returns to the dark kinetics. Under all circumstances the LOG-producing system escapes from P_{fr} control 48 h after sowing (with 25 °C and strictly standardized conditions). It was concluded that the temporal pattern of response must be determined by changes in the system on which P_{fr} acts rather than by P_{fr}. The basic experiment reported in Fig. 10 was repeated over several years with seed samples which differed in origin (different seed growers) as well as in age. While the on-off pattern of sensitivity was always found, the length of time during which the LOG-producing system was sensitive ("competent") for P_{fr} differed (e.g., see Fig. 152 in MOHR 1972). The duration of time of competence became somewhat shorter as the seed material aged.

It should be emphasized, though, that in the case of the LOG-producing system the onset of the period of competence for P_{fr} is late. As an example, competence for P_{fr} in phytochrome-mediated anthocyanin synthesis starts 27 h after sowing under the same growth conditions. It is obvious once again that the onset and duration of the period of competence for P_{fr} is response-specific.

The temporal pattern of competence is not influenced by phytochrome. This important conclusion was already reached during the initial stage of these investigations (OELZE-KAROW and MOHR 1970) when experimental evidence showed that neither the beginning nor the end of the period of sensitivity for P_{fr} of the LOG-producing system could be influenced by phytochrome.

To state the decisive point: In the course of development of the mustard seedling, the sensitivity of the LOG-producing system towards P_{fr} changes abruptly in the manner of an all-or-none response. This is true for the onset as well as for the termination of the period of sensitivity. Hence, not only the LOG response per se shows the characteristics of an all-or-none response; rather the process of acquiring and losing competence for P_{fr} with respect to suppression of LOG synthesis also follows the pattern of an all-or-none response.

7 Concluding Remarks

In the photomorphogenetic, phytochrome-dependent development of a seedling we must keep two steps apart: pattern specification and pattern realization. The specification of the temporal and spatial pattern of development is independent of phytochrome even though the realization of the pattern of development can only occur in the presence of phytochrome. All the examples we have considered support the idea that there is no feedback from pattern realization to pattern specification. In brief, pattern specification with regard to response A is not influenced by the extent of pattern realization with regard to response B. In the forward direction a light action on the future realization response was found only in the sense that the extent of responsivity of the anthocyanin producing cell function towards phytochrome can be determined by light at a time when the cell function itself is not yet competent to light (P_{fr}). But even in this case the starting point and the lag-phase of the response are insensitive to light pre-treatments prior to competence (see Sect. 3.3).

So far approaches to analyze the causalities of pattern specification have not really been successful, either in plants or in animals (for reviews see WOLPERT 1969, SANDER 1976). The concept (as formulated largely by WOLPERT) that cells acquire positional information and respond to it by differentiating in appropriate ways may be very useful in more descriptive treatments of animal developmental physiology (e.g., BRYANT et al. 1977). However, it remains to be seen whether the "field" and "positional information" concepts and the notion of localized information molecules in animal eggs are really fruitful in "future attempts to tackle pattern specification in more molecular terms" (SANDER 1976).

Phyllotaxis, i.e., the specific arrangements of leaf primordia on the apices of growing shoots, is still an outstanding example of pattern formation in plants (BAARSCH-GOLLNAU et al. 1980). The effect of light, exerted through phytochrome (MOHR and PINNIG 1962) on the rate of formation of leaf primordia was investigated at the apex of seedlings of *Sinapis alba* and *Xanthium strumarium*. It was found that light accelerates this rate. On the other hand, no significant light effect was found on the angles of divergence of successive leaves during the transition from the almost decussate leaf position of the cotyledons to the helical phyllotaxis of the stem leaves. In fact, light- and dark-grown plants use the same leaves for the transition from decussate to helical phyllotaxis. Thus, if time is plotted in "biological units" (number of primordia) there is no difference between light and dark-grown plants. Using scanning electron microscope techniques it was found that the "primordia free apical area" enlarges during development. The rate of enlargement is accelerated by light. However, if time is expressed in biological units (number of primordia) no difference between light- and dark-grown plants exists. It is concluded that light accelerates the realization of the apical pattern without interfering with the specification of the pattern. In other words, light accelerates the development of an apex without affecting the temporal and spatial coordination of the events.

Recent models to explain the most common phyllotaxis, the helical pattern of inserted leaves, are based on the diffusion of an inhibitor (VEEN and LINDENMAYER 1977) or of a morphogen (THORNLEY 1975) that is produced in the primordia, decays proportionally with its concentration, and inhibits the development of a new primordium as long as its concentration is above a certain threshold. As VEEN and LINDENMAYER (1977) point out rightly, "leaf determination is basically a two- or three-dimensional process and ... one-dimensional projections of the morphogenetic field will not be sufficient to illuminate the details of its operation". However, in spite of their sophisticated computer study, a convincing relationship between geometric and physiological parameters is not in sight.

Interesting recent suggestions about the part selector genes (GARCIA-BELLIDO 1975) or transposable genetic elements (IS-elements, NEVERS and SAEDLER 1977) may play in pattern specification during development of a eukaryotic multicellular system are still speculative (MARX 1980). Selector genes are considered to operate on the level of differential gene transcription while IS-elements operate on the level of DNA. In any case, since each cell of a multicellular organism possesses the full genome of the species, pattern formation must be related to differential gene action in different cells of the multicellular system. BONNET (1764) stated:

> "if organized bodies are not preformed, then they must be formed every day, in virtue of the laws of a special mechanism. Now, I beg you to tell me what mechanics will preside over the formation of a brain, a heart, a lung and so many other organs?"

The study of photomorphogenesis has greatly enriched our knowledge about the mechanism of pattern *realization* in plant development. In some cases, such as light-mediated flavonoid synthesis (see Chap. 24, this Vol.) or plastidogenesis in angiosperms (see Chap. 12, this Vol.) the molecular mechanism of pattern realization has been elucidated recently on the level of mRNA and enzyme synthesis. In these fields of investigation the thoughtful framing of appropriate questions has in fact been rewarded by molecular answers. However, these answers are still *partial* answers since we do not know yet how physiologically active phytochrome (P_{fr}) causes the pertinent genes to be transcribed. The so far poor understanding of gene function in eukaryotic cells simply has remained a major obstacle in the efforts to unravel the events between the formation of P_{fr} and the onset of transcription.

However, recent findings about "hot spots" (=hypersensitive regions) for DNAse I in its action on chromatin indicate progess (see KOLATA 1981, for a review). It seems that if a gene is capable of being transcribed it must have a hypersensitive region at its 5′ end, although the appearance of the hypersensitive region is not *sufficient* for transcription. The chromatin in a hypersensitive region is in a more relaxed configuration than the rest of the chromatin. "It is reasonable to speculate ... that the hypersensitive regions are sites on chromation that are especially accessible to proteins used to initiate gene transcription" (WEINTRAUB, quoted by KOLATA 1981). The formation of hypersensitive sites might be related to pattern *specification* and concomitant appearance of competence.

References

Barsch-Gollnau S, Ritterbusch A, Mohr H (1980) Photomorphogenesis and phyllotaxis during vegetative growth in *Sinapis alba* and *Xanthium strumarium*. Plant Cell Environ 3:363–370

Berg RL (1959) A general evolutionary principle underlying the origin of developmental homeostasis. Am Nat 93:103–105

Bonnet C (1764) Contemplation de la nature. Quoted by R.A. Raff (1977) The molecular determination of morphogenesis. BioScience 27:394–401

Brüning K, Drumm H, Mohr H (1975) On the role of phytochrome in controlling enzyme levels in plastids. Biochem Physiol Pflanz 168:141–156

Bryant PJ, Bryant SV, French V (1977) Biological regeneration and pattern formation. Sci Am 237:67–81

Bryant SV, French V, Bryant PJ (1981) Distal regeneration and symmetry. Science 212:993–1002

Bühler B (1977) Untersuchungen zur Wechselwirkung von Phytochrom und Äthylen bei der Anthocyansynthese des Senfkeimlings (*Sinapis alba* L.). Doctoral thesis. Univ Freiburg

Bünning E (1948) Entwicklungs- und Bewegungsphysiologie der Pflanze. Springer, Berlin Heidelberg

Drumm H, Mohr H (1973) Control by phytochrome of glutathione reductase levels in the mustard seedling. Z Naturforsch 28c:559–563

Drumm H, Mohr H (1974) The dose-response curve in phytochrome-mediated anthocyanin synthesis in the mustard seedling. Photochem Photobiol 20:127–131

Foyer CH, Halliwell B (1976) The presence of glutathione and glutathione reductase in chloroplasts: a proposed role in ascorbic acid metabolism. Planta 133:21–25

French V, Bryant PJ, Bryant SV (1976) Pattern regulation in epimorphic fields. Science 193:969–981

Frosch S, Bergfeld R, Mohr H (1976) Light control of plastogenesis and ribulose-bisphosphate carboxylase levels in mustard seedling cotyledons. Planta 133:53–56

Frosch S, Drumm H, Mohr H (1977) Regulation of enzyme levels by phytochrome in mustard cotyledons: multiple mechanisms? Planta 136:181–186

Garcia-Bellido A (1975) Genetic control of wing disk development in *Drosophila*. In: Cell Patterning. Ciba Found Symp New Ser 29:161–182

Gehring H, Kasemir H, Mohr H (1977) The capacity of chlorophyll a biosynthesis in the mustard seedling cotyledons as modulated by phytochrome and circadian rhythmicity. Planta 133:295–302

Goodwin BC (1970) Temporal order as the origin of spatial order in embryos. Stud Gen 23:273–282

Goodwin BC, Cohen MH (1969) A phase-shift model for the spatial and temporal organization of development systems. J Theor Biol 25:49–107

Halliwell B, Foyer CH (1978) Properties and physiological function of a glutathione reductase purified from spinach leaves by affinity chromatography. Planta 139: 9–17

Heldt HW, Chou CJ, Lorimer GH (1978) Phosphate requirement for the light activation of ribulose-1,5-bisphosphate carboxylase in intact spinach chloroplasts. FEBS Lett 92:234–240

Hong Y-N, Schopfer P (1981) Control by phytochrome of urate oxidase and allantoinase activities during peroxisome development in the cotyledons of mustard (*Sinapis alba* L.) seedlings. Planta 152:325–335

Jabben M, Deitzer GF (1979) Effects of the herbicide SAN 9789 on photomorphogenic responses. Plant Physiol 63:481–485

Johnson CB (1980) The effect of red light in the high irradiance reaction of phytochrome: evidence for an interaction between P_{fr} and a phytochrome cycling-driven process. Plant Cell Environ 3:45–51

Kasemir H, Huber P, Mohr H (1976) Timing of the initial action of phytochrome with regard to protochlorophyll synthesis in the mustard seedling. Planta 132:157–160

Kolata GB (1981) Genes regulated through chromatin structure. Science 214:775–776

Lewin R (1981) Seeds of change in embryonic development. Science 214:42–44

Marx JL (1980) A moveable feast in the eukaryotic genome. Science 211:153–155

Mohr H (1972) Lectures on photomorphogenesis. Springer, Berlin Heidelberg New York

Mohr H (1974) The role of phytochrome in controlling enzyme levels in plants. In: Paul J (ed) Biochemistry of cell differentiation. MTP Int Rev Sci 9:37–81. Butterworth, London

Mohr H (1978) Pattern specification and realization in photomorphogenesis. Bot Mag Tokyo Spec Issue 1:199–217

Mohr H (1982) Principles in plant morphogenesis. In: Sattler R (ed) Axioms and principles of plant construction. Junk, The Hague, pp. 93–111

Mohr H, Drumm-Herrel H (1981) Interaction between blue/UV light and light operating through phytochrome in higher plants. In: Smith H (ed) Plants and the daylight spectrum. Academic Press, London New York, pp 423–441

Mohr H, Oelze-Karow H (1976) Phytochrome action as a threshold phenomenon. In: Smith H (ed) Light and plant development. Butterworth, London, pp 257–284

Mohr H, Pinnig E (1962) Der Einfluß des Lichts auf die Bildung von Blattprimordien am Vegetationskegel der Keimlinge von *Sinapis alba* L. Planta 58:569–579

Mohr H, Sitte P (1971) Molekulare Grundlagen der Entwicklung. BLV, München

Mohr H, Drumm H, Schmidt R, Steinitz B (1979) The effect of light pre-treatments on phytochrome-mediated induction of anthocyanin and of phenylalanine ammonia-lyase. Planta 146:369–376

Nevers P, Saedler H (1977) Transposable genetic elements as agents of gene instability and chromosomal rearrangements. Nature 268:109–115

Oelze-Karow H, Mohr H (1970) Experiments regarding the problem of differentiation in multicellular systems. Z Naturforsch 25b:1282–1286

Oelze-Karow H, Mohr H (1973) Quantitative correlation between spectrophotometric phytochrome assay and physiological response. Photochem Photobiol 18:319–330

Oelze-Karow H, Mohr H (1974) Interorgan correlation in a phytochrome-mediated response in the mustard seedling. Photochem Photobiol 20:127–131

Quatrano RS, Brawley SH, Hogsett WE (1979) The control of the polar deposition of a sulfated polysaccharide in *Fucus* zygotes. In: Subtelny S, Konigsberg IR (eds) Determinants of spatial organization. Academic Press, London New York, pp 77–96

Rutishauser A (1969) Embryologie und Fortpflanzungsbiologie der Angiospermen. Springer, Wien New York

Sander K (1976) Specification of the basic body pattern in insect embryogenesis. Adv Insect Physiol 12:125–238

Schäfer E, Schmidt W, Mohr H (1973) Comparative measurements of phytochrome in cotyledons and hypocotyl hook of mustard (*Sinapis alba* L.). Photochem Photobiol 18:331–334

Schäfer E, Lassig T-U, Schopfer P (1975) Photocontrol of phytochrome destruction in grass seedlings. The influence of wavelength and irradiance. Photochem Photobiol 22:193–202

Schmidt R, Mohr H (1981a) Lag-phase and rate of synthesis in light-mediated anthocyanin synthesis. Planta 151:541–543

Schmidt R, Mohr H (1981b) Time-dependent changes in the responsiveness to light of phytochrome-mediated anthocyanin synthesis. Plant Cell Environ 4:433–437

Schmidt R, Mohr H (1982) Evidence that a mustard seedling responds to the amount of P_{fr} and not to the P_{fr}/P_{tot} ratio. Plant Cell Environ 5:495–499

Schopfer P (1972) Role of phytochrome in the control of enzyme activity in higher plants: photomodulation and photodetermination of enzyme synthesis. Symp Biol Hung 13:115–126

Schopfer P (1977) Phytochrome control of enzymes. Annu Rev Plant Physiol 28:223–252

Schopfer P, Plachy C (1973) Die organspezifische Photodetermination der Entwicklung von Peroxidaseaktivität im Senfkeimling (*Sinapis alba* L.) durch Phytochrome. Z Naturforsch 28c:296–301

Sharma R, Schopfer P (1982) Sequential control of phytochrome-mediated synthesis de novo of β-amylase in the cotyledons of mustard (Sinapis alba L.) seedlings. Planta 155:183–189

Sinnot E (1960) Plant Morphogenesis. McGraw-Hill, New York

Snow M, Snow R (1934) The interpretation of phyllotaxis. Biol Rev 9:132–137

Steinitz B, Bergfeld R (1977) Pattern formation underlying phytochrome-mediated anthocyanin synthesis in the cotyledons of *Sinapis alba* L. Planta 133:229–235

Steinitz B, Drumm H, Mohr H (1976) The appearance of competence for phytochrome-mediated anthocyanin synthesis in the cotyledons of *Sinapis alba* L. Planta 130:23–31

Steinitz B, Schäfer E, Drumm H, Mohr H (1979) Correlation between far-red absorbing phytochrome and response in phytochrome-mediated anthocyanin synthesis. Plant Cell Environ 2:159–163

Thompson D'Arcy W (1917) On growth and form. Univ Press, Cambridge

Thompson d'Arcy W (1961) On growth and form. (abridged ed, ed Bonner JT). Cambridge Univ Press, London

Thornley JHM (1975) Phyllotaxis I. A mechanistic model. Ann Bot 39:491–507

Turing AM (1952) The chemical basis of morphogenesis. Philos Trans R Soc London B 237:37–72

Veen AH, Lindenmayer A (1977) Diffusion mechanism for phyllotaxis – theoretical physico-chemical and computer study. Plant Physiol 60:127–139

Wagner E, Mohr H (1966) "Primäre" und "sekundäre" Differenzierung im Zusammenhang mit der Photomorphogenese von Keimpflanzen (*Sinapis alba* L.). Planta 71:204–221

Wardlaw CW (1952) Morphogenesis in plants. Methuen, London

Wolk CP (1979) Intercellular interactions and pattern formation in filamentous cyanobacteria. In: Subtelny S, Konigsberg JR (eds) Determinants of spatial organization. Academic Press, London New York, pp 247–266

Wolpert L (1969) Positional information and the spatial pattern of cellular differentiation. J Theor Biol 25:1–47

Woodger JH (1930, 1931) The "concept of organism" and the relation between embryology and genetics. Quart Rev Biol 5:1–22, 438–463; 6:178–207

15 The Control of Cell Growth by Light

V. GABA and M. BLACK

1 Introduction

The photocontrol of growth in higher plants depends on the participation of cell enlargement, as revealed by measurements of cell length, diameter or volume. Although many studies of the photocontrol of growth omit an examination at cellular level, it is nevertheless necessary to include them, sometimes making assumptions about the contribution of cell extension to the overall growth pattern. It is clear, furthermore, that the role of cell division must be considered where necessary. Emphasis in this chapter will be on the growth of selected organs and possible mechanisms of the photocontrol of cell growth. Little or no attention will be given to growth substance/light interactions, effects of light on stem sections, root growth, tropistic movements and seed germination, most of which are covered elsewhere in this and other volumes of this Encyclopedia.

1.1 Growth, Cell Enlargement and Cell Division

Since growth is a dynamic process the pattern can show considerable changes over small time intervals. Growth may also be highly localised, occurring in a restricted region of an organ, or can proceed to different extents in various parts of the organ. Analyses of growth processes ideally should take these features into account. The kinetics of growth should be examined at frequent, preferably equal, time intervals otherwise only gross changes are noted. Some classical work, such as that by BROTHERTON and BARTLETT (1918), for example, in which only one growth point is recorded (end-point analysis), might tell us that light has caused an increase in organ or cell size but conveys no information as to how this was reached. Wherever possible, growth in separate zones of an organ should be followed, which are delimited by artificial markings (e.g. FRANSSEN et al. 1981) or which are naturally indicated (e.g., SILK 1980). Important events in the organ as a whole are obscured if this is not done (HÄCKER et al. 1964, VERBELEN and DE GREEF 1979). Certainly, this approach is essential if a proper insight into localised cellular events is to be achieved, but unfortunately much research which we have had to draw upon in the following account fails to observe these strictures.

Growth is expressed in a variety of ways, including changes in linear dimensions, weight and growth rate, or elementally, in terms of differentials of length, area, volume and weight (see ERICKSON 1977). A recent refinement is the application of fluid flow mechanics (SILK and ERICKSON 1978, 1979, SILK 1980, ERICK-

SON and SILK 1980). Techniques which facilitate the measurement of growth over short time intervals and/or in restricted zones include film analysers and displacement transducers, the latter being especially useful for detecting rapid, even transitory, growth responses (see PENNY and PENNY 1979). At tissue and cell level, the use of a microprocessor-based digitiser system permits rapid and accurate measurements to be made (HARRIS et al. 1981).

It is sometimes stated that organ growth is due to cell division and/or cell expansion. Obviously cell division (sensu stricto) alone cannot generate an increase in volume and to do so it must be accompanied by some degree of cell enlargement (see e.g., WRIGHT 1961, HABER and FOARD 1964, GREEN 1976, 1980, especially the discussion of surface extension and cell partitioning by the last author). Cell enlargement is predominantly involved in a growth process when at least one of the following criteria is satisfied:

a) Where cell size has been measured before and after photostimulation. The occurrence of cell division obviously introduces difficulties here. For example, mean cell size is sometimes determined by dividing the organ volume by the cell number; but clearly, when the rate of volume increase is less than the relative rate of division, the mean cell size diminishes and thus the degree of cell growth becomes obscured.

b) Where an organ or tissue has been proved to consist only of non-dividing cells, as determined by: (1) measurement of cell number, (2) examination for mitotic figures, (3) measurement of DNA quantity, (4) the use of special non-dividing tissues, e.g., gamma plantlets. Certain difficulties arise in some of these. For example, a common way of determining cell number is by counting cells in macerated organs (BROWN and RICKLESS 1949). This gives information for the organ as a whole and not for individual tissues, a serious omission in some cases (e.g., leaves) where tissues develop at different rates (e.g., AVERY 1933, MAKSYMOWYCH 1973).

1.2 Photosystems Involved in the Control of Growth

All the major photosensory systems in plants are involved in the control of cell growth. These are phytochrome (in the low-energy mode, dynamic mode, and high-irradiance reaction), cryptochrome, and photosynthesis. Other photoreceptors have been mooted, for example, for the perception of green light by roots (KLEIN 1979), but there is a paucity of information about them.

2 The Thomson Hypothesis

In his studies of the pea plant, WENT (1941) observed that light inhibits the extension growth of older, elongating internodes but stimulates the younger internodes and leaves. His explanation that this is due to a compensatory mechanism was wrong, since isolated organs also show differential responses to light

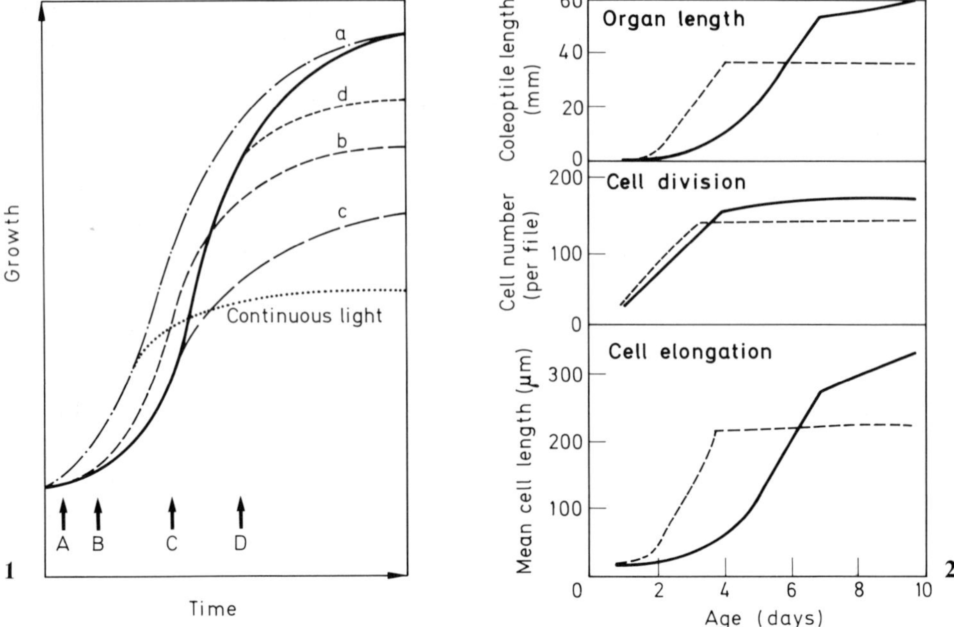

Fig. 1. The relation between age at time of exposure (at points *A–D*) and subsequent response (lines *a–d*) of an etiolated organ to a single period of illumination, and to continuous irradiation. *Solid line* represents growth in continuous darkness. Illumination at times *C* and *D* (lines *c* and *d*) gives rapid inhibition. Light at time *B* produces an intermediate response – promotion then inhibition. (After THOMSON 1950, 1951)

Fig. 2. Time relations of cell division, cell elongation and organ growth of oat coleoptiles, and the effect of continuous light on these processes. *Full line* darkness; *dashed line* light. (After THOMSON 1954)

according to their age (SCHNEIDER 1941, THOMSON 1951). THOMSON (1950) noted that the age (developmental stage) of organs of *Avena sativa* seedlings determines their response to light, the growth of the older coleoptiles and internodes being inhibited while that of young organs is promoted. The generalization that old and young organs respond differently is displayed graphically in Fig. 1. Light is held to accelerate all phases of the developmental sequence of plant cells—division, enlargement and maturation—and therefore promotes the transition from one stage to the next. Thus, an illuminated organ may pass through the enlargement phase so quickly that the full potential growth never occurs. Instead, the maturation phase is reached and hence the effect of light is to prevent the expression of enlargement capacity, i.e., growth relative to that in darkness is inhibited. Thomson also places leaf growth in this scheme but these organs never pass the stage corresponding to the early phase of stem growth (THOMSON and MILLER 1962).

In the oat coleoptile, for example, mean cell length follows a similar path to organ length (Fig. 2). All three parameters—mean cell length, organ length and cell number—show initial stimulation by light and a final inhibition. The relative activity of cell expansion and division as affected by light can be summarised as follows: (a) Light increases the overlap in the times of cell division and elongation. (b) Cell number increases earlier in the light. (c) The attainment of maximum cell size is hastened by light. (d) The times at which division and elongation begin are advanced in the light. The pattern of growth and its response to light is very similar in other cases such as pea internode no. 4 (I_4) (THOMSON 1954). Internodes respond to light in a manner characteristic of their age (THOMSON 1954). Cell division in *Avena* I_1 and *Pisum* I_1 is always inhibited by light, whereas in *Avena* I_2 (and the coleoptile) and *Pisum* I_3 and I_4 it is always promoted by an early irradiation. Exposure when cell division is already in progress hastens the end of this phase and therefore reduces final cell number. Both promotion and inhibition are in fact seen in *Avena* I_2 and *Pisum* I_4 according to the time of illumination (light given late in development causes inhibition) (THOMSON 1959). Several workers have generally confirmed the plausibility of the concepts developed by THOMSON (e.g., FUJII 1957, HÄCKER et al. 1964, ROESEL and HABER 1963, JOSE 1977). Notwithstanding this, doubts have been expressed as to the validity of these concepts (HOCK and MOHR 1965, NAKATA and LOCKHART 1966).

3 The Grass Seedling

Various species in the Gramineae, such as *Avena sativa, Triticum* spp., *Oryza sativa* and *Zea mays* have featured prominently in studies of the photocontrol of plant growth. Although the major component of overall stem elongation of the grass plant occurs during flowering (and hence involves physiological control processes which are outside the scope of this article) it is the juvenile organs, the coleoptile and mesocotyl, which have received most attention with regard to photomorphogenesis. The coleoptile—the sheath surrounding the apical meristem and leaf primordia of the grass embryo—may properly be considered as a leaf (e.g., FAHN 1967) though it is sometimes referred to as a stem (e.g., FIRN and DIGBY 1980). Its growth is, however, mainly polarized (i.e., virtually in one dimension) which distinguishes it from a leaf. The mesocotyl is the first internode—the two terms are interchangeable. Cell division in the mesocotyl occurs in a discrete zone unlike that in the coleoptile where it is dispersed (MER and CAUSTON 1963, HUISINGA 1967).

3.1 The Coleoptile

Much of the experimental work on coleoptiles is complicated by treatment given to the plant material during its preparation. Frequently the imbibed seeds are exposed to dim red light for 20 h or so, intended to inhibit mesocotyl elongation. But the light is also likely to have some effect on the coleoptile, and workers may be mistaken in believing that they are dealing with a dark-grown organ. Indeed, early photoresponses might possibly be saturated by radiation during the last phases of germination and this could account for the apparent failure of young *Avena, Hordeum* and *Oryza* coleoptiles to respond to light (e.g. FURUYA et al. 1969, MUIR and CHEN CHANG 1974).

Illumination is variably effective according to the "age", or more properly, the developmental stage of the coleoptile. Examination of responses of various parts of the coleoptile, an extension of THOMSON'S (1950, 1951) approach, has been carried out by several workers. In dark-grown wheat coleoptiles, in which cell division has ceased, ROESEL and HABER (1963) found that growth finishes first in the basal region and last in the apical region. Brief irradiation with far-red–reversible red, blue or white light stimulates extension in the upper region, but inhibits basal growth, i.e., the growth pattern like that occurring in darkness is exaggerated. There is simultaneous expression of these two effects of light in the middle region of the coleoptile. Thus, overall promotion and inhibition cancel each other out. The dual effect is readily interpretable on the basis of Thomson's hypothesis: closeness to maturity (not cell "age") determines the response to light and here the "immature" cells are stimulated to extend while the mature ones stop elongating. However, long-term blue light (at 1.65 Wm^{-2}) inhibits all parts of the coleoptile. Thus, there are two effects of blue light depending on duration and total energy. Phytochrome and cryptochrome might both be involved. Moreover, at certain ages red-light treatments can greatly alter the distribution of growth among the three regions of the coleoptile. Hence, a brief exposure to red light can cause enhancement of geotropism or a shift in the site of curvature (BLAAUW 1963, HUISINGA 1964). In high-intensity white light the basal rather than the apical portion grows, an opposite effect to that provoked by brief white-light treatment (ROESEL and HABER 1963).

Coleoptiles of a large number of wheat and barley cultivars were investigated by LAWSON and WEINTRAUB (1975), who found the same basic pattern as described above, which is modified with increase in coleoptile length (age). Photostimulation of the apical region ceases and eventually red light inhibits its elongation. Unfortunately, no check was made that cell division had ended, so it is not justified unequivocally to ascribe these effects to changes in cell enlargement.

The degree of the coleoptiles' response to red light, therefore, varies with species, cultivar, the length of the coleoptile, the position of the cells, and their developmental status. LAWSON and WEINTRAUB (1975) suggest that these factors may account for some, if not most, of the variable responses of excised segments to light as recorded in the literature (see also BLAAUW-JANSEN and BLAAUW 1966, PIKE et al. 1979). Failure to find any effect of blue, green, red or far-red light on decapitated *Avena* segments may be due to simultaneous promotion and inhibition in different regions (GENTILE and KLEIN 1964).

In some cases, such as in apical *Avena* segments, coleoptile elongation is due entirely to cell elongation (HOPKINS and HILLMAN 1965). Promotion of cell elongation can occur rapidly (within 46 min) after illumination (PIKE et al. 1979). *Avena* coleoptile length, cell division and cell elongation have been followed over almost the complete time course of growth by MER and CAUSTON (1963). At 3 days, when all three parameters are still increasing, a brief low-irradiance red-light treatment transiently stimulates coleoptile and cell length,

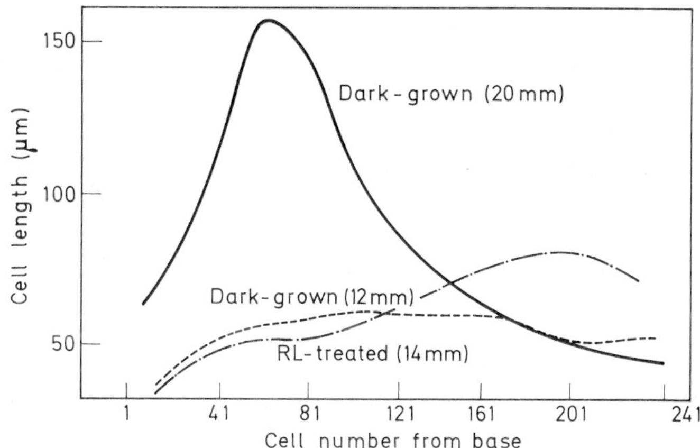

Fig. 3. The growth of inner epidermal cells of etiolated rice coleoptiles. Coleoptiles of 4-day old seedlings (12 mm long) were exposed to red light, and cell length measured after 20 h. After this period the dark controls were 20 mm long, while the irradiated coleoptiles were 14 mm long. (After FURUYA et al. 1969)

both of which are, however, finally reduced relative to the dark control. Cell division stops immediately upon illumination and at the end of the period of measurement cell number is lower than in dark-grown organs.

It is claimed that until rice coleoptiles reach a certain length (>10 mm) photocontrol of elongation is not possible (PJON and FURUYA 1967). However, sensitivity to light is only decreased, not absent, and in certain cultivars coleoptiles of only 3–4 mm length respond (PAUL and FURUYA 1973). Elongation of etiolated 4-day-old rice coleoptiles (10–12 mm long) of the cultivar Aichi Asahi depends on the presence of the tip (which is replaceable by auxin) and occurs in the central region of the organ in the zone 5–6 mm from the tip. Thirty-one percent of the growth is located here and virtually no extension takes place at the base or close to the tip (FURUYA et al. 1969). Elongation is inhibited by red light (54 Jm^{-2}), perception being most effective by the basal portion, but significant amounts also by the tip. Nearly all the cells in the 12-mm long etiolated coleoptile are 50 μm long (Fig. 3). After a further 20 h in darkness, coleoptile length increases to 20 mm and the total cell number remains almost unchanged. Organ extension is due to the elongation of cells 50–80 from the base which reach approximately 150 μm in length. A pulse of red light prevents this cell elongation almost completely but it promotes some extension of the cells close to the tip (cells 160–220 from the base) which do not grow in darkness (Fig. 3). This accounts for the red-light promotion of growth in apical segments (PAUL and FURUYA 1973). This is the typical double response to red light—inhibition and promotion. In this study, measurements of cell size are confined to the inner epidermis of the coleoptile and mean cell lengths are determined for each location in a file from base to tip. Because each cell is assigned a number its fate as a consequence of irradiation can be followed precisely (there are no cell divisions at the stage at which the coleoptiles are used). This type of analysis, first suggested by BINDLOSS

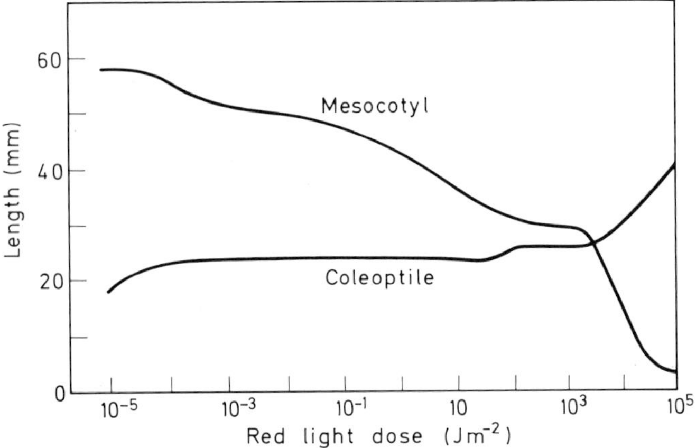

Fig. 4. Multiphasic dose response to red light of the *Avena* coleoptile and mesocotyl. (After BLAAUW et al. 1968)

(1942) for comparing dwarf and tall varieties of plants, is better than the more conventional determination of mean cell length at a certain position in an organ, since in the latter comparisons between different cells must inevitably be made. Whether the restriction of measurements to the inner epidermis is desirable is debatable, however, in view of the possibility that the leaf epidermis may be at a different stage of development from the other tissues in the organ (AVERY 1933).

BLAAUW et al. (1968) studied the fluence-response relationships in 2-day-old intact *Avena* coleoptiles (in which cell division probably had not ended) and identified three steps of growth promotion in the range 10^{-5} to 10^{5} Jm^{-2} of red light (below saturation dosage) (Fig. 4). The first step, extending to approximately 50 Jm^{-2}, gives 26% of the total promotion and is not far-red reversible. The second, minor step produces a further 8% of the growth promotion. The third, final step stimulates growth by another 65%. Comparable steps occur in the inhibition of mesocotyl growth (see Sect. 3.2). Growth of excised coleoptiles is also promoted non-reversibly by red light at dosages similar to those in the first step described by BLAAUW et al. (see PIKE et al. 1979). MANDOLI and BRIGGS (1979) reported similar observations on both coleoptiles and mesocotyls and, like SCHNEIDER (1941), noted that green light (safelight) can stimulate 10%–20% growth according to the age of the coleoptile, though red light is more effective. It is clear, therefore, that the red light dosage is critically important and that a green safelight should be used with great caution. How the multiphasic response changes with the developmental stage of the *Avena* coleoptile, and what are the relative contributions of cell growth in different regions of the coleoptile are important questions.

The growth inhibition observed in 2-day old etiolated rice coleoptiles is directly proportional to the logarithm of the amount of P_{fr} as determined spectrophotometrically (PJON and FURUYA 1968). After a red-light pulse most of the measurable P_{fr} disappears within 6 h (at 27 °C) whereas the response is still photoreversible for at least 9 h.

Most studies with blue light have used only brief pulses which generally act like red light, are often far-red reversible and therefore presumably operate through phytochrome (HUISINGA 1967, PJON and FURUYA 1967, PAUL and FURUYA 1973). Low fluences of blue light (e.g., 1.8–6.4 Jm^{-2}) temporarily reduce the growth rate of *Avena* coleoptiles (BLAAUW 1961, SHEN-MILLER and GORDON 1967), whereas high levels (400 Jm^{-2}) briefly stimulate it (BLAAUW 1961). Intermediate values (54 Jm^{-2}) first stimulate growth, then inhibit it, and finally leave no net effect (BLAAUW 1961). Responses to long-term blue light, possibly mediated by cryptochrome, are noted by ROESEL and HABER (1963) in *Triticum* and by ELLIOT and SHEN-MILLER (1976) in *Avena*. The latter workers report similar action spectra for phototropism and photoinhibition of straight growth with peaks at 360, 440 and 470 nm; prior red light reduces the magnitude of both responses.

So far only the responses of dark-grown coleoptiles have been considered. Some observations have, however, been made on light-grown *Zea mays* (GORTON and BRIGGS 1980). In this case, seedlings are held in darkness for 21 h, then exposed to a light regime consisting of 8 h white light/16 h dark each day for 3 days, with red or far-red immediately following each main photoperiod. Coleoptile growth is stimulated (relative to the red-light control) by the far-red exposure. These effects, presumably on cell elongation, are particularly interesting because they are taking place in de-etiolated coleoptiles, about which there is little information in the case of the other commonly investigated species.

3.2 The Mesocotyl

Growth of this organ is highly sensitive to light, but unlike the responses of the coleoptile, those of the mesocotyl do not seem to be age-dependent. There are three steps in the inhibition of elongation growth by red light (BLAAUW et al. 1968) (Fig. 4): (a) The first step reduces elongation by ca. 15%, is saturated by 10^{-3} Jm^{-2} (at 660 nm) and is far-red irreversible. (b) The second step inhibits elongation approximately 50%, requires red light of 10^{-1} to 10^{2} Jm^{-2}. It is completely far-red reversible and therefore cannot be produced by far-red light. (c) The third step inhibits elongation growth to about 50% of that reached in darkness. The action of red light depends on duration rather than on the total fluence delivered. It is not far-red reversible. MANDOLI and BRIGGS (1979) confirmed the stepwise response and found that reciprocity holds over the red-light fluence range 0.2–10^5 μmol m^{-2} (ca. 0.04–1.8×10^4 Jm^{-2}). Interestingly, a 10-min exposure to a green safelight also inhibits mesocotyl elongation by 15%–30%, even at levels as low as 0.8 μmol m^{-2} (which also, incidentally, promotes coleoptile growth). This confirms an earlier report by MER (1966) that even 30 s of green safelight can reduce *Avena* mesocotyl extension by 25%. SCHNEIDER (1941) noted that green light is much less effective than red light in controlling *Avena* mesocotyl extension, but not inactive.

There are at least two phases of inhibition by red light of *Zea* mesocotyls (VANDERHOEF et al. 1979). The initial, highly photosensitive phase as exhibited by *Avena* seems to be absent in *Zea*, possibly because it has been obscured by the use of green safelights (see MANDOLI and BRIGGS 1979). The first phase in *Zea* (50% of the photoinhibition) (VANDERHOEF et al. 1979) is characterized by (a) a rapid photoresponse, within 20 min of illumination, (b) reciprocity, (c) high levels of FR needed for reversal. The growth rate remains depressed for 36 h after irradiation. The second-phase inhibition is exhibited after 4.5 h and here reciprocity does not hold.

Several workers (WEINTRAUB and PRICE 1947, GOODWIN and OWENS 1948) report a complex series of peaks (670 nm) and shoulders (660, 570 nm) in the action spectra and agree that red light is more effective than blue light in the non-saturating part of the fluence response curve for *Avena*. At high fluence rates, however, blue and red light have similar effectiveness (WEINTRAUB and PRICE 1947, FLINT 1944). Presumably, the low-energy responses are mediated exclusively by phytochrome and the high-energy (irradiance) responses involve cryptochrome. Detailed action spectra show that the first, low-fluence rate inhibition step (15% of inhibition) in *Avena* has a sharp peak at 660 nm (BLAAUW et al. 1968). That all wavelengths have been found to be active (BLAAUW et al. 1968, MER 1966) indicates that either a very low P_{fr}/P_{tot} level is required or that phytochrome is not the only photoreceptor. For the first step in *Zea* there is evidence of a split peak in the red region of the action spectrum (two maxima, at 640 nm and 660 nm) which could suggest the presence of more than one red-light receptor (VANDERHOEF et al. 1979). The potential action of red light is FR-reversible, but FR alone also is inhibitory (DUKE and WICKLIFF 1969). In *Zea* mesocotyls given a 12-h white-light period at between 1.5 d and 4 d, elongation growth is controlled by P_{fr}, with inhibition and P_{fr} levels being linearly related (DUKE et al. 1977). In *Avena*, over part of the inhibition range where red light is effective, the P_{fr} concentration in vivo correlates with the observed growth response (DELINT et al. 1963, EDWARDS and KLEIN 1964). Determinations of action spectra might be confused by the relative magnitudes of each of the two steps up to 50% inhibition. In the mesocotyl, the R/FR reversible system (second step, to 50% inhibition) is of greater magnitude than the first step, but it is possible that the proportions may be altered by the experimental conditions (BLAAUW et al. 1968). In the coleoptile, however, the first step (non-reversible) is of relatively greater magnitude than the second.

Light affects cell division and elongation in the mesocotyl. AVERY et al. (1937) showed in *Avena* that the lowest fluence rates do not affect cell elongation but strongly inhibit division. As the fluence rate increases, cell elongation and division decline in parallel. The major action of light seems to be on division. GOODWIN (1941) also found two phases in the fluence-response relationship in *Avena*. He demonstrated that the very sensitive, first phase includes the suppression of cell division, while the less sensitive, second phase (which actually includes the third inhibition step of BLAAUW et al. 1968) affects elongation. MER and CAUSTON (1963) measured the kinetics of cellular responses to light in the *Avena* mesocotyl. Cell division in the cortex ceases after 4 d in darkness (Fig. 5). At the mesocotyl base, cells are 200–300 μm long, and with increasing distance rapidly attain a length of about 500 μm (Fig. 6). In darkness, most of the mesocotyl is occupied by cells of this size. Near the coleoptilar node,

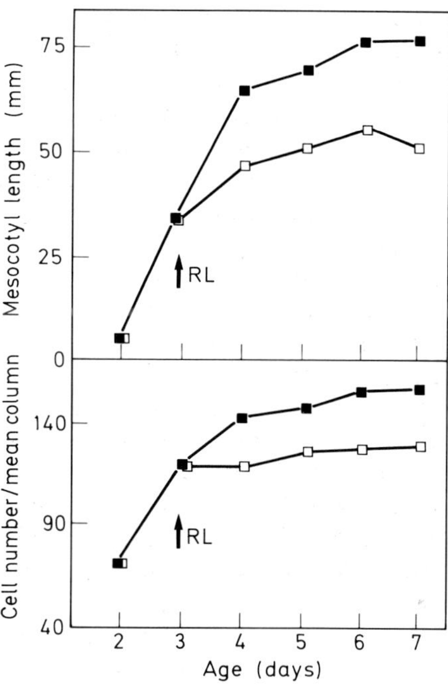

Fig. 5. Effect of a 5-min red light pulse on mesocotyl elongation and cell number. (After Mer and Causton 1963). *Open symbols* red light; *closed symbols* dark

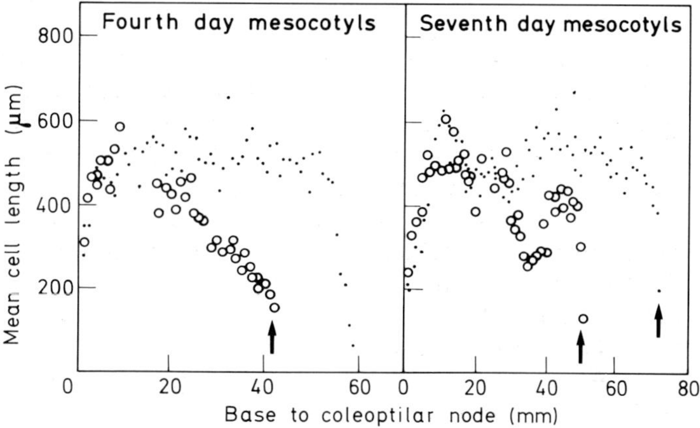

Fig. 6. Variation in mesocotyl cell size with increasing distance from base to node. (Mer and Causton 1963; reprinted by permission of the publisher). *Dots* dark control; *circles* treated with a red light pulse on the third day

on day 4, cell length decreases over a distance of 5 mm to a minimum length of approximately 66 µm. This sub-nodal region is the meristem (Huisinga 1967). At the time of the red-light treatment (day 3: see Fig. 5) none of the cortical cells is mature, but there is a gradient of increasing maturity from coleoptilar node to base, whose growth potential in darkness is dependent on age. Cells near the base greater than 130 µm in length are unaffected by red light and show as much elongation by day 4 in the light as in darkness (Fig. 6). Similar cells further from the base (which later occupy the zone 30–40 mm from the base at day 4) hardly elongate in the light. The rapid arrest of cell division by light (Fig. 5) is also indicated by the size of the topmost cells in the illuminated mesocotyls, as compared with those in the dark controls (130 μm vs. 66 μm—Fig. 6). Four days after irradiation (i.e., after a total of 7 days) the final length of most of the upper cells is slightly less than the mean value of 500 μm reached by the cells of the dark-grown mesocotyl. Situated between 35 and 40 mm (Fig. 6), there is a group of cells so sensitive to light that they can grow to only about 260 μm long.

The timing of inhibition of *Zea* mesocotyls in the two dosage phases (45 min and 4.5 h) indicates that cell elongation is directly affected in both (VANDERHOEF et al. 1979). Also the response curve does not greatly change with age even when cell division can be expected to have ended (MANDOLI and BRIGGS 1979). To clarify this point, a complementary experiment is needed similar to that carried out by MER and CAUSTON (1963) using plants in which cell division has ceased, combined with a determination of their fluence-response relationships. One hour of white light stimulates the swelling of the apical meristematic region of the *Zea* mesocotyl within 24 h, an effect due exclusively to cell enlargement. Ethylene might be involved (CAMP and WICKLIFF 1981).

3.2.1 Photoperception in Mesocotyls

There is evidence that photoinhibition of mesocotyl elongation can be brought about by light perceived directly by the mesocotyl and also by the coleoptile. For example, GOODWIN (1941) showed that irradiation of only the coleoptile tip of *Avena* can cause growth inhibition in the mesocotyl equivalent to about 60% of that occurring when the whole seedling is exposed. The occurrence of light scattering through the coleoptile down to the mesocotyl has not been precluded, however. Even in the intact seedling, irradiation of the coleoptile tip is not needed and maximal inhibition also follows when the tip is covered. Growth of excised *Avena* mesocotyls is reduced by irradiation, so clearly they themselves are photosensitive (SCHNEIDER 1941). Mesocotyls of seedlings from which the whole coleoptile is removed also are photosensitive (KONDO et al. 1969). Confirmation that the coleoptile tip can be involved in photoperception is found in *Zea mays* where, indeed, it appears that the mesocotyl itself has only partial sensitivity (DUKE and WICKLIFF 1969). Stimulation of mesocotyl apical swelling in this species (cell enlargement) is due to light perceived by the mesocotyl (CAMP and WICKLIFF 1981).

4 Growth of Hypocotyls and Stems

Since the appearance of the last issues of the Encyclopedia several general works have covered stem growth, including its photocontrol (e.g., VINCE 1964, SACHS 1965, CUTTER 1971, DORMER 1972, FREDERICQ and DE GREEF 1972, VINCE-PRUE 1975, GREEN 1976, ERICKSON 1977, GOODWIN 1978). Much of the basic photo-

Table 1. Hypocotyl responses of dark- and light-grown seedlings

Species	Dark-grown				Light-grown			References
	R Pulse	R Continuous	B	FR	R Continuous	B	FR	
Cucumis sativus		–	–	–	–	–	0	Black and Shuttleworth (1974), Gaba and Black (1979)
					–	–	+	Thomas and Dickinson (1979)
	0	–						Hillman and Purves (1966)
Lactuca sativa	0	0	–	–				Häcker et al. (1964), Evans et al. (1965), Hartmann (1966), Turner and Vince (1969)
					0	–	+	Evans et al. (1965)
					0	–	0	Thomas and Dickinson (1979)
Sinapis alba	0	–	–	–				Beggs et al. (1980)
					–	0	0	Wildermann et al. (1978a)
Raphanus sativus	0	–	–	–				Jose and Vince-Prue (1977b)
Petunia	0	–	–	–				Evans et al. (1965)
Arabidopsis (wild-type)		–	–	–				Koornneef et al. (1980)
Phaseolus vulgaris		–	–					Fletcher and Zalik (1964)
Lycopersicon esculentum					–	–	0	Thomas and Dickinson (1979)
Ricinus communis					0	0	+	Kigel and Schwartz (1980)

+ =promotion; – =inhibition– 0=little or no response

physiology has been analysed in recent years, sometimes in great detail, but there are still insufficient data on the cellular events (elongation, division) occurring during photocontrol. It is difficult to quantify changes in cell size in a situation where cell number is also changing. Moreover, division and cell enlargement may be differently affected by light. There exists in stems, however, a period after cell division has ended when growth is attributable to cell enlargement alone, and many useful investigations have been carried out on stems in this state, especially in seedlings. Light promotes the transition from cell division to enlargement or from enlargement to maturation (Thomson 1954). But even in those cases where cell enlargement exclusively features in growth, cells of different ages (i.e., states of maturity) occur, and there will, therefore, be multiple effects of light on the cell population as a whole.

4.1 Hypocotyls

4.1.1 Dark-Grown Seedlings

The photoresponses of etiolated hypocotyls have been intensively studied in several species (see summary in Table 1). In all cases, continuous blue and far-red light inhibits elongation through the HIR, characterized by fluence-rate and time dependency. One qualification is that older hypocotyls of some species such as cucumber and lettuce are insensitive to HIR far-red light (EVANS et al. 1965, TURNER and VINCE 1969, BLACK and SHUTTLEWORTH 1974). The HIR acts through different photo-systems: cryptochrome and phytochrome (HARTMANN 1966, MEIJER 1968, MANCINELLI 1980). Evidence for a separate photoreceptor for blue light comes to a large extent from studies of kinetics which show that, in several species, the timing of the inhibition by blue light is extremely rapid compared with that for far-red light (MEIJER 1968, EVANS et al. 1965, TURNER and VINCE 1969, COSGROVE 1981). An interesting genetic analysis of *Arabidopsis* suggests the occurrence of at least four photoreceptors—for UV, blue and two for red light (KOORNNEEF et al. 1980).

The apparent absence of control over hypocotyl growth by red light in lettuce (Table 1) is due to the simultaneous occurrence of two opposing effects—promotion and inhibition—in different zones of the organ, the resultant being no overall response (HÄCKER et al. 1964). All other tested species, however, show unequivocal inhibition in continuous red light, but the induction response is either weak or absent (HILLMAN and PURVES 1966). Repeated inputs of P_{fr}, at least, are required. JOSE and VINCE-PRUE (1977a) suggest that red light functions via phytochrome in two processes, one being an induction process and the other requiring continuous irradiation, which may act through pigment cycling. These have been formalized as a "static mode reaction" which requires a single or intermittent exposure to light and is then controlled by the photostationary equilibrium, and a "dynamic mode reaction" which requires longer irradiation (JOSE and VINCE-PRUE 1978).

4.1.2 De-Etiolated Seedlings

After de-etiolation, hypocotyls are no longer inhibited by prolonged far-red light (see Table 1). The loss of the far-red component of the HIR is itself a phytochrome-mediated response (JOSE and VINCE-PRUE 1977a) and could be due to the substantial reduction in P_{tot} which can follow irradiation with red light (e.g., GRILL and VINCE 1966). The response to blue light via cryptochrome appears to be retained (Table 1), except in *Sinapis alba* in which only phytochrome seems to operate, with a diurnal periodicity (WILDERMANN et al. 1978a, b). A separate blue light receptor was first demonstrated in de-etiolated *Cucumis sativus* by GABA and BLACK (1979) studying growth kinetics, and confirmed for the same species, and tomato and lettuce, by THOMAS and DICKINSON (1979). Inhibition by blue light is rapid (approx. 5 min) and is independent of the phycochrome photoequilibrium. Recovery from inhibition is also fairly quick, taking about 20 min. Inhibition by and recovery from red light in *Cucumis* hypocotyls, on the other hand, takes several hours.

Continuous red light operating through phytochrome again seems to be required (BLACK and SHUTTLEWORTH 1974) although hourly pulses can produce almost the same effect (GABA and BLACK 1979). This "dynamic mode" of phytochrome action has been little investigated, but in green plants it may be more important than induction responses (see JOSE and VINCE-PRUE 1978). *Ricinus communis* hypocotyls are unaffected by red and blue light separately, but require simultaneous irradiation (KIGEL and SCHWARTZ 1981). This is similar to the situation in *Pisum sativum* stem where inhibition is produced only by combined blue (440–470 nm) and yellow light (600 nm) (ELLIOTT 1979). Thus, in some species a synergism between phytochrome and cryptochrome may be necessary to secure inhibition of hypocotyl or stem elongation growth.

The cotyledons of de-etiolated seedlings of several species participate in the inhibition of hypocotyl lengthening (BLACK and SHUTTLEWORTH 1976). In *Cucumis sativus,* a substantial component of the inhibition by red light is due to perception by the cotyledons and transmission to the hypocotyl (BLACK and SHUTTLEWORTH 1974). Direct perception by the hypocotyl also occurs. A similar phenomenon takes place in de-etiolated *Raphanus sativus* (JOSE 1977). Blue light, on the other hand, acts directly upon the hypocotyl, the cotyledons apparently playing no role in inhibition caused by these wavelengths (BLACK and SHUTTLEWORTH 1974, JOSE 1977). Although the "cotyledon effect" is known to be present in seven species (BLACK and SHUTTLEWORTH 1976, JOSE 1977) it is certainly absent in *Ricinus communis* (KIGEL and SCHWARTZ 1981).

4.1.3 Cell Enlargement and Division

The action of light on cell enlargement and cell division has been examined in hypocotyls of only a few species. Hypocotyl photoresponses predominantly concern cell extension, division being of little import, at least in the growth period over which investigations have been made (e.g., in *Sinapis alba*—MOHR and APPUHN 1962, *Lactuca sativa*—HÄCKER et al. 1964). In most cases, however, endpoint determinations have been carried out, so there is little information on the detailed pattern of size changes as revealed by direct cell measurements made at short time intervals.

Although hypocotyl extension is inhibited by light, the growth rate in restricted regions of the organ is actually promoted (HÄCKER et al. 1964). Growth in lettuce was examined between 36 h and 84 h after sowing. Cell division during the early part of this period is at a very low rate and before ceasing completely at 60 h can be reduced by white and far-red light (rather than by red light). Extension growth of excised sections is not markedly affected by γ-irradiations (to stop cell divisions) and remains photoinhibitable (STUART et al. 1977). Far-red and white light strongly inhibit cell enlargement in lettuce, but red light has a reduced effect, and only on cells in the middle region of the hypocotyl. On the other hand, in the upper parts of the organ cell growth is promoted. These two effects of red light on cell enlargement—promotion and inhibition—are of equal magnitude so net elongation of the hypocotyl seems unaffected by red light (HÄCKER et al. 1964). This finding serves to emphasize the importance of measuring growth in discrete regions and not only in the organ as a whole.

Simultaneous promotion and inhibition of cell enlargement has been observed in etiolated hypocotyls of other species, for example *Raphanus sativus* (JOSE 1977). Organ growth between hours 72 and 96 occurs mainly in the uppermost 6 mm and effects of light are therefore on that zone. Red light promotes extension in the extreme 1 mm apical region, but otherwise all light (blue, red and far-red) is inhibitory. However, the magnitude of inhibition by blue light is greater at a distance from the apex, whereas inhibition by red and far-red light is more equitably distributed. After 24 h light, cell length is about 300 μm compared to that in darkness. In far-red light, the largest cells are smaller than those in red light. In blue light, all cells are 200–250 μm long.

Where overall hypocotyl elongation has been recorded at small time intervals we can guess at the underlying changes in cell enlargement reflected therein. Hart and Macdonald (personal communication) marked cress hypocotyls at 1 mm intervals with ion exchange beads and analysed growth by time-lapse photography. In both etiolated and green seedlings, growth is confined to the upper half of the hypocotyl. In darkness, growth of etiolated hypocotyls occurs mostly in the apical region of this zone, but in de-etiolated seedlings it is evenly distributed. The degree of elongation of the marked portions fluctuates with time. With overhead illumination, elongation growth of etiolated hypocotyls is gradually reduced, but the apical segment is least (and last) affected. Inhibition of growth of green seedlings is more abrupt and here the apical segment is the first to respond, followed later by the median region. If these changes in growth pattern are assumed to be due to effects on cell enlargement, the differences are explained in kinetics of different cells, both within a particular hypocotyl and between etiolated and green seedlings.

4.2 Stems of De-Etiolated Plants

Less is known about the photocontrol of stem elongation in green, adult plants (and the underlying cellular events) than about green and dark-grown seedlings. Among the reasons are: (a) the difficulty of relating the photophysiology to photoreceptor content because the latter cannot be measured in green tissue (see Chap. 27, this Vol.), (b) suitable dark controls are difficult to provide, at least for long periods. In recent years, there has been an acceleration in the rate at which knowledge in this area is accumulating.

4.2.1 End-of-Day Effects

The effect of far-red light at the end of the main photoperiod (the end-of-day treatment) provided one of the first demonstrations of phytochrome control of stem growth in green plants (Downs et al. 1957). This FR/R reversible response occurs in plants growing in a variety of daylengths (e.g., Downs et al. 1957, Kasperbauer 1971, Vince-Prue et al. 1976, Frankland and Letendre 1978, Morgan and Smith 1978) and also in seedlings exposed to long-duration illumination (e.g., Wildermann et al. 1978a). Stimulation of growth results from a brief exposure to far-red light and is not increased by prolonged irradiation (e.g., Vince-Prue 1977). Promotion by far-red light is lessened if the exposure is delayed after the end of the main light period (Vince-Prue et al. 1976, Vince-Prue 1977), is reduced with increasing daylength (e.g., Downs et al. 1957), but is enhanced by high fluence rates in the main light period (Lecharny and Jacques 1979). End-of-day far-red light seems to stimulate stem growth fairly rapidly (LECHARNY and JACQUES 1980), presumably because P_{fr} is immediately converted; i.e., the block to subsequent growth in darkness is removed (JOSE and SCHÄFER 1978). There is evidence from *Chenopodium* sp. and *Helianthus* sp. that the far-red irradiation is perceived partly by the leaves as well as by the reacting internodes (GARRISON and BRIGGS 1975, LECHARNY 1979).

Effects of end-of-day far-red light on both cell divisions and cell enlargement occur in several species. Cell divisions in *Chenopodium* are absent in the epidermis of internodes longer than 10 mm, but they continue in the pith until 25 mm length is reached, irrespective of the light conditions. No increase in cell number occurs as a consequence of the far-red treatment (LECHARNY 1979). Far-red light stimulates mean cell number per column of pith cells in *Helianthus annuus* by 60% and mean cell length by 25% (GARRISON and BRIGGS 1975) but it does not alter the time period over which growth takes place. This is also the case in *Phaseolus vulgaris* (NAKATA and LOCKHART 1966) where both cell enlargement and elemental cell division (cell $cell^{-1}$ day^{-1}) are promoted, so that the cell number in far-red-treated stems is higher than in the control. In *Phaseolus acutifolius,* however, the time period over which internodes extend is increased by far-red treatment, which also enhances cell division in the epidermis, cortex and pith (LENOIR 1967). Species appear to differ in their cellular responses but these differences may be due to the experimental techniques and cultural methods.

4.2.2 Effects of Daylength

4.2.2.1 Day-Extension

Green plants display a growth response to days extended by light of low fluence rates. The response becomes enhanced with each successive internode and appears to involve an effect at an early stage of internode development (VINCE-PRUE et al. 1976, VINCE-PRUE 1977). Elongation is promoted by low-intensity tungsten light to saturate at about 200 lux (VINCE-PRUE 1975) when it becomes inhibitory. Extension growth is increased as a linear function of decreasing phytochrome photoequilibrium to a P_{fr}/P_{tot} level of about 0.3 (VINCE-PRUE 1977). That P_{fr} stimulates growth in internodes in an early stage of expansion but inhibits these in a later stage, as in the Thomson Hypothesis, has been suggested by VINCE-PRUE (1977). The contribution made by cell enlargement to these effects is not clear.

4.2.2.2 Light/Dark Cycles

Stem growth in many species is controlled by daylength, in some cases in a truly photoperiodic manner (see VINCE-PRUE 1975). In de-etiolated *Cucumis sativus* seedlings, for example, hypocotyl elongation is greater in short than in long days. Controlling factors appear to be the spectral quality of light (including its blue content), length of the light period, and length of the dark period. The latter is especially important in regulating the extent of recovery in darkness from P_{fr} action, initiated during the light (GABA 1979). The length of endodermal cells (REDINGTON 1929) in the stems of flax, hemp, hop, cotton, *Hibiscus* and *Salvia* varies inversely with the daily light period. However, it is more useful to interpret this effect in terms of night-length, i.e., that cell size is directly related to the length of the dark period (see also Sect. 4.3).

Daily irradiation of dwarf pea plants with red light inhibits internode and cell length, as well as cell number in some internodes. In tall peas, however, the effect is more complex. Both promotive and inhibitory actions on cell size and number occur in different internodes (KÖHLER 1977).

4.2.3 Fixed Daylengths of Restricted Spectral Bands

A frequently used experimental technique to examine the role of light-quality on stem growth employs daylengths of fixed duration, incorporating light of different spectral composition. A complication introduced by this approach is the variable effect on photosynthetic rate.

The experiments of MEIJER (1959) with de-etiolated seedlings illustrate the variation in response among different species and the importance of fluence rate. Generally, red and green light appear to elicit similar responses—inhibition of stem elongation— and far-red light, when it has any effect at all, promotes growth. Blue and red light vary in their relative effectiveness among different species. The fluence-response curves for red light saturate at low fluence rates while the response to blue light is generally linear with fluence rates over a wide range. The marked fluence-rate dependency of the response to blue light is reminiscent of the HIR in lettuce seedlings and in green seedlings of other species (HARTMANN 1966, THOMAS and DICKINSON 1979). Relatively few studies have included an examination of the cellular basis of the growth response to fixed daylengths of different spectral quality. In bean seedling stems, however, red, blue and green light reduce cell length and number (FLETCHER et al. 1965).

4.2.4 Simulated Natural Light Environments

The role of phytochrome in the natural environment is covered by SMITH and MORGAN (see Chap. 19, this Vol.). Much information about this comes from experimental work with simulated natural light environments in which photosynthetically active radiation (PAR) and phytochrome photoequilibrium (φ) value are varied. Blue light also controls plant growth but we have little understanding of how differences in this component of natural light, which are known to occur (HOLMES and SMITH 1977a), can affect growth.

The technical problem of maintaining a constant PAR and φ value while varying other spectral regions of possible physiological activity (such as blue light) has not, as yet, been adequately tackled. WARRINGTON and MITCHELL (1976) found that plants of several species (sorghum, dwarf bean, rye grass and white clover) display a marked inhibition of stem elongation under a spectrum biased in the blue region, compared with plants grown under a more balanced spectrum with no great differences in PAR or φ. When grown in an environment deficient in blue light, enhancement of stem elongation is found.

There are suggestions that differences in growth habit due to spectral composition become less pronounced at high fluence rates (MEIJER 1971, WARRINGTON and MITCHELL 1976). This may be of importance in the natural environment where there is a concomitant reduction in mean fluence rate with decreasing ζ (and φ) during progressive natural shading (FRANKLAND and LETENDRE 1978).

4.2.5 The Effect of Fluence Rate

The role of fluence rate was found to be central in controlling the growth and development of shade plants. One problem associated with experimental variation of the fluence rate arises out of possible effects on photosynthesis, and sometimes it is difficult to disentangle photomorphogenetic effects from photosynthetic ones, especially when treatments last for extended time periods (e.g., POPP 1926b). Clearly, the products of photosynthesis promote growth, and the availability of photosynthates for stem growth may even be regulated by P_{fr} (MEIJER 1959, LECHARNY and JACQUES 1979).

A comparison of seedling establishment among species of different degrees of shade tolerance reveals that a reduction of fluence rate induces rapid growth of shade-intolerant grassland species (GRIME and JEFFREY 1964). Extension growth of shade-intolerant species has been suggested to be determined by two radiation-dependent processes: (a) the supply of photosynthates, and (b) an inhibition process, both directly related to the fluence rate (GRIME 1966). The studies of WARRINGTON and MITCHELL (1976) indicate that more extension growth occurs in a selection of species in the long term, in a variety of spectra, at low fluence rates than at high values. This is in spite of the higher levels of carbohydrate made under the latter conditions. The differences between treatments become larger as the duration of the treatment increases, showing that these effects operate continuously during the growth period.

The detector pigment for fluence rate has been suggested to be phytochrome, chlorophyll or cryptochrome (e.g., FRANKLAND and LETENDRE 1978, MCLAREN and SMITH 1978). The marked fluence-rate-dependent response to blue light has often been noted (MEIJER 1959, HARTMANN 1966, GABA 1979, GABA and BLACK 1979, THOMAS and DICKINSON 1979, BEGGS et al. 1980).

One of the few analyses of fluence-rate effects in cellular terms is by BUTLER (1963) on *Vicia faba*. Here, a sharp decrease in internode length and cell number occur at very low fluence rates. In fact, at very low levels the cell size in the youngest internodes may be greater than in darkness, while cell number is reduced. At 1.1 lux mean cell length is not reduced although cell number is; and at 108 lux and above, cell number suffers a further decrease, but cell size becomes more strongly affected.

4.3 The Role of Darkness in Stem and Hypocotyl Elongation

Transfer of light-grown plants to darkness permits an increase in the growth rate of the stems. This also occurs in plants growing in daily light–dark cycles, where growth rates may be higher at night (see SACHS 1882). In green seedlings of *Cucumis sativus*, for example, phytochrome-controlled inhibition of extension growth begins to diminish after 8 to 10 h of darkness (GABA and BLACK 1979). At about this time, hypocotyls of cucumber seedlings transferred from white light to darkness start to show a rise in mean cell length (KADOURI and ATSMON 1974). Where inhibition of growth normally continues into the dark period because of the remaining P_{fr}, it can be relieved by end-of-day FR.

Blue light operating through cryptochrome also participates in the control of stem elongation so it should be considered if this contributes any carry-over effects into the dark period. The recovery from blue-light inhibition in many species seems to be relatively rapid, taking less than 30 min (GABA and BLACK 1979, COSGROVE 1981). Hence, any continuation of photoinhibition by white light into the dark period is presumably due to P_{fr} and not to the action of cryptochrome. Where high fluence-rate white light has been experienced and thus, where inhibition during the light may be imposed largely by the blue component, the *only* expression of P_{fr} would be in the succeeding dark period.

Although of obvious importance in many (perhaps most) species the continuation of photoinhibition into darkness is not universal. In pea stems, for example, inhibition may be lost as soon as plants are darkened (ELLIOT 1979, NAUNOVIĆ and NEŠKOVIĆ 1979).

5 Hook Opening

The opening of the plumular (apical) hook in response to light involves differential cell enlargement, those cells on the inner (concave) surface of the hook expanding relatively more than those on the outer (convex) surface. Cells of

Table 2. Cellular responses in light-mediated hook opening: a static approach

Species	RL treatment time: fluence rate (Wm^{-2})	Response measured after (h)	Intact/ Detached
1. Bean (probably *P. vulgaris*)	20 h: 0.01	20	Intact
2. *Phaseolus vulgaris*	20 h: 0.375	20	Detached
3. *Phaseolus mungo*	5 min: 27	24	Detached
4. *Cucumis sativus*	6×1 min: 3.9 over 2 h	24	Intact
5. *Lactuca sativa*	Hook opening response		Intact
	Hook closing response		Intact

the inner surface generally increase in size twofold, the cells of the outside apparently being unaffected (KLEIN 1959, KANG and RAY 1969, PORATH and ATSMON 1977) (Table 2). Differential cell division rates are not involved (e.g., KLEIN 1959, MOHR and HAUG 1962). In one well-known case—*Lactuca sativa*—plumular hook formation occurs only upon irradiation, experimentally with continuous red light, and opening is induced by far-red or blue light (MOHR and NOBLE 1960) and fluorescent white light (SILK 1980). This property of lettuce seedlings has permitted important advances to be made in understanding hook physiology (SILK and ERICKSON 1978, 1979, SILK 1980, ERICKSON and SILK 1980).

The determination of cell size after a fixed period following illumination (e.g., Table 2) may be described as a static approach because it fails to take full account of the dynamic nature of the hook. This structure is situated in a region of very rapid growth yet maintains its position relative to the apex. In the bean (*P. vulgaris*) for example, as growth in darkness proceeds the hypocotylar hook moves into the epicotyl (KLEIN 1959). In *P. vulgaris* (RUBINSTEIN 1971) the hook continuously tends to open even in darkness by basal growth at the inner (concave) region, but this is counteracted by greater growth at the apical portion of the convex side. In lettuce, also, the plumular hook retains its form even though its components are continually changing (MOHR and HAUG 1962). Therefore, the hook is always present and appears to move up the hypocotyl. When seedlings, or isolated hooks, are exposed to red light the growth of these areas of the hook, which are already expanding in the dark, is stimulated further, but to a much greater extent on the concave side.

The methods of fluid flow characterization have been used to analyse the movement of tissues through the hook region of lettuce (SILK and ERICKSON 1978, SILK 1980). Spatial (local) changes, refer to a fixed position in space, and material changes of the physical elements to the hook's cellular components,

Table 2 (continued)

Tissue(s) measured	Cell size change		Degree of opening	Reference
	Convex outside	Concave inside		
Epidermal cells	0	×2 increase	Complete	KLEIN (1959)
Cortical cells	0	×2.5 increase	Complete	KANG and RAY (1969)
Cortical cells	0?	×1.5–2 increase	Incomplete (180° – 50° change)	GEE and VINCE-PRUE (1976)
Second subepidermal layer	0	×2 increase	Complete	PORATH and ATSMON (1977)
Cells in 3rd and 5th	0	50% increase	Incomplete 170° – 90°	MOHR and HAUG (1962)
layers from the epidermis	0	20% decrease	90° – 180°	

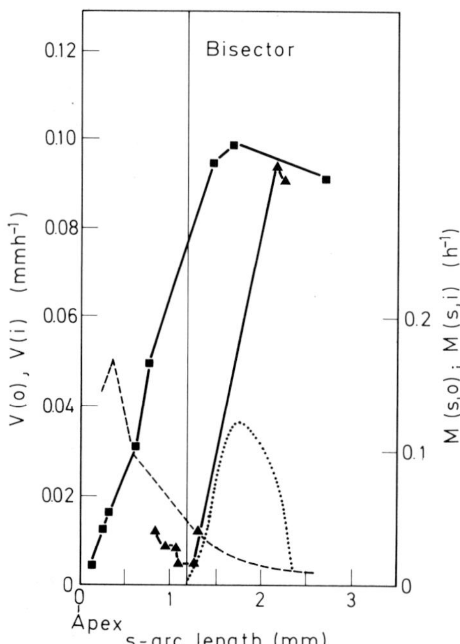

Fig. 7. A kinematic analysis of hook maintenance in lettuce seedlings, showing the velocities of displacement of points on the outer edge, v(o) *squares,* and inner edge, v(i) *triangles,* and relative elemental rates of elongation at points on the outer edge, M(s, o) *dashed line,* and points on the inner edge M(s, i) *dotted line.* (SILK and ERICKSON 1978)

i.e., groups of cells. While closed, the hook maintains a zero spatial (local) rate of change though there is a marked flow of cells through the hook, i.e., variable material or elemental rates of change of movement. Epidermal hairs on the hook can be used as natural markers, so the material rates of change can be followed when the spatial or local changes are zero (i.e., when the hook remains closed) and when they occur (i.e., when the hook opens). Both the displacement and expansion of the material elements can be traced (Fig. 7). When the hook is closed, there are certain points on the arc, extending from the apex through the hook bisector and into the hypocotyl, where elements on the outside (o) move faster than those on the inside (i). Kinematic analysis shows that the velocity of points on the outer v(o) and inner v(i) surfaces differ in this way and that rapid elongation of the hypocotyl as a whole starts after the hook bisector, there being very little displacement of the inside until this point is passed. The relative elemental rate of elongation (i.e., rate of cell growth) of the outside M(s, o) is highest just after the apex and declines towards the bisector. Just after the bisector, the elemental rate of elongation of the inner surface M(s, i) shows a burst of activity until both sides have equal displacement velocities. When plotted as differences in growth rate between the upper and lower surface [i.e., M(s, o)–M(s, i)], these variations extending over the arc are revealed (Fig. 8). In summary, during the maintenance of the hook "the growth rate of the outer edge, M(s, o), exceeds the inner growth rate, M(s, i), when the element is on the apical side of the hook. After an element is displaced past the hook, its inner growth rate exceeds its outer growth rate and the element straightens". During hook opening, provoked by light rich in the blue wavelengths, the growth velocities of the inner and outer edges

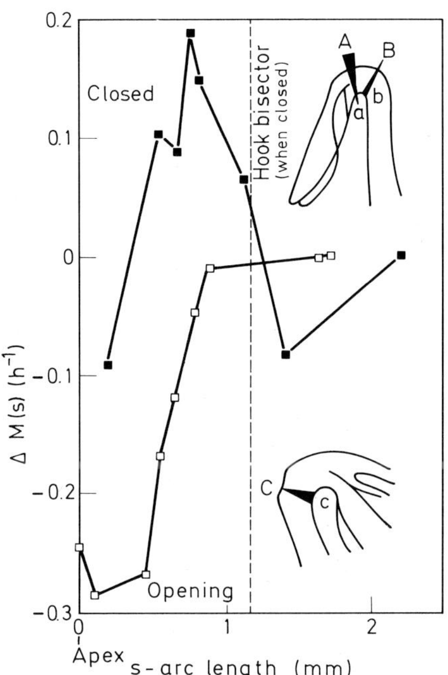

Fig. 8. The difference between the relative elemental rates of elongation [ΔM(s)] at points on the outer edge [M(s, o)] and inner edge [M(s, i)], i.e., M(s, o)–M(s, i), plotted against s, the arc length distance from the apex of the stem, for lettuce seedlings; "closed" data from cv. "Grand Rapids", "opening" data is from cv. "Arctic". *Closed symbols* M(s) when closed; *open symbols* M(s) 14–16th after the start of hook opening. (Data of SILK and ERICKSON 1978). *Inset diagrams* show the growth rate gradients which cause curvatures. *Upper drawing* shows that during hook maintenance the relative elemental growth rates must decrease between *A* and *a,* and *b* and *B*. *Lower diagram* shows that during hook opening the growth rate on the inside of the hook increases relative to the outside. The width of the dark wedges indicates the magnitude of the local stretch rate. (SILK 1980)

change (Fig. 9). The inner edge elongates throughout the straightening process while the outer edge, surprisingly, shrinks perceptibly but resumes elongation when straightening is over. Estimates of ΔMs during opening are displayed in Fig. 8 for comparison with the pattern in darkness. A diagrammatic representation of the events occurring during hook maintenance and opening is also shown in Fig. 8. In hook opening, the sharpest differential growth rate is within 0.5 mm of the apex and the magnitude of ΔMs decreases until it reaches zero at 0.9 mm from the apex. Light (white or blue) causes the replacement of the two gradients shown diagrammatically in Fig. 8 (A, a and B, b) by a single gradient (C, c).

The complex differences in cell elongation existing over just a few millimeters should be explicable in terms of growth-regulating factors such as wall extensibility, sensitivity to hormones, etc., but at present there is no understanding of the basic control processes that are taking place. The 5–12 h lag period for the start of rapid opening (RUBINSTEIN 1971, GEE and VINCE-PRUE 1976, JANES et al. 1976) may not be the time for these control factors to begin to operate but rather the time for a cell near the apex to reach the hook center (SILK and ERICKSON 1978, SILK 1980).

During its formation the hook of lettuce migrates towards the apex, and movements of the hook are always in the cotyledonary plane (MOHR and HAUG 1962). Bending (i.e., hook closure) occurs when the cells on the potentially outer (convex) side of the future hook increase their rate of elongation. Histological examination reveals no differences between the cells of the prospective inner and outer regions. In hook formation, the median cell length decreases in the inside and outside of the hook, but then increases on the outside. This is because

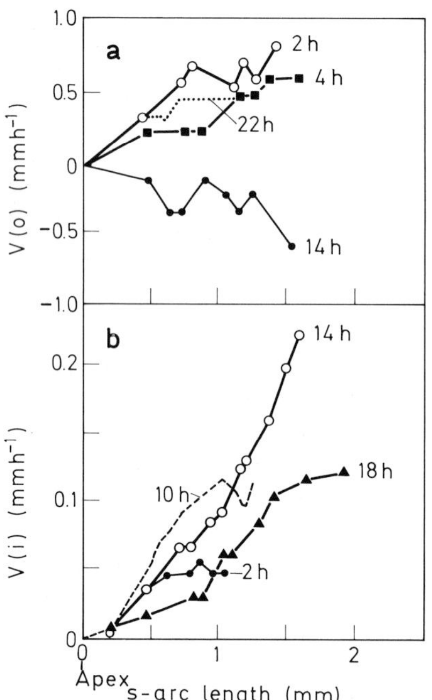

Fig. 9 a, b. A kinematic analysis of hook opening in lettuce seedlings. **a** Velocities of displacement along the outer edge, v(o), plotted against s (arc length) for various times. **b** Similar analysis for points along inner edge, v(i). (After SILK and ERICKSON 1978)

as the hook closes it moves closer to the apex and in this direction the cells are smaller ("material" changes—ERICKSON and SILK 1980).

5.1 Photobiology of Hook Opening

Hook opening is generally a phytochrome-controlled response. It can be elicited by low-fluence rate red light in a wide range of species and cultivars (POWELL and MORGAN 1980) (see Table 2). A single, far-red reversible red pulse is frequently effective, e.g., in *Phaseolus mungo* and *P. vulgaris* (CAUBERGS and DE GREEF 1975, GEE and VICE-PRUE 1976). In *Cucumis sativus,* intermittent red light pulses are needed and far-red light is also partially effective (PORATH and ATSMON 1977). Hook closure is induced by P_{fr} in lettuce, while opening responds to blue and far-red light (MOHR and NOBLE 1960). Blue light overrides the action of P_{fr} so that lettuce hooks are open in white fluorescent light (SILK 1980).

Photoperception by the hook itself is known in several species, such as *Cucumis sativus* (PORATH et al. 1980) and *Phaseolus mungo* (GEE and VINCE-PRUE 1976). In the latter, the basal limit of red-light sensitivity is about 5 mm below the hook, corresponding roughly with the edge of the zone of short cells on the inner side. The portion of the hypocotyl just below the hook seems to be the major photoperceptive region in *Phaseolus vulgaris* while the hook itself is only slightly sensitive (KLEIN 1965). The stimulus for hook opening can also be transmitted from the cotyledons in cotton and other species (POWELL and

MORGAN 1970, 1980) and in *C. sativus* far-red light is effective only through the cotyledons (whereas red light is more effective on the hook itself) (PORATH et al. 1980). Inter-organ effects have been studied in detail in *P. vulgaris* by CAUBERGS and DE GREEF (1975) and DE GREEF et al. (1976), where there is evidence for a very rapidly transmitted phytochrome action. Full red/far-red reversibility is obtained either when the whole plant is treated or when the whole plant is irradiated with red light, defoliated, and then exposed to far-red light. Far-red is ineffective upon the leaves.

Organs other than the hypocotyl have a role in hook opening, quite apart from photoperception. In nine different species and in three cultivars of cotton varying degrees of participation of cotyledons, apical bud and roots are apparent (POWELL and MORGAN 1980). In general, the cotyledons and the apical region of the shoot beyond the hook inhibit light-mediated opening, except in cotton where these organs are stimulatory. The roots variously promote, have no effect, or are effective only in combination with the cotyledons. The organs have less effect in the dark than in the light, except in one cultivar of cotton where the cotyledons are always inhibitory. The shank of the hypocotyl is important in hook opening in some species (e.g., in *P. vulgaris:* KLEIN 1965). These various effects suggest a role for transmissible factors, possibly hormonal.

5.2 Concluding Remarks

Hook opening is a complex phenomenon involving not just an increase in cell size on the underside of the hook. It is a dynamic process in which the flow of component elements occurring during hook maintenance is profoundly affected by light. The data of SILK and ERICKSON (1978) and SILK (1980) make valuable contributions to our understanding of these events, though it is regrettable that the maintenance and opening processes are described only for different cultivars of lettuce. Important questions remain to be answered: (a) How is the growth rate gradient across the hypocotyl maintained? Layers of cells expand at different rates, but the nature of the operative controls is still obscure. There is some evidence that cells from the inside of bean hooks respond more to IAA and to red light than do the "outside" cells (RUBINSTEIN 1971) indicating the existence of fundamental physiological difference between them. Variation in bioelectric potentials in response to blue light have been found at points on the bean hook, again suggesting that localized differences occur (HARTMANN and SCHMID 1980). However, no histological differences in potentially inside and outside cells of lettuce hooks are detectable (MOHR and HAUG 1962). (b) How does blue light act to convert the double gradient of growth rate differential in the lettuce hook into a single gradient at a different point? (c) Which cells perceive and translate the stimulus? (d) Must cells reach a certain position in the hook before they can begin to react? (e) Does phytochrome-controlled hook opening have similar kinematics to the blue-controlled opening of lettuce? We know from THOMSON'S work that the growth of young cells at the apical parts of an organ (i.e., nearest to the apical meristem) is promoted by light. How this fits into the dynamics of hook opening is not understood, but some evidence suggests that the photophysiology of hypocotyl elongation and hook opening are not closely related (MOHR and NOBLE 1960, JANES et al. 1976, PORATH et al. 1980).

6 Growth of Leaves

Though of great physiological importance and interest, the action of light on leaf growth has not been extensively reviewed since HUMPHRIES and WHEELER (1963). Monographs on their own work by MAKSYMOWYCH (1973) and WILLIAMS (1975) give detailed analyses of leaf ontogeny and development in selected systems. Several contributions on fundamental aspects can be found in MILTHORPE (1956) and EVANS (1972). MILTHORPE and MOORBY (1974) discuss leaf development and its environmental control while GOODWIN (1978) covers hormonal aspects. DALE (1976) deals with cell division. Development and growth of grass leaves receives attention from LANGER (1979) and general aspects of developmental anatomy are considered in standard texts such as ESAU (1960) and CUTTER (1971).

6.1 Leaf Development and Growth—an Outline

Shortly after its inception, the leaf primordium produces by apical growth an elongating structure projecting from the apical meristem. The whole leaf at this stage is meristematic. Divisions cease first at the tip, then progressively down the leaf, and longitudinal growth continues by means of an intercalary meristem. Marginal meristems arise laterally on the primordium at a stage and size characteristic for each species. These develop the lamina and establish the number of cell layers (mesophyll). Plate (subapical) meristems appear early on, expanding the lamina by anticlinical divisions, the entire blade then functioning as a plate meristem. Usually the palisade cells differentiate from the adaxial layer of the mesophyll formed by the marginal meristem, and spongy mesophyll from the abaxial layers and sometimes parts of the middle layer. In grasses, the marginal growth is limited and is restricted to the early phases of leaf development. Most of the grass leaf cells are initially produced by a subapical meristem in longitudinal files, but as the leaf ages the meristem becomes restricted to the leaf base as an intercalary meristem.

Cessation of growth occurs initially at the tip and proceeds basipetally. In tobacco, cell division ends first in the epidermis, next in the spongy mesophyll and lastly in the palisade layer, but cell enlargement continues in the epidermis after it has stopped in the mesophyll (AVERY 1933). Similar patterns have been observed (or implied) in other cases such as *Xanthium* sp. (MAKSYMOWYCH 1963), *Trifolium repens* (DENNE 1966) and *Phaseolus vulgaris* (DALE 1964, VERBELEN and DE GREEF 1979). Mathematical analyses of this pattern have been made by SILK and ERICKSON (1978) and ERICKSON and SILK (1980).

It is often stated that leaf growth and development are composed of two separate phases—a cell division phase followed by a cell enlargement phase. (A third process—formation of intercellular spaces—also contributes significantly to the growth of the leaf). Of course, cell division alone cannot account for an increase in organ size and enlargement of the daughter cells must occur during organ growth (see Sect. 1.1). An early, restricted cell division phase has, however, been reported in some cases. For example, in *Ipomoea* sp. cell division in the epidermis finishes when the tissue has reached only 5% of its final area, and thereafter growth is solely by cell enlargement (ASHBY and WANGERMAN 1950). Events in the epidermis are not typical of all leaf tissues and it would be unwise to make generalizations based on its behavior. Two phases in leaf growth certainly cannot be delineated in other species. SUNDERLAND (1960) found in both *Lupinus albus* and *Helianthus annuus* that cell division generally continues until a large proportion of the final leaf area has been achieved. In other species, e.g., *Vicia faba* (BUTLER 1963), *Spinacia oleracea* (SAURER and POSSINGHAM 1970), *Medicago sativa* (KOEHLER 1973), *Phaseolus vulgaris* (DALE 1964, VERBELEN and DE GREEF 1979) and *Trifolium subterraneum* (WILLIAMS and RIJVEN 1970), cell division occurs in the later stages of leaf growth. In some (e.g., *Capsicum frutescens:* STEER 1971) it continues until leaf growth stops (see DALE 1976, for a discussion of the misconceptions about two phases in leaf growth). Cell divisions are always accompanied by cell enlargement though the latter may be hidden by a very high division rate which keeps mean cell size down to a low value, or even reduces it (e.g., MAKSYMOWYCH 1973, VERBELEN and DE GREEF 1979). That is to say, mean cell size is the resultant of total cell enlargement (organ volume) divided by cell number (see GREEN 1976). We should note that primordial formation and leaf growth to normal dimensions can occur even when there are no cell divisions, e.g., in gamma-irradiated wheat (HABER 1962, FOARD 1971).

6.2 The Effect of Light

The rate of cell division, the duration of division, and final cell size are all affected by light. Leaf tissue can respond differentially to light, so to gain a full appreciation of the photocontrol of leaf growth all tissue types must be taken into account. The comprehensive study by VERBELEN and DE GREEF (1979) of primary leaves of *Phaseolus vulgaris* held in continuous white light and darkness illustrates these points.

a) Surface Generation. This increases in the light by about 800 fold and is almost complete by day 14. In darkness, on the other hand, the increase in area of about 50 times is mostly complete by day 7.

b) Leaf Thickness. This increases by about 3.4 times in the light over 14 days and about 2.2 times in the dark. Under both conditions there is an initial two fold increase in leaf thickness due simply to hydration of the young leaf within the seed.

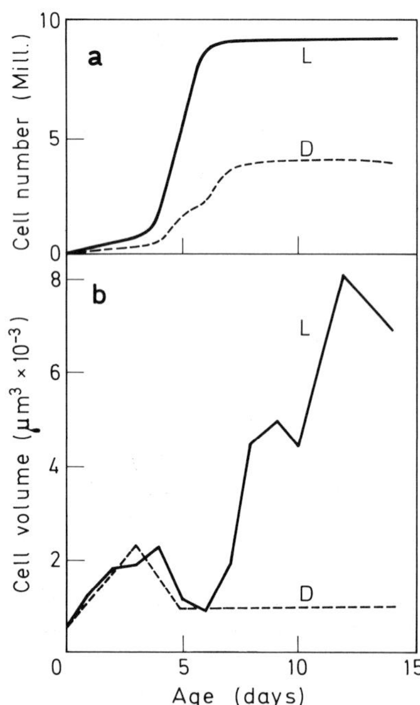

Fig. 10. The effect of white light (*L*) and darkness (*D*) on **a** cell number and **b** cell volume in the palisade parenchyma of *Phaseolus vulgaris* L. (Data of VERBELEN and DE GREEF 1979)

c) Cell Division. This continues for 4 days in both light and darkness, but in the light it starts and ends 1 day earlier than in darkness, at day 3 and 6, respectively (Fig. 10). The total cell number reached in illuminated leaves is about twice that in leaves maintained in darkness, but differences between tissues occur. In the palisade parenchyma, for example, cell number in the light is about double that in darkness (Fig. 10), whereas in the upper epidermis it is about 25% greater.

d) Cell Size. With hydration of the tissues mean cell size rises appreciably, and to the same extent in light and darkness. When cell divisions begin, cell volume decreases dramatically. When the period of intense cell division is over, mean cell volume starts to increase in the light, but not in darkness (Fig. 10). Palisade cells elongate less in darkness than in light. Therefore, the very large white light-mediated increases in cell volume and leaf surface area take place in the latter part of leaf growth. In summary: (a) During the first 2 days, imbibition and hydration cause cell enlargement, leading to an increase in volume and area. There are no cell divisions. (b) There follows a long period of surface expansion stimulated by white light. Accompanying cell division begins and ends earlier in the light than in the dark. (c) Surface expansion continues in the light, not in the dark. It is not accompanied by cell division.

6.2.1 Timing of Light Action

The above pattern is for plants exposed to white light from the start of the seed's imbibition. A delay in the application of white light to *Phaseolus vulgaris* leads to a decrease in final leaf area, mean cell number and cell volume, as compared with plants irradiated continuously. When exposures are made on day 6, increasing durations of illumination cause increasingly greater final cell volumes. One single white light pulse is not as effective as two pulses (of equivalent total duration) separated by a period of darkness. Long-term red, blue and far-red light all promote cell number and cell volume by about twofold over the dark control. Therefore, the initial effects of light are photomorphogenetic, but continuous white light may act also by a photosynthetic component (DALE and MURRAY 1968, 1969).

6.3 Effect of Light Quantity

The quantity of radiation — duration and fluence rate — has marked effects on leaf growth. In many species, leaf expansion is minimal at very low and very high fluence rates and maximal at intermediate values (NJOKU 1956, FRIEND et al. 1962, HUMPHRIES and WHEELER 1963, SCHWABE 1963, LOACH 1970, WARRINGTON and MITCHELL 1976, NOBEL 1977, FRANKLAND and LETENDRE 1978, MCLAREN and SMITH 1978, GRIME 1979). In *P. vulgaris*, for example, a parabolic curve relates total leaf area per plant and individual leaf area to the total energy received over the range 0.4–4.2 $MJm^{-2}\ d^{-1}$, (DALE 1965). (However, this is not true for leaf discs — see below). Variations in light fluence obtained by altering the photoperiod produce similar effects and in *P. vulgaris* the maximum leaf area is reached at a photoperiod of 16 h. Photoperiod and fluence rate also affect the specific leaf area (SLA: $cm^2\ g^{-1}$, hence an indicator of leaf thickness) which decreases linearly with an increase in photoperiod (DALE 1965) and fluence rate (e.g., FRIEND et al. 1962), i.e., $SLA \propto \frac{1}{\text{fluence}}$. Leaf thickness $\left(\propto \frac{1}{SLA}\right)$ is known in several species to increase with a rise in the integrated light fluence, i.e., photoperiod and/or fluence rate (NOBEL 1976, 1977, CHABOT et al. 1979). Apart from the values for thickness itself, and for specific leaf area, other parameters which express this change are the specific leaf weight (fresh weight/unit area), the ratio of mesophyll area to leaf area (A^{mes}/A) (NOBEL 1976, 1977), percent mesophyll and mesophyll cell volume.

The increase in thickness caused by high fluence rates is due in part to increased palisade thickness, including the number of layers of mesophyll cells. Hence, cell division is promoted by high fluence rates, in some cases leading to palisade formation on both sides of the leaf. The additional layers are produced early in leaf development (see CUTTER 1971, DALE 1976). Cell number in the first pair of leaves in *P. vulgaris* is unaffected by high fluence rates, but in the trifoliate leaves it is approximately doubled (DALE 1968). It is not possible from the literature to quantify the amount of light needed for the promotion of an extra layer of mesophyll and whether or not this is a threshold response. But cell number is not the only parameter affected by light quantity. For example, similar cell numbers are reached in primary leaves of *P. vulgaris* exposed to photoperiods of 6, 12 and 18 h, though leaf area is at a maximum at 12 h and the thickness increases with the increase in photoperiod. Thinness is due to the palisade mesophyll cells remaining about twice as long as broad under 6 h photoperiods while being three to five times as long under 12 h and 18 h daily light periods (DALE 1968). The effects of light intensity on leaf size after cell division has ceased are, of course, through an action on cell expansion (VAN VOLKENBURGH and CLELAND 1979).

6.4 Leaf Discs

The response pattern of isolated leaf discs to light differs from that of intact leaves. Growth of leaf discs is promoted by red light and is reversible by far-red light (e.g., LIVERMAN et al. 1955, KLEIN and WANSOR 1963). The promotion of growth may be due solely to enhanced cell division rates with no increase in mean cell size (POWELL and GRIFFITHS 1960, 1963, HUMPHRIES and WHEELER 1960, KLEIN and WANSOR 1963) but some reports indicate that red light stimulates division and enlargement equally (DALE and MURRAY 1968). Interestingly, promotion of growth in darkness by kinetin seems to be by cell enlargement and also occurs in gamma-irradiated tissues (POWELL and GRIFFITHS 1960, 1963, HUMPHRIES and WHEELER 1960). Thus, light and cytokinin can act on discs through different mechanisms. Cell size of discs held in darkness prior to illumination is equally promoted by white light and blue light, and less so by red light. But in discs cut from light-grown leaves, cell enlargement is stimulated most by white, less by blue and least by red light (POSSINGHAM 1973). High intensity light is effective and discs do not show the reduced growth response to high light fluxes which is observed in many intact leaves (DALE 1966, POSSINGHAM and SMITH 1972) (see Sect. 6.3).

Because disc growth (e.g., in spinach) is promoted by light of different spectral qualities, more than one photoreceptor may be involved, and possibly two photocontrolled growth processes exist – a low-energy one which is easily saturated (phytochrome?) and a high-energy one activated by white (fluorescent) or blue light. In some of these cases, and also in intact leaves (e.g., WARRINGTON and MITCHELL 1976) blue light may be acting through cryptochrome (see SENGER 1980). Blue-biased spectra have been found, for some species, to induce larger leaves than do "balanced" or "red-biased" spectra of similar PAR and φ values (WARRINGTON and MITCHELL 1976).

6.5 Leaves of the Gramineae

After the plate meristem stage, cell divisions become limited to an intercalary meristem at the leaf base. This region is divided into two zones by a band of parenchymatous cells coinciding with the appearance of the ligule, which develops from an epidermal outgrowth.The upper part of this meristem gives rise to the lamina and the lower part to the sheath. Cell division and enlargement cause the lamina to move up within the sheath of the older leaves, with cell enlargement being restricted to the regions protected by the sheath. On emergence of the lamina, cell enlargement ceases in the part thus exposed. When

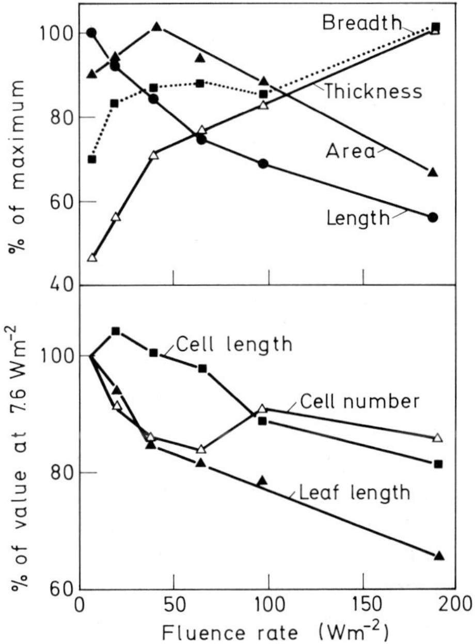

Fig. 11a, b. The effect of fluence rate on wheat leaf development. **a** Leaf length, breadth, area and thickness expressed as a percentage of the maximum obtained. (MILTHORPE and MOORBY 1974, using the data of FRIEND et al. 1962). **b** Leaf length, cell length and cell number expressed as a percentage of the response to the lowest fluence rate used ($7.6\ Wm^{-2}$). (After FRIEND and POMEROY 1970)

the ligule is differentiated, meristematic activity in the lamina stops, but cell division and enlargement elongate the sheath until the ligule is exposed. Thus, leaf growth occurs in two regions, sheath and lamina, and in the latter only in the part enclosed by the older sheaths. Phytochrome may be responsible for the abrupt cessation of cell extension in the lamina as it emerges from the sheath (BEGG and WRIGHT 1962), but this has not yet been proved, nor is the photophysiology of sheath growth clear. Phytochrome does control the unrolling of the lamina, an event in which gibberellin has been implicated (REID et al. 1968, BEEVERS et al. 1970).

Fluence rate is important in determining various growth parameters of grass leaves. At low fluence rates leaf area and length are greater, and leaf thickness and breadth are smaller than at high energy levels. Indeed, maximum area is reached at relatively low fluence rates (FRIEND et al. 1962, MILTHORPE and MOORBY 1974, LANGER 1979, see Fig. 11a). Leaf area of some species may increase with lengthening daily photoperiod, an effect which can occur relatively quickly and which is due to an enhancement of both cell size and cell number (LANGER 1979). Reduction of lamina length at high fluence rates, for example in wheat, is accompanied by lower numbers and lengths of bulliform cells (i.e., epidermis and mesophyll) of the leaf (Fig. 11b). While lamina breadth is enhanced, epidermal cell breadth is altered but little. Instead, the cell number increases considerably. Lamina thickness also increases greatly at high fluence rates, a feature matched by a rise in mesophyll thickness (FRIEND and POMEROY 1970), but it is difficult to assess the change in cell layers because of irregular packing. Fluence rate effects on *Dactylis* sp. and *Lolium* sp. are in general similar to those on wheat (FORDE 1966).

Note, however, that light controls leaf growth through photosynthesis, as well as by photomorphogenetic effects. The early phase of leaf growth depends on the provision of photosynthates from the older leaves, while the latter phase is affected directly by light acting morphogenetically (e.g., FRIEND et al. 1962).

6.6 Sun and Shade Leaves

Environmental effects on leaf size, especially the phenomena of sun and shade leaves, have been investigated by physiologists and ecologists for well over a century (see Chap. 19, this Vol.), an early general account being given by HABERLANDT (1914) and later by BURKHOLDER (1936), SHIRLEY (1945), SHIELDS (1950), HUMPHRIES and WHEELER (1963), ETHERINGTON (1975), BOARDMAN (1977), and GRIME (1979). The sun/shade syndrome is shown at several different levels: (a) Sun and shade species, colonizing open and shady habitats respectively (e.g., BOARDMAN 1977). (b) Experimental plants grown in shade produced by vegetation canopies or by neutral density "filters", under light in which quality and energy level can be accurately controlled. Leaves of both sun and shade species can thus be handled experimentally. (c) Leaves in the crown vs. outer leaves of a plant.

Shading generally produces larger, thinner leaves (see Sect. 6.3). At extremely low fluence rates, however, leaf area becomes reduced. In ten species studied by WYLIE (1951), reduced thickness is due to shortening of the palisade cells, the spongy mesophyll at the same time expanding laterally. The increased thickness under higher fluence rates is generally associated with greater palisade development, vertical elongation taking place at the expense of laterally expanding mesophyll (SHIELDS 1950). In very bright light additional palisade layers may occur (CHABOT et al. 1979), even on both sides of the leaf, with the spongy mesophyll much reduced. Since palisade development begins in the bud (even in darkness), when also the number of mesophyll cell layers is determined (HABERLANDT 1914, AVERY 1933, SHIELDS 1950), these effects of light are early in leaf development. Cells tend to be thinner-walled in the shade, and epidermal cells are longer and thinner (SHIELDS 1950). Large, intercellular air spaces occur in shade leaves. Detailed accounts of the comparative anatomy of sun and shade leaves, especially in rainforest species, may be found in GOODCHILD et al. (1972) and BOARDMAN (1977).

A quantitative expression of the anatomical differences between sun and shade leaves employs the ratio of internal leaf area to external leaf area (TURRELL 1936) now referred to as the ratio of mesophyll area to leaf area (A^{mes}/A) (e.g., NOBEL et al. 1975). This ratio gives no information on the number of cell layers. TURRELL (1936) showed that in a number of dicot species the ratio is low for shade leaves, intermediate for mesomorphic leaves and high for xeromorphic, "sun" leaves. In the latter, the greater development of palisade-type mesophyll exposes more surface (per unit volume) than the spongy type by a factor of 1.6–3.5. Over a wide range of leaf thickness, there is an almost linear relationship between thickness and A^{mes}/A (NOBEL et al. 1975). Using leaves of *Hyptis emoryi*, NOBEL (1976) found that the increase of A^{mes}/A from

13 (shade) to 40 (sun) is due rather to cell proliferation in the mesophyll rather than to shape changes, and that the palisade and spongy mesophyll make equal contributions.

Thus, total, daily photon fluence rather than some instantaneous, peak photon flux value determines the anatomical characteristics of the leaf. In *Fragaria vesca,* for example, as the daily photon fluence rises there is an increase in leaf thickness, specific leaf weight (fresh weight/unit area) mesophyll cell volume, percent mesophyll and A^{mes}/A (CHABOT et al. 1979). The relationship between daily photon fluence and thickness does not hold in all cases, however. It is absent in *Plectanthrus parviflorus,* for example, for which daily CO_2 uptake rather than daily PAR is more important (NOBEL and HARTSTOCK 1981). Similarly, high ambient CO_2 concentrations cause an increase in thickness. In this species, leaf thickness seems to be more sensitive to fluence rate early rather than late in development though fluence rate is influential throughout. However, in these studies on *Plectanthrus* possible photoperiodic effects were not precluded as they were by the use of low background lighting in the work on *F. vesca*. Photoperiodic control of leaf area has indeed been reported in several cases (e.g., VINCE-PRUE 1975, FRANKLAND and LETENDRE 1978). Changes in leaf thickness are physiologically important since rates of photosynthesis at light saturation are affected, thicker leaves tending to have the higher rates. This reflects the high A^{mes}/A value and as the latter rises there is a corresponding increase in the area available for diffusion of CO_2 into the mesophyll cells and hence also in mesophyll conductance. If stomatal conductance remains unchanged the increase in leaf thickness in response to high fluence rates (PAR?) during development results in leaves with an enhanced photosynthetic capability.

Phytochrome evidently plays a role in some effects of shade and the P_{fr} remaining at the end of the light period stimulates leaf growth (KASPERBAUER 1971, HOLMES and SMITH 1977b, FRANKLAND and LETENDRE 1978, see SMITH and MORGAN Chap. 19, this Vol.). But to what extent phytochrome is concerned in the effects of different fluence rates is not yet clear. Another system besides phytochrome may measure fluence rates (McLAREN and SMITH 1978), one possibility being cryptochrome. Blue light can affect the control of leaf growth. In the absence of blue light, for example, leaf anatomy is similar to that of shade leaves (POPP 1926a, SHIELDS 1950) and high fluence-rate blue light promotes the formation of xeromorphic leaves (WASSINK and STOLWIJK 1956).

7 Mechanisms of Photocontrol of Cell Growth

The physiology of the photocontrol of organ and cell growth is complex. As far as its mediation by phytochrome in the low-energy mode is concerned, light can promote or inhibit growth according to the developmental stage of the cells which are irradiated. There is some evidence that phytochrome acting in its high-energy mode can also sometimes stimulate growth (e.g., of cotyledons – MOHR 1966), though its effects are generally inhibitory. Blue light seems always to inhibit cell expansion in stems, except in hook opening in some species, but it promotes cell enlargement in leaves. The mechanism of all these effects cannot yet be understood, but some notable advances have been made in the past few years with respect to the action of phytochrome and to the photocontrol of leaf expansion.

The growth rate of a plant cell is determined by a number of factors (see LOCKHART 1965, RAY et al. 1972):

$$\frac{dv}{dt} = Lp \cdot W_{ex} \frac{(\Psi_{ext} - \Psi_{int} - Y)}{Lp + W_{ex}}$$

where $\frac{dv}{dt}$ = growth rate:

Lp = hydraulic conductivity:
W_{ex} = wall extensibility:
Ψ_{ext} = external water potential:
Ψ_{int} = internal water potential (osmotic potential):
Y = wall yield stress.

Changes in the growth rate such as those affected by light would be brought about by alterations in the factors in this equation. Hence, an increase in any one of W_{ex}, Lp or $(\Psi_{ext} - \Psi_{int})$, or a decrease in Y would cause a rise in growth rate.

An observation by GESSNER (1934) on *Helianthus* hypocotyls suggested that inhibitory light causes a decrease in extensibility (plasticity) of the cell walls, confirmed by LOCKHART (1960) using pea stems. Red light depresses the growth of etiolated pea stems within about 120 min, and measurements made 180 min after illumination reveal a decrease in plasticity. No change occurs in osmotic concentration of the pea stem cells. A decrease in wall extensibility has also been found in several other cases where light inhibits growth, such as in *Avena* coleoptiles (here it is a rather long-term effect of red light – BLAAUW-JANSEN and BLAAUW 1966), in rice coleoptiles (MASUDA et al. 1970, FURUYA et al. 1972) and in lettuce hypocotyls (STUART and JONES 1977). But whether light directly affects wall properties in these cases, or whether extensibility is indirectly affected cannot yet be decided.

On the other hand, where light promotes growth, such as in apical sections of *Avena* coleoptile (BLAAUW-JANSEN and BLAAUW 1966), subapical sections of *Zea mays* coleoptiles (WARNER et al. 1981, WARNER and ROSS 1981) (red light in both cases) and leaves of *Phaseolus vulgaris* (VAN VOLKENBURGH and CLELAND 1980, 1981) (high fluence-rate white light) an increase in wall extensibility follows irradiation within 90 min, 20 min and 2 h, respectively. The kinetics of these changes are sufficiently rapid to suggest that wall extensibility is a factor affected very early in the action of light. Interestingly, in both *Avena* and *Zea*, the potentiation by P_{fr} of growth and enhanced wall extensibility appears to be very fast, since reversal by far-red light, given very shortly after red light, is non-existent or weak (BLAAUW-JANSEN and BLAAUW 1966, PIKE et al. 1979, WARNER and ROSS 1981). Far-red light is effective, however, when it is applied simultaneously with red light, to reduce the φ value to approx. 0.25 (WARNER and ROSS 1981). In the leaf of *P. vulgaris*, the wall extensibility seems to be the only parameter affected by light. Yield stress, osmotic potential, and hydraulic conductivity remain virtually unchanged (VAN VOLKENBURGH and CLELAND 1981). However, any alteration in the osmotic potential of the epidermis, the cell layers which may control the ability of the leaf to expand, may not be detected. Hydraulic conductivity of pea epicotyl, *Avena* coleoptile, and *P. vulgaris* buds is unaffected by red light (PIKE 1976).

Little is known concerning the action of blue light on the parameters upon which cell growth depends. However, inhibition of cucumber hypocotyl elongation by blue light may be due to its effect on cell wall properties (COSGROVE 1981), (see note added in proof).

Much evidence has been presented in recent years that increased wall extensibility, especially in response to auxin, results from the acidification of the cell wall by protons excreted from the protoplast (see CLELAND 1977). Proton excretion has also been detected in response to growth-promoting light, in apical sections of *Avena* coleoptiles (in the presence and absence of auxin) and in *P. vulgaris* leaves (PIKE and RICHARDSON 1977, VAN VOLKENBURGH and CLELAND 1980, 1981). In both of these cases, the acidification is not caused by enhanced respiration, but seems to be a specific response to light. The kinetics of proton excretion in *P. vulgaris* leaves appears to match those for light-stimulated growth rather well, since acidification is simultaneous with or just precedes cell enlargement. But in *Avena* the latent period for proton excretion is longer than that for the promotion of growth (PIKE et al. 1979). How far this is due only to relative insensitivity of the measuring techniques is not yet clear. In sub-apical sections of *Avena* coleoptiles where, after 4 h, red light inhibits cell expansion, there is a reduction in acidification, and red light also decreases the magnitude of auxin-induced proton excretion (LURSSEN 1976). On the whole then, there is a fair correlation in *Avena* between the red light-controlled growth and the enhancement or inhibition of hydrogen ion excretion. The correlation does not hold in pea stems, however, where effects of red light on acidification seem to be absent even though growth is inhibited (PIKE and RICHARDSON 1979).

If control of proton excretion is indeed the basis for the photoregulation of cell growth, how might this be achieved? A proton pump may be mediated by ATPase located in or associated with the plasma membrane. Since P_{fr} might interact with the plasma membrane, say, of *Avena* coleoptile (see NEWMAN and BRIGGS 1972), an effect on the pump mechanism becomes a possibility. The rapidity with which blue light inhibits cell enlargement (MEIJER 1968, GABA and BLACK 1979, COSGROVE 1981) might suggest that here, too, a direct effect on the membrane-associated pump could be implicated. Certainly, effects of blue light on membrane properties are well known (HARTMANN and SCHMIDT 1980). It has been suggested, however, that in bean leaves where high-fluence rate white light promotes expansion, the ATPase-dependent pump may be affected by intracellular ATP produced during photosynthesis (VAN VOLKENBURGH and CLELAND 1980). In certain organs (stems, coleoptiles) P_{fr} promotes or inhibits growth according to the developmental stage of the cell. Studies on acidification suggest that the proton pump is also promoted or inhibited according to cell age or position in the tissue. It is difficult to understand, however, what determines whether light should have a positive or negative effect on pump activity.

Much attention has been given to the possibility that light regulates growth through effects on plant hormones (see BLACK and VLITOS 1972 and Chap. 16, this Vol).

Acknowledgements. The authors would like to thank all those colleagues who sent unpublished data, and who drew their attention to much useful information; Mrs. MONIKA SHAFFER-FEHRE for translations; and the S.E.R.C. for supporting V.G. as a postdoctoral assistant. Also appreciation is given to JOAN HAJ-SHAFI for typing the manuscript.

Note Added in Proof: After the completion of the manuscript, a report by COSGROVE and GREEN appeared (Plant Physiol 1981, 68:1447–1453) which confirmed that blue light inhibits selongation growth of etiolated cucumber and sunflower seedlings by rapidly decreasing the yielding properties of the cell wall.

References

Ashby E, Wangermann E (1950) Studies in the morphogenesis of leaves. V. A note of the origin of differences in cell size among leaves of different levels of insertion on the stem. New Phytol 49:189–192

Avery GS Jr (1933) Structure and development of the tobacco leaf. Am J Bot 20:565–592

Avery GS Jr, Burkholder PB, Creighton HB (1937) Polarised growth and cell studies in the first internode and coleoptile of *Avena* in relation to light and dark. Bot Gaz 99:125–143

Beevers L, Loveys B, Pearson JA, Wareing PF (1970) Phytochrome and hormonal control of expansion and greening of etiolated wheat leaves. Planta 90:286–294

Begg JE, Wright MJ (1962) Growth and development of leaves from intercalary meristems in *Phalaris arundinacea* L. Nature 194:1097–1098

Beggs CJ, Holmes MG, Jabben M, Schäfer E (1980) Action spectra for the inhibition of hypocotyl growth by continuous irradiation in light- and dark-grown *Sinapis alba* L. seedlings. Plant Physiol 66:615–618

Bindloss EA (1942) A developmental analysis of cell length as related to stem length. Am J Bot 29:179–188

Blaauw OH (1961) The influence of blue, red and far red light on geotropism and growth of the *Avena* coleoptile. Acta Bot Neerl 10:397–450

Blaauw OH (1963) Effects of red light on geotropism of *Avena* and their possible relations to phototropic phenomena. Acta Bot Neerl 12:424–432

Blaauw-Jansen G, Blaauw OH (1966) Effect of red light on irreversible and reversible expansion of *Avena* coleoptile sections. Planta 71:291–304

Blaauw OH, Blaauw-Jansen G, Van Leeuwen WJ (1968) An irreversible red-light-induced growth response in *Avena*. Planta 82:87–104

Black M, Shuttleworth JE (1974) The role of the cotyledons in the photocontrol of hypocotyl extension in *Cucumis sativus* L. Planta 117:57–66

Black M, Shuttleworth JE (1976) Inter-organ effects in the photocontrol of growth. In: Smith H (ed) Light and plant development. Butterworth, London, pp 317–331

Black M, Vlitos AJ (1972) Possible interrelationships of phytochrome and plant hormones. In: Mitrakos K, Shropshire W Jr (eds) Phytochrome. Academic Press, London New York, pp 517–550

Boardman NK (1977) Comparative photosynthesis of sun and shade plants. Annu Rev Plant Physiol 28:355–377

Brotherton WB, Bartlett HH (1918) Cell measurement as an aid in the analysis of quantitative variation. Am J Bot 5:192–206

Brown R, Rickless P (1949) A new method for the study of cell division and cell extension with some preliminary observations on the effect of temperature and of nutrients. Proc R Soc London B 136:110–125

Burkholder PR (1936) The role of light in the life of plants. II. The influence of light upon growth and differentiation. Bot Rev 2:97–172

Butler RD (1963) The effect of light intensity on stem and leaf growth in broad bean seedlings. J Exp Bot 14:142–152

Camp PJ, Wickliff JL (1981) Light or ethylene treatments induce transverse cell enlargement in etiolated maize mesocotyls. Plant Physiol 67:125–128

Caubergs R, De Greef JA (1975) Studies in hook opening in *Phaseolus vulgaris* L. by selective R/FR pretreatments of embryonic axis and primary leaves. Photochem Photobiol 22:139–144

Chabot BF, Jurik TW, Chabot JF (1979) Influence of instantaneous and integrated light-flux density on leaf anatomy and photosynthesis. Am J Bot 66:940–945

Cleland RE (1977) The control of cell enlargement. In: Jennings DH (ed) Integration of activity in the higher plant. Cambridge Univ Press, pp 101–115

Cosgrove D (1981) Rapid suppression of growth by blue light. Occurrence, time course and general characteristics. Plant Physiol 67:584–590

Cutter EG (1971) Plant anatomy: experiment and interpretation. Part 2. Organs. Arnold, London

Dale JE (1964) Leaf growth in *Phaseolus vulgaris*. I. Growth of the first pair of leaves under constant conditions. Ann Bot NS 28:579–589

Dale JE (1965) Leaf growth in *Phaseolus vulgaris*. II. Temperature effects and the light factor. Ann Bot NS 29:293–308

Dale JE (1966) The effect of nutritional factors and certain growth substances on the growth of disks cut from the young leaves of *Phaseolus*. Physiol Plant 19:385–396

Dale JE (1968) Cell growth in expanding primary leaves of *Phaseolus*. J Exp Bot 19:322–332

Dale JE (1976) Cell division in leaves. In: Yeoman MM (ed) Cell division in higher plants. Academic Press, London New York, pp 315–345

Dale JE, Murray D (1968) Photomorphogenesis, photosynthesis, and early growth of primary leaves of *Phaseolus vulgaris*. Ann Bot NS 32:767–780

Dale JE, Murray D (1969) Light and cell division in primary leaves of *Phaseolus*. Proc R Soc London B 173:541–555

De Greef JA, Caubergs R, Verbelen JP, Moerells E (1976) Phytochrome-mediated interorgan dependence and rapid transmission of the light stimulus. In: Smith H (ed) Light and plant development. Butterworth, London, pp 295–316

DeLint PJAL, Edwards JL, Klein WH (1963) The red, far-red system and phytochrome. Plant Physiol 38:Suppl 24

Denne PM (1966) Leaf development in *Trifolium repens*. Bot Gaz 127:202–210

Dormer KJ (1972) Shoot organisation in vascular plants. Chapman and Hall, London

Downs RJ, Hendricks SB, Borthwick HA (1957) Photoreversible control of elongation of Pinto beans and other plants under normal conditions of growth. Bot Gaz 118:199–208

Duke SO, Wickliff JL (1969) *Zea* shoot development in response to red light interruption of the dark-growth period. I. Inhibition of the first internode elongation. Plant Physiol 44:1027–1030

Duke SO, Naylor AW, Wickliff JL (1977) Phytochrome control of longitudinal growth and phytochrome synthesis in maize seedlings. Physiol Plant 40:59–68

Edwards JL, Klein WH (1964) Relationship of phytochrome concentration to physiological responses. Plant Physiol 39:Suppl 50

Elliot WM (1979) Control of leaf and stem growth in light-grown pea seedlings by two high irradiance responses. Plant Physiol 63:833–836

Elliot WM, Shen-Miller J (1976) Similarity in dose responses, action spectra, and red light responses between phototropism and photoinhibition of growth. Photochem Photobiol 23:195–199

Erickson RO (1977) Modeling of plant growth. Annu Rev Plant Physiol 27:407–434
Erickson RO, Silk WK (1980) The kinematics of plant growth. Sci Am 242 (5):102–113
Esau K (1960) Anatomy of seed plants. Wiley and Sons, New York
Etherington JR (1975) Environment and plant ecology. Wiley and Sons, New York
Evans LT (ed) (1972) Environmental control of plant growth. Academic Press, London New York
Evans LT, Hendricks SB, Borthwick HA (1965) The role of light in suppressing hypocotyl elongation in lettuce and *Petunia*. Planta 64:201–218
Fahn A (1967) Plant anatomy. Pergamon, Oxford
Firn RD, Digby J (1980) The establishment of tropic curvatures in plants. Annu Rev Plant Physiol 31:131–148
Fletcher RA, Zalik S (1964) Effect of light quality on growth and free indoleacetic acid content in *Phaseolus vulgaris*. Plant Physiol 39:328–331
Fletcher RA, Peterson RL, Zalik S (1965) Effect of light quality on elongation, adventitious root production and the relation of cell number and cell size to bean seedling elongation. Plant Physiol 40:541–548
Flint LH (1944) Light and the elongation of the mesocotyl in corn. Plant Physiol 19:537–543
Foard DE (1971) The initial protrusion of a leaf primordium can form without concurrent periclinal cell divisions. Can J Bot 49:1601–1603
Forde BJ (1966) Effects of various environments on the anatomy and growth of perennial ryegrass and cocksfoot. NZ J Bot 4:455–468
Frankland B, Letendre RJ (1978) Phytochrome and effects of shading on growth of woodland plants. Photochem Photobiol 27:223–230
Franssen JM, Cooke SA, Digby J, Firn RD (1981) Measurements of differential growth causing phototropic curvature of coleoptiles and hypocotyls. Z Pflanzenphysiol 103:207–216
Fredericq H, De Greef JA (1972) Control of vegetative growth by red, far-red reversible photoreactions in higher and lower plant-systems. In: Mitrakos K, Shropshire W Jr (eds) Phytochrome. Academic Press, London New York, pp 319–346
Friend DJC, Pomeroy ME (1970) Changes in cell size and number associated with the effects of light intensity and temperature on the leaf morphology of wheat. Can J Bot 48:85–90
Friend DJC, Helson VA, Fisher JE (1962) Leaf growth in Marquis wheat, as regulated by temperature, light intensity and daylength. Can J Bot 40:1299–1311
Fujii R (1957) Effect of light on the growth of higher plants red and near infra-red on the *Vigna* seedling. Seiri Seitai 7:79–86
Furuya M, Pjon C-J, Fujii T, Ito M (1969) Phytochrome action in *Oryza sativa* L. III. The separation of photoreceptive site and growing zone in coleoptiles, and auxin transport as effector system. Dev Growth Diff 11:62–76
Furuya M, Masuda Y, Yamamoto R (1972) Effects of environmental factors on mechanical properties of the cell wall in rice coleoptiles. Dev Growth Differ 14:95–105
Gaba VP (1979) An analysis of photoinhibition of hypocotyl elongation in de-etiolated seedlings of *Cucumis sativus* L. PhD Thesis, Univ London
Gaba V, Black M (1979) Two separate photoreceptors control hypocotyl elongation in green seedlings. Nature 278:51–54
Garrison R, Briggs WR (1975) The growth of internodes in *Helianthus* in response to far-red light. Bot Gaz 136:353–357
Gee H, Vince-Prue D (1976) Control of the hypocotyl hook angle in *Phaseolus mungo* L.: the role of the parts of the seedling. J Exp Bot 27:314–323
Gentile AC, Klein RM (1964) Absence of effect of visible radiation on elongation of decapitated *Avena* coleoptile segments. Physiol Plant 17:299–300
Gessner F (1934) Wachstum und Wanddehnbarkeit am Helianthus-Hypokotyl. Jahrb Wiss Bot 80:143–168
Green PB (1976) Growth and cell pattern formation on an axis: critique of concepts, terminology, and modes of study. Bot Gaz 137:187–202
Green PB (1980) Organogenesis – a biophysical view. Annu Rev Plant Physiol 31:51–82

Grill R, Vince D (1966) Photocontrol of anthocyanin formation in turnip seedlings. III. The photoreceptors involved in the responses to prolonged radiation. Planta 70:1–12

Grime JP (1966) Shade avoidance and shade tolerance in flowering plants. In: Bainbridge R, Evans GC, Rackham O (eds) Light as an ecological factor. Blackwell, Oxford, pp 187–207

Grime JP (1979) Plant Strategies and vegetation processes. Wiley and Sons, New York

Grime JP, Jeffrey DW (1964) Seedling establishment in vertical gradients of sunlight. J Ecol 53:621–642

Goodchild DJ, Björkman O, Pyliotis NA (1972) Chloroplast ultrastructure, leaf anatomy, and content of chlorophyll and soluble protein in rainforest species. Carnegie Inst Wash Year Book 71:102–107

Goodwin PB (1978) Phytohormones and growth and development of organs of the vegetative plant. In: Letham DS, Goodwin PB Higgins TJV (eds) Phytohormones and related compounds – a comprehensive treatise Vol II. Elsevier, Amsterdam, pp 31–173

Goodwin RH (1941) On the inhibition of the first internode of *Avena* by light. Am J Bot 28:325–332

Goodwin RH, Owens OVH (1948) An action spectrum for inhibition of the first internode of *Avena* by light. Bull Torrey Bot Club 75:18–21

Gorton HL, Briggs WR (1980) Phytochrome responses to end-of-day irradiations in light-grown corn in the presence and absence of Sandoz 9789. Plant Physiol 66:1024–1026

Haber AH (1962) Non-essentiality of concurrent cell divisions for degree of polarisation of leaf growth. I. Studies with radiation-induced mitotic inhibition. Am J Bot 49:583–589

Haber AH, Foard DE (1964) Interpretations concerning cell division and growth. In: Régulateurs naturels de la croissance végétale. (5th Int Conf on Plant Growth Substances). Cent Nat Rech Sci, pp 491–503

Haberlandt (1914) Physiological plant anatomy. Translated by M. Drummond from Physiologische Pflanzenanatomie (1884). McMillan, London

Häcker M, Hartmann KM, Mohr H (1964) Zellteilung und Zellwachstum im Hypokotyl von *Lactuca sativa* L. unter dem Einfluß des Lichtes. Planta 63:253–268

Harris PJ, Lowry KH, Chapas LC (1981) Comparison of methods for measuring tissue areas in sections of plant organs. Ann Bot 47:151–154

Hartmann E, Schmid K (1980) Effects of UV and blue light on the biopotential changes in etiolated hooks of dwarf beans. In: Senger H (ed) The blue light syndrome. Springer, Berlin Heidelberg New York, pp 221–237

Hartmann KM (1966) A general hypothesis to interpret 'high energy phenomena' of photomorphogenesis on the basis of phytochrome. Photochem Photobiol 5:349–366

Hillman WS, Purves WK (1966) Light responses, growth factors and phytochrome transformations of *Cucumis* seedling tissues. Planta 70:275–284

Hock B, Mohr H (1965) Eine quantitative Analyse von Wachstumsvorgängen im Zusammenhang mit der Photomorphogenese von Senfkeimlingen (*Sinapis alba* L.). Planta 65:1–16

Holmes MG, Smith H (1977a) The function of phytochrome in the natural environment. I. Characterisation of daylight for studies in photomorphogenesis and photoperiodism. Photochem Photobiol 25:533–538

Holmes MG, Smith H (1977b) The function of phytochrome in the natural environment. IV. Light quality and plant development. Photochem Photobiol 25:551–557

Hopkins WG, Hillman WS (1965) Response of excised *Avena* coleoptile segments to red and far-red light. Planta 65:157–166

Huisinga B (1964) Influence of light on growth, geotropism and guttation of *Avena* seedlings grown in total darkness. Acta Bot Neerl 13:445–487

Huisinga B (1967) Influence of irradiation on the distribution of growth in dark-grown *Avena* seedlings. Acta Bot Neerl 16:197–201

Humphries EC, Wheeler AW (1960) The effects of kinetin, gibberellic acid, and light on expansion and cell division in leaf disks of dwarf bean (*Phaseolus vulgaris*). J Exp Bot 11:81–85

Humphries EC, Wheeler AW (1963) The physiology of leaf growth. Annu Rev Plant Physiol 14:385–410

Janes HW, Loercher L, Frenkel C (1976) Effect of red light and ethylene on growth of etiolated lettuce seedlings. Plant Physiol 57:420–423

Jose AM (1977) Photoreception and photoresponses in the radish hypocotyl. Planta 136:125–129

Jose AM, Schäfer E (1978) Distorted phytochrome action spectra in green plants. Planta 138:25–28

Jose AM, Vince-Prue D (1977a) Light-induced changes in the photoresponses of plant stems; the loss of a high irradiance response to far-red light. Planta 135:95–100

Jose AM, Vince-Prue D (1977b) Action spectra for the inhibition of growth in radish hypocotyls. Planta 136:131–134

Jose AM, Vince-Prue D (1978) Phytochrome action: a re-appraisal. Photochem Photobiol 27:209–216

Kadouri A, Atsmon D (1974) The effects of various light regimes on chloroplast DNA synthesis and replication. In: Bieleski RL, Ferguson AR, Cresswell MM (eds) Mechanisms of regulation of plant growth. R Soc NZ Bull 12:339–343

Kang BG, Ray PM (1969) Role of growth regulators in the bean hypocotyl hook opening response. Planta 87:193–205

Kasperbauer MJ (1971) Spectral distribution of light in a tobacco canopy and effects of end-of-day light quality on growth and development. Plant Physiol 47:775–778

Kigel J, Schwartz A (1981) Cooperative effects of blue and red light in the inhibition of hypocotyl elongation of de-etiolated castor bean. Plant Sci Lett 21:83–88

Klein RM (1965) Photomorphogenesis of the bean plumular hook. Physiol Plant 18:1026–1033

Klein RM (1979) Reversible effects of green and orange-red radiation on plant cell elongation. Plant Physiol 63:114–116

Klein RM, Wansor J (1963) Effects of non-ionizing radiation on expansion of disks from leaves of dark-grown bean plants. Plant Physiol 38:5–10

Klein WH (1959) Interaction of growth factors with photoprocess in seedling growth. In: Withrow RB (ed) Photoperiodism and related phenomena in plants and animals. Am Assoc Adv Sci, Washington, DC, pp 207–215

Koehler PG (1973) The roles of cell division and cell expansion in the growth of alfalfa leaf mesophyll. Ann Bot NS 37:65–68

Köhler D (1977) Cell division and cell enlargement as influenced by gibberellic acid and red light in dwarf and tall peas. Z Pflanzenphysiol 82:125–136

Kondo N, Fujii T, Yamaki T (1969) Effect of light on auxin transport and elongation of *Avena* mesocotyl. Dev Growth Diff 11:46–61

Koornneef M, Rolff E, Spruit CJP (1980) Genetic control of light-inhibited hypocotyl inhibition in *Arabodopsis thaliana* (L.) HEYNH. Z Pflanzenphysiol 100:147–160

Larger RHM (1979) How grasses grow 2nd edn. Arnold, London

Lawson VR, Weintraub RL (1975) Effects of red light on the growth of intact wheat and barley coleoptiles. Plant Physiol 56:44–50

Lecharny A (1979) Phytochrome and internode elongation in *Chenopodium polyspermum* L. Sites of perception. Planta 145:405–409

Lecharny A, Jacques R (1979) Phytochrome and internode elongation in *Chenopodium polyspermum* L. The light fluence rate during the day and the end-of-day effect. Planta 146:575–577

Lecharny A, Jacques R (1980) Light inhibition of internode elongation in green plants. Planta 149:384–388

LeNoir WC Jr (1967) The effect of light on the cellular components of polarised growth in bean internodes. Am J Bot 54:876–887

Liverman JL, Johnson MP, Starr L (1955) Reversible photoreaction controlling expansion of etiolated bean-leaf disks. Science 121:440–441

Loach K (1970) Shade tolerance in tree seedlings. II. Growth analysis of plants raised under artificial shade. New Phytol 69:273–286

Lockhart JA (1960) Intracellular mechanism of growth inhibition by radiant energy. Plant Physiol 35:129–135
Lockhart JA (1965) Analysis of irreversible plant cell elongation. J Theor Biol 8:264–275
Lurssen K (1976) Counteraction by phytochrome to the IAA-induced hydrogen-ion excretion in *Avena* coleoptile cylinders. Plant Sci Lett 6:389–399
Maksymowych R (1963) Cell division and cell elongation in leaf development of *Xanthium pennsylvanicum*. Am J Bot 50:891–901
Maksymowych R (1973) Analysis of leaf development. Cambridge Univ Press
Mancinelli AL (1980) The photoreceptors of the high irradiance responses of plant photomorphogenesis. Photochem Photobiol 32:853–857
Mandoli DF, Briggs WR (1979) Growth characteristics of etiolated *Avena sativa* L. (cv. Lodi) in response to red and green light. Carnegie Inst Wash Year Book 78:140–144
Masuda Y, Pjon C-J, Furuya M (1970) Phytochrome action in *Oryza sativa* L. V. Effects of decapitation and red and far-red light on cell wall extensibility. Planta 90:236–242
McLaren JS, Smith H (1978) Phytochrome control of the growth and development of *Rumex obtusifolius* under simulated canopy light environments. Plant Cell Environ 1:61–67
Mer CL (1966) The inhibition of cell division in the mesocotyl of etiolated oat plants by light of different frequencies. Ann Bot NS 30:17–23
Mer Cl, Causton DR (1963) Carbon dioxide: a factor influencing cell division. Nature 199:360–362
Meijer G (1959) The spectral dependence of flowering and elongation. Acta Bot Neerl 8:189–246
Meijer G (1968) Rapid growth inhibition of gherkin hypocotyls in blue light. Acta Bot Neerl 17:9–14
Meijer G (1971) Some aspects of plant irradiation. Acta Hortic 22:103–108
Milthorpe FL (ed) (1956) The growth of leaves. Butterworth, London
Milthorpe FL, Moorby J (1974) An introduction to crop physiology. Cambridge Univ Press
Mohr H (1966) Differential gene activation as a mode of action of phytochrome 730. Photochem Photobiol 5:469–483
Mohr H, Appuhn U (1962) Die Steuerung des Hypocotylwachstums von *Sinapis alba* L. durch Licht und Gibberellinsäure. Planta 59:49–67
Mohr H, Haug A (1962) Die histologischen Vorgänge während der lichtabhängigen Schließung und Öffnung des Plumulahakens bei den Keimlingen von *Lactuca sativa* L. Planta 59:151–164
Mohr H, Noble A (1960) Die Steuerung der Schließung und Öffnung des Plumula-Hakens bei Keimlingen von *Lactuca sativa* durch sichtbare Strahlung. Planta 55:327–342
Morgan DC, Smith H (1978) The relationship between phytochrome photoequilibrium and development in light-grown *Chenopodium album* L. Planta 142:187–193
Muir RM, Chen Chang K (1974) Effect of red light on coleoptile growth. Plant Physiol 54:286–288
Nakata S, Lockhart JA (1966) Effects of red and far-red radiation on cell division and elongation in the stem of Pinto bean seedlings. Am J Bot 53:12–20
Naunović G, Něsković M (1979) Rapid responses to light and gibberellic acid in etiolated pea stems. Photochem Photobiol 29:1173–1175
Newman IA, Briggs WR (1972) Phytochrome-mediated electric potential changes in oat seedlings. Plant Physiol 50:687–693
Njoku E (1956) Studies in the morphogenesis of leaves. XI. The effect of light intensity on leaf shape in *Ipomea caerulea*. New Phytol 55:91–110
Nobel PS (1976) Photosynthetic rates of sun versus shade leaves of *Hyptis emoryi* Torr. Plant Physiol 58:218–223
Nobel PS (1977) Internal leaf area and cellular CO_2 resistance: photosynthetic implications of variations with growth conditions and plant species. Physiol Plant 40:137–144
Nobel PS, Hartstock TL (1981) Development of leaf thickness for *Plectranthus parviflorus* – influence of photosynthetically active radiation. Plant Physiol 51:163–166
Nobel PS, Zaragoza LJ, Smith WK (1975) Relation between mesophyll cell surface area,

photosynthetic rate, and illumination level during development of leaves of *Plectranthus parviflorus* Henckel. Plant Physiol 55:1067–1070

Paul R, Furuya M (1973) Phytochrome action in *Oryza sativa* L. VI. Red far-red reversible effect on early development of coleoptiles. Bot Mag 86:203–211

Penny P, Penny D (1978) Rapid responses to phytohormones. In: Letham DS, Goodwin PB, Higgins TJV (eds) Phytohormones and related compounds – a comprehensive treatise Vol II. Elsevier/North-Holland Biomedical Press, Amsterdam, pp 537–597

Pike CS (1976) Lack of influence of phytochrome on membrane permeability to tritiated water. Plant Physiol 57:185–187

Pike CS, Richardson A (1977) Phytochrome-controlled hydrogen ion excretion by *Avena* coleoptiles. Plant Physiol 59:615–617

Pike CS, Richardson AE (1979) Red light and auxin effects on ^{86}rubidium uptake by oat coleoptile and pea epicotyl segments. Plant Physiol 63:139–141

Pike CS, Richardson AE, Weiss ER, Aynari JM, Grushow J (1979) Short term phytochrome control of oat coleoptile and pea epicotyl growth. Plant Physiol 63:440–443

Pjon C-J, Furuya M (1967) Phytochrome action in *Oryza sativa* L. I. Growth responses of etiolated coleoptiles to red, far-red and blue light. Plant Cell Physiol 8:709–718

Pjon C-J, Furuya M (1968) Phytochrome action in *Oryza sativa* L. II. The spectrophotometric versus the physiological status of phytochrome in coleoptiles. Planta 81:303–313

Popp HW (1926a) A physiological study of the effect of light of various ranges of wavelength on the growth of plants. Am J Bot 13:706–737

Popp HW (1926b) Effect of light intensity on growth of soy beans and its relation to the autocatalyst theory of growth. Bot Gaz 82:306–320

Porath D, Atsmon D (1977) Hook opening in cucumber seedlings by narrow-band red and far-red light. Plant Sci Lett 8:217–222

Porath D, Atsmon D, Raviv J (1980) Hook opening in cucumber seedlings: difference in perception in red and far-red light demonstrated using light-conducting fibres. Plant Sci Lett 17:311–316

Possingham JV (1973) Effect of light quality on chloroplast replication in spinach. J Exp Bot 24:1247–1260

Possingham JV, Smith JW (1972) Factors affecting chloroplast replication in spinach. J Exp Bot 23:1050–1059

Powell RD, Griffith MM (1960) Some anatomical effects of kinetin and red light on disks of bean leaves. Plant Physiol 35:273–275

Powell RD, Griffith MM (1963) Some effects of kinetin, red light, and gamma radiation on growth of disks of bean leaves. Bot Gaz 124:274–278

Powell RD, Morgan PW (1970) Factors involved in the opening of the hypocotyl hook of cotton and beans. Plant Physiol 45:548–552

Powell RD, Morgan PW (1980) Opening of the hypocotyl hook in seedlings as influenced by light and adjacent tissues. Planta 148:188–191

Ray PM, Green PB, Cleland RE (1972) Role of turgor in plant cell growth. Nature 239:163–164

Redington G (1929) Effect of the duration of light upon plant growth and development. Biol Rev Camb Philos Soc 4:180–208

Reid DM, Clements JB, Carr DJ (1968) Red light induction of gibberellin synthesis in leaves. Nature 217:580–582

Roesel HA, Haber AH (1963) Studies of effects of light on growth pattern and of gibberellin sensitivity in relation to age, growth rate, and illumination in intact wheat coleoptiles. Plant Physiol 38:523–532

Rubinstein B (1971) The role of various regions of the bean hypocotyl on red-light-induced hook opening. Plant Physiol 48:183–186

Sachs J von (1882) Lectures on the physiology of plants (Transl. Ward HM) Clarendon, Oxford

Sachs RM (1965) Stem elongation. Annu Rev Plant Physiol 16:73–96

Saurer W, Possingham JV (1970) Studies on the growth of spinach leaves (*Spinacia oleracea*). J Exp Bot 21:151–158

Schneider CL (1941) Effect of red light on growth of the *Avena* seedling with special reference to the first internode. Am J Bot 28:878–886

Schwabe WW (1963) Morphogenetic responses to climate. In:Evans LT (ed) Environmental control of plant growth. Academic Press, London New York, pp 311–316

Senger H (ed) (1980) The blue light syndrome. Springer, Berlin Heidelberg New York

Shen-Miller J, Gordon SA (1967) Gravitational compensation and the phototropic response of oat coleoptiles. Plant Physiol 42:352–360

Shields LM (1950) Leaf xeromorphy as related to physiological and structural influences. Bot Rev 16:399–447

Shirley HL (1945) Light as an ecological factor and its measurement II. Bot Rev 11:497–532

Silk WK (1980) Growth rate patterns which produce curvature and implications for the physiology of the blue light response. In: Senger H (ed) The blue light syndrome. Springer, Berlin Heidelberg New York, pp 643–655

Silk WK, Erickson RO (1978) Kinematics of hypocotyl curvatures. Am J Bot 65:310–319

Silk WK, Erickson RO (1979) Kinematics of plant growth. J Theor Biol 76:481–501

Steer BT (1971) The dynamics of leaf growth and photosynthetic capacity in *Capsicum frutescens* L. Ann Bot NS 35:1003–1015

Stuart DA, Jones RL (1977) Roles of extensibility and turgor in gibberellin- and dark-stimulated growth. Plant Physiol 59:61–68

Stuart DA, Durnam DJ, Jones RL (1977) Cell elongation and cell division in elongating lettuce hypocotyl sections. Planta 135:249–255

Sunderland N (1960) Cell division and expansion in the growth of the leaf. J Exp Bot 11:68–80

Thomas B, Dickinson HG (1979) Evidence for two photoreceptors controlling growth in de-etiolated seedlings. Planta 146:545–550

Thomson BF (1950) The effect of light on the rate of development of *Avena* seedlings. Am J Bot 37:284–291

Thomson BF (1951) The relation between age at time of exposure and response of parts of the *Avena* seedling to light. Am J Bot 38:635–638

Thomson BF (1954) The effect of light on cell division and cell elongation in seedlings of oats and peas. Am J Bot 41:326–332

Thomson BF (1959) Far-red reversal of internode-stimulating effect of red light on peas. Am J Bot 46:740–742

Thomson BF, Miller PM (1962) The role of light in histogenesis and differentiation in the shoot of *Pisum sativum*. II. The leaf. Am J Bot 49:303–310

Turner MR, Vince D (1969) Photosensory mechanisms in the lettuce seedling hypocotyl. Planta 84:368–382

Turrell FM (1936) The area of the internal exposed surface of dicotyledon leaves. Am J Bot 23:255–264

Vanderhoef LN, Quail PH, Briggs WR (1979) Red light-inhibited mesocotyl elongation in maize seedlings. II. Kinetic and spectral studies. Plant Physiol 63:1052–1067

Van Volkenburgh E, Cleland RE (1979) Separation of cell enlargement and division in bean leaves. Planta 146:245–247

Van Volkenburgh E, Cleland RE (1980) Proton excretion and cell expansion in bean leaves. Planta 148:273–278

Van Volkenburgh E, Cleland RE (1981) Control of light-induced bean leaf expansion: role of osmotic potential, wall yield stress and hydraulic conductivity. Planta 153:572–577

Verbelen JP, De Greef JA (1979) Leaf development of *Phaseolus vulgaris* L. in light and darkness. Am J Bot 66:970–976

Vince D (1964) Photomorphogenesis in plant stems. Biol Rev 39:506–536

Vince-Prue D (1975) Photoperiodism in plants. McGraw-Hill, London

Vince-Prue D (1977) Photocontrol of stem elongation in light-grown plants of *Fuchsia hybrida*. Planta 133:144 – 156

Vince-Prue D, Gutteridge CG, Buck MW (1976) Photocontrol of petiole elongation in light-grown strawberry plants. Planta 131:109–114

Warner T, Ross JD (1981) Phytochrome control of maize coleoptile section elongation: the role of cell wall extensibility. Plant Physiol 68:1024–1026

Warner T, Ross JD, Coombs J (1981) Phytochrome control of maize coleoptile section elongation. Plant Physiol 67:355–357

Warrington IJ, Mitchell KJ (1976) The influence of blue- and red-biased light spectra on the growth and development of plants. Agric Met 16:247–262

Wassink EC, Stolwijk JAJ (1956) Effects of light quality on plant growth. Annu Rev Plant Physiol 7:373–400

Weintraub RL, Price L (1947) Developmental physiology of the grass seedling. II. Inhibition of mesocotyl elongation in various grasses by red and violet light. Smithson Misc Collect 106:No 21

Went FW (1941) Effect of light on stem and leaf growth. Am J Bot 28:83–95

Wildermann A, Drumm H, Schäfer E, Mohr H (1978a) Control by light of hypocotyl growth in de-etiolated mustard seedlings. I. Phytochrome as the only photoreceptor pigment. Planta 141:211–216

Wildermann A, Drumm H, Schäfer E, Mohr H (1978b) Control by light of hypocotyl growth in de-etiolated mustard seedlings. II. Sensitivity for newly-formed phytochrome after a light to dark transition. Planta 141:217–223

Williams RF (1975) The shoot apex and leaf growth. Cambridge Univ Press

Williams RF, Rijven AHGC (1970) The physiology of growth in subterranean clover. 2. The dynamics of leaf growth. Aust J Bot 18:149–166

Wright STC (1961) A sequential growth response to gibberellic acid, kinetin and indolyl-3-acetic acid in the wheat coleoptile (*Triticum vulgare* L.). Nature 190:699–700

Wylie RB (1951) The principles of foliar organisation shown by sun-shade leaves from ten different species of deciduous dicotyledenous trees. Am J Bot 38:355–361

16 Photomorphogenesis and Hormones

J.A. DE GREEF and H. FREDERICQ

1 Introduction

This review is restricted to phytochrome-mediated light effects and to the generally recognized five classes of natural hormones, auxins, gibberellins, cytokinins, abscisic acid and ethylene. Other environmental conditions having morphogenic effects in relation to phytochrome and hormone actions are included when they are relevant to our objective. Aspects of flower induction are mentioned when an interaction between phytohormones and phytochrome is evident.

2 Germination Studies: the *Lactuca* System and Some Other Light-Requiring Seeds

The *Lactuca* seed has been a preferred material of many studies related to germination mediated by phytochrome and plant hormones.

The effects of both light and growth regulators on the germination process, in detail, are exceedingly complicated and the available literature is overwhelming. Seeds have the capacity to assess information about environmental cues such as light, moisture and temperature. This information is integrated into a time-frame, determining whether or not germination is favorable. Counteractive systems appear to be present, the one system promoting germination, the other inducing dormancy. The appropriate level of active phytochrome, growth regulators (such as GA[1], ethylene and cytokinins), optimal temperature range, and imbibition conditions promote germination. Improper poising of the phytochrome system, ABA and growth inhibitors, sub- and supra-optimal temperature ranges, constraint of covering structures and osmotic stress are inhibitory, suppressive or dormancy inducing. Each of these factors may interact. Therefore, seeds exhibit any degree of response from full germination to complete dormancy.

When the balance between promotive and suppressive systems is controlled by phytochrome (see Chap. 17, this Vol.), the seeds are light-sensitive. Besides phytochrome regulation and the particular state of the seed in question, lettuce seed germination is also influenced by endogenous growth regulators. Whether

1 *Abbreviations.* ABA: abscisic acid; AMO 1618: 2′-isopropyl-4′-(trimethylammonium chloride)-5′-methylphenyl piperidine-1-carboxylate; BA: benzyladenine; CCC: 2-chloroethyltrimethyl ammonium chloride; DCMU: 3-(3,4-dichlorophenyl)-1,1-dimethylurea; DNP: 2,4-dinitrophenol; GA: gibberellic acid; HBA: o-hydroxybenzyl adenosine; IAA: indol-3yl-acetic acid; W: white light

or not the expression of phytochrome in germination behaviour reflects interactions with hormonal regulatory systems, is still a matter of debate.

2.1 Can a Phytochrome Treatment Be Replaced by Phytohormones?

In several photoblastic seeds P_{fr} action can be mimicked by exogenous GA (see Evenari 1965). Kinetin slightly promotes germination in darkness while it strongly accelerates germination induced by R or GA (Ikuma and Thimann 1963). The R effect is nullified by short FR in the presence of kinetin, but not in the presence of GA. Light or GA act as primary stimulus in the hypocotyl axis, while exogenous kinetin has a secondary action which facilitates bursting of the seed coat by expansion of the cotyledons. When GA acts by initiating one of the chemical reactions which result from the light reaction, its end product is the same as that produced by light.

In several studies cytokinin activity in light-sensitive seeds appears to be under phytochrome control. Van Staden and Wareing (1972) are the first authors to present evidence for a rapid R/FR reversible increase of cytokinins in the butanol and aqueous extracts obtained from imbibed *Rumex obtusifolius* L. seeds. When *Lactuca* seeds are irradiated with R or imbibed with GA in darkness, the level of water-soluble cytokinins decreases and that of the butanol-soluble ones increases (Van Staden 1973). However, FR does not completely reverse this effect. These results are also confirmed with dormant celery seeds by Thomas et al. (1978). The authors stress the importance of hormone balances as a possible mechanism for phytochrome action. In light-promoted *Picea* seeds similar results are obtained (Taylor and Wareing 1979).

ABA can negate GA-induced dark germination and this inhibition is removed when cytokinin is added (Khan 1968). The nature of the ABA inhibition remains obscure since excess GA cannot overcome the inhibitory effect at 25 °C. Studying the interactions of GA, ABA and cytokinins in intact lettuce seeds, Khan (1971) suggests that the kinetin effect may be regarded as permissive to the action of GA in the presence of ABA.

Ethylene cannot reverse ABA inhibition, but it promotes synergistically the kinetin reversal of ABA inhibition of both light and GA-induced germination (Rao et al. 1975).

From these studies it is obvious that the stimulatory or inhibitory action of light through phytochrome can be replaced by hormones in light-sensitive seeds. In most cases exogenous hormones are used to elicit the physiological response, while phytochrome-mediated changes of endogenous hormone levels correlated with germination behaviour are poorly documented.

2.2 Is There Evidence that Phytochrome and Phytohormones Interact During Germination Processes?

Kahn (1960) has shown that the promotive effects of GA and phytochrome are additive when R is given 4 h after the beginning of a GA treatment, while they become synergistic when R is administered after 20 h. These facts may indicate that part of the GA action in lettuce either increases the responsiveness of the seeds to R or maintains R sensitivity at its maximum level. When GA is applied with an increasing delay after imbibition, a decreasing rate and lower final percentage of germination are observed. In contrast, R given with the same delay influences the rate but not the final germination percentage (LEWAK and KHAN 1977). They conclude that R and GA seem to have a different mode of action.

BEWLEY et al. (1968) find a strong synergism between a very rapid P_{fr} action and suboptimal concentrations of exogenous GA in lettuce seeds. The same kind of synergism between P_{fr} and kinetin is found by BLACK et al. (1974). The parallelism with the promotive effect of kinetin in flower induction of *Pharbitis* seedlings is remarkable (NAKAYAMA et al. 1962). When these seedlings are sprayed with BA after a short, suboptimal R irradiation, flowering is drastically enhanced (OGAWA and KING 1979).

Recently, CARPITA and NABORS (1981) have shown that the synergism described by BEWLEY et al. (1968) between R and exogenous GA does exist when germination of intact lettuce seeds is examined, but this interaction is only additive when the growth of the axes of excised embryos is measured. Dose-response curves demonstrate quantitative increases in the growth response of the latter after R or GA treatments insufficient to induce germination of whole seeds, indicating that a threshold growth potential must be achieved by the embryonic axes of intact seeds before the endosperm and the fruit coat can be punctured. This co-action is apparently a case of two additive subthreshold stimuli, surpassing the threshold potential when applied simultaneously to intact seeds. The nature of this type of interaction will only be elucidated when it becomes technically possible to measure kinetically whether or not phytochrome-mediated changes of endogenous GA levels are correlated with the germination response.

These findings are consistent with earlier observations that the covering structures of the seed restrict embryo growth (NABORS and LANG 1971) and that cytokinin without the presence of GA can reverse ABA inhibition in isolated embryos (BEWLEY and FOUNTAIN 1972).

In several types of imposed seed dormancy interactions between light and hormones have been demonstrated. Skotodormant lettuce seeds (maintained in prolonged darkness in the imbibed state at 20 °C) lose their GA responsiveness prior to that to light. When GA and light are applied together, they stimulate germination at times when each treatment alone is ineffective (VIDAVER and HSIAO 1974).

These results are confirmed by BEWLEY (1980). Embroys dissected from skotodormant seeds germinate and are as capable of radicle expansion in an osmoticum as are freshly imbibed seeds. Hence, skotodormancy is an endosperm-imposed dormancy as is primary dormancy, but it is different from the latter because it is not relieved by R or by GA alone. Skotodormancy does not reside within the embryo as an inherent block to germination processes, but it is due to the inability of the intact seed to respond to the stimulation of either R or the hormone.

Photodormant seeds (intact seeds treated with FR during imbibition) display also a persistent loss of GA sensitivity. When these seeds are punctured or treated with a GA solution at low pH, germination is restored. Prolonged FR treatment in the presence of GA suspends only GA action, but it does not prevent GA-stimulated germination during a subsequent dark incubation (BURDETT 1972).

GA can be substituted for by ethylene in its synergistic action with R in breaking secondary dormancy, but ethylene cannot substitute for the phytochrome-mediated action (SPEER et al. 1974). In a similar way the establishment of secondary dormancy in seeds of *Chenopodium bonus-henricus* can be prevented by daily brief R or by the combination of kinetin + ethylene + GA and, to a lesser extent, by GA alone. Following the establishment of secondary dormancy the hormone combination is relatively more active than light or GA in removing dormancy. High levels of exogenous ABA do not prevent the breaking of dormancy by light (KHAN and KARSSEN 1980).

At higher temperature (>27 °C) GA is much less effective in promoting germination when the seeds are kept in the light than in the dark (REYNOLDS and THOMPSON 1973). Temperature cut-off points which have been observed in seed germination (THOMPSON 1970) appear to be of overriding importance. Attention to this fact in future research of thermodormancy may unify many scattered studies on germination behavior and could explain a number of inconsistencies.

In overcoming thermodormancy (35 °C) of lettuce, light action through phytochrome or GA action requires the addition of ethylene plus CO_2 (NEGM et al. 1973). Exogenous ethylene partially reverses dormancy induced by either high temperature (32 °C), ABA, or an osmoticum, but only in the presence of GA or light. However, thermodormancy at 36 °C is not alleviated by either GA + C_2H_4 or light + C_2H_4, while it is reversed by the addition of kinetin to both of these mixtures. Dormancy imposed by ABA is also reversed by kinetin, and faster in the light than in darkness. Osmodormancy is not reversed by kinetin and its reversal by ethylene or C_2H_4 + GA is inhibited by kinetin but only in the light. These observations can be interpreted that kinetin in the presence of light stimulates cotyledonary growth, acting as a metabolic sink which leads to competition with radicle protrusion for critical metabolites and thus reduced germination. In thermodormancy and osmodormancy kinetin and ethylene seem to regulate common events leading to germination, but through mechanisms unique to each respective growth regulator. In germination the regulatory role of ethylene is absolutely dependent upon an interaction with GA and/or light (DUNLAP and MORGAN 1977a, b).

With respect to the site(s) of phytochrome and hormone actions the experimental data give rise to very different views. In *Spergula arvensis* seeds cytokinin production is under control of a combination of light and ethylene as is germination (VAN STADEN et al. 1973). They assume that the observed changes in hormone content are probably only part of the process initiated by light and ethylene, since cytokinin on its own has no effect. The data indicate at least two sites of hormonal and light action.

During germination of *Chenopodium album* seeds two sites of hormonal action can be discerned (KARSSEN 1976). The induction phase is primarily controlled by a high level of P_{fr} which can be substituted in darkness by GA at low pH and by ethylene to a lesser extent. During the growth stage proper a second site of regulation is apparent. In this phase the growth of the radicle is promoted by low P_{fr} and inhibited by ABA which is antagonized by GA, zeatin, kinetin and ethylene. The primary inducers act additionally rather than synergistically.

In the photomorphogenesis of *Sinapis* seedlings BAJRACHARYA et al. (1975) find no essential role of ABA. The competence to this hormone and to phytochrome are separated in time during embryo development.

Studying the interactions of phytochrome and exogenous GA on germination of *Lamium amplexicaule* TAYLORSON and HENDRICKS (1976) conclude that, when secondary dormancy is not a factor, P_{fr} increases the effectiveness of GA stimulation, indicating similar sites of action.

In other seeds, light cannot be replaced by GA treatment, although germination is obviously influenced by light–GA interactions. Seeds of *Begonia evansiana* do not germinate in complete darkness, even when GA is administered to the external medium. However, the light requirement for germination is markedly reduced by GA treatment (NAGAO et al. 1959). It is suggested that the latter intensifies the light action or substitutes for a part of the light requirement. GA cannot induce germination of *Kalanchoë blossfeldiana* in darkness, but it enhances the effect of suboptimal photoperiods (BÜNSOW and VON BREDOW 1958).

Phytochrome involvement has been proven with the cv. Feuerblüte at 20 °C. Water-imbibed seeds fully germinate when they are exposed daily to 1 min R or W (ELDABH et al. 1974). This effect is annihilated by a subsequent FR treatment which is ineffective on its own. When GA is present in the medium, the number and the duration of the effective light treatments is strongly reduced. After an initial lag phase of 2 days (needed for GA-uptake and primary differentiation of the embryo) two 1-min R exposures and even two 5-s broadband FR pulses given on 2 consecutive days, are saturating for full germination. Consequently, there is no longer R/FR reversibility. Moreover, there is a strong synergism between GA and FR, while each of both treatments is completely ineffective for germination (FREDERICQ et al. 1976) (Table 1).

Preliminary action spectra taken in the presence and absence of GA strongly indicate phytochrome involvement (Fig. 1). The light sensitivity of the seeds is enhanced drastically in the presence of GA (at 660 nm the fluence required for 50% germination becomes at least 5000 times smaller). It seems thus that a low threshold level of P_{fr} is sufficient for germination in the presence of GA, assuming that the efficiency of $P_r \rightleftharpoons P_{fr}$ phototransformations is not altered by the hormone.

In fully imbibed, but not germinated seeds the effect of one (or two) short R or FR irradiation(s) is maintained over a very long dark period (several weeks) at 20 °C, when exogenous GA is present. A further light exposure (almost ineffective in itself), given after the dark period, strongly promotes germination

Table 1. Germination (%) of *Kalanchoë blossfeldiana* (cv. Feuerblüte) in water or in GA_3 under indicated light treatments

		Water	GA_3
A. Daily light treatment	2 min W	94	94
	2 min W+15 s FR	6	93
	2 min W+12 h FR	0	85
	15 s FR	0	78
	12 h FR	0	89
	Dark control	0	0
B. Light treatment on day 5 and 6	5 s FR	0	89
	30 s FR	0	86
	Dark control	0	0

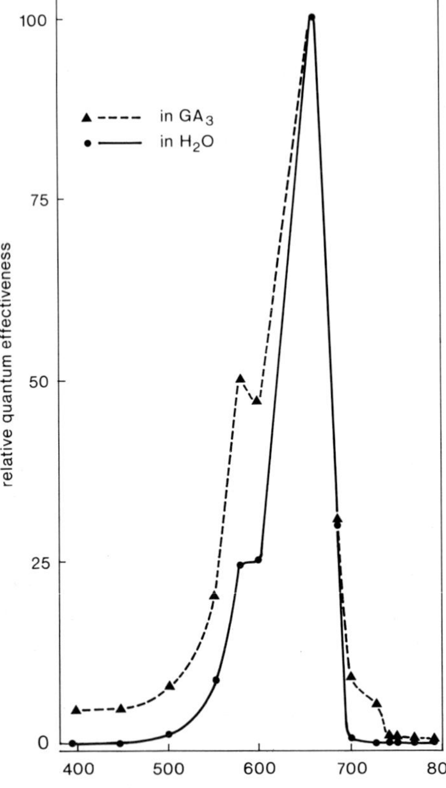

Fig. 1. Action spectra for the induction of *Kalanchoë* seed germination in the presence and in the absence of GA_3. Quantum effectiveness for the induction of 50% germination. For both spectra the response at 660 nm is normalized to 100%. Corrections have been made for the transmittance of the seed coat at the wavelengths used (DE PETTER et al. unpublished)

(RETHY et al. 1976). This memory effect is temperature-dependent as demonstrated, e.g., on the energy metabolic level (FREDERICQ et al. 1980). The irradiation(s) inducing the physiological memory effect, cause(s) a long-lasting rise of the endogenous ATP level in the ungerminated seeds (above the dark con-

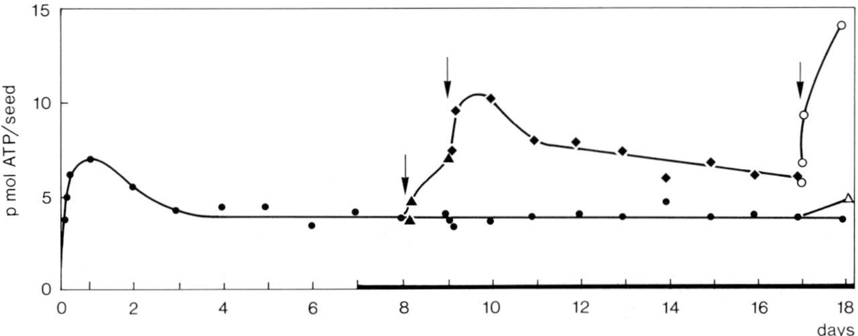

Fig. 2. Kinetics of ATP levels in *Kalanchoë* seeds. After 6 days in darkness the secondary dormant seeds are transferred to GA_3 at day 7 and exposed to 2-min R at the times indicated (*arrows*). ●: dark controls; ▲: after R at day 8; ◆: after an additional R exposure at day 9; ○: after a third R exposure at day 17; △: after a R exposure at day 17 only

trols). The light exposure after the prolonged dark period again provokes a considerable ATP rise and induces germination (Fig. 2).

In screening experiments concerning the effects of light and GA on the germination of different seed species there is a correlation between light sensitivity and seed weight. Most of the absolutely light-requiring seeds are very small, hence making analytical and biochemical work very difficult (DEDONDER et al. 1980).

2.3 Mode of Action at the Metabolic Level

2.3.1 General Metabolic Effects

In the early stages of germination of photoblastic seeds and well before radicle protrusion, respiration is enhanced by R and this effect is FR-reversible (see EVENARI 1965). A comparison of the effects of chemical and light treatments on respiration of lettuce seeds indicates that chemical and photocontrol mechanisms have a different temporal pattern (WOODSTOCK and TOOLE 1976). Short R increases the respiratory rates within 2 to 3 h and the differences between R- and FR-irradiated seeds become more pronounced with time. In darkness, at 25 °C, GA + kinetin stimulates respiration slightly by the 6th hour and markedly by the 12th and 24th hour. ABA inhibits germination and represses respiration of both light-induced and GA + kinetin-treated seeds. The delay of respiratory effects between light and chemical treatments could be due to a different site of action, although the chemical effect may be retarded by a diffusion barrier.

2.3.2 Enzyme Studies

In cotyledons of *Sinapis alba* seedlings amylase activity can be induced by phytochrome, but this light effect cannot be replaced by GA (DRUMM et al. 1971).

The authors visualize that P_{fr} and hormones act as independent factors in connection with regulation of enzyme synthesis.

In lettuce seeds fat bodies disappear during the early stage of imbibition and this process ends before radicle protrusion. R and GA treatment are promotive, FR, ABA treatment and thermodormancy are inhibitory. A correlation with a thermolabile factor is suggested (TAKEBA and MATSUBARA 1977). In dark-grown apple embryos the activity of alkaline lipase is stimulated by R and GA, but not by FR or in the presence of AMO 1618. Germination is influenced in a way similar to the alkaline lipase activity. This correlation suggests that light stimulates the germination of apple embryos by promoting GA biosynthesis via the phytochrome system. GA increases the activity of lipase hydrolyzing the storage lipids (SMOLENSKA and LEWAK 1974).

JACOBSEN and PRESSMAN (1979) find that radicle emergence is preceded by the breakdown of endosperm cells near the root cap. This breakdown in celery seed is dependent on a stimulus from the embryo in response to light. Indirect evidence indicates that the stimulus may be a gibberellin released from the embryo which causes the endosperm cells to produce degrading enzymes. R- and GA-induced lettuce seed germination is followed by the induction of hydrolytic enzymes affecting the endosperm cell wall and the stored reserves in the cotyledons. Promotion of endo-β-mannanase activity in the endosperm is controlled by the combined actions of the axis and the cotyledon. It involves the production of hormones in the axis and the overcoming of an inhibitor in the endosperm (BEWLEY and HALMER 1980). A very rapid P_{fr} action on the induction of α-galactosidase, prior to the completion of germination, is also shown by LEUNG and BEWLEY (1981). While the induction of the enzyme is finished very quickly, the enzyme activity arises only 3 h after light exposure, indicating that P_{fr} sets in motion a sequence of metabolic events. It is suggested that this lag phase is required for the passage of a diffusible promotor from the site of light perception (embryonic axis) to the site of enzyme production (cotyledons). GA, and even better GA + BA can replace the requirement for the axis. FR can inhibit germination at times when enzyme activity is unaffected, indicating two distinct phytochrome-controlled reactions.

Some lines of evidence can be found for the existence of interactions between different molecular modulators on the one hand, and for the existence of phase sequences in germination events, affected by environmental and hormonal factors, on the other.

When only primary dormancy is involved, exogenous GA and phytochrome interact synergistically in intact seeds. This means that both modulator molecules build up to a threshold growth potential which must be achieved before the embryonic axis can overcome the restraint of the covering structures of the seed. The interaction of GA and P_{fr} is additive when the growth of isolated axes is measured, but this response is insufficient to cause germination of intact seeds. Isolated embryonic axes also respond to much lower GA concentrations than intact seeds, since the covering tissues are barriers to GA uptake. Considering the facts that the induction of the growth potential increase by P_{fr} or GA occurs at the same physiological time and that P_{fr} increases the effectiveness of GA stimulation and vice versa, both growth effectors appear to have a common or, at least, similar site(s) of action.

In seeds where GA can substitute for P_{fr} the final percentage of germination is not changed when the R exposure is given with greater delay, while the same delay of GA application results in a lower germination percentage. This observation is in favor of the hypothesis that P_{fr} is a more primary inducer than GA and may control the amount of active GA molecules in the seed tissues. On the other hand, GA seems to play a more prominent part in light-controlled growth than the other hormones.

Since exogenous kinetin strongly accelerates germination induced by R or GA and is less effective in darkness, its mode and site of action are different in the sequence of germination events. This idea is also strengthened by the fact that ABA inhibition of both light- and GA-induced germination are reversed by kinetin, while an excess of GA cannot overcome the inhibitory effect. This kinetin reversal is synergistically promoted by ethylene, but ABA inhibition cannot be negated by ethylene alone. Thus this compound affects the pathway of the germination process at a point co-acting with kinetin but different from ABA, when secondary dormancy is not a factor.

More attention should be paid to the phase sequences of the pattern specification and realization in germination studies and in studies of photomorphogenesis in general (see Chap. 14, this Vol.). It is conceivable that the germination response depends on the tissue sensitivity defined by the developmental state at the moment when P_{fr} is formed and/or hormones are either applied to or endogenously formed in the seed. Many puzzling phenomena in photomorphogenic studies can be resolved if the concept of sensitivity variation as the determinant of growth patterns (=developmental acquisition) is taken into account. Sensitivity as a growth-limiting factor refers to the competence of the tissue to respond to modulator molecules. Although the molecular basis of this concept is not known, it is a current view that specific receptor molecules (proteins) are present in particular tissues. During the initial lag time necessary for physiological and biochemical changes induced by imbibition, the seeds are rather insensitive to light and hormones. Thereafter the time lapse of sensitivity varies for the different modulator molecules. This is very conspicuous when the seeds enter secondary dormancy. In this case, seeds lose their GA responsiveness prior to that to light. The deeper the dormant state of the seed the more complex the mix of modulator molecules has to be in order to re-establish germination conditions. It is clear that the latent period of suspended germination processes is correlated with varying degrees of responsiveness that can be evolved by the seed tissues in the presence of the dormancy agent used.

3 Studies Related to Vegetative Development

3.1 Basic Observations Concerning Light and GA Action

The light-induced inhibition of pea stem growth can be reversed by GA (LOCKHART 1956). SCOTT and LIVERMAN (1957) observed that R stimulated expansion of bean leaf discs can be simulated by GA treatment. BRIAN (1958) has drawn attention to the relationship between the effects of exogenous GA and long days, and has postulated that GA biosynthesis is controlled in vivo by the phytochrome system. However, it has not yet been delineated whether light exerts its effect directly or indirectly on endogenous gibberellins.

3.2 Leaf Growth and Light-Controlled Changes in Endogenous GA Content

Unequivocal evidence of a rather rapid, but transient increase of GA-like activity is shown by Reid et al. (1968). Etiolated barley leaves exposed to 30 min R contain higher amounts of extractable GA, peaking after 15 min of subsequent darkness. Afterwards there is a rapid decline. Since CCC and AMO 1618 reduce considerably GA activity in the leaf segments, the authors conclude that the rise in GA activity must be largely the result of R-induced synthesis rather than the release of GA from a storage site. The R response is blocked by chloramphenicol and actinomycin D, implicating the need of active RNA and protein synthesis (Reid and Clements 1968). A similar observation is made by Beevers et al. (1970), who find that unrolling of etiolated wheat leaf segments is stimulated by GA and also by kinetin in the dark and by R, while ABA and FR counteract this response. These authors also confirm the rapid and transient increase of GA activity: a maximum is found 10 min after the end of a 5-min R exposure. The rapid increase and rapid subsequent decline in GA levels indicate a high rate of GA turnover.

Feeding homogenates of etiolated barley leaves with ^{3}H-GA_9, Reid et al. (1972) demonstrate that the incorporation of radioactivity into other GAs occurs to a greater extent in R than in darkness.

Loveys and Wareing (1971) have compared their results on wheat with the barley system. In wheat a 5-min R pulse induces a transient GA increase, but this response is not affected by AMO 1618. In barley, however, a 30-min R exposure causes a subsequent GA increase which can be significantly reduced by the inhibitors of GA biosynthesis. R stimulated increase of extractable GA activity in wheat is mirrored by a corresponding decrease in the level of bound GA. From these data it seems that R-stimulated GA production is due to release from a bound form as a short term effect, while GA synthesis may be involved as a long-term effect.

Cooke and coworkers (1975, 1975a, b) report that plastids are a major site of rapid phytochrome controlled GA production in etiolated wheat leaves. They also suggest a close relationship between R stimulated GA activity and leaf unrolling. Evans and Smith (1976a), working with preparations of partially purified etioplasts from barley leaves, bring even more convincing evidence for phytochrome involvement in GA production by using the α-amylase assay which is much more sensitive than the lettuce hypocotyl bioassay used by Cooke and Saunders (1975a). Sonication of intact etioplasts in the dark yield increased amounts of extractable GA-like activity (Cooke and Kendrick 1976). In both wheat and barley it is suggested that phytochrome is located in or on the etioplast envelope (Evans and Smith 1976b, Cooke and Kendrick 1976). These authors present alternative interpretations to explain phytochrome control of GA production in etioplasts: alteration in the permeability of the envelope following phytochrome phototransformation, R-induced release, and interconversion of acidic GA-like substances from the bound site within the envelope. STODDART (1968) also reports that more GA-like material can be extracted from chloroplast suspensions after sonication. However, since phytochrome of these plastids would be in the P_{fr} form already, they should not need sonication, if the situations are comparable. BROWNING and SAUNDERS (1977) suggest that there are at least two fractions of extractable gibberellins sequestered within the wheat chloroplast membranes—one released by methanol, another by Triton. Recently, HILTON and SMITH (1980) show that a purified etioplast fraction exhibits a phytochrome mediated regulation of the levels of GA-like substances extractable into aqueous methanol, while fractions enriched with mitochondria do not.

The idea of compartmentalization of endogenous plant growth regulators coupled with environmentally mediated release mechanisms may offer new approaches for a better understanding of plant development.

3.3 Control of Stem Growth and Root Formation

Based on the first observations of LOCKHART (1956), many studies have been undertaken to correlate light effects with GA levels. The results are often contradictory with regard to GA metabolism, the rate of turnover and the active forms of GA. The complexity of the relationships between GA effects, light inhibition of growth and responsiveness of the tissues is generally recognized, but leads to very different interpretations (GRAEBE and ROPERS 1978).

Also apparent light–auxin interactions related to subsequent photomorphogenic events do not fit a simple model for the light control of development via auxins. R stimulates auxin transport through the inside half of the bean hypocotyl shank and increases the sensitivity of this tissue to IAA. This suggests that endogenous auxin-like substances participate in controlling the unbending of the hypocotyl hook of *Phaseolus vulgaris* and may be related to light-induced unbending of the hook (RUBINSTEIN 1971). SHERWIN and FURUYA (1973) demonstrate a phytochrome-mediated effect on IAA-uptake by intact etiolated rice coleoptiles. The absorption is biphasic and each phase shows R/FR reversibility. R causes an initial inhibition of the rate of IAA uptake followed by a large promotion. R affects neither the destruction nor the immobilization of the applied IAA. The authors suggest that phytochrome controls the amount of auxin diffusing from the coleoptile tips (permeability).

In *Lupinus albus* the inhibition of hypocotyl lengthening is mediated through phytochrome in its low energy mode. This effect is reversed by exogenous GA. Inhibition by HIR is largely overcome by IAA treatment, but not by GA-like substances. Application of IAA to etiolated seedlings causes growth inhibition (ACTON and MURRAY 1974). It is concluded that light antagonizes hormone-regulated growth through a complex relationship between hormone balance and visible radiation. The R-inhibited growth of excised coleoptiles is only reversed by FR in the presence of IAA (LIVERMAN and BONNER 1953). R-treated *Avena* tissue contains reduced amounts of extractable auxins relative to dark-grown material (BLAAUW-JANSEN 1959). Application of high concentrations of exogenous IAA (HILLMAN 1959) or treatment of *Avena* coleoptiles with tryptophan, a precursor of IAA (MUIR and CHANG 1974) prevents the R-induced growth inhibition. FLETCHER and ZALIK (1964, 1965) demonstrate a direct relationship between IAA content after one photoperiod of 8 h R and the height of bean plants after seven cycles. Dark controls are the tallest and contain the most IAA. R-treated plants are the shortest and have the least IAA. R contaminated with FR increases both height and IAA-content. Feeding the plants with ^{14}C-IAA the auxin seems to be transformed into three distinct fractions, their relative proportions being related to light quality. Both effects, stem elongation and auxin metabolism, provide evidence for phytochrome control. Further support of these observations is found in the work of RUSSEL and GALSTON (1969). R inhibits the growth of etiolated internodes of dwarf pea and causes a shift to higher IAA concentrations in the dose-response curve for excised sections. Similarly, the IAA content of the subapical internode is greatly reduced in light-treated dwarf pea varieties, relative to light-treated normal varieties or dark-grown dwarfs (MUIR 1970).

Also in apical dominance confusing interactions of hormones are met when physiological responses to light quality (phytochrome control) are examined (Table 2).

Studying the differential effect of plant hormones and R in different growing zones of the bean hypocotyl GOTÔ and ESASHI (1974, 1976) show that the greatest promotive effect is in the order cytokinin, R, GA, CO_2, ethylene and auxin going from the apical to the basal part. ABA seems to prevent the stem from excessive elongation caused by both GA- and R-dependent responses of the immature region during the light period, and the auxin-dependent response of the more mature region during the night. Exogenous IAA promotes the elongation of the older shank portion and FR acts synergistically. When the hormone is applied to the younger stem portion, FR nullifies the IAA-promoting effect (GOTÔ and SUZUKI 1980). Recently it was observed that R- or W-induced growth inhibition of the subapical internode in bean is overcome by exogenous GA, while R+FR abolishes the GA effect. Thus phytochrome controls GA action (DE GREEF et al. unpublished).

With respect to root formation the available information is much less coherent. HUMPHRIES (1961) suggests that root formation may be under phytochrome control. In isolated, decapitated pea roots the phytochrome controlled lateral root initiation requires the presence of exogenous IAA (FURUYA and TORREY 1964). Other chemical constituents in the nutrient medium, however, can also become limiting factors for root formation. The regeneration of adventitious roots in mustard seedlings is promoted by phytochrome (PFAFF and SCHOPFER 1974). These authors suggest that phytochrome is needed to produce a hormonal rooting factor in the cotyledons. PFAFF and SCHOPFER (1980) find also that primordium formation in dark-grown or FR-treated rest seedlings is not promoted by exogenous IAA, GA_3, kinetin, ABA or ethylene. The application of these hormones is either ineffective or inhibitory in the rooting response.

Table 2. Effects of phytochrome in relation to hormones in apical dominance

Species	Treatment	Physiological response (outgrowth of lateral bud)	Hormone response	References
Xanthium strumarium	W	+		Tucker and Mansfield (1972)
	W+FR	−	Accumulation of ABA and cytokinin	
Solanum lycopersicum	W	+		Tucker (1976)
	W+FR	−	Increase of IAA and ABA	
	IAA application to apical bud	+		Tucker (1977)
	ABA application to lateral bud	−		
	ABA application to apical bud	+		
	W+FR+kinetin application to lateral bud	+ (Slight and short lived)		
	W+hadacidin application to lateral bud	− (During treatment)		
Phaseolus vulgaris	W	+	No ABA involvement	White and Mansfield (1978)
	W+FR	+	id.	

+: stimulation; −: inhibition

It is concluded that phytochrome does not operate through changes of hormone levels.

When segment physiology is used in studies of light and hormone action, the results have to be evaluated carefully. Decapitation or excision of sections may cause a strong growth inhibition effect due to the absence of the hormone source. Under these conditions normal growth can be recovered by hormone application vastly in excess of that present in the intact plant. During the process of differentiation both in light and dark the same tissues of intact seedlings develop successive phases of different sensitivity to photoreceptors and hormones. Consequently different responses are obtained according to that specific sensitivity state. In general, green shoot tissues are much more sensitive to GA than to IAA, while the opposite is true in etiolated seedlings. This means that light may also change the sensitivity of the tissue to the hormone. On the other hand, it is remarkable that green tissues contain only a small fraction of the phytochrome present in etiolated tissues.

3.4 Cytokinin Effects and Studies on Endogenous Cytokinin Levels

The reports of Miller (1956) and Scott and Liverman (1956) concerning the similarity of R and cytokinin effects on etiolated bean leaf discs have been followed by a number of studies supporting these observations. The substitution of kinetin for R in short term experiments is confirmed by Hillman (1957) for the frond multiplication rate of *Lemna minor*. After 3 weeks in darkness, however, the growth level of kinetin and non-kinetin cultures is about the same, while 5 min R every 24 h in the presence of kinetin maintains a much higher growth rate (Rombach 1971). Therefore, the effect of kinetin on growth in darkness is not substitutive for, but synergistic with that of short P_{fr} treatment. Growth of sterile cultures of *Spirodela oligorrhiza* can also be stimulated in the dark either with cytokinins or intermittent red light (McCombs and Ralph 1972).

BA pretreatment of excised cotyledons of etiolated cucumber stimulated chlorophyll formation during subsequent illumination, but the hormone seems to act independently of R (Dei and Tsuji 1978). In cotyledons of intact mustard seedlings only P_{fr} is active, while kinetin has no significant effect on chlorophyll accumulation. When excised cotyledons are used, both P_{fr} and kinetin contribute to the elimination of the lag phase of chlorophyll production. When the two factors are applied simultaneously, their effects are always additive (Ford et al. 1981). The authors conclude that the enhanced chlorophyll production by kinetin application cannot be considered as evidence that P_{fr} action is mediated through changes in endogenous cytokinin activity. Analogous results are obtained by Kochhar et al. (1981) studying the action of light and kinetin on betalain synthesis in *Amaranthus* seedlings. The two factors act additively irrespective of what factor is varied. However, it seems that a pretreatment with kinetin increases the effectiveness of P_{fr}.

Cytokinin levels of mature poplar leaves are rapidly influenced by short R. When detached green leaves are exposed to 5 min R at the end of the night, there is a transient increase of butanol-soluble cytokinins after 30 min of darkness. The response is dependent on the physiological conditions of the leaves. The cytokinin content shows diurnal changes with a pronounced peak of activity at daybreak (Hewett and Wareing 1973). During the main photoperiod the leaves contain a complex of several cytokinin-active fractions decreasing to low levels in the night. Exposed to 5-min R and harvested after 10 min of darkness the leaves contain a single main peak of cytokinin activity, which was identified as o-hydroxybenzyl adenosine (HBA) (Jorgan et al. 1973). The levels of HBA in leaves of plants growing in the field are low during the night, they rise rapidly at dawn and then they decline in a manner similar to the changes observed following a period of 5 min R. Later, during the day the decline is followed by a slow rise in HBA (Thompson et al. 1975). Wareing and Thompson (1976) also demonstrate R/FR reversibility and they find that the escape time for reversibility is between 5 to 10 min. As is the case for gibberellins (Cooke et al. 1975), chloroplast preparations contain significant amounts of HBA, but neither R nor sonication has any effect upon the distribution of this compound between chloroplast and ambient medium. Thus the data reported for gibberellins in etioplasts are not concomitant with HBA present in green chloroplasts.

These changes of endogenous cytokinin levels seem not to be correlated yet with any growth response of the mature leaf. Of course, the site of hormone synthesis can be spatially different from the site of physiological action. From the systems listed above it seems that cytokinin can promote division and expansion in immature leaves and induce dormant lateral bud outgrowth, indicating that the acquisition of cytokinin sensitivity precedes the onset of cell division. For root formation auxin may have an equivalent position since it can induce this process in young and old shoot cells.

3.5 Is Xanthoxin More Involved in Phytochrome-Mediated Growth Inhibition than ABA?

Dwarfism and light-induced stem extension inhibition have been considered as a consequence of changed ABA levels in plant tissues. R-treated dwarf peas contain high levels of an unknown inhibitor of stem extension (KÖHLER and LANG 1963), but no significant differences in ABA content of light- or dark-grown pea shoots could be found (KENDE and KAYS 1971, BARNES 1972, DÖRFFLING 1973).

Leaf blades of dwarf and tall rice plants do not contain different ABA levels. However, extracts made from the actively growing parts of dwarfs (stem apex and basal parts of leaves) contain up to 50% more ABA than those from tall plants. Dwarf and tall plants respond differently to exogenous ABA (TIETZ 1979). In *Spinacea* leaves (ZEEVAART 1974) and *Betula* seedlings (LOVEYS et al. 1974) the ABA content seems to be controlled by the length of the photoperiod, but there is no apparent relationship between measured ABA levels and photoperiodic control of growth. The phytochrome-mediated unrolling response of barley leaf segments in light can be inhibited by ABA (POULSON and BEEVERS 1970). Examining the effect of exogenous ABA on the development of mustard seedlings grown in the presence and absence of P_{fr}, BAJRACHARYA et al. (1975) conclude that their results do not support the hypothesis that ABA interacts with phytochrome-mediated morphogenesis.

When the levels of both ABA and xanthoxin, another potent growth inhibitor, in etiolated and R-treated dwarf pea seedlings are measured, ABA concentrations are the same, but R-treated material contains five times as much xanthoxin as dark-grown material (on a dry weight basis). Concomitantly, lutein and violaxanthin, two major xanthophylls in pea seedlings, increase markedly after R treatment. It is suggested that the R effect may involve enhancement of xanthoxin levels by increasing the availability of the precursor violaxanthin (BURDEN et al. 1971). The results on xanthoxin levels are confirmed by ANSTIS et al. (1975). They also demonstrate that the R induced xanthoxin increase is paralleled by a decrease in growth rate. Since xanthoxin and ABA are chemically related, DÖRFFLING (1978) suggests that a light-induced transformation could occur between both inhibitors.

Table 3. Data in favor or against interactions between light and ethylene

Plant species	Response	Treatment	Effect on		Inter-action	References
			Response	Ethylene production		
Pisum sativum	Plumular expansion	Ethylene	–		A	Goeschl et al. (1967)
		R	+	–		
Phaseolus vulgaris	Hook opening	R	+	–		Kang et al. (1967)
		R + ethylene	–		A	Kang and Ray (1969)
Pisum sativum	Hook development	Hypobaric	–		A	Kang and Burg (1972a, b, c)
		CO_2	–		A	
		dark + ethylene	+		A	
	Geotropic bending of subapical stem sections	R	+	–		
		ethylene	–		A	
		hypobaric	+			
	Carotenogenesis	R	+			
		R + ethylene	–			
		dark + hypobaric				
Lactuca sativa	Hook closure	R	+	No effect	B	Janes et al. (1976)
		R + hypobaric	+		B	
	Hypocotyl elongation	Ethylene	–			
		R	No effect		B	
Oryza sativa	Coleoptile growth (intact)	Etiolated + ethylene	+			Suge et al. (1971)
		R + ethylene	+		B	Imaseki et al. (1971)
	(apical segment)	R	+	–		
		R + ethylene	No effect		A	
		etiolated + ethylene	No effect			
Glycine max	Growth of apical hypocotyl part and prevention of lateral swelling	Etiolated + R	+	–	A	Samimy (1978)

Sorghum vulgare		Etiolated + FR		+	A	Craker et al. (1973)
		etiolated + B		+		
Sinapis alba	Primordium formation of adventitious roots	Ethylene	No effect			Pfaff and Schopfer (1980)
		dark + FR	+		A/B	
		ethylene removal	+			
Sorghum vulgare	Anthocyanin synthesis	Light	+		A	Craker and Wetherbee (1973)
		light + ethylene	−			
Brassica oleracea	Anthocyanin synthesis	R	+			Kang and Burg (1973)
		R + ethylene	−			
		IAA	−	+	A	
Sinapis alba	Anthocyanin synthesis	P_{fr}	+			Bühler et al. (1978a, b)
		P_{fr} + ethylene	Partial inhibition		B	
	Enzyme levels and chlorophyll synthesis	P_{fr}	+			
		P_{fr} + ethylene	Partial or no inhibition			

− : inhibition; + : promotion; A: in favor; B: against

From these data it is impossible to discern any meaningful correlation between light-induced changes of ABA content and light control of growth. Whether phytochrome is involved in changing levels or metabolism of growth-inhibiting substances (ABA or related compounds) remains an open question. Neither is there any evidence that phytochrome modified the sensitivity of tissues to ABA.

3.6 Ethylene

From basic observations in several studies suggestions have been made that light-mediated changes of endogenous ethylene production might be correlated with photomorphogenic responses. Light-grown pea seedlings treated with ethylene show a reduced rate of leaf expansion similar to that of etiolated seedlings. Brief R exposure of these etiolated seedlings induces plumular expansion and a corresponding transient decrease of ethylene production (Goeschl et al. 1967). Both effects are FR-reversible to the level achieved by FR alone. Kang et al. (1967) have found that the R-induced opening of the hypocotyl bean hook is accompanied with a decreased ethylene production.

Does ethylene play a mediator role in the causal sequence between phytochrome and the photoresponse?

In Table 3 a number of papers concerning this problem are summarized. When kinetic measurements in an open flow system are performed to examine light-induced ethylene production related to photomorphogenic responses, serious difficulties arise with regard to the current notion that ethylene acts as a trigger and as a controlling factor. Experiments with green *Marchantia* thalli indicate that the ethylene release is dramatically influenced when basic metabolic activities are altered by changing either O_2 or CO_2 concentrations in the ambient atmosphere, by varying the light intensity during the photoperiod and by administering inhibitors such as DNP and DCMU (De Greef et al. 1980, 1981). From these studies it is concluded that energy-producing systems, namely respiration and photosynthesis, are closely related to the ethylene release of the thalli.

Growth and the orientation of thalli are modified when a short FR exposure is given at the end of a W photoperiod. Since this effect is fully reversed by R, phytochrome involvement is proven (Fredericq and De Greef 1966). Terminal FR drastically changes the pattern of ethylene release in the subsequent photoperiod (Veroustraete et al. 1982). In the presence of CO_2 the ethylene production is inhibited by FR, whereas it is stimulated in the absence of CO_2. Both FR effects are R reversible. The fluence response curves for the stimulation of the CO_2-independent component, the elimination of epinasty (Fig. 3), and those for R reversal show the same saturation levels. When dark intervals of increasing length are inserted between the end of the photoperiod and the FR treatment, there is an immediate escape of the phytochrome control of epinasty, but the related stimulation of ethylene production lasts for several hours (Fig. 4). The escape curves show that FR-stimulated ethylene production rapidly disappears when the morphological expression of the phytochrome treatment van-

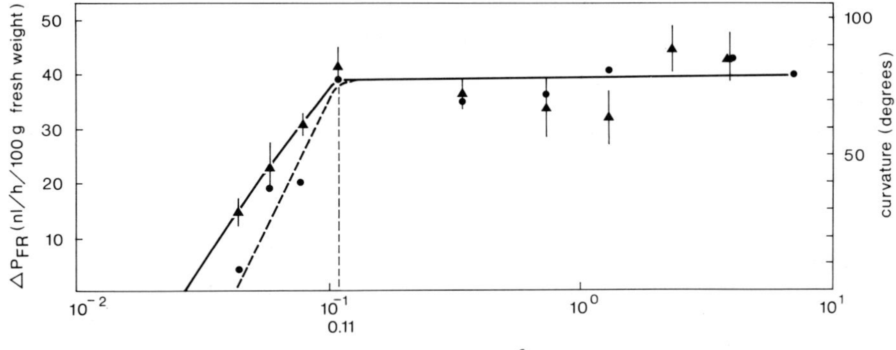

Fig. 3. Fluence-response curves of the stimulation of light-induced ethylene production (●) and the elimination of epinasty (▲) in CO_2-free air, when FR is given at the end of the preceding photoperiod. (VEROUSTRAETE et al. 1982)

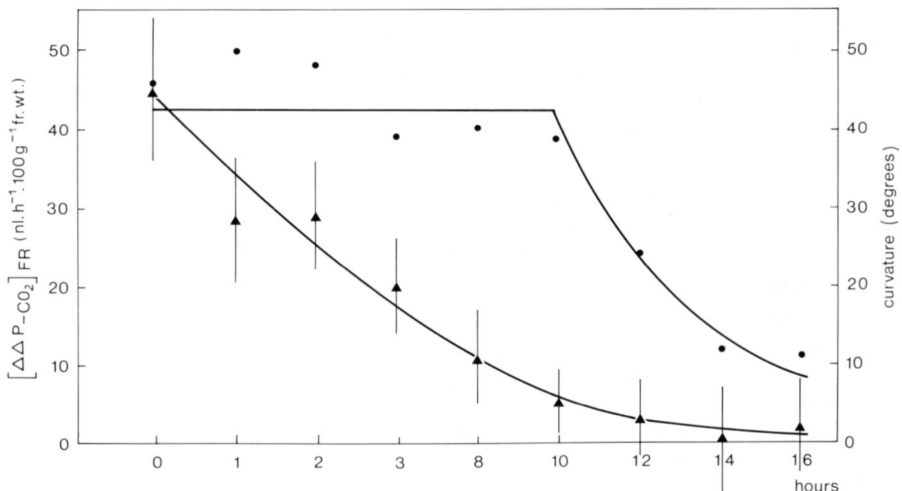

Fig. 4. Escape from reversibility of FR-induced ethylene stimulation by the CO_2-independent component ($[\Delta\Delta P_{-CO_2}]_{FR}$, ●) and elimination of epinasty (degrees curvature of thalli in CO_2-free air, ▲). (VEROUSTRAETE et al. 1982)

ishes. On the other hand, exogenous ethylene cannot substitute for the light effect.

Since the phytochrome induced change of thallus morphology is accompanied with increased growth vigor and changed metabolic activity, the CO_2-independent increase of ethylene release reflects the modifications of both metabolism and growth. This places ethylene production quite definitely as a concomitant of the changed growth pattern rather than as a precedent. Hence ethylene evolution is an excellent indicator of changed metabolic activities induced by phytochrome through the environmental stimulus (light) and results in a modified growth pattern. In this case no hormonal action can be attributed to endogenous ethylene, which is merely a by-product of metabolic activity.

4 Concluding Remarks

Despite massive research efforts and the collection of many data dealing with the effects of phytochrome and hormones on different tissues throughout the plant's life cycle, there have been no breakthroughs in the quest for understanding the action and/or interaction of these modulator molecules. Scientists who are familiar with the current state of this research field in plant physiology, will be fully aware of a lot of confusion, puzzling interactions and apparent uncertainties concerning supposedly established facts.

It is the author's conviction that many inconsistencies of experimental data have to be imputed to the incomparability of the methods and the plant materials (physiological age and developmental evolution of sensitivity to modulator molecules) used. A large body of experiments concerning the linkage between light effects on growth and the role of growth substances has no decisive value. Many statements are only superficially descriptive and have no mechanistic relevance. In many cases separate mechanisms of action related to different competences of the tissues are compared. This state of affairs is in sharp contrast to what has been achieved in the equivalent area of animal physiology. Studies on the mechanism of steroid hormone action continue to make headlines with fundamental discoveries in receptor site and gene regulation.

All of us are convinced that growth, differentiation and morphogenesis are regulated by very fine mechanisms of sequential adjustments to obtain an orderly arranged expression of the intrinsic developmental program in the successive phases of the life cycle of lower and higher plants. It is easy to say that plant development involves a continuous re-routing of metabolic pathways under the influence of environmental and endogenous cues, but is seems very difficult to come to grips with the problem of what these changes are.

So, what are the barriers to achieving knowledge of the role of modulator molecules in plant photomorphogenesis?

If developmental events are controlled by changes in growth substance concentrations and if these changes are influenced in their turn by environmental cues such as light for example, then correlated variations in growth substance level must occur, commensurate with the onset of developmental stages in the intact plant. However, in a recent encyclopedia work GOODWIN et al. (1978) established from literature data that "there is generally a poor correlation between growth rate and endogenous phytohormone levels in developing roots, stems, leaves, cambial tissues and vegetative food storage organs".

There is a triangle of certainties for each phenomenal event in plant morphogenesis: environmental signal, developmental stage, and physiological response (Fig. 5). In between environment and developmental stage the mediating role of sensor molecules and hormones is still a matter of debate. At the level of

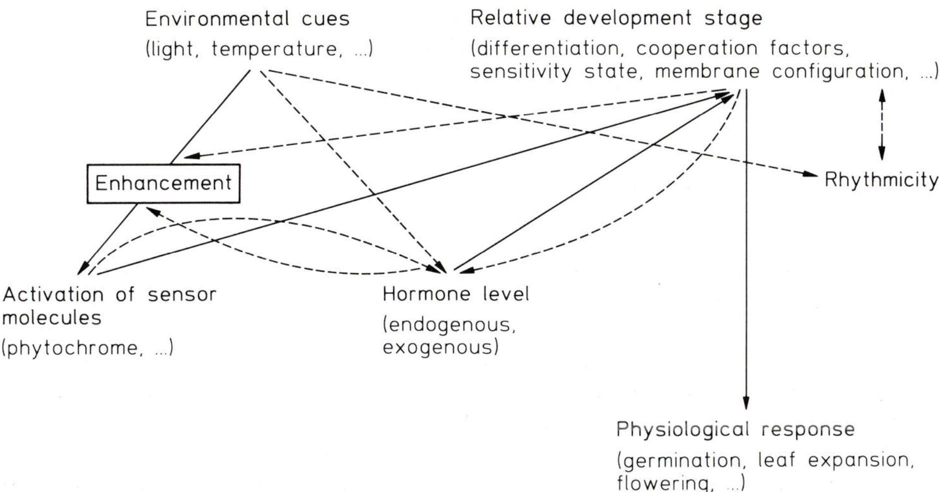

Fig. 5. Certainties (——) and uncertainties (– – –) concerning phytochrome and hormone interactions in photomorphogenesis

possible interactions there are many uncertainties. A great deal of our ignorance of interactions is presented in Fig. 5 by dashed lines. There are observations that light through daylength or phytochrome may change hormone concentrations or balances between hormones or between different forms of the same hormone, but there are no direct correlations made with responsiveness at the levels of sensitivity and morphogenesis. There are observations that hormone levels may vary with the developmental stage of the plant without any obvious change in environmental signal, but it is not clear what kind of specificity can be attributed to the hormonal activity for integrated plant growth. Light can increase the sensitivity of tissues to hormones, and inversely, hormone application may enhance the sensitivity to light action, while a low P_{fr} content may abolish the hormone action in other cases. P_{fr} and hormones often appear to speed up reaction rates, on the one hand, and to synchronize mutually dependent processes, on the other. This leads to the unsolved problem of the fate of applied hormones and the multiplicity of the effects of hormones and phytochrome. What is the real meaning of interaction, alleviation, substitution, additive and synergistic effects of exogenous hormones with regard to phytochrome-mediated responses in the molecular chain of events? In studies of the mechanisms of phytohormone action various kinds of inhibitors have been used. In most cases hormone-induced growth is more or less unspecifically suppressed by any kind of inhibitor. When the interactions between light and levels of endogenous hormones are examined, the light action often involves short- and long-term effects, suggesting a continuity of light control throughout the completion of the photomorphogenic response. The observed light effects on changes in growth substance concentrations may be the result of conversion between different forms, biosynthesis, transport, the number of binding sites, etc. There are also many indications that phytochrome exerts its effect on membranes and/or membrane-associated structures. These alterations may intrinsically affect the sensitivity of the target tissues to endogenous hormone action.

Are there any particular areas of underdevelopment in photomorphogenic and hormonal research?

In a perfect model of plant developmental processes all factors concerned should have to be taken into account to explain the observed physiological response. Hence it is of crucial importance to verify at all times whether the observed changes of hormone levels are functionally and quantitatively related to the ultimate growth response.

Such a multi-factor analysis of any growth phenomenon is impossible at the present. Most methods are destructive or unphysiological since they are based either on extraction or accumulation. In this way kinetic studies in the same individual are excluded. Within the pool of the same extract it is very difficult to make any distinction between the free and conjugated forms with regard to their physiological action. Great achievements in identification and quantification of modulator molecules and in the definition of their function are not possible without an appropriate, analytical instrumentation of high performance allowing kinetic studies of endogenous hormone levels. At the present such analytical tools are available to determine accurately ABA and ethylene at physiological levels. For auxins, gibberellins and cytokinins there is a varying degree of reliability. Only for ethylene can continuous and quantitative studies be performed on the same plant material. Radio-immunochemical assays seem to offer a new outcome for the localization and estimation of the phytochrome molecules at the cellular level.

As long as there is no insight into the complexity of the primary actions of phytochrome and hormones at the subcellular level, many missing links will remain in the overall reaction chain leading from initiation to expression of a photomorphogenic response.

References

Acton GJ, Murray PB (1974) The roles of auxin and gibberellin in reversing radiation inhibition of hypocotyl lengthening. Planta 117:219–226

Anstis PJP, Friend J, Gardner DCJ (1975) The role of xanthoxin in the inhibition of pea seedling growth by red light. Phytochemistry 14:31–35

Bajracharya D, Tong W-F, Plachy C, Schopfer P (1975) On the role of abscisic acid in phytochrome-mediated photomorphogenesis. Biochem Physiol Pflanz 168:421–432

Barnes MF (1972) Abscisic acid in pea shoots. Planta 104:182–184

Beevers L, Loveys B, Pearson JA, Wareing PF (1970) Phytochrome and hormonal control of expansion and greening of etiolated wheat leaves. Planta 90:286–294

Bewley JD (1980) Secondary dormancy (skotodormancy) in seeds of lettuce (*Lactuca sativa* cv. Grand Rapids) and its release by light, gibberellic acid and benzyladenine. Physiol Plant 49:277–280

Bewley JD, Fountain DW (1972) A distinction between the actions of abscisic acid, gibberellic acid and cytokinins in light-sensitive lettuce seed. Planta 102:368–371

Bewley JD, Halmer P (1980) Embryo-endosperm interactions in the hydrolysis of lettuce seed reserves. Isr J Bot 29:118–132

Bewley JD, Negbi M, Black M (1968) Immediate phytochrome action in lettuce seeds and its interaction with gibberellins and other germination promotors. Planta 78:351–357

Blaauw-Jansen G (1959) The influence of red and far-red light on growth and phototropism of the *Avena* seedling. Acta Bot Neerl 8:1–39

Black M, Bewley JD, Fountain D (1974) Lettuce seed germination and cytokinins: their entry and formation. Planta 117:145–152

Brian PW (1958) Role of gibberellin-like hormones in regulation of plant growth and flowering. Nature 181:1122–1123

Browning G, Saunders PF (1977) Membrane localized gibberellins A_9 and A_4 in wheat chloroplasts. Nature 265:375–377

Bühler B, Drumm H, Mohr H (1978a) Investigations on the role of ethylene in phytochrome-mediated photomorphogenesis. I. Anthocyanin synthesis. Planta 142:109–117

Bühler B, Drumm H, Mohr H (1978b) Investigations on the role of ethylene in phytochrome-mediated photomorphogenesis. II. Enzyme levels and chlorophyll synthesis. Planta 142:119–122

Bünsow R, von Bredow K (1958) Wirkung von Licht und Gibberellin auf die Samenkeimung der Kurztagpflanze *Kalanchoë blossfeldiana.* Biol Zentr 132–141

Burden RS, Firn RD, Hiron RWP, Taylor HF, Wright STC (1971) Induction of plant growth inhibitor xanthoxin in seedlings by red light. Nature New Biol 234:95–96

Burdett AN (1972) Two effects of prolonged far-red light on the response of lettuce seeds to exogenous gibberellin. Plant Physiol 49:531–534

Carpita NC, Nabors MW (1981) Growth physics and water relations of red light-induced germination in lettuce seeds. V. Promotion of elongation in the embryonic axes by gibberellins and phytochrome. Planta 152:131–136

Cooke RJ, Kendrick RE (1976) Phytochrome-controlled gibberellin metabolism in etioplast envelopes. Planta 131:303–307

Cooke RJ, Saunders PF (1975a) Phytochrome-mediated changes in extractable gibberellin activity in a cell-free system from etiolated wheat leaves. Planta 123:299–302

Cooke RJ, Saunders PF (1975b) Photocontrol of gibberellin levels as related to the unrolling of etiolated wheat leaves. Planta 126:151–160

Cooke RJ, Saunders PF, Kendrick RE (1975) Red-light-induced production of gibberellin-like substances in homogenates of etiolated wheat leaves and in suspensions of intact etioplasts. Planta 124:319–328

Craker LE, Wetherbee PJ (1973) Ethylene, light and anthocyanin synthesis. Plant Physiol 51:436–438

Craker LE, Abeles FB, Shropshire W (1973) Light-induced ethylene production in *Sorghum.* Plant Physiol 51:1082–1083

Dedonder A, Rethy R, De Petter E, Fredericq H, De Greef J (1980) Preliminary screening experiments on the effects of light and GA_3 on the germination of different seed species. In: De Greef J (ed) Photoreceptors and plant development. Antwerpen Univ Press, Antwerpen, pp 431–435

De Greef J, Veroustraete F, Fredericq H, Van Wiemeersch L (1980) Study on the interaction of light and limiting physiological factors on the ethylene production by green *Marchantia polymorpha* thalli. In: De Greef J (ed) Photoreceptors and plant development. Antwerpen Univ Press, Antwerpen, pp 423–429

De Greef J, De Proft M, Veroustraete F, Fredericq H (1981) Case studies of ethylene release in higher and lower plant systems. In: Jeffcoat B (ed) Aspects and prospects of plant growth regulators, Monograph 6. Brit Plant Growth Regulator Group, Wantage, pp 9–18

Dei M, Tsuji H (1978) The influence of various plant hormones on the stimulatory action of red light on greening in excised etiolated cotyledons of cucumber. Plant Cell Physiol 19(8):1407–1414

Dörffling K (1973) Die Regulation des Internodienwachstums von Erbsenkeimlingen durch Licht und Abscisinsäure. Z Pflanzenphysiol 70:131–137

Dörffling K (1978) The possible role of xanthoxin in plant growth and development. Phil Trans R Soc London B 284:499–507

Drumm H, Elchinger I, Möller J, Peter K, Mohr H (1971) Induction of amylase in mustard seedlings by phytochrome. Planta 99:265–274

Dunlap JR, Morgan PW (1977a) Reversal of induced dormancy in lettuce by ethylene, kinetin, and gibberellic acid. Plant Physiol 60:222–224

Dunlap JR, Morgan PW (1977b) Characterization of ethylene/gibberellic acid control of germination in *Lactuca sativa* L. Plant Cell Physiol 18:561–568

Eldabh R, Fredericq H, Maton J, De Greef J (1974) Photophysiology of *Kalanchoë* seed germination. I. Interrelationship between photoperiod and terminal far-red light. Physiol Plant 30:185–191

Evans A, Smith H (1976a) Localization of phytochrome in etioplasts and its regulation in vitro of gibberellin levels. Proc Natl Acad Sci USA 73:138–142

Evans A, Smith H (1976b) Spectrophotometric evidence for the presence of phytochrome in the envelope membranes of barley etioplasts. Nature 259:323–325

Evenari M (1965) Light and seed dormancy. In: Ruhland W (ed) Encyclopedia of plant physiology Vol XV/2. Springer, Berlin Göttingen Heidelberg, pp 804–847

Fletcher RA, Zalik S (1964) Effect of light quality on growth and free indoleacetic acid content in *Phaseolus vulgaris*. Plant Physiol 39:328–331

Fletcher RA, Zalik S (1965) Effect of light of several spectral bands on the metabolism of radioactive IAA in bean seedlings. Plant Physiol 40:549–552

Ford MJ, Kasemir H, Mohr H (1981) The influence of phytochrome and kinetin on chlorophyll accumulation in mustard cotyledons: a two-factor analysis. Ber Dtsch Bot Ges 94:35–41

Fredericq H, De Greef J (1966) Red (R), far-red (FR) photoreversible control of growth and chlorophyll content in light-grown thalli of *Marchantia polymorpha* L. Naturwissenschaften 53:337

Fredericq H, De Greef J, Rethy R (1976) Kinetic analysis of the synergism between gibberellic acid and the physiologically active form of phytochrome. Arch Int Physiol Biochim 84:160–162

Fredericq H, Rethy R, Dedonder A, De Greef J, De Petter E (1980) Photocontrol of *Kalanchoë blossfeldiana* seed germination. In: De Greef J (ed) Photoreceptors and plant development. Antwerpen Univ Press, Antwerpen, pp 367–374

Furuya M, Torray JG (1964) The reversible inhibition by red and far-red light of auxin-induced lateral root initiation in isolated pea roots. Plant Physiol 39:987–991

Goeschl JD, Pratt HK, Bonner BA (1967) An effect of light on the production of ethylene and the growth of the plumular portion of etiolated pea seedlings. Plant Physiol 42:1077–1080

Goodwin PB, Gollinow BI, Letham DS (1978) Phytohormones and growth correlations. In: Letham DS, Goodwin PB, Higgins TJ (eds) Phytohormones and related compounds – a comprehensive treatise Vol II. Elsevier/North-Holland Biomedical Press, Amsterdam, pp 215–249

Gotô N, Esashi Y (1974) Differential hormone responses in different growing zones of the hypocotyl. Planta 116:225–241

Gotô N, Esashi Y (1976) Aging progression involving dwarfism and its acceleration by red light in bean hypocotyls. Plant Physiol 57:547–552

Gotô N, Suzuki M (1980) Far-red light control of elongation in light-grown bean hypocotyl: effect on different age zones and interaction with indole-3-acetic acid. Plant Cell Physiol 21:105–113

Graebe JE, Ropers HJ (1978) Gibberellins. In: Letham DS, Goodwin PB, Higgins TJ (eds) Phytohormones and related compounds – a comprehensive treatise Vol I. Elsevier/North-Holland Biomedical Press, Amsterdam, pp 107–204

Hewett EW, Wareing PF (1973) Cytokinins in *Populus* x *robusta* (Schneid): Light effects on endogenous levels. Planta 114:119–129

Hillman WS (1957) Nonphotosynthetic light requirement in *Lemna minor* and its partial satisfaction by kinetin. Science 126:165–166

Hillman WS (1959) Interactions of growth substances and photoperiodically active radiations on the growth of pea internode sections. In: Withrow RB (ed) Photoperiodism and related phenomena in plant and animals. Am Assoc Adv Sci, Washington, pp 181–196

Hilton JR, Smith H (1980) The presence of phytochrome in purified barley etioplasts and its in vitro regulation of biologically-active gibberellin levels in etioplasts. Planta 148:312–318

Horgan R, Hewett EW, Purse J, Wareing PF (1973) A new cytokinin of *Populus robusta*. Tetrahedron 30:2827–2828

Humphries EC (1961) Effects of quality of light on development of roots on dwarf bean hypocotyls in presence and absence of boron. Nature 190:701–703

Ikuma H, Thimann KV (1963) Action of kinetin on photosensitive germination of lettuce seed as compared with that of gibberellic acid. Plant Cell Physiol 4:113–128

Imaseki H, Pjon CJ, Furuya M (1971) Phytochrome action in *Oryza sativa* L. IV. Red and far-red reversible effect on the production of ethylene in excised coleoptiles. Plant Physiol 48:241–244

Jacobsen JV, Pressman E (1979) A structural study of germination in celery (*Apium graveolens* L.) seed with emphasis on endosperm breakdown. Planta 144:241–248

Janes HW, Loercher L, Frenkel C (1976) Effects of red light and ethylene on growth of etiolated lettuce seedlings. Plant Physiol 57:420–423

Kahn A (1960) Promotion of lettuce seed germination by gibberellin. Plant Physiol 35:333–339

Kang BG, Burg SP (1972a) Ethylene as a natural agent inducing plumular hook formation in pea seedlings. Planta 104:275–281

Kang BG, Burg SP (1972b) Relation of phytochrome-enhanced geotropic sensitivity to ethylene production. Plant Physiol 50:132–135

Kang BG, Burg SP (1972c) Involvement of ethylene in phytochrome-mediated carotenoid synthesis. Plant Physiol 49:631–633

Kang BG, Burg SP (1973) Role of ethylene in phytochrome-induced anthocyanin synthesis. Planta 110:227–235

Kang BG, Ray PM (1969) Ethylene and carbon dioxide as mediators in the response of the bean hypocotyl hook to light and auxins. Planta 87:206–216

Kang BG, Yocum CS, Burg SP, Ray PM (1967) Ethylene and carbon dioxide: mediation of hypocotyl hook opening response. Science 156:958–959

Karssen CM (1976) Two sites of hormonal action during germination of *Chenopodium album* seeds. Physiol Plant 36:264–270

Kende H, Kays SE (1971) The level of (+)-abscisic acid in dwarf pea shoots. Naturwissenschaften 58:524–525

Khan AA (1968) Inhibition of gibberellic acid-induced germination by abscisic acid and reversal by cytokinins. Plant Physiol 43:1463–1465

Khan AA (1971) Cytokinins: permissive role in seed germination. Science 171:853–859

Khan AA, Karssen CM (1980) Induction of secondary dormancy in *Chenopodium bonus-henricus* L. seeds by osmotic and high temperature treatments and its prevention by light and growth regulators. Plant Physiol 66:175–181

Kochhar VK, Kochhar S, Mohr H (1981) Action of light and kinetin on betalain synthesis in seedlings of *Amaranthus caudatus*: a two-factor analysis. Ber Dtsch Bot Ges 94:27–34

Köhler D, Lang A (1963) Evidence for substances in higher plants interfering with gibberellin responses. Plant Physiol 37:555–560

Leung DWM, Bewley JD (1981) Red-light and gibberellic-acid-enhanced α-galactosidase activity in germinating lettuce seeds, cv. Grand Rapids. Control by the axis. Planta 152:436–441

Lewak S, Khan AA (1977) Mode of action of gibberellic acid and light on lettuce seed germination. Plant Physiol 60(4):575–577

Liverman JL, Bonner J (1953) The interaction of auxin and light in the growth responses of plants. Proc Natl Acad Sci USA 39:905–916

Lockhart JA (1956) Reversal of the light inhibition of pea stem growth by the gibberellins. Proc Natl Acad Sci USA 42:841–848

Loveys BR, Wareing PF (1971) The red-light-controlled production of gibberellin in etiolated wheat leaves. Planta 98:109–116

Loveys BR, Leopold AC, Kriedemann PE (1974) Abscisic acid metabolism and stomatal physiology in *Betula lutea* following alteration in photoperiod. Ann Bot 38:85–92

McCombs PJA, Ralph RK (1972) Cytokinin control of growth of *Spirodela oligorrhiza* in darkness. Planta 107:97–109
Miller CO (1956) Similarity of some kinetin and red light effects. Plant Physiol 31:318–319
Muir RM (1970) The control of growth by the synthesis of IAA and its conjugates. In: Carr DJ (ed) Plant growth substances. Springer, Berlin Heidelberg New York, pp 96–101
Muir RM, Chang KC (1974) Effect of red light on coleoptile growth. Plant Physiol 54:286–288
Nabors MW, Lang A (1971) The growth physics and water relations of red-light-induced germination in lettuce seeds. I. Embryos germinating in osmoticum. Planta 101:1–25
Nagao M, Esashi Y, Tanaka T, Kumagai T, Fukumoto S (1959) Effects of photoperiod and gibberellin on the germination of seeds of *Begonia evansiana* Andr. Plant Cell Physiol 1:39–47
Nakayama S, Tobita H, Okumura FS (1962) Antagonism of kinetin and far-red light or β-indoleacetic acid in flowering of *Pharbitis* seedlings. Phyton 19:43–48
Negm FB, Smith OE, Kumamoto J (1973) The role of phytochrome in an interaction with ethylene and carbon dioxide in overcoming lettuce seed thermodormancy. Plant Physiol 51:1089–1094
Ogawa Y, King RW (1979) Establishment of photoperiodic sensitivity by benzyladenine and a brief red irradiation in dark-grown seedlings of *Pharbitis nil* Chois. Plant Cell Physiol 20(1):115–122
Pfaff W, Schopfer P (1974) Phytochrome-induzierte Regeneration von Adventivwurzeln beim Senfkeimling (*Sinapis alba* L.). Planta 117:269–278
Pfaff W, Schopfer P (1980) Hormones are no causal links in phytochrome-mediated adventitious root formation in mustard seedlings (*Sinapis alba* L.). Planta 150:321–329
Poulson R, Beevers L (1970) Effects of light and growth regulators on leaf unrolling in barley. Plant Physiol 46:509–514
Rao VS, Sankhla N, Khan AA (1975) Additive and synergistic effects of kinetin and ethrel on germination, thermodormancy, and polyribosome formation in lettuce seeds. Plant Physiol 56:263–266
Reid DM, Clements JB (1968) RNA and protein synthesis: prerequisites of red-light-induced gibberellin synthesis. Nature 219:607–609
Reid DM, Clements JB, Carr DJ (1968) Red light induction of gibberellin synthesis in leaves. Nature 217:580–582
Reid DM, Tuing MS, Durley RC, Railton ID (1972) Red-light-enhanced conversion of tritiated gibberellin A_9 into other gibberellin-like substances in homogenates of etiolated barley leaves. Planta 108:67–75
Rethy R, Fredericq H, De Greef J, Van Onckelen H, Maton J (1976) Long-lasting light effects in secondary dormant seeds of *Kalanchoë blossfeldiana* (cv. Feuerblüte) treated with gibberellic acid (GA_3). Arch Int Physiol Biochim 84:1102–1104
Reynolds T, Thompson PA (1973) Effects of kinetin, gibberellins and ($\pm$) abscisic acid on the germination of lettuce (*Lactuca sativa*). Physiol Plant 28:516–522
Rombach J (1971) On the interaction of kinetin and phytochrome in *Lemna minor* growing in the dark. Acta Bot Neerl 20:636–645
Rubinstein B (1971) Auxin and red light in the control of hypocotyl hook opening in beans. Plant Physiol 48:187–192
Russell DW, Galston AW (1969) Blockage by gibberellic acid of phytochrome effects on growth, auxin responses, and flavonoid synthesis in etiolated pea internodes. Plant Physiol 44:1211–1216
Samimy C (1978) Effect of light on ethylene production and hypocotyl growth of soybean seedlings. Plant Physiol 61:772–774
Scott RA Jr, Liverman JL (1956) Promotion of leaf expansion by kinetin and benzylaminopurine. Plant Physiol 31:321–322
Scott RA, Liverman JL (1957) Control of etiolated bean leaf disc expansion by gibberellins and adenine. Science 126:122–124
Sherwin JE, Furuya M (1973) A red-far-red reversible effect on uptake of exogenous indoleacetic acid in etiolated rice coleoptiles. Plant Physiol 51:295–298

Smolenska G, Lewak S (1974) The role of lipases in the germination of dormant apple embryos. Planta 116:361–370
Speer HL, Hsiao AI, Vidaver W (1974) Effects of germination-promoting substances given in conjunction with red light on the phytochrome-mediated germination of dormant lettuce seeds (*Lactuca sativa* L.). Plant Physiol 54:852–854
Stoddart JL (1968) The association of gibberellin-like activity with the chloroplast fraction of leaf homogenates. Planta 81:106–112
Suge H, Katsura N, Inada K (1971) Ethylene–light relationship in the growth of the rice coleoptile. Planta 101:365–368
Takeba G, Matsubara S (1977) Rapid disappearance of small fat bodies during the early stage of imbibition of lettuce seeds. Plant Cell Physiol 18:1067–1075
Taylor JS, Wareing PF (1979) The effect of light on the endogenous levels of cytokinins and gibberellins in seeds of sitka spruce (*Picea sitchensis* Carriere). Plant Cell Environ 2:173–179
Taylorson RB, Hendricks SB (1976) Interactions of phytochrome and exogenous gibberellic acid on germination of *Lamium amplexicaule* L. seeds. Planta 132:65–70
Thomas TH, Biddington NL, Palevitch D (1978) The role of cytokinins in the phytochrome-mediated germination of dormant imbibed celery (*Apium graveolens*) seeds. Photochem Photobiol 27(2):231–236
Thompson PA (1970) The characterization of the germination response to temperature of species and ecotypes. Nature 225:827–831
Thompson AG, Horgan R, Heald JK (1975) A quantitative analysis of cytokinin using ion-current monitoring. Planta 124:207–210
Tietz A (1979) Zwergwuchs und Abscisinsäure. Biochem Physiol Pflanz 174:499–503
Tucker DJ (1976) Effects of far-red light on the hormonal control of side shoot growth in the tomato. Ann Bot 40:1033–1042
Tucker DJ (1977) Hormonal regulation of lateral bud outgrowth in the tomato. Plant Sci Lett 8:105–111
Tucker DJ, Mansfield TA (1972) Effects of light quality on apical dominance in *Xanthium strumarium* and the associated changes in endogenous levels of abscisic acid and cytokinins. Planta 102:140–151
Van Staden J (1973) Changes in endogenous cytokinins of lettuce seed during germination. Physiol Plant 28:222–227
Van Staden J, Wareing PF (1972) The effect of light on endogenous cytokinin levels in seeds of *Rumex obtusifolius*. Planta 104:126–133
Van Staden J, Olatoye ST, Hall MA (1973) Effect of light and ethylene upon cytokinin levels in seeds of *Spergula arvensis*. J Exp Bot 24:662–666
Veroustraete F, Fredericq H, van Wiemeersch L, De Greef J (1982) Specific photoregulation by phytochrome of the epinasty and the light-induced ethylene production in *Marchantia polymorpha*. Photochem Photobiol 35:261–264
Vidaver W, Hsiao AI (1974) Actions of gibberellic acid and phytochrome on the germination of Grand Rapids lettuce seeds. Plant Physiol 53:266–268
Wareing PF, Thompson AG (1976) Rapid effects of red light on hormone levels. In: Smith H (ed) Light and plant development. Butterworth, London, pp 285–294
White JC, Mansfield TA (1978) Correlative inhibition of lateral bud growth in *Phaseolus vulgaris* L. – Influence of the environment. Ann Bot 42:191–196
Woodkstock LW, Toole VK (1976) Respiration of Grand Rapids lettuce (*Lactuca sativa* L.) seeds in relation to chemical and photocontrol of germination. Plant Cell Physiol 18:1–8
Zeevaart JAD (1974) Levels of (+)-abscisic acid and xanthoxin in spinach under different environmental conditions. Plant Physiol 53:644–648

17 Light Control of Seed Germination

B. FRANKLAND and R. TAYLORSON

1 Introduction

Seed germination in many plant species is influenced by light. Promotive effects of light were first documented in the second half of the 19th century, but later inhibitory effects were also described. Historical reviews are given by BARTON and CROCKER (1948) and by EVENARI (1956, 1965). There are also general reviews on light and seed germination by BLACK (1969), ROLLIN (1972), TOOLE (1973), KENDRICK (1976) and VIDAVER (1977).

In several species, germination can be induced by exposing seeds to a single, brief pulse of light. However, experiments with prolonged irradiation of seeds first contributed to the discovery of the photoreceptor regulating plant development (FLINT and MCALISTER 1935, 1937; MEISCHKE 1936). In 1952 BORTHWICK et al. showed that the promoting effect of a brief red irradiation on *Lactuca sativa* seeds could be reversed by a subsequent brief far-red irradiation. There was a repeatable red/far-red reversibility and on this basis proposed the photoreceptor existed in two interconvertible forms, an inactive red absorbing P_r form and an active far-red absorbing P_{fr} form. The photoreceptor was subsequently named *phytochrome* and shown also to be involved in many other plant responses to light.

2 Definition and Events of Germination

2.1 Definition of Germination and Dormancy

Seeds with no dormancy germinate after imbibition of water and exposure to adequate temperature and oxygen. Germination begins with enlargement of the embryonic axis and ends with root protrusion through the seed coat. When a viable seed fails to germinate under conditions favorable for seedling growth, it is said to be dormant. The dormancy may be referred to as innate or primary in freshly matured seeds, or as induced or secondary in imbibed seeds exposed to unfavorable germination conditions, or by other terms (HARPER 1977, VILLIERS 1972). A dormant seed requires some special stimulus to germinate. One such stimulus is light. Other kinds of seed dormancies are described elsewhere (KOLLER and MAYER 1962, TAYLORSON and HENDRICKS 1977).

2.2 Events Preceding and During Germination

It is necessary, although difficult, to distinguish between events in germinating versus dormant seeds. Dormant seeds may lack some essential metabolic process or a coupling between metabolism and cell extension in the embryonic root (Simpson 1978). There are several reviews of the metabolic events that accompany or precede germination (e.g., Roberts 1969, Mayer and Polyakoff-Mayber 1975, Ching 1972, Jann and Amen 1977, Mayer 1977, Bewley and Black 1978).

In many seeds germination occurs by cell enlargement and not by cell division, which begins after germination. Thus, elongation of the embryonic root involves cell water relations. Naked embryos from dormant seeds often germinate, but may be prevented from doing so by replacing the restricting effect of the seed coat by an osmotic stress (Scheibe and Lang 1965). The "growth potential" or "germination potential" of the embryo may be quantified by the water potential of the external medium which just suppresses germination.

3 Photostimulation of Germination

3.1 Relationship Between Light Fluence and Germination Response

It is important to note that germination of an individual seed is an all-or-nothing event. Germination of a population of seeds may be quantified by the proportion which germinate under a given set of conditions. Germination data are usually presented as final (maximum, total) percentage germination or as percentage germination a specified time after sowing. However, seeds in a given population do not all germinate at the same point in time. Full information is provided by a germination time course curve, that is, a plot of accumulated percentage germination against time. The sigmoid shape of this curve reflects the variation in time to germination of the individual seeds in the population.

In some seeds, a single pulse of light induces maximum germination. Within certain limits there is reciprocity between light fluence rate and irradiation time, that is, the same fluence (fluence rate × time) produces the same germination response. There is usually a sigmoid relationship between percentage germination and logarithm of light fluence which reflects a normal distribution among individual seeds in their light requirements for germination (Frankland 1976, Duke 1978a). The linear relationship between probit of germination and log fluence has been used by others (e.g., Evenari and Neumann 1953, Borthwick et al. 1954, Toole et al. 1955a, Taylorson and Hendricks 1973) in analysis of germination data.

The fluence giving 50% germination represents that required for the average seed in the population. For example, Toole et al. (1955a) reported values for red light induction of germination of *Lepidium virginicum, L. densiflorum* and

Lactuca sativa of 1400, 6 and 20 Jm^{-2} respectively. Attempts to classify light-requiring seeds by degree of photosensitivity (ISIKAWA and SHIMOGAWARA 1954) are questionable because of variation within a species. In *Sinapis arvensis,* for instance, three seed batches gave fluences for 50% germination of 2, 50 and 1,000 Jm^{-2} (FRANKLAND 1976).

If the variation in a population of seeds with respect to light requirement does not show a normal distribution, then the fluence-response curve will not be symmetrically sigmoid and there will be a non-linear relationship between probit of germination and log fluence. A biphasic fluence-response curve was obtained by BLAAUW-JANSEN and BLAAUW (1975) and SMALL et al. (1979a) for red light induction of germination in thermodormant lettuce seeds. There appeared to be at least two response types, one being induced by red light of about 10^{-3} Jm^{-2} and the other by 10 Jm^{-2}. For some seed batches the fluence-response curves were even more complex.

3.2 Relationship Between Wavelength and Germination Response

One way to examine wavelength dependence of the germination response is to provide equal levels of irradiations over a range of wavelengths. However, to investigate the nature of the photoreceptor one must determine a true action spectrum. The biological response at a given wavelength will be proportional to $Nt\varepsilon\Phi$ (photon fluence rate × time × extinction coefficient × quantum yield). Therefore the extinction coefficient will be proportional to the reciprocal of the photon fluence giving some standard response (SHROPSHIRE 1972), assuming no other interfering pigments are in the tissue. The first true action spectrum

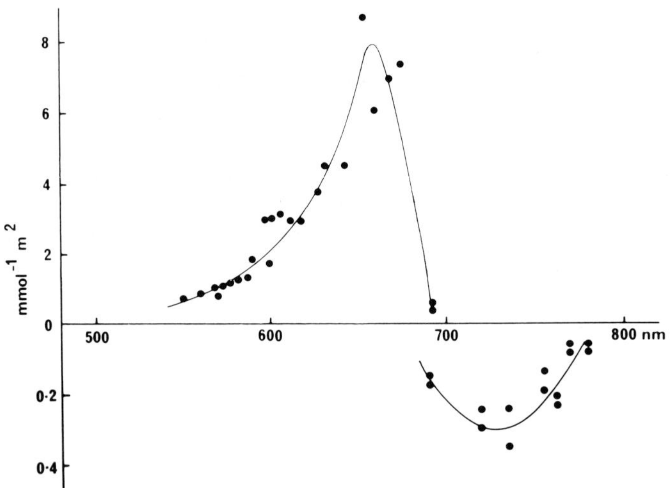

Fig. 1. Action spectra for stimulation of seeds of *Lactuca sativa* cv. Grand Rapids to 50% germination (*upper graph*) and for inhibition of red light-induced seeds to 50% (*lower graph*). Vertical axis reciprocal of photon fluence. (After BORTHWICK et al. 1952)

for photostimulation of seed germination was determined by BORTHWICK et al. (1952, 1954). Their data are plotted in Fig. 1 as reciprocal of photon fluence required to stimulate 50% germination in seeds of *Lactuca sativa* cv. Grand Rapids; also plotted is the reciprocal of photon fluence required to inhibit germination of red light-induced seeds to 50%. The two action peaks at about 660 and 730 nm correspond to the absorption peaks of the P_r and P_{fr} forms of phytochrome, respectively. Similar action peaks have been obtained for seeds of *Lepidium virginicum* (TOOLE et al. 1955a) and *Arabidopsis thaliana* (SHROPSHIRE et al. 1961). Although few detailed action spectra have been published red/far-red reversibility has been widely demonstrated (see list in TOOLE 1973).

SMALL et al. (1979a) stressed care in interpreting fine details of action spectra for processes involving photoreversible pigments, such as phytochrome, with partially overlapping absorption spectra (see Chap. 6, this Vol.). The action spectra cannot be identical to the absorption spectra of P_r and P_{fr} because of the overlapping nature of the latter. For instance, the action spectrum for photostimulation will fall away more sharply on the far-red side as the amount of light required to produce a given amount of P_{fr} is affected by light absorption by both forms of the pigment.

3.3 Quantitative Aspects of Phytochrome-Controlled Germination

Red/far-red reversibility of seed germination confirms phytochrome as the photoreceptor (TOOLE 1973). It is assumed that a fluence-response curve reflects variation among individual seeds in their requirement for a particular minimum proportion or minimum amount of P_{fr} phytochrome (FRANKLAND 1976, DUKE 1978a). Differences in respose to specified fluences have also been explained in terms of variation in total phytochrome (HARTMANN 1966, MANCINELLI et al. 1966, TAYLORSON and HENDRICKS 1971, YANIV and MANCINELLI 1967). Because phytochrome has been difficult to detect spectrophotometrically in seeds, phytochrome behavior in seeds has been usually inferred from studies on dark-grown seedlings or seedling extracts (see reviews by BUTLER 1972, FRANKLAND 1972, KENDRICK and SPRUIT 1977, PRATT 1978, SMITH and KENDRICK 1976).

The transformations of phytochrome may be summarized as follows, where k_1 and k_2 are the wavelength and fluence rate-dependent rate constants for the photochemical transformations (they are the products of N, the photon fluence rate, and σ_1 and σ_2, the wavelength dependent photoconversion cross-sections), and k_0, k_3 and k_4 are the rate constants for P_r synthesis, P_{fr} dark reversion to P_r and P_{fr} destruction respectively:

$$\xrightarrow{k_0} P_r \underset{k_2\,(=N\sigma_2)}{\overset{k_1\,(=N\sigma_1)}{\rightleftarrows}} P_{fr} \xrightarrow{k_4} \qquad P_{fr} \xrightarrow{k_3} P_r$$

The photochemical transformations are rapid relative to the dark transformations and the ratio of P_{fr} to P_r is simply equal to σ_1/σ_2. At low photon fluence rates or wavelengths which ineffectively photoconvert phytochrome, one must account for the dark transformations. Consequently, responses to long, saturating irradiations do not always reflect the expected P_{fr}/P_r ratio, with red light remaining more effective than, say, green light (see early pseudo-action spectrum of FLINT and MCALISTER 1935). In seeds there is no strong evidence for P_r synthesis and P_{fr} destruction, so one may simply consider only thermal reversion of P_{fr} to P_r. At photoequilibrium the proportion of phytochrome as P_{fr} (the P_{fr}/P ratio) is equal to $\sigma_1/(\sigma_1+\sigma_2)$. Allowing for dark reversion of P_{fr} to P_r, P_{fr}/P is equal to $N\sigma_1/(N\sigma_1+N\sigma_2+k_3)$. Under polychromatic light P_{fr}/P may be calculated by using known photoconversion rate constants and integrating $N_\lambda\sigma_{1\lambda}$ and $N_\lambda\sigma_{2\lambda}$ across the spectrum. The rate constants could be those for phytochrome in vitro (e.g., BUTLER et al. 1964) or, more appropriately, those for phytochrome in vivo (e.g., PRATT and BRIGGS 1966, see Chap. 9, this Vol.). The amount of P_{fr} formed by a nonsaturating fluence of red light can be calculated as $0.75\ (1-e^{-\sigma Nt})$. This assumes that the maximum proportion of phytochrome as P_{fr} is 0.75 (PRATT 1975). Alternatively, P_{fr} may be determined spectrophotometrically in dark-grown seedling tissue placed under the same light conditions as the seeds being studied. All these methods assume that phytochrome in seeds has the same properties as phytochrome in seedlings. Also, light will be attenuated in passing through the seed coats and into the denser seed tissue.

DUKE (1978a) showed that the probit of germination in *Rumex crispus* could be expressed in terms of P_{fr} as follows: probit $y=5+\frac{1}{\sigma}\ (\ln P_{fr}-\mu)$, where μ is the natural logarithm of the P_{fr} level required for 50% germination and σ is the standard deviation of $\ln P_{fr}$. The concentration of P_{fr} was calculated from the photon fluence of red light using known photoconversion constants, assuming that total phytochrome was constant and assuming that the seed tissue had a spectral attenuation factor of 0.5. This justified the assumption that there is a logarithmic normal distribution in P_{fr} requirement for germination around a mean level. The P_{fr} requirement of the average seed will vary among seed batches and may be altered by a variety of conditions. In other light-requiring seeds there is a good correlation between percentage germination and estimated P_{fr}/P ratio (e.g., KARSSEN 1970d). In Fig. 2 germination data are presented for *Sinapis arvensis* seeds after various dichromatic (red/far-red) irradiations, compared with calculated P_{fr}/P and measured P_{fr}/P in seedling tissue after the same irradiations. In general, germination follows the change in P_{fr}, although as no allowance has been made for the differential transmission of the seed coats and tissues the P_{fr}/P ratios may be overestimated.

3.4 Effects of Short Irradiation with Far-Red and Blue Light

A short irradiation with far-red light reverses the red inductive effect by photoconversion of P_{fr} to P_r. Short far-red can also reduce germination below the dark control in light-requiring species such as *Lactuca sativa* (BORTHWICK et al. 1954) and in a few dark-germinating species such as *Lycopersicon esculentum* (MANCINELLI et al. 1966). This is evidence for some P_{fr} in seeds sown in dark

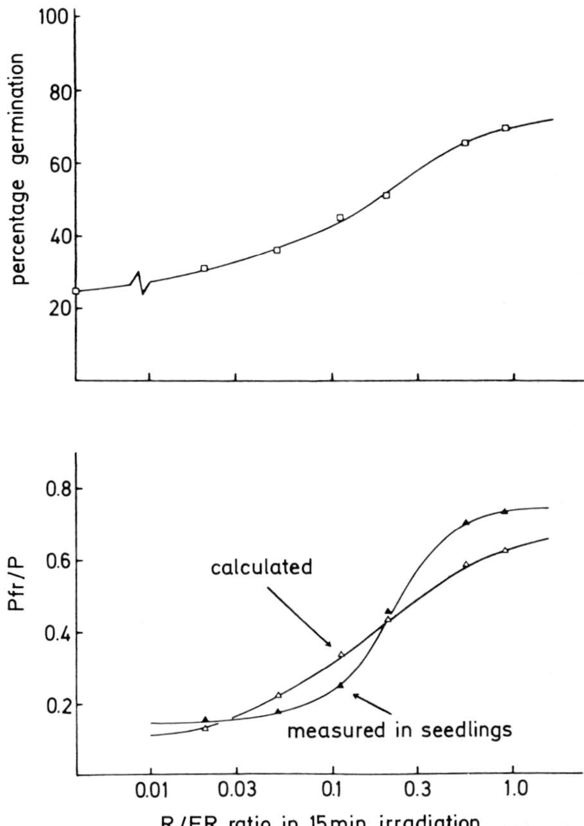

Fig. 2. Relationship, for a range of red/far-red ratios, between percentage germination of *Sinapis arvensis* seeds (*upper graph*) and Pfr/P ratio (*lower graph*) calculated or measured in seedling tissue. (After WADDOUPS 1976)

causing germination without light. Failure of far-red to inhibit germination could be for several reasons; germination may not be under phytochrome control or may have progressed beyond phytochrome control; the low P_{fr}/P ratio established by the far-red light may allow germination to proceed; or possibly, P_{fr} can reappear from some intermediate form following far-red irradiation.

Far-red irradiation can also promote germination, as in bromeliads such as *Wittrockia superba* (DOWNS 1964) and in *Eragrostis curvula* (TOOLE and BORTHWICK 1968). Full germination is commonly not induced by far-red as only a small amount of P_{fr} is formed (P_{fr}/P ratio of about 0.03). Such stimulation is inevitably associated with apparent incomplete reversal by far-red of a previous red irradiation.

Many effects of blue light on plants may involve a photoreceptor, possibly a flavoprotein, other than phytochrome (SENGER 1980). However, conclusive evidence for a photomorphogenic photoreceptor other than phytochrome in seeds is lacking. As blue light produces a P_{fr}/P ratio intermediate between those produced by red light and far-red light it may promote or inhibit seed germination (BORTHWICK et al. 1954. Also see EVENARI et al. 1957, BLACK and WAREING

1957, WAREING and BLACK 1958, MALCOSTE 1968a). SMALL et al. (1979b) found that thermodormant *Lactuca* seeds would germinate after blue fluences of 0.1 Jm^{-2}, whereas far-red dormant seeds required 1,000 Jm^{-2}. Red light sensitivity was parallel, although both types of seed were about 100 times more sensitive to red light. This is consistent with phytochrome being the photoreceptor. However, the action spectrum, with peaks at 422 and 446 nm, resembled that for phototropism rather than a phytochrome-controlled response. Heavily pigmented seed coats absorbing blue light more than red light would result in the sensitivity to blue light being less than predicted from the absorption properties of phytochrome. In *Amaranthus retroflexus,* the absence of a blue light effect was attributed to seed coats which do not transmit these wavelengths (KADMAN-ZAHAVI 1960). Other visible light bands can influence seeds that have a very high sensitivity to P_{fr}. Green light, often used as a safelight, may induce germination (BASKIN and BASKIN 1979, BLOM 1978).

3.5 Escape from Far-Red Reversibility

"Escape" describes the increasing failure of the red/far-red reversion effect as the dark interval between the two irradiations increases. It indicates that P_{fr} action has been completed in seeds so that reversion to P_r will not affect subsequent germination. Variations among seeds in P_{fr} required for germination in part reflects variation in P_{fr} action required. Escape may be presented as the time of P_{fr} action required in half the population. The rate of escape may vary according to the pre-treatment prior to irradiation.

DUKE et al. (1977) postulated that P_{fr} interacts with some "reaction partner" or "co-effector" X. Escape curves can then indicate seed variation in threshold levels of P_{fr} X complex required to induce germination. The amount of $P_{fr}X$ will be affected for instance, by the amount of X and the rate of disappearance of P_{fr}.

3.6 Requirement for Repeated or Prolonged Irradiation

Many species require relatively long periods of light exposure to induce germination. In such cases, germination can usually be induced by repeated brief irradiations. For example, seeds of *Paulownia tomentosa* (BORTHWICK et al. 1964) require 48 h exposure to light but a short irradiation each day for 3 days is equally effective. Effects of repeated daily irradiations have been taken as evidence for "photoperiodism" in germination (see review by EVENARI 1965, BLACK and WAREING 1955). However, this response may not be analogous to photoperiodic control of flowering. Most apparent "photoperiodic" effects might only reflect need for intermittent irradiation to maintain some level of P_{fr} over a long period of time as in *Begonia evansiana* (NAGAO et al. 1959) and *Epilobium cephalostigma* (ISIKAWA and YOKOHAMA 1962).

In some species only a proportion of seeds germinate after a single pulse of light, but more germinate after prolonged irradiation. Examples include

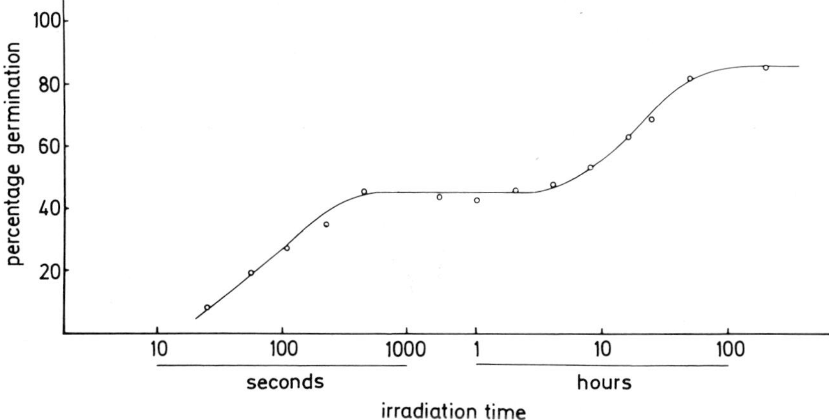

Fig. 3. Effect of red light irradiation time on seed germination in *Chenopodium album*. (After KARSSEN 1970d)

Chenopodium album (KARSSEN 1970d), *Portulaca oleracea* (VAN ROODEN et al. 1970) and *Plantago major* (FRANKLAND 1981). In *Chenopodium* seeds (Fig. 3) 45% germinated after an irradiation of less than 10 min, whereas a further 40% required an irradiation of from 8 h to 48 h. Intermittent irradiation 15 min every 8 h gave as much germination as 48 h continuous light (KARSSEN 1970d). The long period of P_{fr} action required suggests thermal reversion of P_{fr} to P_r probably limits the effects of a single light pulse.

4 Changes in Responsivity to Light with Time

4.1 Increased Responsivity During Imbibition

Most dry seeds (6%–8% water content) are unaffected by irradiation. The proportion of photoresponsive seeds usually increases with increasing dark imbibition time. In *Lactuca sativa* (BORTHWICK et al. 1954) and *Sinapis arvensis* (FRANKLAND 1976) responsivity reaches a maximum within a few hours after sowing, with a few seeds becoming responsive within minutes. It is supposed that the increased responsivity results from rehydration of phytochrome molecules in the dry seed. In *Lactuca,* the correlation between responsivity and seed water content has been studied using imbibition at various temperatures (IKUMA and THIMANN 1964), various osmotica (BERRIE et al. 1974) and water-saturated atmospheres (HSIAO and VIDAVER 1971a, b). Full red/far-red reversibility could be observed in seeds with 15% to 20% water content, well below complete imbibition. In *Amaranthus retroflexus* seeds, phytochrome is fully hydrated at half their maximum water content (TAYLORSON and HENDRICKS 1972). At 15% water content phytochrome may be transformed, but cannot act.

Often, maximum responsivity requires 24 h or more of imbibition. Some explanations involve phytochrome synthesis (TAYLORSON and HENDRICKS 1971), decrease in inhibitor levels (BURKART and SANCHEZ 1969) and increase in the factor (X) interacting with P_{fr} (KOLLER et al. 1964). The latter explanation was used by KARSSEN (1967, 1970c) for *Chenopodium album,* although increased responsivity was mainly attributed to phytochrome rehydration. DUKE et al. (1977) also concluded that changes in the amount or receptivity of a component (X), which interacts with P_{fr} changed responsivity. The component could be membrane-related. Rapid increases in responsivity, as in *Lactuca* seeds, could be associated with membrane hydration. Where full responsivity requires considerable time, there is clear evidence of involvement of an unknown temperature-dependent process.

4.2 Decreases in Responsivity

In *Oryzopsis miliacea* (KOLLER and NEGBI 1959) and *Amaranthus retroflexus* (TAYLORSON and HENDRICKS 1969) responsivity to a saturating light pulse remains high for long periods. In other species responsivity is progressively lost after one or more days, as in *Lactuca sativa* (BORTHWICK et al. 1954, IKUMA and THIMANN 1964) and *Chenopodium album* (KARSSEN 1970c). This may be regarded as a type of secondary dormancy or dark-induced dormancy (KARSSEN 1980a).

After reaching a peak, responsivity may decline immediately as in *Portulaca oleracea* (DUKE et al. 1977) or remain high for several days before declining as in *Rumex crispus* (TAYLORSON and HENDRICKS 1973). The rate of decline is markedly affected by temperature. In *Sinapis arvensis,* for instance, maximum responsivity is reached after 6 h imbibition at both 25° and 15 °C. At 25 °C an immediate decline begins that makes seeds unresponsive after 24 h, but at 15 °C high responsivity persists for about 24 h before a slow decline over several days (FRANKLAND 1976). *Lactuca* seeds held in a water saturated atmosphere at 20 °C shift from being light-independent to light-requiring before becoming unresponsive (HSIAO and VIDAVER 1973).

DUKE et al. (1977) argued that loss in responsivity was related to declining levels of the P_{fr} reaction partner X because levels of total phytochrome and rate of P_{fr} dark reversion to P_r was constant. Similar conclusions were reached by KARSSEN (1970c) and TAYLORSON and HENDRICKS (1973). Incubation in nitrogen gas will prevent loss of light responsivity (secondary dormancy) in *Lactuca* (VIDAVER and HSIAO 1975) and *Rumex* (LE DEUNFF 1973).

4.3 Effects of Light on Dry Seeds

While irradiation of dry seeds is normally ineffective, germination in *Pinus banksiana* can be stimulated by red irradiation of dry seeds (ORLANDINI and BULARD 1972) and MCARTHUR (1978) reported that some *Lactuca* seed batches

Table 1. Effects of irradiating dry seeds with far-red or red light. Data as percentage germination 2 days after sowing in the dark at 25 °C (in %)

		D	R	FR	FR-R	R-FR
Lactuca sativa	Normal seeds	30	29	12	27	10
	High P_{fr} seeds	87	88	46	83	48
Sinapis arvensis	Normal seeds	10	9	5	9	6
	High P_{fr} seeds	48	47	30	48	34

with only 6% water content were influenced by red and far-red irradiation. However, whole seed water content may not reflect hydration of phytochrome.

Some seeds show reduced germination if irradiated with far-red immediately before sowing. This far-red inhibitory effect can be reversed by red light in both *Lactuca sativa (*KENDRICK and RUSSELL 1975) and *Sinapis arvensis* (FRANKLAND 1976). Far-red/red reversible effects on dry seeds can be seen in seeds with high levels of P_{fr} (see Table 1), produced by irradiating imbibed seeds with red and then immediately re-drying. Persistance of P_{fr} in dry seeds is prolonged (VIDAVER and HSIAO 1972, TAYLORSON and HENDRICKS 1972, LOERCHER 1974), and can induce germination after subsequent imbibition.

Far-red effects on dry seeds may arise from occurrence of intermediates in phytochrome interconversions in the dehydrated tissue (see Sect. 9.2). Thus, red light cannot convert dehydrated phytochrome to P_{fr}, but far-red light can remove existing P_{fr} by forming the intermediate meta-Fa (see scheme, Sect. 9.2), which can only generate P_{fr} by subsequent red irradiation. This could be a red-light-induced physiological response where P_r is not the light absorber. Preparation of "high P_{fr}" seeds by drying under continuous red light makes them less sensitive to inhibition by far-red irradiation, probably because much phytochrome is "trapped" as meta-Ra (KENDRICK and RUSSELL 1975). This is unaffected by far-red light but the P_{fr} generated by rehydrating the phytochrome can be removed by far-red light. Far-red fails to inhibit germination if given several hours before sowing; probably by slow regeneration of P_{fr} from meta-Fa.

Other work suggesting involvement of phytochrome intermediates by SCHEUERLEIN (1980) showed higher germination in *Lactuca* after two flashes of 1 ms red light separated by a few seconds dark than a single flash of the same total energy. Possibly, this involves dark reactions of the P_r to P_{fr} transformation.

4.4 Changes During Post-Harvest Storage

Freshly harvested seeds often germinate poorly even when incubated in light. After dry storage dormancy may decrease. In *Sinapis arvensis* germination in dark and after red light increased to a maximum value after 30 weeks dry storage at 20 °C (WADDOUPS 1976). These changes reflect greater responsivity to P_{fr}, not changes in phytochrome. Changes in *Lactuca sativa* seeds to light and temperature during after-ripening are shown in Table 2 (SUZUKI et al. 1980).

Table 2. Effect of dry storage at 23 °C on germination of seeds of *Lactuca sativa* cv. Grand Rapids at three temperatures in response to red or far-red irradiation after 8 h imbibition (in %)

		Period of after-ripening in months						
		0	1	2	3	4	5	6
20 °C	D	78	82	89	86	85	83	88
	R	95	92	93	93	94	90	92
	FR	12	19	29	37	88	80	86
25 °C	D	1	0	5	1	25	28	43
	R	76	67	89	76	78	88	86
	FR	0	2	0	2	8	4	3
30 °C	D	0	0	0	0	0	0	0
	R	0	0	0	0	0	45	66
	FR	0	0	0	0	0	0	0

5 Photoinhibition of Germination

5.1 Wavelength Dependence of Photoinhibition

Seeds of *Amaranthus caudatus, Nemophila insignis* and *Phacelia tanacetifolia* germinate readily in the dark but are inhibited by prolonged irradiation with white light. Such photoinhibition is also observed in many light-requiring species such as *Amaranthus retroflexus* (KADMAN-ZAHAVI 1960), *Oryzopsis miliacea* (NEGBI and KOLLER 1964) and *Atriplex dimorphostegia* (KOLLER 1970). Here, a dual action of light is seen with short periods of light promoting germination and long periods of white light inhibiting. In *Oryzopsis* and *Amaranthus,* seeds germinate on transfer back to darkness whereas *Nigella sativa* seeds (ISIKAWA 1957) no longer germinate in darkness and are termed photodormant.

Greatest inhibition of germination is by far-red radiation (710–720 nm) with blue light (ca. 470 nm) being slightly less inhibitory. However, in *Phacelia tanacetifolia* red light can also be inhibitory (SCHULZ and KLEIN 1963, ROLLIN and MAIGNAN 1967). Prolonged far-red effectively inhibits germination in many species, e.g., *Lactuca sativa, Lamium amplexicaule, Nemophila insignis, Lycopersicon esculentum, Cucumis sativus* and *Amaranthus caudatus* (JONES and BAILEY 1956, HENDRICKS et al. 1959, MANCINELLI and BORTHWICK 1964, YANIV et al. 1967, KENDRICK and FRANKLAND 1969). Periods of far-red insufficient to reduce the final germination percentage may have marked delaying effects. Prolonged far-red may also induce a red light requirement and can inhibit a late stage of the germination process in light-requiring seeds. This latter effect occurs after completion of P_{fr} action as shown by escape from reversibility to a short far-red irradiation (MOHR and APPUHN 1963, ROLLIN 1963, HARTMANN 1966, KARSSEN 1970c). Also, NEGBI et al. (1968) showed that far-red irradiated *Lactuca* seeds become insensitive to a subsequent application of gibberellin.

Photoinhibition is dependent on fluence rate, particularly in white light, although far-red inhibits at low fluence rates. This suggested that photoinhibition was similar to the high energy reaction or high irradiance reaction (HIR) observed in seedling photomorphogenesis (see review by MANCINELLI and RABINO 1978). HARTMANN (1966) suggested that the high irradiance reaction operated through phytochrome. He found that the action peak at 720 nm for photoinhibition of *Lactuca* germination could be nullified if mixed with otherwise inactive light, either from the red or from the infra-red side of 720 nm. Maximum photoinhibition in both *Lactuca* (HARTMANN 1966), and *Nemophila insignis* (ROLLIN et al. 1970) was obtained with a light mixture giving a P_{fr}/P ratio of about 0.10. Similar effects were obtained for *Poa pratensis* and *Amaranthus arenicola* (HENDRICKS et al. 1968, BORTHWICK et al. 1969) although the action peak remained at 720 nm even when simultaneous red light significantly increased the proportion of P_{fr}. In *Amaranthus caudatus* the inhibitory effect of 720 nm light could be nullified by mixing it with either 655 or 755 nm light (FRANKLAND 1976).

Prolonged far-red irradiation maintains a low proportion of phytochrome as P_{fr} (ca. 0.03). While this accounts for its inhibitory action, it also implies that P_{fr} must be formed over a long period. In *Amaranthus caudatus,* for instance, a far-red period of 48 h is required to suppress germination fully (KENDRICK and FRANKLAND 1969). There is good evidence that far-red acts by maintaining a low P_{fr}/P ratio. For instance, intermittent far-red is just as effective as continuous far-red. There is also a good inverse correlation between degree of photoinhibition and the P_{fr}/P photoequilibrium established by the light source used (KENDRICK and FRANKLAND 1969, KARSSEN 1970d). However, there are several important reasons why the above cannot fully explain photoinhibition. For instance, light giving high P_{fr}/P ratios can also inhibit if the fluence rate is also high. Secondly, the 720 nm action peak does not yield the lowest possible P_{fr}/P ratio. Also, photoinhibition can occur after the apparent completion of P_{fr} action.

5.2 Some Explanations of the High Irradiance Response

In seedling photomorphogenesis the high irradiance reaction acts in the same direction as the inductive low energy reaction. In seed germination, the time and fluence rate-dependent photoreaction inhibits, which is opposite to the low energy reaction. Dependence on fluence rate suggests that rate of interconversions of P_r and P_{fr} is important. The P_{fr}/P ratio is little affected by fluence rate whereas the photostationary flux is directly proportional to fluence rate. JOHNSON and TASKER (1979) suggested conversion of P_r to P_{fr} drives some essential reaction. Alternatively, the effector molecule for the high irradiance reaction is (1) a short-lived activated species of P_{fr} which is more reactive than stable P_{fr} (HARTMANN 1966), (2) a short-lived $P_{fr}X$ complex which changes to $P_{fr}X'$ the effector for the inductive low energy reaction (SCHÄFER 1975, MANCINELLI and RABINO 1978) and (3) an intermediate in the P_r to P_{fr} or P_{fr} to P_r pathways. In some models P_{fr} itself is the effector for the high irradiance reaction (HART-

MANN 1966, SCHÄFER 1975), whereas in others it only interacts synergistically with some product of the high irradiance reaction (JOHNSON and TASKER 1979). Photoinhibition by prolonged irradiation after apparent completion of P_{fr} action was explained by BORTHWICK et al. (1969) in terms of $P_{fr}X$ having an absorption maximum at 720 nm and dissociating to P_r or P_{fr} and X under high fluence rate conditions. The explanation implies depletion of either P_{fr} or its active complex, $P_{fr}X$. Under conditions of rapid cycling, lifetime of the active complex may be too short to achieve physiological action. However, as photoinhibition can operate at a late stage in germination an effect independent of depletion of the low energy reaction effector complex is suggested.

5.3 Two Points of Action in Photocontrol of Germination

The dual effect of light on seed germination can be visualized by photocontrol at two points in events leading to cell elongation in the embryonic root (ROLLIN 1966, KENDRICK 1976, FRANKLAND 1981).

$$\begin{array}{cccccc} & P_{fr} & & H & & \\ & \text{promotes} & & \text{inhibits} & & \\ \rightarrow & \rightarrow & \rightarrow & \rightarrow & \rightarrow & \text{germination} \end{array}$$

Induction of elongation is the first step and is dependent upon P_{fr}. Another step, designated as H, is inhibited by a time and fluence rate dependent photoreaction. It can be interpreted as an effect of prolonged phytochrome interconversion and can be quantified in terms of photostationary flux or cycling rate

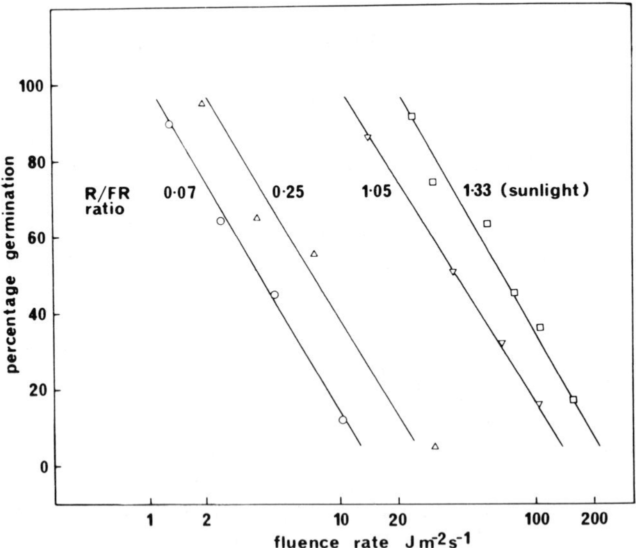

Fig. 4. Seed germination in *Lactuca sativa* cv. Vanguard at various fluence rates and various red/far-red ratios using filtered sunlight. (After GORSKI and GORSKA 1979)

($H = N\sigma_1 P_r = N\sigma_2 P_{fr}$). The photoreaction H can act late and after the completion of P_{fr} action. However, by blocking reactions leading to germination it can lead to loss of P_{fr} induction so that the first, P_{fr} controlled, step has to be re-initiated. After prolonged irradiation at low P_{fr}/P, germination can be re-induced by establishing a high P_{fr}/P. Seeds which become photodormant lose responsivity to even high P_{fr}/P during prolonged irradiation. An antagonistic interaction between P_{fr} and H can be seen in photoinhibition of dark-germinating *Lactuca* seeds by sunlight (GORSKI and GORSKA 1979). These data show a negative correlation between germination and fluence rate although much lower fluence rates cause germination with low red/far-red ratios (Fig. 4).

6 Effect of Temperature on Responsivity to Light

6.1 Effects of Constant Germination Temperatures

The effect of light on seed germination is strongly conditioned by temperature (see review by TOOLE 1973). Seeds of *Brassica juncea* germinate best in both light and dark at relatively low temperatures (15 °C) whereas *Sisymbrium officinale* germinates best at 25 °C (TOOLE et al. 1955b), and *Portulaca oleracea* needs high temperature (35–40 °C). The temperature giving the greatest difference between germination in dark and light is not necessarily optimal. In *Lactuca sativa,* there is high germination in both dark and light below 15 °C and no germination under either condition above 35 °C, with the greatest difference being at about 25 °C. In contrast, *Lepidium virginicum* does not germinate in the dark over a wide range of temperatures. These temperature effects result from changes in germination potential or degree of dormancy of the seeds that reflect an altered P_{fr} requirement for germination. Increased dark germination is caused by response to the small amount of P_{fr} present in the seed. In *Amaranthus retroflexus,* germination in prolonged white light is decreased at 20 °C but increased at 30 °C (KADMAN-ZAHAVI 1960). Short irradiations increase germination more at the higher temperature. The seeds may have a higher potential for germination at high temperature and thus are less susceptible to photoinhibition.

6.2 Effects of Pre-Incubation at Low or High Temperature

Germination at one temperature may be altered by pre-incubation at another. Several hours to a few days incubation at low temperature enhances germination of some light-requiring seeds. Pre-chilling may act by delaying thermal reversion of pre-existing P_{fr} to P_r (SCHEIBE and LANG 1965), an explanation used for enhanced *Amaranthus retroflexus* germination after pre-chilling (TAYLORSON and HENDRICKS 1969).

VANDERWOUDE and TOOLE (1980) studied the effects of chilling on *Lactuca* seeds in which P_{fr} was reduced to very low levels by far-red. Chilling did not increase dark germination, confirming P_{fr} dependence, but rendered the seeds responsive to low P_{fr} levels.

It was suggested that the chilling effect arose from altered membrane properties. Interaction between chilling and phytochrome in *Betula papyrifera* seeds has also been interpreted as membrane changes affecting availability or sensitivity of P_{fr} receptor sites (BEVINGTON and HOYLE 1981).

In seeds where optimum germination is at low temperatures, preincubation at high temperatures may inhibit subsequent germination. *Lactuca* seeds which germinate in darkness at 15 °C require light to germinate at 25 °C (BORTHWICK et al. 1954). This is often termed thermodormancy. An explanation involving enhanced dark reversion of P_{fr} is unlikely since some seeds lose responsivity to light after a high temperature pre-treatment. Also unlikely is a loss of phytochrome or the P_{fr} reaction partner X. Light responsivity in thermodormant *Rumex crispus* seeds can be reinstated by low temperature treatment (TAYLORSON and HENDRICKS 1973).

6.3 Effects of Fluctuating Temperatures

Germination of light-requiring species may be stimulated by fluctuating temperatures. Where no dark germination occurs, fluctuating temperatures enhance the light effect but where some dark germination occurs at constant temperature, fluctuating temperatures may give high germination and appear to replace the light effect (TOOLE et al. 1955b). Alternating temperatures increased germination in *Rumex obtusifolius* and *R. cirspus,* provided the difference is 5 °C or more and provided one temperature is above 15 °C and one temperature is below 25 °C (TOTTERDELL and ROBERTS 1980). More cycles of daily alternation were necessary for maximum germination in darkness than in the light. In *Rumex obtusifolius* seeds studied by TAKAKI et al. (1981) a single temperature perturbation stimulated high germination. Only a few minutes of high temperature stimulated germination and there was close interaction between temperature and P_{fr}. Short periods of 30 °C, following 40 °C pre-incubations and formation of small amounts of P_{fr}, stimulate germination in *Amaranthus retroflexus* (HENDRICKS and TAYLORSON 1978). Since formation of P_{fr} before the down shift was more effective than after, changes in phytochrome action that depended on membrane organization were suggested.

7 Effects of Stimulants and Other Factors on Germination

7.1 Effects of Gibberellins

The naturally occurring gibberellins affect more species and types of dormancy than other chemicals known to induce germination, suggesting a possible role in induction of germination (AMEN 1968). However, there is little evidence for changes in endogenous gibberellins following light treatment. JONES and STODDART (1977) suggest the effects of applied gibberellins are "pharmacological" and unrelated to the natural mechanisms of germination induction.

CARPITA et al. (1979b) found that exogenous gibberellin had no effect on the elongation of *Lactuca* embryonic axes exposed to red light which supports the hypothesis that P_{fr} stimulates the release or synthesis of endogenous gibberellins. However, although applied gibberellins induce germination in the dark their effect may depend on pre-existing P_{fr}. This suggests that germination depends on both P_{fr} and gibberellin rather than a P_{fr} control of synthesis or release of endogenous gibberellins. Prolonged far-red irradiation prevented response to gibberellin (NEGBI et al. 1968) but the far-red effect may not be due to lowering the P_{fr} level. Hydrated *Lactuca* seeds kept in dark lose responsivity to gibberellin before losing responsivity to red light (VIDAVER and HSIAO 1974, BEWLEY 1980). Combining gibberellin and red light stimulates germination even after loss of responsivity to red light, confirming the synergistic interaction of gibberellin and P_{fr}. Synergism between P_{fr} and gibberellin has also been noted in *Kalanchoë blossfeldiana* seeds (FREDERICQ et al. 1976). In this case, either stimulant alone fails to act, but a combination elicits full response.

7.2 Effects of Other Substances

Although cytokinins are not widely effective as germination stimulants, there is evidence for rapid changes in extractable cytokinins following irradiation of seeds (e.g., *Apium graveolans*) (THOMAS et al. 1978). However, stimulation by cytokinins often requires light (KHAN and TOLBERT 1965) which argues that light acts by increasing endogenous cytokinin levels. Ethylene stimulates germination in some light-requiring seeds (e.g., OLATOYE and HALL 1973, SPEER et al. 1974, SCHONBECK and EGLEY 1980) although there is no evidence that P_{fr} controls ethylene production.

Germination of light-sensitive species can also be induced by chemicals such as thiourea, nitrate and cyanide, and other compounds (see reviews by MAYER and POLJAKOFF-MAYBER 1975, TAYLORSON and HENDRICKS 1977). These may act by influencing metabolism (ROBERTS 1973, HENDRICKS and TAYLORSON 1975). Ethanol interacts with P_{fr} to stimulate germination in seeds of *Panicum dichotomiflorum* and other light-requiring species (TAYLORSON and HENDRICKS 1979).

7.3 Effects of Water Stress

Since germination involves cell enlargement and water uptake, media of high osmotic concentration (low water potential) can inhibit germination, or cause a light requirement in seeds which normally germinate in dark (SCHEIBE and LANG 1967), or increase susceptibility to photoinhibition by prolonged white light. For instance, non-dormant seeds of *Chenopodium album* sown on 0.7 M mannitol germinated 45% in the dark but were increased to 70% by red light or decreased to 5% by prolonged white light (KARSSEN 1970b). Seeds of *Zea mays,* normally regarded as light insensitive, may be brought under phytochrome control by osmotic stress (THANOS and MITRAKOS 1979).

When dark-germinating seeds are maintained under osmotic stress for 1 or 2 weeks, a light requirement may develop, as shown in *Lactuca sativa* (BERRIE et al. 1974) and *Chenopodium bonus-henricus* (KHAN and KARSSEN 1980). The dormancy may involve reversion of preexisting P_{fr} to P_r under conditions where P_{fr} action cannot proceed. However, some light-requiring seeds may actually lose their responsiveness to light during osmotic stress indicating that more profound changes occur. Seeds maintained in conditions which preclude germination (e.g., *Lactua* seeds in a water-saturated atmosphere) eventually fail to germinate in either light or dark (HSIAO and VIDAVER 1973). In *Rumex crispus,* water stress inhibited interaction of P_{fr} with its reaction partner X more than the dark reversion of P_{fr} to P_r (DUKE 1978b).

7.4 Effects of Pre-Harvest Conditions

Seed batches of a species often vary greatly in degree of dormancy and response to light. Some variation arises from genetic differences, as in *Nicotiana tabacum* (KASPERBAUER 1968), *Cucumis sativus* (SUZUKI and TAKAHASHI 1969) and *Lactuca sativa* (GLOBERSON et al. 1974). However, differences among batches of genetically similar seeds arise from environmental effects during formation and maturation.

Light effects during seed ripening may be on the seed's phytochrome or indirectly through the parent plant. The photoperiod during seed maturation may influence subsequent germination. In most cases seeds from plants grown in short days are less dormant than those from plants in long days, as shown in *Ononis sicula* (EVENARI et al. 1966), *Chenopodium album* (KARSSEN 1970a) and *Portulaca oleracea* (GUTTERMAN 1974). Photoperiodic effects may arise through changes in seed coat thickness, permeability of seed coat to water, and inhibitor levels. However, in *Chenopodium* photoperiodic effects are partly phytochrome-mediated.

The spectral quality of the light received by the parent plant can influence subsequent seed germination. Plants of *Arabidopsis thaliana* grown under incandescent light, which is rich in far-red, produce light-requiring seeds (MCCULLOUGH and SHROPSHIRE 1970, SHROPSHIRE 1973). Similar effects were noted in *Cucumis prophetarium* and *C. sativus* (GUTTERMAN and PORATH 1973). The effect is likely to result from changes in the proportion of phytochrome as P_{fr}. In *Arabidopsis,* phytochrome in developing seeds may be interconverted until the seeds are nearly dehydrated and is then stabilized in the form last established (HAYES and KLEIN 1974).

Even though seeds ripen in sunlight, mature seeds of light-requiring species apparently contain low levels of P_{fr}. Possibly, thermal reversion of P_{fr} continues after the water content decreases below that needed for photoreversibility (ca. 18%), or the P_r to P_{fr} photoreaction pathway is blocked before the P_{fr} to

P_r pathway as dehydration occurs (see discussion by KENDRICK 1976), or chlorophyll in the covering structures results in light of relatively high far-red/red ratio reaching the developing seed (CRESSWELL and GRIME 1981).

8 Mode of Action of P_{fr} and Later Events in Germination

Action of P_{fr} in seeds may occur over many hours as shown by the prolonged period of escape. While phytochrome action is more rapid in other plant systems, initial events may nonetheless be similar. Considerable evidence from plants indicates that an early action of P_{fr} may involve association with something causing a change in membrane(s) function (MARMÉ 1977, BROWNLEE et al. 1979, HENDRICKS and VANDERWOUDE 1982, ROUX et al. 1981). Subsequent early events are not known and may vary depending upon the specific ultimate display and the period of action (HENDRICKS and VANDERWOUDE 1982). In seeds, some later events of phytochrome action may be indistinguishable from early events of germination.

Action of P_{fr} in seeds may be associated with the action of temperature and hormones, which also act on germination and possibly also by affecting membrane(s) (HENDRICKS and TAYLORSON 1976, 1978, 1979). Temperature may act by increasing responsivity of seeds to low levels of P_{fr} (VANDERWOUDE and TOOLE 1980) or by direct interaction with P_{fr} at a common site of action (TAKAKI et al. 1981). Hormonal involvement in P_{fr} action is likely but not clearly shown (see Sect. 7.1). CARPITA and NABORS (1981) interpret findings on additivity of gibberellin and P_{fr} growth responses in isolated embryos versus synergism in intact seeds as suggesting P_{fr} may induce gibberellin production. New evidence for GA increases in *Lactuca* following irradiation (BIANCO and BULARD 1981) support this interpretation.

Changes in respiratory metabolism (ROBERTS and SMITH 1977) and in nucleic acid and protein synthesis (TAO and KHAN 1977) are implicated in induction of germination. Also, formation of polyribosomes in *Portulaca* seeds is markedly increased by light (REGER et al. 1975). However, it is uncertain that P_{fr} action involves direct effects on any of these processes.

9 Properties and Localization of Phytochrome in Seeds

9.1 Detection of Phytochrome in Seeds by Spectrophotometry

Spectrophotometric detection of phytochrome in seeds is very difficult since levels are very low. Phytochrome is easier to detect in large seeds such as *Pisum sativum* (MCARTHUR and BRIGGS 1971), although these usually do not show germination responses to light. A very sensitive dual wavelength spectrophotometer has been used to study phytochrome in some relatively large light-sensitive seeds such as *Cucumis sativus* (MANCINELLI and TOLKOWSKY 1968, SPRUIT and MANCINELLI 1969, MALCOSTE and BOISARD 1970, GUTTERMAN and PORATH 1975), *Curcurbita pepo* and *C. maxima* (MALCOSTE et al. 1970, BOISARD and MALCOSTE 1970, ZOUAGHI et al. 1972). Phytochrome has also been detected in seeds of *Pinus nigra* (ORLANDINI and MALCOSTE 1972), and in some relatively small-seeded species such as *Lactuca sativa, Nemophila insignis, Sinapis alba* (BOISARD et al. 1968), *Raphanus sativus* (MALCOSTE 1969), *Amaranthus caudatus* (KENDRICK et al. 1969) and *Rumex alpinus* (BIANCO 1973).

Phytochrome is not usually detectable until after 3 to 5 h of imbibition. Following the initial increase, the amount of phytochrome remains constant for about 5 h in *Amaranthus* (KENDRICK et al. 1969) or 11 h in *Lactuca* (BOISARD et al. 1968) and *Cucurbita* (MALCOSTE et al. 1970, ZOUAGHI et al. 1972). Another larger phytochrome increase follows which is usually associated with germination and seedling growth. The first increase represents rehydration of preexisting phytochrome. The second increase probably represents synthesis of phytochrome as P_r and is usually prevented by low temperature and is blocked by inhibitors of protein synthesis.

9.2 Properties of Phytochrome and Intermediates in Seeds

Phytochrome in dry seeds is not fully photoreversible and hence not detected by normal photometric assay. Phytochrome in dry seeds of *Cucumis* appeared to have a modified difference spectrum following far-red irradiation compared to that of phytochrome in seedlings (SPRUIT and MANCINELLI 1969). A similar difference spectrum was observed in freeze-dried *Pisum* seedling tissue (KENDRICK and SPRUIT 1974). An understanding of these spectral characteristics comes from properties of dehydrated phytochrome and its intermediates (KENDRICK 1976). A simplified version of the scheme of KENDRICK and SPRUIT (1977) shows the intermediates involved in the P_r and P_{fr} interconversions (dashed lines indicate photoinduced steps and continuous lines dark steps:

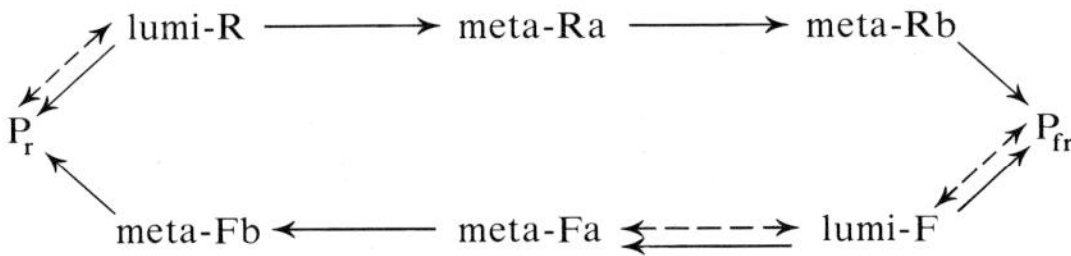

The initial changes involve the phytochrome chromophore and can proceed in the dehydrated state whereas the later steps involve protein conformation and require hydrated phytochrome. In dehydrated tissue red light converts P_r to lumi-R which cannot be transformed to P_{fr} and rapidly reverts back to P_r in the dark. However, far-red light converts P_{fr} to meta-Fa which can revert slowly to P_{fr} in the dark. The difference spectrum for dry seeds therefore represents the interconversion of P_{fr} and meta-Fa. Note that phytochrome can be detected in dehydrated tissue by red/far-red reversibility of the 730/800 nm absorbance difference provided that it is initially present as P_{fr}. As hydration begins, the red/far-red reversible difference spectra result from a mixture of reactions of dehydrated and partially hydrated phytochrome and after adequate imbibition resemble typical spectra of seedling tissue (SPRUIT and MANCINELLI 1969).

9.3 Appearance of Phytochrome P_{fr} in Dark-Imbibed Seeds

In several dark-imbibed seeds, considerable phytochrome is found as P_{fr} (BOISARD et al. 1968, SPRUIT and MANCINELLI 1969, KENDRICK et al. 1969), which could account for high dark germination. Appearance of P_{fr} during imbibition could be from rehydration of pre-existing P_{fr} or from the relatively long-lived intermediate meta-Rb trapped during maturation. The appearance of P_{fr} in dry *Cucumis* seeds following far-red irradiation is probably from intermediates such as meta-Fa in the P_{fr} to P_r pathway. The kinetics of this P_{fr} appearance (SPRUIT and MANCINELLI 1969) is similar to that found in dehydrated seedling tissue (KENDRICK and SPRUIT 1974, SPRUIT et al. 1975).

9.4 Localization of Phytochrome in Seeds

Light requirements are often lost if the seed coat and other structures surrounding the embryo are removed. In *Lactuca,* the endosperm furnishes the restriction. The site of photoreception is in the embryo, thus naked embroys under osmotic stress often show light sensitivity (SCHEIBE and LANG 1965). The restricting effect of the seed covers may be related to mechanical restraint, reduced oxygen penetration and presence of inhibitors. Light increases the growth potential of the embryo to overcome the restraint. This increase in growth potential can be quantified in terms of water potential. Effects of red light on elongation of *Lactuca* embryonic axes in osmotica (NABORS and LANG 1971 a, b, CARPITA et al. 1979 a) indicate P_{fr} decreased water potential of the axes by 4 to 5 bar.

In *Lactuca* seed IKUMA and THIMANN (1959) concluded that the hypocotyl was the light-sensitive region, and VANDERWOUDE and PRATT (1978) showed phytochrome is located in the cortical cells of the radicle and hypocotyl. BOISARD and MALCOSTE (1970) indicated photoreception in *Cucumis sativus* seeds is mainly in or near the embryonic axis. However, phytochrome is detectable in both the cotyledons and embryonic axis of *Cucurbita pepo* (ZOUAGHI et al. 1972). Root cap removal fails to affect photosensitivity of *Lactuca* embryonic axes maintained under osmotic stress (CARPITA et al. 1979 b).

10 Ecological Significance of Light-Controlled Germination

10.1 In Relation to Soil Burial

In seeds, phytochrome is basically a light detector enabling germination when seeds are near the soil surface. It also provides a mechanism for maintaining a supply of dormant weed seeds within the soil which may germinate only after light exposure accompanying disturbance of the surface soil (SAUER and STRUIK 1964, WESSON and WAREING 1967, 1969a, TAYLORSON 1970, 1972, ROBERTS 1972). If seeds of *Sinapis arvensis* and *Plantago major* are sown at different depths and subsequently exposed to light, germination progressively declines with depth reaching the dark control level at 6 to 9 mm (FRANKLAND and POO 1980, FRANKLAND 1981). In well-aerated soil, the decline in germination correlates with the progressive decrease in light fluence rate reaching the seeds. Also, light quality changes due to differential light scattering by the soil particles (WELLS 1959). For instance, the fluence rate at 2 mm depth in finely sieved compost was 0.5% of that at the soil surface and the ratio of red light energy to far-red was 0.6 compared to 1.2 in direct sunlight. However, the apparent low P_{fr}/P ratio in buried seeds may arise mainly because low fluence rates convert P_r to P_{fr} slower than the dark reversion of P_{fr} (FRANKLAND 1981).

Highly dormant freshly shed seeds may be buried before developing an ability to respond to light. Conversely, seeds initially capable of dark germination may develop a light requirement during burial in soil (WESSON and WAREING 1969b). Dark reversion of P_{fr} may occur when germination is prevented, but induction of secondary dormancy may also involve other changes that are unrelated to the absence of light, particularly in wet and poorly aerated soils. Both increases and decreases in dormancy take place during burial in soil (COURTNEY 1968, TAYLORSON 1970, 1972, ROBERTS and LOCKETT 1978, BASKIN and BASKIN 1978, KARSSEN 1980b). Often there are cyclical changes in dormancy associated with temperature effects on induction and removal of secondary dormancy.

10.2 Leaf Shading Effects

Many seeds fail to germinate when shaded by foliage. Phytochrome may detect the shading through its red/far-red reversible properties. Since chlorophyll absorbs red light more effectively than far-red light, shade light will have a relatively low ratio of red to far-red light; about 0.1 compared to 1.2 for sunlight (HOLMES and SMITH 1975, FRANKLAND and LETENDRE 1978, FRANKLAND and POO 1980). The possible biological significance of this on germination was recognized by MEISCHKE (1936) and numerous studies demonstrate phytochrome involvement in inhibitory effects of shade (CUMMING 1963, MALCOSTE 1968b, TAYLORSON and BORTHWICK 1969, BLACK 1969, VAN DER VEEN 1970, STOUTJESDIJK 1972, SMITH 1973, GORSKI 1975, KING 1975, VAZQUEZ-YANES 1976). However, THOMPSON et al. (1977) argue that fluctuating temperature could also lead to shade detection; being much greater in open as opposed to shaded areas. Germination of light-requiring seeds of *Sinapis arvensis* and *Plantago major* sown on the soil surface under various degrees of plant shade decreased as far-red/red ratio of incident light increased, reflecting progressive decreases of phytochrome as P_{fr} (FRANKLAND and POO 1980, FRANKLAND 1981).

GORSKI et al. (1977), GORSKI et al. (1978), and SILVERTOWN (1980) found germination of many species is inhibited by shade light. The effects are due partly to the low P_{fr}/P ratio established but also partly to an inhibitory photoreaction (see Sect. 5.3). Prolonged shade light may cause seeds of some species to lose responsivity to light while in others a light requirement may be induced. Effects of shade light are likely to be of adaptive significance although no precise role in the plants' life strategy at the seed stage has been clearly shown.

References

Amen RD (1968) A model of seed dormancy. Bot Rev 34:1–31

Barton LV, Crocker W (1948) Twenty years of seed research. Faber and Faber, London

Baskin JM, Baskin CC (1978) Seasonal changes in the germination response of *Cyperus inflexus* seeds to temperature and their ecological significance. Bot Gaz 139:231–235

Baskin JM, Baskin CC (1979) Promotion of germination of *Stellaria media* seeds by light from a green safe lamp. New Phytol 82:381–383

Berrie AMM, Paterson J, West HR (1974) Water content and the responsivity of lettuce seeds to light. Physiol Plant 31:90–96

Bevington JM, Hoyle MC (1981) Phytochrome action during prechilling-induced germination of *Betula papyrifera* Marsh. Plant Physiol 67:705–710

Bewley JD (1980) Secondary dormancy (skotodormancy) in seeds of lettuce (*Lactuca sativa* cv. Grand Rapids) and its release by light, gibberellic acid and benzyladenine. Physiol Plant 49:277–280

Bewley JD, Black M (1978) Physiology and biochemistry of seeds in relation to germination. I. Development, germination and growth. Springer, Berlin Heidelberg New York

Bianco J (1973) Phytochrome et germination des semences de *Rumex*. Physiol Plant 28:61–66

Bianco J, Bulard C (1981) Influence of light treatment on gibberellin (GA_4, GA_7 and GA_9) content of *Lactuca sativa* L. cv. Grand Rapids achenes. Z Pflanzenphysiol 101:189–194

Blaauw-Jansen G, Blaauw OH (1975) A shift of the response threshold to red irradiation in dormant lettuce seeds. Acta Bot Neerl 24:199–202

Black M (1969) Light-controlled germination of seeds. Symp Soc Exp Biol 23:193–217

Black M, Wareing PF (1955) Growth studies in woody species VII. Photoperiodic control of germination in *Betula pubescens* EHRH. Physiol Plant 8:300–316

Black M, Wareing PF (1957) Sensitivity of light-inhibited seeds to certain spectral regions. Nature 180:395

Blom CWPM (1978) Germination, seedling emergence and establishment of some *Plantago* species under laboratory and field conditions. Acta Bot Neerl 27:257–271

Boisard J, Malcoste R (1970) Le photocontrole de la germination des graines de *Cucurbita pepo* L. (Courge) et le site de la photosensibilité. CR Acad Sci Paris 271:304–307

Boisard J, Spruit CJP, Rollin P (1968) Phytochrome in seeds and an apparent dark reversion of P_r to P_{fr}. Meded Landbouwhogesch Wageningen 68–17:1–5

Borthwick HA, Hendricks SB, Parker MW, Toole EH, Toole VK (1952) A reversible photoreaction controlling seed germination. Proc Natl Acad Sci 38:662–666

Borthwick HA, Hendricks SB, Toole EH, Toole VK (1954) Action of light on lettuce-seed germination. Bot Gaz 115:205–225

Borthwick HA, Toole EH, Toole VK (1964) Phytochrome control of *Paulownia* seed germination. Isr J Bot 13:122–133

Borthwick HA, Hendricks SB, Schneider MJ, Taylorson RB, Toole VK (1969) The high-energy light action controlling plant responses and development. Proc Natl Acad Sci 64:479–486

Brownlee C, Roth-Bejerano N, Kendrick RE (1979) The molecular mode of phytochrome action. Sci Prog 66:217–229

Burkart S, Sanchez RA (1969) Interaction between an inhibitor present in the seeds of *Datura ferox* L. and light in the control of germination. Bot Gaz 130:42–47

Butler WL (1972) Photochemical properties of phytochrome in vitro. In: Mitrakos K, Shropshire W (eds) Phytochrome. Academic Press, London New York, pp 185–192
Butler WL, Hendricks SB, Siegelman HW (1964) Action spectra of phytochrome in vitro. Photochem Photobiol 3:521–527
Carpita NC, Nabors MW (1981) Growth physics and water relations of red-light-induced germination in lettuce seeds. V. Promotion of elongation in the embryonic axes by gibberellins and phytochrome. Planta 152:131–136
Carpita NC, Nabors MW, Ross CW, Petretic NL (1979a) The growth physics and water relations of red-light-induced germination in lettuce seeds. III. Changes in the osmotic and pressure potential in the embryonic axes of red- and far-red-treated seeds. Planta 144:217–224
Carpita NC, Ross CW, Nabors MW (1979b) The influence of plant growth regulators on the growth of the embryonic axes of red- and far-red-treated lettuce seeds. Planta 145:511–516
Ching TM (1972) Metabolism of germinating seeds. In: Kozlowski TT (ed) Seed biology II. Academic Press, London New York, pp 103–218
Courtney AD (1968) Seed dormancy and field emergence in *Polygonum aviculare*. J Appl Ecol 5:675–684
Cresswell EG, Grime JP (1981) Induction of a light requirement during seed development and its ecological consequences. Nature 291:583–585
Cumming BG (1963) The dependance of germination on photoperiod, light quality and temperature in *Chenopodium* spp. Can J Bot 41:1211–1233
Downs RJ (1964) Control of germination of seeds of the Bromeliaceae. Phyton 21:1–6
Duke SO (1978a) Significance of fluence-response data in phytochrome-initiated seed germination. Photochem Photobiol 28:383–388
Duke SO (1978b) Interactions of seed water content with phytochrome-initiated germination of *Rumex crispus* L. seeds. Plant Cell Physiol 19:1043–1049
Duke SO, Egley EH, Reger BJ (1977) Model for variable light sensitivity in imbibed dark-dormant seeds. Plant Physiol 59:244–249
Evenari M (1956) Seed germination. In: Hollaender A (ed) Radiation biology: visible and near visible light III. McGraw-Hill, New York, pp 519–549
Evenari M (1965) Light and seed dormancy. In: Ruhland W (ed) Encyclopedia of plant physiology Vol XV/2. Springer, Berlin Göttingen Heidelberg, pp 804–847
Evenari M, Neumann G (1953) The germination of lettuce seeds. III. The effect of light on germination. Bull Res Counc Isr 3:136–144
Evenari M, Neumann G, Stein G (1957) Action of blue light on the germination of seeds. Nature 180:609–610
Evenari M, Koller D, Gutterman Y (1966) Effects of the environment of the mother plant on the germination by control of seed-coat permeability to water in *Ononis sicula* Guss. Aust J Biol Sci 19:1007–1016
Flint LH, McAlister ED (1935) Wave lengths of radiation in the visible spectrum inhibiting the germination of light-sensitive lettuce seed. Smithson Misc Collect 94:1–11
Flint LH, McAlister ED (1937) Wave lengths of radiation in the visible spectrum promoting the germination of light-sensitive lettuce seed. Smithson Misc Collect 96:1–8
Frankland B (1972) Biosynthesis and dark transformations of phytochrome. In: Mitrakos K, Shropshire W (eds) Phytochrome. Academic Press, London New York, pp 195–225
Frankland B (1976) Phytochrome control of seed germination in relation to the light environment. In: Smith H (ed) Light and plant development. Butterworth, London, pp 477–491
Frankland B (1981) Germination in shade. In: Smith H (ed) Plants and the daylight spectrum. Academic Press, London New York, pp 187–204
Frankland B, Letendre RJ (1978) Phytochrome and effects of shading on growth of woodland plants. Photochem Photobiol 27:223–230
Frankland B, Poo WK (1980) Phytochrome control of seed germination in relation to shading. In: De Greef J (ed) Photoreceptors and plant development. Antwerpen Univ, Antwerpen, pp 357–366
Fredericq H, De Greef J, Rethy R (1976) Kinetic analysis of the synergism between

gibberellic acid and the physiologically active form of phytochrome. Arch Int Physiol Biochim 84:160–162

Globerson D, Kadman-Zahavi A, Ginzburg C (1974) A genetic method for studying the role of the seed coats in the germination of lettuce. Ann Bot 38:201–203

Gorski T (1975) Germination of seeds in the shadow of plants. Physiol Plant 34:342–346

Gorski T, Gorska K (1979) Inhibitory effects of full daylight on the germination of *Lactuca sativa* L. Planta 144:121–124

Gorski T, Gorska K, Nowicki J (1977) Germination of seeds of various herbaceous species under leaf canopy. Flora 166:249–259

Gorski T, Gorska K, Rybicki J (1978) Studies in the germination of seeds under leaf canopy. Flora 167:289–299

Gutterman Y (1974) The influence of the photoperiodic regime and red-far red light treatments of *Portulaca oleracea* L. plants on the germinability of their seeds. Oecologia 17:27–38

Gutterman Y, Porath D (1973) Phytochrome and germination of *Cucumis prophetarum* L. and *Cucumis sativus* L. seeds, influenced by fruit storage under different light conditions. Isr J Bot 22:200

Gutterman Y, Porath D (1975) Influences of photoperiodism and light treatments during fruit storage on the phytochrome and on the germination of *Cucumis prophetarum* L. and *Cucumis sativus* L. seeds. Oecologia 18:37–43

Harper JL (1977) Population biology of plants. Academic Press, London New York

Hartmann KM (1966) A general hypothesis to interpret "high energy phenomena" of photomorphogenesis on the basis of phytochrome. Photochem Photobiol 5:349–366

Hayes RG, Klein WH (1974) Spectral quality influence of light during development of *Arabidopsis thaliana* plants in regulating seed germination. Plant Cell Physiol 15:643–654

Hendricks SB, Taylorson RB (1975) Breaking of seed dormancy by catalase inhibition. Proc Natl Acad Sci USA 72:306–309

Hendricks SB, Taylorson RB (1976) Variation in germination and amino acid leakage of seeds with temperature related to membrane phase change. Plant Physiol 58:7–11

Hendricks SB, Taylorson RB (1978) Dependence of phytochrome action in seeds on membrane organization. Plant Physiol 61:17–19

Hendricks SB, Taylorson RB (1979) Dependence of thermal responses of seeds on membrane transitions. Proc Natl Acad Sci USA 76:778–781

Hendricks SB, Toole VK, Borthwick HR (1968) Opposing actions of light in seed germination of *Poa pratensis* and *Amaranthus arenicola*. Plant Physiol 43:2023–2028

Hendricks SB, Toole EH, Toole VK, Borthwick HA (1959) Photocontrol of plant development by the simultaneous excitations of two interconvertible pigments. III. Control of seed germination and axis elongation. Bot Gaz 121:1–8

Hendricks SB, VanDerWoude W (1982) How phytochrome acts – a continuing quest. In: Shropshire W, Mohr H (eds) Photomorphogenesis. Encyclopedia of plant physiology new ser vol 16. Springer, Berlin Heidelberg New York

Holmes MG, Smith H (1975) The function of phytochrome in plants growing in natural environment. Nature 254:512–514

Hsiao AI, Vidaver W (1971a) Seed water content in relation to phytochrome-mediated germination of lettuce seeds (*Lactuca sativa* L. var. Grand Rapids). Can J Bot 49:111–115

Hsaio AI, Vidaver W (1971b) Water content and phytochrome-induced potential germination responses in lettuce seeds. Plant Physiol 47:186–188

Hsiao AI, Vidaver W (1973) Dark reversion of phytochrome in lettuce seeds stored in a water-saturated atmosphere. Plant Physiol 51:459–463

Ikuma H, Thimann KV (1959) Photosensitive site in lettuce seeds. Science 130:568–569

Ikuma H, Thimann KV (1964) Analysis of germination processes of lettuce seed by means of temperature and anaerobiosis. Plant Physiol 39:756–767

Isikawa S (1957) Interaction of temperature and light in the germination of *Nigella* seeds II. Bot Mag 70:271–275

Isikawa S, Shimogawara G (1954) Effects of light upon the germination of forest tree seeds. I. Light sensitivity and its degree. J Japan Forest Soc 36:318–323

Isikawa S, Yokohama Y (1962) Effect of "intermittent irradiations" on the germination of *Epilobium* and *Hypericum* seeds. Bot Mag 75:127–132

Jann RC, Amen RD (1977) What is germination? In: Khan AA (ed) The physiology and biochemistry of seed dormancy and germination. Elsevier/North-Holland, Amsterdam, pp 7–28

Johnson CB, Tasker R (1979) A scheme to account quantitatively for the action of phytochrome in etiolated and light-grown plants. Plant Cell Environ 2:259–265

Jones MB, Bailey LF (1956) Light effects on the germination of seeds of henbit (*Lamium amplexicaule* L.). Plant Physiol 31:347–349

Jones RL, Stoddart JL (1977) Gibberellins and seed germination. In: Khan AA (ed) The physiology and biochemistry of seed dormancy and germination. Elsevier/North-Holland, Amsterdam, pp 77–109

Kadman-Zahavi A (1960) Effects of short and continuous illuminations on the germination of *Amaranthus retroflexus* seeds. Bull Res Counc Isr 9:1–20

Karssen CM (1967) The light promoted germination of the seeds of *Chenopodium album* L. I. The influence of the incubation time on quantity and rate of the response to red light. Acta Bot Neerl 16:156–160

Karssen CM (1970a) The light-promoted germination of the seeds of *Chenopodium album* L. III. Effect of the photoperiod during growth and development of the plants on the dormancy of the produced seeds. Acta Bot Neerl 19:81–94

Karrsen CM (1970b) The light-promoted germination of the seeds of *Chenopodium album* L. IV. Effects of red, far-red and white light on non-photoblastic seeds incubated in mannitol. Acta Bot Neerl 19:95–108

Karssen CM (1970c) The light-promoted germination of the seeds of *Chenopodium album* L. V. Dark reactions regulating quantity and rate of the response to red light. Acta Bot Neerl 19:187–196

Karssen CM (1970d) The light-promoted germination of the seeds of *Chenopodium album* L. VI. P_{fr} requirement during different stages of the germination process. Acta Bot Neerl 19:297–312

Karssen CM (1980a) Environmental conditions and endogenous mechanisms involved in secondary dormancy of seeds. Isr J Bot 29:43–64

Karssen CM (1980b) Patterns of change in dormancy during burial of seeds in soil. Isr J Bot 29:65–73

Kasperbauer MJ (1968) Germination of tobacco seeds. I. Inconsistency of light sensitivity. Tob Sci 12:19–24

Kendrick RE (1976) Photocontrol of seed germination. Sci Prog 63:347–367

Kendrick RE, Frankland B (1969) Photocontrol of germination in *Amaranthus caudatus*. Planta 85:326–339

Kendrick RE, Russell JH (1975) Photomanipulation of phytochrome in lettuce seeds. Plant Physiol 56:332–334

Kendrick RE, Spruit CJP (1974) Inverse dark reversion of phytochrome: An explanation. Planta 120:265–272

Kendrick RE, Spruit CJP (1977) Phototransformations of phytochrome. Photochem Photobiol 26:201–214

Kendrick RE, Spruit CJP, Frankland B (1969) Phytochrome in seeds of *Amaranthus caudatus*. Planta 88:293–302

Khan AA, Karssen CM (1980) Induction of secondary dormancy in *Chenopodium bonus henricus* L. seeds by osmotic and high temperature treatments and its prevention by light and growth regulators. Plant Physiol 66:175–181

Khan AA, Tolbert NE (1965) Reversal of inhibitors of seed germination by red light plus kinetin. Physiol Plant 18:41–43

King TJ (1975) Inhibition of seed germination under leaf canopies in *Arenaria serpyllifolia*, *Veronica arvensis* and *Cerastium holosteoides*. New Phytol 75:87–90

Koller D (1970) Analysis of the dual action of white light on germination of *Atriplex dimorphostegia* (Chenopodiaceae) L. Isr J Bot 19:499–516

Koller D, Mayer AM (1962) Seed germination. Annu Rev Plant Physiol 13:403–436

Koller D, Negbi M (1959) The regulation of germination in *Oryzopsis miliacea*. Ecology 40:20–36

Koller D, Sachs M, Negbi M (1964) Spectral sensitivity of seed germination in *Artemisia monosperma*. Plant Cell Physiol 5:79–84

Le Deunff Y (1973) Phytochrome et choc thermique dans la germination des semences de *Rumex crispus* L. CR Acad Sci Ser D 276:2443–2446

Loercher L (1974) Persistence of red light induction in lettuce seeds of varying hydration. Plant Physiol 53:503–506

Malcoste R (1968a) Un probleme écologique: la germination des semences dans les conditions naturelles. Ann Bot 7:241–274

Malcoste R (1968b) L'induction de la germination des graines de *Nigella damascena* L. par la lumière. CR Acad Sci Paris 267:613–616

Malcoste R (1969) Étude par spectrophotometrie in vivo du phytochrome de quelques semences. CR Acad Sci Paris 269:701–703

Malcoste R, Boisard J (1970) Étude du phytochrome dans les graines de *Cucumis sativus* (cornichon, var. Vert de Paris) par spectrophotométrie differentielle in vivo. CR Acad Sci Paris 270:331–334

Malcoste R, Boisard J, Spruit CJP, Rollin P (1970) Phytochrome in seeds of some Cucurbitaceae: In vivo spectrophotometry. Meded Landbouwhogesch Wageningen 10–16:1–16

Mancinelli AL, Borthwick HA (1964) Photocontrol of germination and phytochrome reaction in dark-germinating seeds of *Lactuca sativa* L. Ann Bot (Rome) 28:9–24

Mancinelli AL, Borthwick HA, Hendricks SB (1966) Phytochrome action in tomato-seed germination. Bot Gaz 127:1–5

Mancinelli AL, Rabino I (1978) The "high irradiance responses" of plant morphogenesis. Bot Rev 44:129–180

Mancinelli AL, Tolkowsky A (1968) Phytochrome and seed germination. V. Changes of phytochrome content during the germination of cucumber seeds. Plant Physiol 43:489–494

Marmé D (1977) Phytochrome: Membranes as possible sites of primary action. Annu Rev Plant Physiol 28:173–197

Mayer AM (1977) Metabolic control of germination. In: Khan AA (ed) The physiology and biochemistry of seed dormancy and germination. Elsevier/North-Holland, Amsterdam, pp 357–384

Mayer AM, Poljakoff-Mayber A (1975) Germination of seeds, 2nd edn. Pergamon, New York

McArthur JA (1978) Light effects upon dry lettuce seeds. Planta 144:1–5

McArthur JA, Briggs WR (1971) In vivo phytochrome reversion in immature tissue of the Alaska pea seedling. Plant Physiol 48:46–49

McCullough JM, Shropshire W Jr (1970) Physiological predetermination of germination responses in *Arabidopsis thaliana* L. Heynh. Plant Cell Physiol 11:139–148

Meischke D (1936) Über den Einfluß der Strahlung auf Licht- und Dunkelkeimer. Jahrb Wiss Bot 83:359–405

Mohr H, Appuhn U (1963) Die Keimung von *Lactuca*-Achänen unter dem Einfluß des Phytochromsystems und der Hochenergiereaktion der Photomorphogenese. Planta 60:274–288

Nabors MW, Lang A (1971a) The growth physics and water relations of red-light-induced germination in lettuce seeds. I. Embryos germinating in osmoticum. Planta 101:1–25

Nabors MW, Lang A (1971b) The growth physics and water relations of red-light-induced germination in lettuce seeds. II. Embryos germinating in water. Planta 101:26–42

Nagao M, Esashi Y, Tanaka T, Kumaga T, Fukumoto S (1959) Effects of photoperiod and gibberellin on the germination of seeds of *Begonia evansiana* Andr. Plant Cell Physiol 1:39–47

Negbi M, Koller D (1964) Dual action of white light in the photocontrol of germination of *Oryzopsis miliacea*. Plant Physiol 39:247–253

Negbi M, Black M, Bewley JD (1968) Far-red sensitive dark processes essential for light- and gibberellin-induced germination of lettuce seed. Plant Physiol 43:35–40

Olatoye ST, Hall MA (1973) Interaction of ethylene and light on dormant weed seeds. In: Heydecker W (ed) Seed ecology. Butterworth, London, pp 233–249

Orlandini M, Bulard C (1972) Photosensibilité des graines de *Pinus banksiana* Lamb. Biol Plant 14:260–268

Orlandini M, Malcoste R (1972) Étude de phytochrome des graines de *Pinus nigra* Arn. par spéctrophotométrie bichromatique in vivo. Planta 105:310–316

Pratt LH (1975) Re-examination of photochemical properties and absorption characteristics of phytochrome using high molecular weight preparations. In: Smith H (ed) Light and plant development. Butterworth, London, pp 19–30

Pratt LH (1978) Molecular properties of phytochrome. Photochem Photobiol 27:81–105

Pratt LH, Briggs WR (1966) Photochemical and non-photochemical reactions of phytochrome in vivo. Plant Physiol 41:467–474

Reger BJ, Egley GH, Swanson C (1975) Polysome formation in light sensitive common purslane seeds. Plant Physiol 55:928–931

Roberts EH (1969) Seed dormancy and oxidation processes. Symp Soc Exp Biol 23:161–192

Roberts EH (ed) (1972) Viability of seeds. Chapman and Hall, London

Roberts EH (1973) Oxidative processes and the control of seed germination. In: Heydecker W (ed) Seed ecology. Butterworth, London, pp 189–218

Roberts EH, Smith RD (1977) Dormancy and the pentose phosphate pathway. In: Khan AA (ed) The physiology and biochemistry of seed dormancy and germination. Elsevier/North-Holland, Amsterdam, pp 385–411

Roberts HA, Lockett PM (1978) Seed dormancy and periodicity of seedling emergence in *Veronica hederifolia* L. Weed Res 18:41–48

Rollin P (1963) Observations sur la différence de nature de deux photoréactions controlant la germination des akènes de *Lactuca sativa* var. Reine de Mai. CR Acad Sci 257:3642–3645

Rollin P (1966) The influence of light upon seed germination. Possible interpretations of data. Photochem Photobiol 5:367–371

Rollin P (1972) Phytochrome control of seed germination. In: Mitrakos K, Shropshire W (eds) Phytochrome. Academic Press, London New York, pp 229–254

Rollin P, Maignan G (1967) Phytochrome and the photoinhibition of germination. Nature 214:741–742

Rollin P, Malcoste R, Eude D (1970) Le role du phytochrome dans la germination des graines de *Nemophila insignis* (L.). Planta 91:227–234

Roux SJ, McEntire K, Slocum RD, Cedel TE, Hale CC (1981) Phytochrome induces photoreversible calcium fluxes in a purified mitochondrial fraction from oats. Proc Natl Acad Sci USA 78:283–287

Sauer J, Struik G (1964) A possible ecological relation between soil disturbance, light-flash and seed germination. Ecology 45:884–886

Schäfer E (1975) A new approach to explain the "high irradiance responses" of photomorphogenesis on the basis of phytochrome. J Math Biol 2:41–56

Scheibe J, Lang A (1965) Lettuce seed germination: Evidence for a reversible light-induced increase in growth potential and for phytochrome mediation of the low temperature effect. Plant Physiol 40:483–492

Scheibe J, Lang A (1967) Lettuce seed germination: a phytochrome mediated increase in the growth rate of lettuce seed radicles. Planta 72:348–354

Scheuerlein R (1980) Short-term reactions of phytochrome: Flash induction of seed germination in *Lactuca sativa*. In: De Greef J (ed) Photoreceptors and plant development. Antwerpen Univ, Antwerpen, pp 375–380

Schonbeck MW, Egley GH (1980) Effects of temperature, water potential, and light on germination responses of redroot pigweed seeds to ethylene. Plant Physiol 65:1149–1154

Schulz MR, Klein RM (1963) Effects of visible and ultraviolet radiation on the germination of *Phacelia tanacetifolia*. Am J Bot 50:430–434

Senger H (ed) (1980) The blue light syndrome. Springer, Berlin Heidelberg New York

Shropshire W Jr (1972) Action spectroscopy. In: Mitrakos K, Shropshire W Jr (eds) Phytochrome. Academic Press, London New York, pp 161–181

Shropshire W Jr (1973) Photoinduced parental control of seed germination and the spectral quality of solar radiation. Sol Energy 15:99–105

Shropshire W Jr, Klein WH, Elstad VB (1961) Action spectra of photomorphogenic induction and photoinactivation of germination in *Arabidopsis thaliana*. Plant Cell Physiol 2:63–69

Silvertown J (1980) Leaf-canopy-induced seed dormancy in a grassland flora. New Phytol 85:109–118

Simpson GM (1978) Metabolic regulation of dormancy in seeds – a case history of the wild oat (*Avena fatua*). In: Clutter ME (ed) Dormancy and developmental arrest. Academic Press, London New York, pp 167–220

Small JGC, Spruit CJP, Blaauw-Jansen G, Blaauw OH (1979a) Action spectra for light-induced germination in dormant lettuce seeds. I. Red region. Planta 144:125–131

Small JGC, Spruit CJP, Blaauw-Jansen G, Blaauw OH (1979b) Action spectra for light-induced germination in dormant lettuce seeds. II. Blue region. Planta 144:133–136

Smith H (1973) Light quality and germination: Ecological implications. In: Heydecker W (ed) Seed ecology. Butterworth, London, pp 219–231

Smith H, Kendrick RE (1976) The structure and properties of phytochrome. In: Goodwin TW (ed) Chemistry and biochemistry of plant pigments Vil I. Academic Press, London New York, pp 377–424

Speer HL, Hsiao AI, Vidaver W (1974) Effects of germination-promoting substances given in conjunction with red light on the phytochrome-mediated germination of dormant lettuce seeds (*Lactuca sativa* L.). Plant Physiol 54:852–854

Spruit CJP, Mancinelli AL (1969) Phytochrome in cucumber seeds. Planta 88:303–310

Spruit CJP, Kendrick RE, Cooke RJ (1975) Phytochrome intermediates in freeze-dried tissue. Planta 127:121–132

Stoutjesdijk PH (1972) Spectral transmission curves of some types of leaf canopies with a note on seed germination. Acta Bot Neerl 21:185–191

Suzuki Y, Takahashi N (1969) Red and far-red reversible photoreactions on seed germination of *Cucumis sativa*. Plant Cell Physiol 44:148–150

Suzuki Y, Soejima Y, Matsui T (1980) Influence of after-ripening on phytochrome control of seed germination in two varieties of lettuce (*Lactuca sativa* L.). Plant Physiol 66:1200–1201

Takaki M, Kendrick RE, Dietrich SMC (1981) Interaction of light and temperature on the germination *Rumex obtusifolius* L. Planta 152:209–214

Tao K-L, Khan AA (1977) Hormonal regulation of nucleic acids and proteins in germination. In: Khan AA (ed) The physiology and biochemistry of seed dormancy and germination. Elsevier/North-Holland, Amsterdam, pp 413–433

Taylorson RB (1970) Changes in dormancy and viability of weed seeds in soils. Weed Sci 18:265–269

Taylorson RB (1972) Phytochrome controlled changes in dormancy and germination of buried weed seeds. Weed Sci 20:417–422

Taylorson RB, Borthwick HA (1969) Light filtration by foliar canopies: Significance for light-controlled weed seed germination. Weed Sci 17:48–51

Taylorson RB, Hendricks SB (1969) Action of phytochrome during prechilling of *Amaranthus retroflexus* L. seeds. Plant Physiol 44:821–823

Taylorson RB, Hendricks SB (1971) Changes in phytochrome expressed by germination of *Amaranthus retroflexus* L. seeds. Plant Physiol 47:619–622

Taylorson RB, Hendricks SB (1972) Rehydration of phytochrome in imbibing seeds of *Amaranthus retroflexus* L. Plant Physiol 49:663–665

Taylorson RB, Hendricks SB (1973) Phytochrome transformation and action in seeds of *Rumex crispus* L. during secondary dormancy. Plant Physiol 52:475–479

Taylorson RB, Hendricks SB (1977) Dormancy in seeds. Annu Rev Plant Physiol 28:331–354

Taylorson RB, Hendricks SB (1979) Overcoming dormancy in seeds with ethanol and other anesthetics. Planta 145:507–510

Thanos CA, Mitrakos K (1979) Phytochrome-mediated germination control of maize caryopses. Planta 146:415–417

Thomas TH, Biddington NL, Palevitch D (1978) The role of cytokinin in the phytochrome-mediated germination of dormant imbibed celery (*Apium graveolens*) seeds. Photochem Photobiol 27:231–236

Thompson K, Grime JP, Mason G (1977) Seed germination in response to diurnal fluctuations of temperature. Nature 267:147–149

Toole EH, Toole VK, Borthwick HA, Hendricks SB (1955a) Photocontrol of *Lepidium* seed germination. Plant Physiol 30:15–21

Toole EH, Toole VK, Borthwick HA, Hendricks SB (1955b) Interaction of temperature and light in germination of seeds. Plant Physiol 30:473–478

Toole VK (1973) Effects of light, temperature and their interactions on the germination of seeds. Seed Sci Technol 1:339–396

Toole VK, Borthwick HA (1968) The photoreaction controlling seed germination in *Eragrostis curvula*. Plant Cell Physiol 9:125–136

Totterdell S, Roberts EH (1980) Characteristics of alternating temperatures which stimulate loss of dormancy in seeds of *Rumex obtusifolius* L. and *Rumex crispus* L. Plant Cell Environ 3:3–12

Van der Veen R (1970) The importance of the red-far-red antagonism in photoblastic seeds. Acta Bot Neerl 19:809–812

VanDerWoude WJ, Pratt LH (1978) Localization of phytochrome in photodormant lettuce embryos. Plant Physiol Suppl 61:17

VanDerWoude WJ, Toole VK (1980) Studies of the mechanism of enhancement of phytochrome-dependent lettuce seed germination by prechilling. Plant Physiol 66:220–224

Van Rooden J, Akkermans LMA, van der Veen R (1970) A study on photoblastism in seeds of some tropical weeds. Acta Bot Neerl 19:257–264

Vazquez-Yanes C (1976) Seed dormancy and germination in secondary vegetation tropical plants: the role of light. Comp Physiol Ecol 1:30–34

Vidaver W (1977) Light and seed germination. In: Khan AA (ed) The Physiology and biochemistry of seed dormancy and germination. Elsevier/North-Holland, Amsterdam, pp 181–192

Vidaver W, Hsiao AI (1972) Persistence of phytochrome-mediated germination control in lettuce seeds for 1 year following a single monochromatic light flash. Can J Bot 50:687–689

Vidaver W, Hsiao AI (1974) Actions of gibberellic acid and phytochrome on the germination of Grand Rapids lettuce seed. Plant Physiol 53:266–268

Vidaver W, Hsiao AI (1975) Secondary dormancy in light-sensitive lettuce seeds incubated anaerobically or at elevated temperature. Can J Bot 53:2557–2560

Villiers TA (1972) Seed dormancy. In: Kozlowski TT (ed) Seed biology II. Academic Press, London New York, pp 220–281

Waddoups DR (1976) Studies on variation in phytochrome-controlled seed germination in *Sinapis arvensis*. PhD Thesis, Univ London

Wareing PF, Black M (1958) Similar effects of blue and infra-red radiation on light-sensitive seeds. Nature 181:1420–1421

Wells PV (1959) Ecological significance of red-light sensitivity in germination of tobacco seed. Science 129:41–42

Wesson G, Wareing PF (1967) Light requirements of buried seeds. Nature 213:600–601

Wesson G, Wareing PF (1969a) The role of light in the germination of naturally occuring populations of buried weed seeds. J Exp Bot 20:402–413

Wesson G, Wareing PF (1969b) The induction of light sensitivity in weed seeds by burial. J Exp Bot 20:414–425

Yaniv Z, Mancinelli AL (1967) Phytochrome and seed germination. II. Changes of P_{fr} requirement for germination in tomato seeds. Plant Physiol 42:1147–1148

Yaniv Z, Mancinelli AL, Smith P (1967) Phytochrome and seed germination. III. Action of prolonged far-red irradiation on the germination of tomato and cucumber seeds. Plant Physiol 42:1479–1482

Zouaghi M, Malcoste R, Rollin P (1972) Étude du phytochrome détéctable, in vivo dans less graines de *Cucurbita pepo* L. au cours des différentes phases de la germination. Planta 106:30–43